Electrical Machines and Automatic Control System

Other Book of Interst

S.N.	ISBN	Title Name	Author	Pgs.
1.	9789381068939	Linear Integrated Circuits	Rishabh Anand	620
2.	9789380016955	Electronics Analog and Digital	R Singh, I Singh, B Singh	639
3.	9789380016108	Electrical Machines & Automatic Control System	A Ambikapathy	568
4.	9789380016818	Power System Analysis	A Ambikapathy	292
5.	9789380016931	Elements of Power System	A Ambikapathy	260
6.	9789380016948	Control System	A Ambikapathy	769
7.	9788187522089	Networks and Systems (Second Edition)	Ashfaq Husain	1160
8.	9788190645133	Network Analysis and Synthesis	Ashfaq Husain, Haroon Ashfaq	630
9.	9789380016078	Switchgear and Protection (Second Edition)	Haroon Ashfaq	260
10.	9789381068342	Communication Systems [Analog and Digital]	Rishabh Anand	1100
11.	9789382609445	Digital Electronics (Second Edition)	Rishabh Anand	732
12.	9789380016160	Electromagnetic Field Theory	Rishabh Anand	755
13.	9788190645102	Combinatorics and Graph Theory (Third Edition)	SB Singh	480
14.	978-8187522058	Audio Video Systems	SP Bali, Rajeev Bali	518
15.	9789380016894	Analog and Digital Electronics	Sonveer Singh	435
16.	9789381068922	Advance Microprocessor	AK Gautam	420
17.	9789381068915	Digital Electronics	AK Gautam	608
18.	9789381068830	Non-Conventional Resources (Second Edition)	Ashish Chandra, Taru Chandra	340
19.	9789381068854	Intelligent Instrumentation for Engineers	Rishabh Anand	120
20.	9789382609179	Fundamentals of Microprocessor Programming	Mansih Kumar Ghodki	304
21.	9789382609216	Network Analysis and Synthesis	NK Sharma	536
22.	9789382609506	Modern Concepts of Communication Systems	BMK Prasad	424
23.	9789382609650	Electric Machines	PS Bimbhra	488

Electrical Machines and Automatic Control System

Second Edition

A.AMBIKAPATHY
Phd. (P)
Assistant Professor
Electronics and Instrumentation Engineering
Galgotia College of Engineering and Technology
Greater Noida

KHANNA BOOK PUBLISHING CO. (P) LTD.
Publisher of Engineering and Computer Books
4C/4344, Ansari Road, Darya Ganj, New Delhi-110002
Phone: 011-23244447-48 **Mobile:** +91-99109 09320
E-mail: contact@khannabooks.com, sales@khannapublishers.com
Website: www.khannabooks.com

Dear Readers,

To prevent the piracy, this book is secured with HIGH SECURITY HOLOGRAM on the front title cover. In case you don't find the hologram on the front cover title, please write us to at contact@khannabooks.com or whatsapp us at +91-99109 09320 and avail special gift voucher for yourself.

Specimen of Hologram on front Cover title: 

Moreover, there is a SPECIAL DISCOUNT COUPON for you with EVERY HOLOGRAM.

How to avail this SPECIAL DISCOUNT:

Step 1: Scratch the hologram
Step 2: Under the scratch area, your "coupon code" is available
Step 3: Logon to www.khannabooks.com
Step 4: Use your "coupon code" in the shopping cart and get your copy at a special discount
Step 5: Enjoy your reading!

Thanks,

Khanna Publishing House

Price: ₹ 350.00

Electrical Machines & Automatic Control System

A. Ambikapathy

ISBN: 978-93-82609-79-7

Second Edition: 2016
Reprint: 2021

Published by:

Khanna Book Publishing Co. (P) Ltd.
4C/4344, Ansari Road, Darya Ganj, New Delhi-110 002
Phone: 011-23244447-48 **Mobile:** +91-9910909320
E-mail: contact@khannabooks.com

Printed by: SPS Printers

Preface

A text book on "Electrical Machines and Automatic Control" has been written for the undergraduate students of mechanical department. The material of this book is very useful to the students for understanding the concepts as well as from examination point of view. A number of examples are solved in a detailed manner in every unit so that the students can practise more on numerical part. Every unit contains exercise problems with answers which help the students to prepare for their exams. The content of this book is explained in a simple language so that understanding will be much easy. I will be thankful to the students and teachers who pinpoint the errors and suggestions for the improvement of the book.

A.AMBIKAPATHY

Brief Table of Contents

Detailed Table of Contents

CHAPTER

1

Single Phase Transformer

1.1 INTRODUCTION

Transformer is a static Piece of a device which transforms electrical Power from one Circuit to another Circuit without change in frequency.

Let we see the Symbolic representation of transformer.

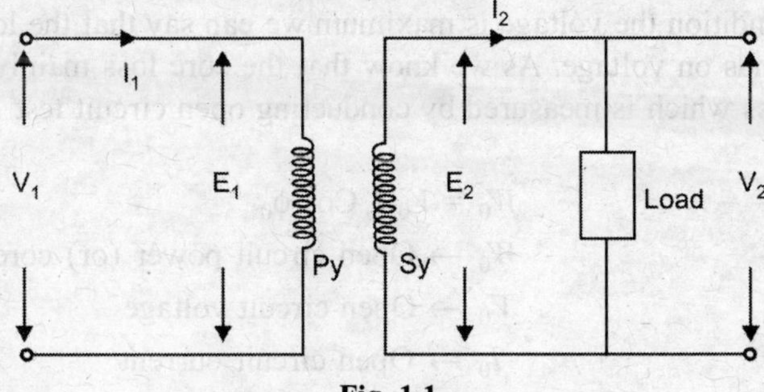

Fig. 1.1

It has two windings

(i) Primary winding

(ii) Secondary

According to the type of transformer the number of turns in the primary and secondary get varied.

Step down transformer will have less number of turns in secondary winding as compared to primary winding and **step up** transformer will have more number of turns in secondary winding as compared to primary winding.

Note: Since two basic Topic of transformer are covered in first year syllabus, in this we will be switching directly to testing, efficiency, regulation etc.

1.2 OPEN CIRCUIT AND SHORT CIRCUIT TESTS

The main objective of conducting these tests in transformer is to determine the total loss in the transformer so as to determine the efficiency and voltage regulation of the transformer.

(i) Open Circuit test : [To determine Core loss (or) Iron loss (or) Constant loss]

The objective of this test is to find out core loss. Let we see the circuit diagram for conducting open circuit test.

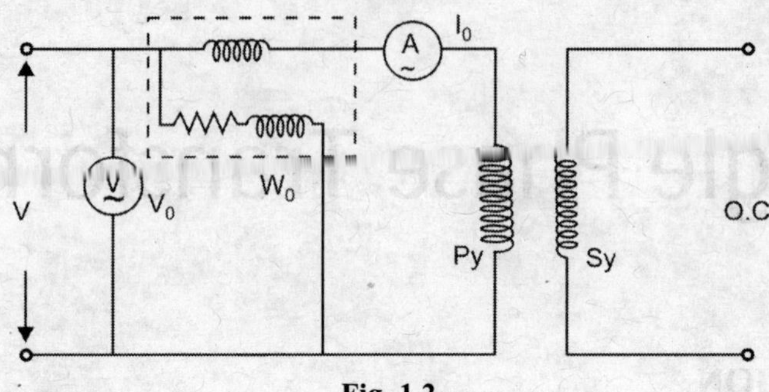

Fig. 1.2

By Leaving the Secondary terminal open, connect Voltameter, Ammeter and Wattmeter in the primary to measure open circuit voltage (V_0), open circuit Current (I_0) and open circuit power (W_0).

Watt meter measures (W_0) core loss of the transformer. Generally core loss depends on voltage. Since, in open circuit condition the voltage is maximum we can say that the loss which is measured by watt meter mainly depends on voltage. As we know that the core loss mainly depends on voltage, we can conclude that the loss which is measured by conducting open circuit test is core loss (or) Iron loss (or) Constant loss.

Let
$$W_0 = V_0 I_0 \text{ Cos } \phi_0 \tag{1.1}$$

where,

$W_0 \rightarrow$ Open circuit power (or) core loss

$V_0 \rightarrow$ Open circuit voltage

$I_0 \rightarrow$ Open circuit current

From Eq. (1.1)
$$\cos \phi_0 = \frac{V_0 I_0}{W_0} \tag{1.2}$$

and
$$\phi_0 = \cos^{-1}\left[\frac{V_0 I_0}{W_0}\right] \tag{1.3}$$

and also
$$\sin \phi_0 = \sin\left[\cos^{-1}\left[\frac{V_0 I_0}{W_0}\right]\right] \tag{1.4}$$

The above Eqs. (1.1) to (1.4) can be used for calculation purposes.

From equations (1.2) and (1.4) we can find out the equivalent open circuit parameters R_0 and X_0.

i.e.,
$$I_\omega = I_0 \cos \phi_0$$
$$I_\mu = I_0 \sin \phi$$

and
$$R_0 = \frac{V}{I_\omega}$$
$$X_0 = \frac{V}{I_\mu}$$

Note: All the meters are connected in low voltage winding and high voltage winding is left open. Copper loss is negligibly small in primary and zero in secondary. Hence, the watt meter reading represents core loss under open circuit test.

(ii) Short Circuit Test

The objective of this test is to find out copper loss.

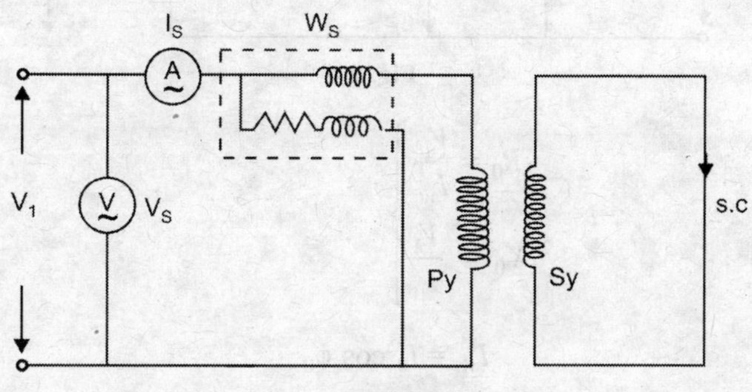

Fig. 1.3

By short circuiting the secondary terminals, true readings of Voltmeter, Ammeter and wattmeter are noted.

where,

$W_s \rightarrow$ Short circuit Power (or) copper loss

$V_s \rightarrow$ Short Circuit Voltage

$I_S \rightarrow$ Short Circuit current

As we know that, in short circuit the value of Current is high and the Voltage is low, the power which is measured by watt meter (W_s) mainly depends on short circuit Current (I_s). So, we can conclude that, the power which is measured by wattmeter is the copper loss. Because, copper loss mainly depends on current.

$$\therefore \quad W_s \rightarrow I_s^2 R_{01} \tag{1.5}$$

$$R_{01} = \frac{W_s}{I_s^2} \tag{1.6}$$

By using short circuit voltage and short circuit current we can say

$$Z_{01} = \frac{V_s}{I_s} \tag{1.7}$$

$$\therefore X_{01} = \sqrt{Z_{01}^2 - R_{01}^2} \tag{1.8}$$

Note: By using the Eqs. (1.1) to (1.8) we can determine the equivalent circuit parameters.

Now we will see how to find equivalent circuit parameters using o.c and s.c test results.

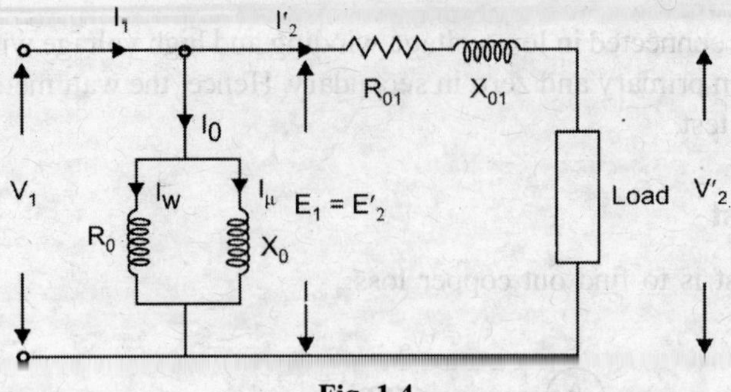

Fig. 1.4

From Fig. (1.4)

$$\left.\begin{aligned} R_0 &= \frac{V_1}{I_w} \\ X_0 &= \frac{V_1}{I_\mu} \end{aligned}\right\} \tag{1.9}$$

and

where,

$$I_w = I_0 \cos \phi_0 \tag{1.10}$$

$$I_\mu = I_0 \sin \phi_0 \tag{1.11}$$

From Eq. (1.9) it is understood that, to find out R_0 and X_0 (equivalent open circuit parameter) the open circuit test data's are used.

From short circuit test,

$$R_{01} = \frac{W_s}{I_s^2}$$

$$Z_{01} = \frac{V_s}{I_s}$$

$$X_{01} = \sqrt{Z_{01}^2 - R_{01}^2}$$

In Fig. 1.4, to find short circuit parameters R_{01} and X_{01} we can consider the data obtained from short circuit test data.

Conclusion

Equivalent circuit Parameters are used to calculate efficiency and regulation of transformer. Where the equivalent circuit parameters can be determined using open circuit and short circuit test results.

1.3 EFFICIENCY

The efficiency of a transformer can be defined as the ratio of output power obtained to the input power given to it.

$$\text{(ie) efficiency, } \eta = \frac{\text{Output Power}}{\text{Input Power}} \tag{1.12}$$

$$\text{(or) Percentage efficiency } \eta = \frac{\text{Output Power}}{\text{Input Power}} \times 100\% \tag{1.13}$$

Transformer is a highly efficient piece of device, it has very small loss. Since, it does not have rotating part, it has better efficiency as compared to other rotating mechines

$$\text{efficiency} = \frac{\text{Output Power}}{\text{Output Power + Power losses}} \quad (1.14)$$

The Eq. (1.14) is the Eq. for efficiency in terms of output.

$$\text{efficiency} = \frac{\text{Input} - \text{losses}}{\text{Input}} \quad (1.15)$$

The Eq. (1.15) is the Eq. for efficiency in terms of input.
where,

$$\text{Losses} = \text{Cu loss + Iron loss (or) Variable loss + Constant loss}$$

From Eq. (1.15) $$\eta = 1 - \frac{\text{Losses}}{\text{Input}} \quad (1.16)$$

Condition for maximum efficiency

$$\text{Copper loss, } W_{cu} = I_1^2 R_{01}$$

$$\text{Iron loss, } W_i = \text{hystersis loss + eddy current loss} = W_h + W_e$$

considering primary,
Primary input power = $V_1 I_1 \cos\phi_1$

$$\eta = \frac{V_1 I_1 \cos\phi_1 - Losses}{V_1 I_1 \cos\phi_1} = \frac{V_1 I_1 \cos\phi_1 - I_1^2 R_{01} - W_i}{V_1 I_1 \cos\phi_1}$$

$$\eta = 1 - \frac{I_1^2 R_{01}}{V_1 I_1 \cos\phi_1} - \frac{W_i}{V_1 I_1 \cos\phi_1}$$

Differentiating both sides w.r. to I_1 we got

$$\frac{d\eta}{dI_1} = 0 - \frac{R_{01}}{V_1 \cos\phi_1} + \frac{W_i}{V_1 I_1^2 \cos\phi_1}$$

For η to be maximum

$$\frac{d\eta}{dI_1} = 0$$

Hence the above Eq. becomes

$$\frac{R_{01}}{V_1 \cos\phi_1} = \frac{W_i}{V_1 I_1^2 \cos\phi_1} \Rightarrow W_i = \frac{\cancel{V_1} I_1^2 \cancel{\cos\phi_1} R_{01}}{\cancel{V_1} \cancel{\cos\phi_1}}$$

Hence, $$W_i = I_1^2 R_{01}$$

Therefore Cu Loss = Iron loss.

(or) Variable loss = Constant loss.

The output current corresponding to maximum efficiency is, $I_2 = \sqrt{\left(\frac{W_i}{R_{02}}\right)}$

The efficiency at any load is given by,

$$\eta = \frac{(x \times \text{Full load } KVA \times \text{power factor})}{(x \times \text{Full load } KVA \times \text{power factor}) + W_{cu} + W_i} \times 100$$

Where,

$x \rightarrow$ ratio of actual to full load KVA

$W_i \rightarrow$ Iron loss in KW

$W_{cu} \rightarrow$ Cu loss in KW

EXAMPLE 1.1: In a 20 KVA, 1000/100V, single phase transformer, the iron and full-load copper losses are 200W and 250 W respectively. Calculate the efficiency at unity power factor on (i) full load (ii) half load.

Given data:

$$\text{Power} = 20KVA$$
$$V_1 = 1000V$$
$$V_2 = 100V$$
$$W_i = 200W$$
$$W_{cu} = 250W$$

Solution: (i) At full load

$$\text{Total Losses} = 200 + 250 = 450W$$

Full load output at unity power factor in KW

$$= KVA \times P.f$$
$$= 20 \times 1 = 20KW$$

$$\text{efficiency, } \eta = \frac{\text{output}}{\text{input}} = \frac{\text{output}}{\text{output} + W_i + W_{cu}}$$

$$= \frac{20 \times 1000W}{20{,}000W + 200W + 250W}$$

$$\eta = 0.977$$
$$\%\,\eta = .977 \times 100$$
$$\%\,\eta = 97.7\%$$

(ii) At half load:

$$\text{Cu loss at half load} = \left(\frac{1}{2}\right)^2 \times \text{Full load cu loss}$$

$$= \frac{1}{4} \times 250 = 62.5\ W$$

Iron loss remains constant.

$$\therefore \quad \text{Total loss} = W_i + W_{cu}$$

$$= 200 + 62.5$$

$$\boxed{\text{Total loss} = 262.5W}$$

at half load, the output at unity power factor is,

$$= \frac{\text{(out put power)}}{2} \times \text{power factor}$$

$$= \left(\frac{20}{2}\right) \times 1 = 10\ KW$$

$$\therefore \quad \eta = \frac{0/P}{0/P + \text{losses}} = \frac{10{,}000W}{10{,}000W + 262.5W}$$

$$\eta = .974$$

$$\%\ \eta = .974 \times 100 = 97.4\%$$

$$\boxed{\%\eta = 97.4\%}$$

Ans. $\%\ \eta$ at full load $= 97.7\%$

$\%\ \eta$ at half load $= 97.4\%$

EXAMPLE 1.2: If P_1 and P_2 be the iron and copper losses of a transformer on full-load, find the ratio of P_1 and P_2 such that the maximum efficiency occurs at 60% of full load.

Solution: If P_2 is the Cu loss at full load, its value at 60% of full load is $= P_2(.60)^2 = (i.e,\ .36i^2)$. At maximum efficiency, it equals the iron loss P_1 which remains constant throughout. Hence at maximum efficiency

$$P_1 = .36P_2$$

$$\frac{P_1}{P_2} = .36$$

Ans. At maximum efficiency the ratio between P_1 and P_2 is .36.

EXAMPLE 1.3: A 13000/250 V, 150 KVA, single phase 50 HZ transformer has core loss of 2.2 KW and full load Cu load of 2 KW. Determine.

(i) *KVA* for maximum efficiency and value of maximum efficiency at unity power factor.

(ii) The efficiency at half load at 0.8 power factor lagging.

Solution:

(i) Load *KVA* corresponding to maximum efficiency is

$$= \text{Full load } KVA \times \sqrt{\frac{\text{iron loss}}{\text{Full load Cu loss}}}$$

$$\text{KVA at maximum efficiency} = 150 \times \sqrt{\frac{2.2}{2}} = 157.32 \text{ KVA}$$

We know at maximum efficiency

$$\text{Cu loss} = \text{Iron loss}$$

$$\therefore \quad \text{Total loss} = 2\,KW + 2\,KW = 4\,KW$$

Output power in kw at maximum efficiency is

$$= (157.32) \times p.f$$

$$= 157.32 \times 1 = 157.32\,KW \qquad [\therefore \text{at unity p.f}]$$

$$\therefore \quad \eta = \frac{\text{out put in KW}}{\text{0/p in KW + losses in KW}}$$

$$\%\,\eta = \frac{157.32}{157.32 + 4} \times 100\%$$

$$\boxed{\%\eta = 97.52\%}$$

(*ii*) $\text{Cu loss at half load} = (2) \times \left(\frac{1}{2}\right)^2$

$$W_{cu} = 0.5 \text{ KW}$$

$$\therefore \quad \text{Total loss} = W_{cu} + W_i = 0.5 + 2.2$$

$$= 2.7 \text{ KW}$$

Out put power in KW at 0.8 power factor

$$= \left(\frac{\text{Power in KVA}}{2}\right) \times p.f$$

$$= \left(\frac{150}{2}\right) \times 0.8$$

$$= 60 \text{ KW}$$

$$\%\,\eta = \frac{0/P}{0/P + \text{losses}} \times 100 = \frac{60}{60 + 2.7} \times 100$$

$$\boxed{\%\eta = 95.69\%}$$

Ans. (i) 97.52% (ii) 95.69%

EXAMPLE 1.4: A 200 KVA transformer has an efficiency of 98% at full load. If the maximum efficiency occurs at three quarters of full load, calculate the efficiency at half load. Assume negligible magnetizing current and power factor 0.8 at all loads.

Solution:

$$\text{Full load output} = 200 \times 0.8 = 160\,KW$$

$$\text{Full load input} = \frac{160}{.98} = 163.265$$

Note: $$\eta = \frac{O/P}{I/P}$$

$$\therefore \quad I/P = \frac{O/P}{\eta} = \frac{160}{.98}$$

$$\text{Full load losses} = 163.265 - 160 = 3.265 \text{ KW}$$

This loss consists of full load Cu loss (x) and iron loss (y).

$$\therefore \quad x + y = 3.265 \text{ KW} \qquad (1)$$

It is also given that maximum efficiency occurs at three quarters of full load when Cu loss becomes equal to iron loss.

$$\therefore \quad \text{Cu loss at 75\% of full load} = x\left(\frac{3}{4}\right)^2 = \frac{9x}{16}$$

Since y remains constant, hence $\frac{9x}{16} = y$ (2)

Substituting the value of y in Eq. (1) we get,

$$x + \frac{9x}{16} = 3.265$$

or

$$16x + 9x = 52.24$$

$$25x = 52.24$$

$$\boxed{x = 2.089 \text{ KW}}$$

Substitute x in Eq. (1)

$$2.089 + y = 3.265$$

$$\boxed{y = 1.1754 \text{ KW}}$$

Half Load Unity Power Factor

$$\text{Copper loss} = 2089 \times \left(\frac{1}{2}\right)^2$$

$$\text{Copper loss} = 522.25\ W$$

$$\text{total loss} = 522 + 1175 = 1697\ W$$

$$\text{output} = 100 \times 0.8 = 80 \text{ KW}$$

$$\eta = \frac{80}{81.697} = .979$$

or

$$= 97.9\%$$

Ans. $\%\eta = 97.9\%$

EXAMPLE 1.5: A 25 KVA, single phase transformer of 2200/220V, has a primary resistance of 1Ω and a secondary resistance of .01Ω. Find the equivalent secondary resistance and the full load efficiency at 0.8 p.f. if the iron loss of the transformer is 80% of the full load Cu loss.

Solution: Turns ratio $= \dfrac{\text{Secondary voltage}}{\text{Primary voltage}} = \dfrac{220}{2200} = \dfrac{1}{10}$

$$R_{02} = R_2 + K_2 R_1$$

$$= 0.01 + 1(0.1)^2 = .02\Omega$$

Full load current, $I_2 = 25000/220$

$$= 113.6A$$

Full load Cu loss $= I_2^2\, R_{02} = 113.6^2 \times 0.02$

$$= 258\ W$$

Iron loss $= 80\%$ of $258 = 206.4\ W$

Total loss $= 258 + 206.4 = 464.4\ W$

Full load output $= 25 \times 0.8 = 20$ KW

$$= 20{,}000W$$

∴ Full load $\eta = \dfrac{20000 \times 100}{(20{,}000 + 464.4)} = 97.7\%$ **Ans.** $\boxed{\%\eta = 97.7\%}$

EXAMPLE 1.6: Consider a 5 KVA, 300/500V single-phase transformer supplying full load current at 0.8 lagging power factor.

The *o.c/s.c* test results are as follows.

o.c test :	300 *v*,	0.9 *A*	80 *W*	(*L.v* side)
s.c test :	30 *v*,	11 *A*	70 *W*	(*h.v* side)

Calculate efficiency.

Solution: Full load secondary current $= \dfrac{\text{Power in } KVA}{\text{Secondary voltage}}$

$$I_2 = \frac{5000}{500} = 10A$$

Total losses $= W_i + W_{cu} = 80 + 70$

$$= 150\ W$$

Full load output in kw = output in $KVA \times P.f$

$$= 5 \times 0.8 = 4 \text{ KW}$$

$$= 4000\ W$$

∴ Full load efficiency, $\eta = \dfrac{O/P}{I/P} = \dfrac{4000}{4000 + 150} \times 100$

$$\boxed{\%\eta = 96.3\%}$$

EXAMPLE 1.7: A 500 KVA, single phase transformer has an efficiency of 94% at full load and also at half load. Determine its efficiency when it operates at unity power factor and 60% of full load.

Solution: At full load

$$\text{output} = 500\ kw \qquad (\because \text{ take } \cos\phi = 1)$$

$$\text{input} = \frac{500}{.94} \qquad \left(\because \eta = \frac{O/P}{I/P} \Rightarrow I/P = \frac{O/P}{\eta}\right)$$

$$= 531.91\ kw$$

$$\text{Total loss} = 531.91 - 500$$

$$= 31.91 \text{ KW}$$

Let,

$$x = \text{iron loss} \rightarrow \text{it is constant}$$

$$y = \text{Cu loss} \rightarrow \text{it is proportional to } (\text{KVA})^2$$

$$\therefore \quad x + y = 31.91 \text{ KW} \qquad (1)$$

At Half Load

$$\text{output} = \frac{500}{2} = 250 \text{ KW}$$

$$I/P = \frac{250}{.94} = 265.95 \text{ KW}$$

$$\text{losses} = 265.95 - 250 = 15.95 \text{ KW}$$

Since Cu loss becomes one fourth of its full load value, hence

$$x + \frac{y}{4} = 15.95 \text{ KW} \qquad (2)$$

Solving Eq. (1) and (2) we get,

$$\cancel{x} + y = 31.91$$

$$\cancel{x} + \frac{y}{4} = 15.95$$

$$.75y = 15.96$$

$$\boxed{y = 21.28}$$

Substituting y in Eq. (1) we get,

$$x + 21.28 = 31.91$$

$$x = 31.91 - 21.28$$

$$\boxed{x = 10.63}$$

$$\therefore \quad \text{iron loss, } x = 10.63 \text{ KW}$$

$$\text{Cu loss, } y = 21.28 \text{ KW}$$

At 60% of full load

$$\text{Cu loss} = 0.60 \times 21.28 = 12.768 \text{ KW}$$

$$\text{Total losses} = 10.63 + 12.768 = 23.398 \text{ KW}$$

$$\text{output} = 500 \times .6 = 300 \text{ KW}$$

$$\%\eta = \frac{O/P}{I/P} \times 100\% = \frac{300}{300 + 23.398} \times 100$$

$$\boxed{\%\eta = 92.76\%}$$ **Ans** (i) $\%\eta = 92.76\%$

EXERCISE

1 In a 25 KVA, 2000/200 *V*, single phase transformer, the iron and full load copper losses are 350 and 400 *W* respectively. Calculate the efficiency at unity power factor on (i) full load (ii) half load. [**Ans.** (i) 97% (ii) 94.3%]

2 If P_1 and P_2 be the iron and copper losses of a transformer on full load, find the ratio of P_1 and P_2 such that maximum efficiency occurs at 75% full load.

[**Ans.** $\frac{P_1}{P_2} = \frac{9}{16} = .5625$]

3 A 1100/230*V*, 150*KVA*, single phase, 50 *HZ* transformer has core loss of 41.4 kw and full load Cu loss of 1.6 KW. Determine .

(i) The *KVA* load for maximum efficiency and value of maximum efficiency at unity power factor.

(ii) The efficiency at half full load 0.8 *P.f* leading. [**Ans.** (i) $\eta_{max} = 98.2\%$ (ii) $\eta = 97\%$]

4 A 4 *KVA*, 200/400 *V*, single phase transformer has equivalent resistance and reactance referred to low voltage side equal to .5Ω and 1.5Ω respectively. Find the terminal voltage on the high-voltage side when it supplies 3/4 full-load at power factor of 0.8, the supply voltage being 220*V*. Hence find the output of the transformer and its efficiency if the core losses are 100*W*.

[**Ans.** $\eta = 91.87\%$]

5 Consider a 4 *KVA*, 200/400 *V* single phase transformer supplying fall load current at 0.8 lagging power factor. The *o.c/sc.* test results are as follows:

o.c test : 200*v* 0.8 *A* 70*w* (*L.v* side)

s.c test : 20 *v* 10 *A* 60 *w* (*h. v.* side)

Calculate the load at unity power factor corresponding to maximum efficiency.

[**Ans.** Load = 4.32 KW]

6 A 600 *KVA*, single phase transformer when working at *U.P.f.* has an efficiency of 92% at full load and also at half-load. Determine its efficiency when it operates at *U.P.f* and 60% of full load. [**Ans.** $\eta = 96.5\%$]

1.4 VOLTAGE REGULATION

It is defined as the change in magnitude of the secondary terminal voltage, expressed as percentage of the secondary voltage, when load at a given power factor is reduced to zero. With primary voltage held constant.

If, $V_2 \rightarrow$ Secondary terminal voltage at any load

$E_2 \rightarrow$ Secondary terminal voltage at no load.

Then at given power factor and specified load, the voltage regulation is given by

$$\text{Voltage regulation} = \frac{E_2 - V_2}{\text{Secondary rated voltage}}$$

where, secondary no load voltage $= E_2$

$$\therefore \quad \text{Voltage regulation} = \frac{E_2 - V_2}{E_2}$$

$$\%\ \text{Voltage regulation} = \frac{E_2 - V_2}{E_2} \times 100$$

Regulation can also be defined as, the change in secondary terminal voltage from no load to full load, expressed as percentage of the secondary voltage.

Now we will see vector diagram under lagging Power factor.

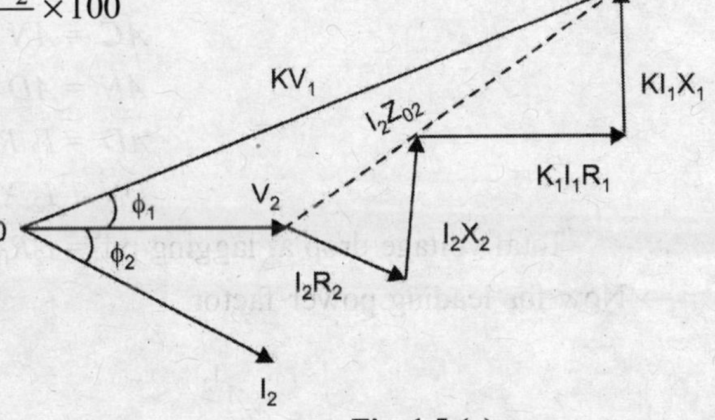

Fig. 1.5 (a)

From Fig. 1.5 (a),

$$V_2 = KV_1 - I_2 Z_{01}$$

The Fig. 1.5 (a) can be reduced as

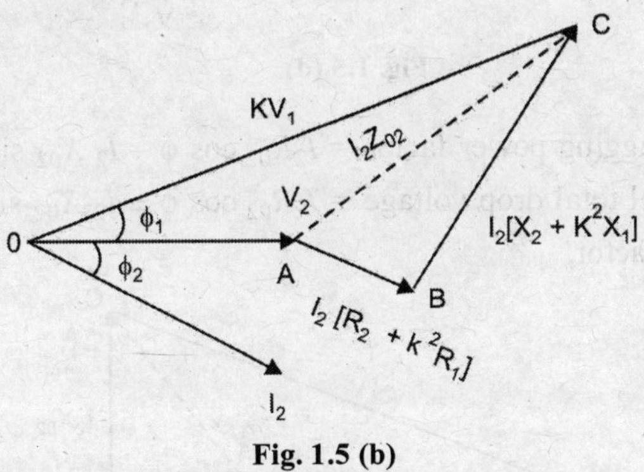

Fig. 1.5 (b)

If $\phi_1 = \phi_2 = \phi$

Then the Fig. 1.5 (b) can be drawn as,

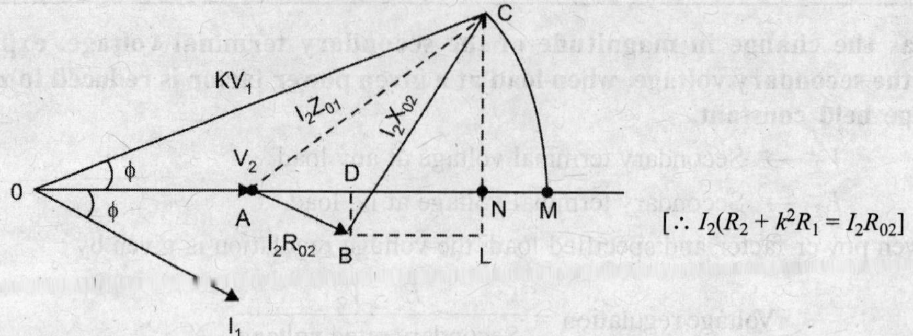

$[\therefore I_2(R_2 + k^2R_1 = I_2R_{02}]$

Fig. 1.5 (c)

Total voltage drop $I_2Z_{01} = AC$

$AC = AM =$ Total voltage drop

Approximately $AM = AN$

$\therefore$ $AC = AN =$ Total voltage drop

$AN = AD + DN$

$AD = I_2 R_{02} \cos \phi$

$DN = I_2 X_{02} \sin \phi$

$\therefore$ Total voltage drop at lagging p.f $= I_2R_{02} \cos \phi + I_2 X_{02} \sin \phi$

→ Now for leading power factor.

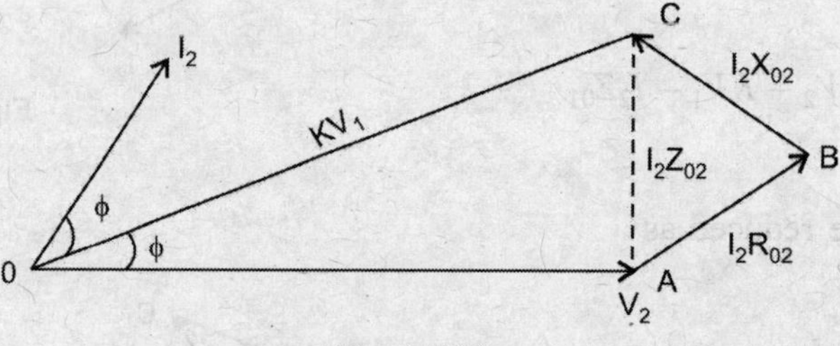

Fig. 1.5 (d)

Total voltage drop at lagging power factor $= I_2R_{02} \cos \phi - I_2 X_{02} \sin \phi$

In general total drop voltage $= I_2R_{02} \cos \phi \pm I_2 X_{02} \sin \phi$ (1.22)

→ Now for unity power factor,

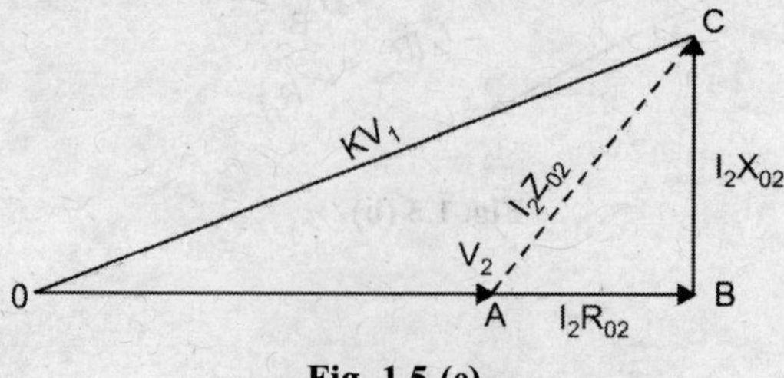

Fig. 1.5 (e)

Total voltage drop at unity power factor $= I_2R_{02}$

$\therefore$ at $u.p.f \cos\phi = 1$

i.e., $\phi = 0$

Substituting $\phi = 0$ in Eq. (1.22) we get

Total voltage drop $= I_2R_{02}$.

Conclusion

Regulation at lagging power factor

$$\therefore \quad \% \text{ Reg} = \frac{I_2R_{02}\cos\phi + I_2X_{02}\sin\phi}{E_2}\times 100 \qquad (1.17)$$

Regulation at leading power factor

$$\therefore \quad \% \text{ Reg} = \frac{I_2R_{02}\cos\phi - I_2X_{02}\sin\phi}{E_2}\times 100 \qquad (1.18)$$

Regulation at unity power factor

$$\therefore \quad \% \text{ Reg} = \frac{I_2R_{02}}{E_2}\times 100 \qquad (1.19)$$

Condition For Zero Voltage Regulation

It is clearly understood that voltage regulation varies with load power factor. If load power factor is varied with constant values of load current and secondary emf, then zero voltage Regulation will occur when.

$$I_2R_{02}\cos\phi + I_2X_{02}\sin\phi = 0$$

1.5 THREE PHASE TRANSFORMER CONNECTION

Years ago, it was a common practice to use interconnected three single-phase transformers instead of a single 3-phase transformer. But now a days, due to improvement in design and manufacturing single 3-phase transformers are used. As compared to a bank of single-phase transformers, the main advantages of a 3-phase transformer are that it occupies less floor space for equal rating, weighs less, cost about 15% less and further, that only one unit is to be handled and connected.

There are various methods available for transforming 3-phase voltages to higher to lower 3-phase voltages, *i.e,* for handling a considerable amount of power. The most common connections are

(i) $Y - Y$

(ii) $\Delta - \Delta$

(iii) $Y - \Delta$

(iv) $\Delta - Y$

(v) Open delta (or) $V - V$

(vi) Scott connection (or $T - T$ connection)

(i) **Y – Y** ***Connection***

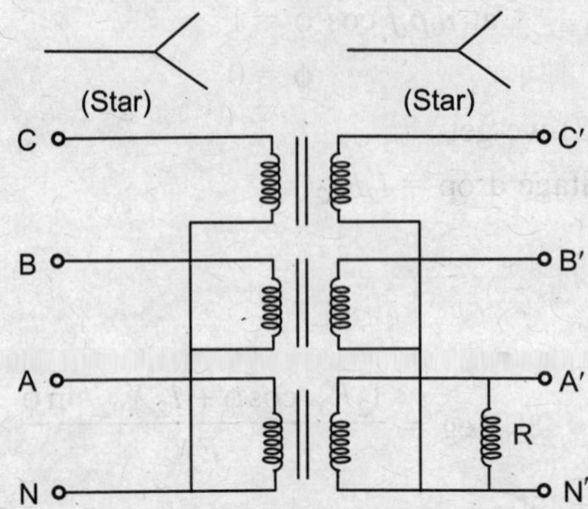

Fig. 1.6: Star-Star connection with 0° displacement

This connection is most economical for small, high voltage transformers because the number of turns/phase and amount of insulation required is minimum.

Fig. (1.6) shows a bank of 3 transformers connected in star (Y) on both sides.

The ratio of line voltages on the primary and secondary sides is the same as the transformation ratio of each transformer.

This connection works efficiently only at balanced load. With unbalanced load, the neutral point shifts so, that line to neutral voltages of three phases become unequal.

- **This type of connections are rarely used in practice because it is not balanced under loading condition due to absence of neutral.**
- **Presence of third harmonics in magnetising current also disturb the satisfactory operation of this type of transformers.**
- **By providing tertiary winding (or) solidly grounding neutral point, the third hormonics can be eliminated.**

The following point can be noted, from Y – Y connection.

1. Line voltage is $\sqrt{3}$ times the phase voltage
2. Phase current is equal to line current.
3. Phase voltage is $\frac{1}{\sqrt{3}}$ times the line voltage.
4. $\frac{\text{Primary phase vg}}{\text{Secordary phase vg}} = K$
5. $\frac{\text{Primary phase ct}}{\text{Secondary phase ct}} = \frac{1}{K}$
6. Primary phase voltage, $V_{\text{ph, py}} = \frac{V_L}{\sqrt{3}}$
7. Secondary phase voltage, $V_{\text{ph, sy}} = \frac{V_L}{\sqrt{3}}$

Where, "a" is a turns ratio.

(ii) Δ – Δ Connection

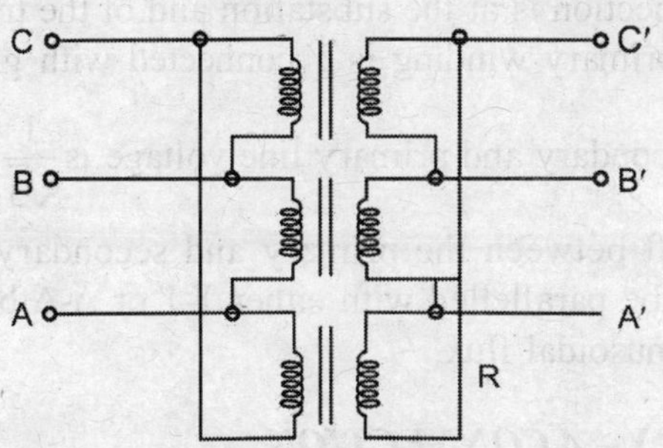

Fig. 1.7: Delta-Delta connection with 0° angular displacement

This connection is economical for large, low-voltage transformers in which insulation problem is not big issue, because it increases the number of turns/phase. The ratio of transformation between primary and secondary line voltage is exactly same as that of each transformer. There is no angular displacement between primary and secondary and there is no internal phase shift between phase and line voltages as in case of Y–Y connection.

- There is no third harmonic problem in this connection, because, third harmonic component of magnetising current can flow in Δ connected transformer without flowing in the line wires.
- There is no unbalanced condition under load. Three phase voltages remain practically constant regardless of load impedance.
- The main advantage of this type of connection is that, it can operate without one phase (or) it can operate with two phase also with reduced capacity of 58%

The Following Points are Noted from Δ – Δ Connection

1. Phase currents are $\frac{1}{\sqrt{3}}$ times the line currents
2. Phase and line voltages are same on primary and secondary side.
3. $\dfrac{\text{Line voltage of primary}}{\text{Line voltage of secondary}} = a$
4. $\dfrac{\text{Line current of primary}}{\text{Line current of secondary}} = \dfrac{1}{a}$

(iii) Y – Δ connection

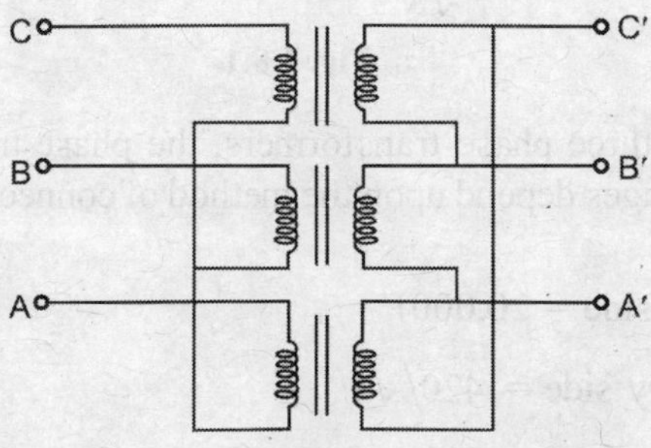

Fig. 1.8: Y – Δ connection with 30° Angular displacement.

The main use of this connection is at the substation and of the transmission line where the voltage is to be stepped down. The primary winding is Y connected with grounded neutral as shown in Fig. 1.8. The ratio between the secondary and primary line voltage is $\frac{1}{\sqrt{3}}$ times the transformation ratio of each other. There is 30° shift between the primary and secondary line voltages which means that Y-Δ transformer bank cannot be parallelled with either Y-Y or Δ-Δ bank. Also, third harmonic current flows in the Δ to provide a sinusoidal flux.

POINTS TO BE NOTED IN Y – Δ CONNECTION

1. Primary line voltage is $\sqrt{3}$ times the phase voltage.
2. Secondary line voltage is equal to the secondary phase voltage.
3. The voltage Ratio of each phase is a

i.e., $$\frac{\text{phase voltage of primary}}{\text{phase voltage of secondary}}$$

The above Eq. is applicable to all the phases (such as A, B & C)

EXAMPLE 1.8: A 3 phase, 50 Hz transformer has a delta connected primary and star connected secondary the line voltage being 20,000V and 420V respectively. The secondary has a star connected balanced load at 0.85 power factor lagging. The line current on the primary side is 6A. Determine the current in each coil of the primary and in each secondary line. What is the output of the transformer in KW.

Solution:

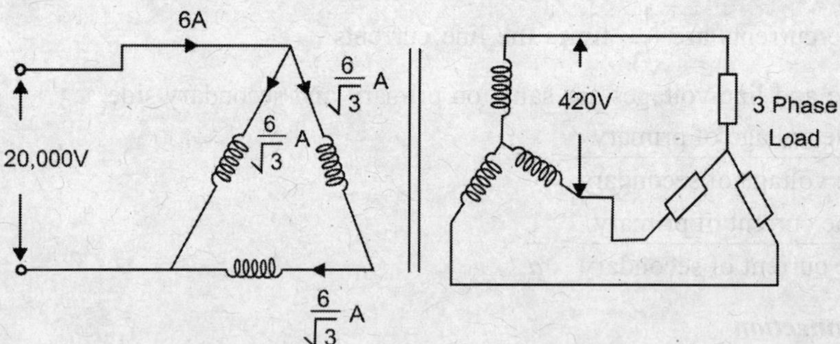

Fig. 1.8.1

It should be noted that in three phase transformers, the phase transformation ratio is equal to the turns ratio but the terminal voltages depend upon the method of connection employed. The Δ/Y connection is shown in Fig (1.8.1).

Phase voltage on primary side = 20,000V

Phase voltage on secondary side = $420/\sqrt{3}$

$$K = \frac{420}{20000} \times \sqrt{3} = 0.021 \times \sqrt{3} = .0363$$

Primary phase current = $\frac{6}{\sqrt{3}}$ A.

Secondary phase current = $\frac{\text{Primary phase current}}{\text{Turns ratio (k)}}$

$$\therefore \text{ Secondary phase current } = \frac{6\sqrt{3}}{\left(\frac{420}{20000}\right)\times\sqrt{3}} = \frac{3.464}{.0363} = 95.432 \text{ A}$$

In secondary (Y Connection), phase current = line current

∴ Secondary line current = 95.432 A

$$\therefore \text{ Output } = \sqrt{3}\, V_L I_L \cos\phi = \sqrt{3}\times 420\times(95.432)\times 0.85$$

$$= 59\ KW$$

Ans. output = 59 KW

EXAMPLE 1.9: A 100 KVA, 3 phase, 50 HZ, 3300/400 transformer is Δ connected on the h.v side and Y connected on $L.V$ side. The resistance of the $h.v$ winding is 3.5 per phase and that of the L.V winding 0.02 Ω per phase. Calculate the iron losses of the transformer at normal voltage and frequency if its full load efficiency is 95.8% at 0.8 p.f (Lag).

Solution: output in KW = output in KVA × p.f = 100 × .8 = 80 KW

We know, $\eta = \frac{O/P}{I/P} \Rightarrow \text{I/P} = \frac{O/P}{\eta} = \frac{80}{.968} = 83.6$ KW.

Losses = $I/P - O/P$ = 83.5 – 80 = 3.5 KW

Let us find full-load Cu losses,

$$K = \frac{\text{Secondary } V_g \text{ / phase}}{\text{primary } Vg \text{ / phase}} = \frac{400/\sqrt{3}}{3300} = \frac{4}{33\sqrt{3}}$$

$$\therefore \quad R_{02} = R_2 + k^2 \cdot R_1$$

$$R_{02} = .02 + \left(\frac{4}{\sqrt{3}}\times 33\right)^2 \times 3.5$$

$$R_{02} = .037\ \Omega$$

Full load secondary phase current is

$$I_2 = \frac{100,000}{\sqrt{3}\times 400}$$

$$I_2 = 144.1A$$

$$\text{Total cu loss } = 3I_2^2\, R_{02}$$

$$= 3 \times (144.1)^2 \times .037 = 2305 \text{ W}$$

Iron loss = Total loss – *F. L* Cu loss

$= 3500 - 2305 = 1{,}195$ W **Ans.** Iron loss = $1{,}195W$.

EXAMPLE 1.10: A 3 phase Δ/Y connected 11000/440V, 50 Hz transformer takes a line current of 5A, when secondary load of 0.8 lagging p. f is connected. Determine each coil current and output of transformer.

Solution: Primary coil rating $= 11{,}000V$

$$\text{Secondary coil rating} = \frac{440}{\sqrt{3}} = 254\ V.$$

$$\text{Primary coil current} = \frac{5}{\sqrt{3}} = 2.887\ A$$

Each coil is delivering equal volt-amps.

$$\text{Since three phase voltages} = 3 \times 1100 \times \frac{5}{\sqrt{3}}$$

$$= 95266$$

$$\text{Volt amps/phase} = 31735$$

This corresponds to the Secondary coil-current of I_2, is given by

$$I_2 = \frac{31755}{250} = 125\ A$$

Total output of transformer in KVA = 31755

Since, the *P.f.* is given as 0.8 Lagging

The total output power in kw = 31,755 × 0.80

= 25.4 KW **Ans.** *O/P* = 25.KW

1.6 OPEN DELTA (OR) V-V CONNECTION

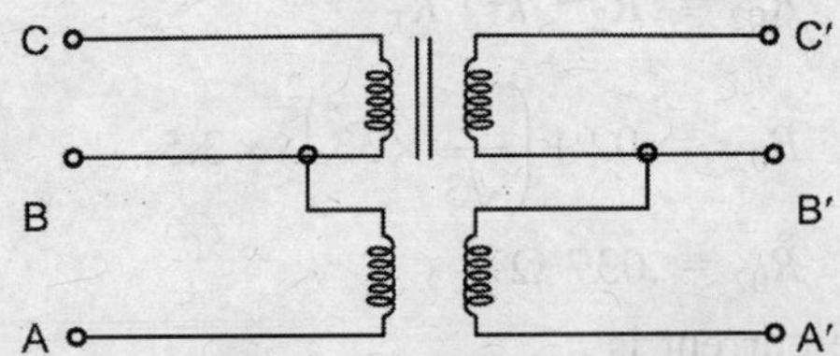

Fig. 1.9: Connection diagram of open delta connection.

If one of the transformers of a Δ-Δ connection is removed and 3-phase supply is connected to the primaries as shown in Fig. (1.9), transforming 3 phase power by using 2 transformers is called open Δ(or) *V-V* connection.

At what situation this kind of transformers are needed?

- When the 3-phase load is too small

- When one phase (or) one of the single phase transformer in Δ-Δ connection is disconnected. So that, there arrangement is done.

The load capacity of this kind of transformer is 57.8% of original rating

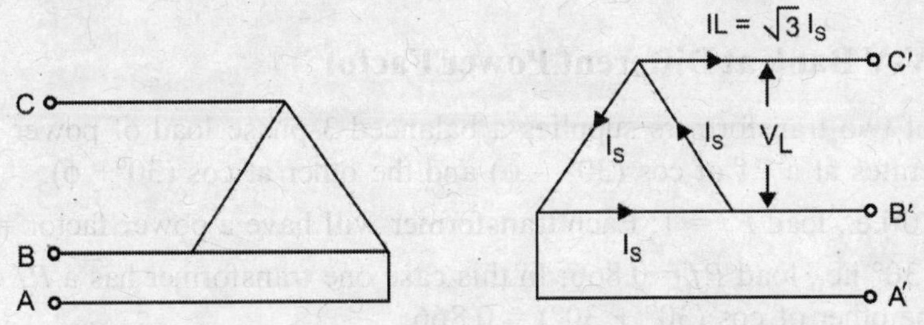

Fig. 1.10 (a): Closed Δ

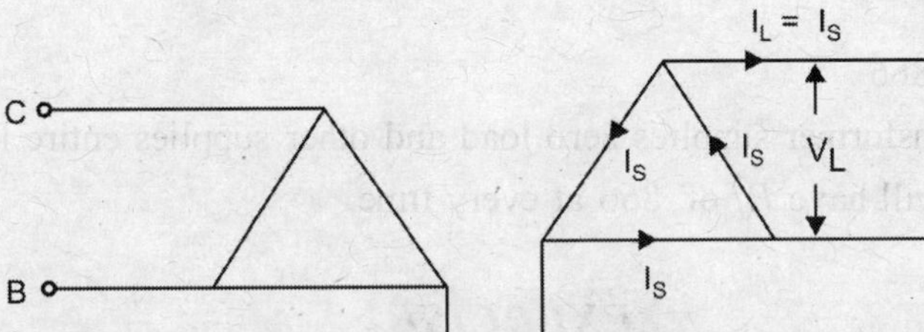

Fig. 1.10 (b) Open Δ

From Fig. 1.10 (a)

$$\Delta - \Delta \text{ capacity} = \sqrt{3}\, V_L I_L$$

$$= \sqrt{3}\, V_L\,(\sqrt{3} I_S) = 3 V_L I_S$$

From Fig. 1.10 (b)

When Δ – Δ connection becomes *V-V* bank, the secondary line current I_L becomes equal to the secondary phase current I_S.

$$\therefore \quad V\text{-}V \text{ Capacity} = \sqrt{3}\, V_l I_l = \sqrt{3}\, V_L I_S$$

$$\frac{V - V \text{ Capacity}}{\Delta - \Delta \text{ Capacity}} = \frac{\sqrt{3}\, V_l I_s}{3 V_l I_s} = \frac{1}{\sqrt{3}} = 0.577 \text{ (or) } 58\%$$

when 3 transformers are making Δ-Δ connection and if one transformer becomes disable, then there is an overloading for other two transformers which is 73.2%. Because,

$$\frac{\text{Total Load in } V - V}{\text{NA-Transformer}} = \frac{\sqrt{3}\, V_l I_S}{V_l I_S} = \frac{1}{\sqrt{3}} = 1.732$$

It has following disadvantages:

1. This kind of connection will have low power factor.

2. Secondary terminal voltage tend to become unbalanced to a great extent when the load is increased, this happens when the load is perfectly balanced.
3. Voltage regulation get differs for different transformers which remain in the supply.

Power Factor of V-V Bank at Different Power Factor

When a *V-V* bank of two transformers supplies a balanced 3-phase load of power factor cos ϕ, then one transformer operates at a P.f of cos $(30° - \phi)$ and the other at cos $(30°+ \phi)$.

(i) when $\phi = 0$ i.e., load *P.f* = 1: Each transformer will have a power factor = cos 30° = 0.866.

(ii) when $\phi = 30°$ i.e., load *P.f* = 0.866: In this case one transformer has a *P.f* of cos (30° – 30°) = 1 and the other of cos (30° + 30°) = 0.866.

(iii) when $\phi = 60°$ i.e., *P.f* = 0.5: In this case one transformer will have a *P.f.* = cos (30° – 60°) =

cos (– 30°) = 0.866

(iv) It means one transformer supplies zero load and other supplies entire load at 0.866 *P.f.*

Conclusion: *V – V* bank will have *P.f* of .866 at every time.

EXERCISE

7. *A* 3 phase, 50 Hz transformer has a delta-connected Primary and star connected secondary, the line voltages being 22,000*V* and 400*V* respectively. The secondary has a star connected balanced load at 0.8 power factor lagging. The line current on the primary side is 5*A*. Determine the current in each coil of the primary and in each secondary line. What is the output of the transformer in KW? **[Ans.** 15.24 KW]

8. A 500 *KVA*, 3 phase, 50 *Hz* transformer has a voltage ratio (line voltages) of 33/11*KV* and is delta/star connected. The resistances per phase are high voltage 35Ω, low voltage 876Ω and the iron loss is 3050W. Calculate the value of efficiency at full load and one half of full load respectively

(a) at unity *P. f* and (b) 0.8 *P. f* **[Ans.** (a) 98.20%; (b) 98%]

EXAMPLE 1.11: What should be the KVA rating of each transformer in a V-V bank when the 3-phase balanced load is 60 KVA? If a third similar transformer is connected for operation, what is the rated capacity? What percentage increase in ratings affected in this way?

Solution: We know the *KVA* rating of each transformer has to be 15% greater.

$$\therefore \quad KVA/\text{transformer} = \left(\frac{60}{2}\right) \times 1.15 = 34.50$$

$$\Delta\text{–}\Delta \text{ bank rating} = 34.50 \times 3 = 103.5$$

$$\therefore \quad \text{increase} = 103.5 - 60 = 43.5$$

$$\%\text{age increase} = \frac{43.5}{60} \times 100 = 72.5\%$$

Ans. 1. rated capacity = 103.5 KVA;

2. % age increase = 72.5%

EXAMPLE 1.12: A Δ–Δ bank consisting of three 30 KVA, 3300/330 transformers supplies a load of 60 KVA. If one transformer is removed, find for the resulting V–V connection.

1. *KVA* load carried by each transformer
2. %age rated load carried by each transformer
3. Total rating of *V–V* bank
4. Ratio of the *V–V* bank to Δ–Δ bank transformer ratings.

Solution: We know,

$$\frac{\text{Total load in V–V connection}}{\text{VA transformer}} = \sqrt{3}$$

$$\therefore \text{ KVA load supplied by each of the two transformers} = \frac{60}{\sqrt{3}} = 34.64\ KVA$$

Obviously,

each transformer in *V–V* bank carry 57.7.1% of the original load.

(ii) $$\%\text{age rated load} = \frac{\text{KVA load/transformer}}{\text{KVA rating/ transformer}}$$

$$= \frac{34.64}{30} = 1.154 = 115.4\%$$

In this case transformer is over loaded till 115.4%

(iii) Rating of the *V–V* bank = 2 × 30 × 0.866
= 51.96 *KVA*

One transformer rating is 30 *KVA*

Two transformer rating is 2 × 30

∴ *KVA* rating of *V – V* = 2 × 30 × .866 = 51.96 *KVA*

(iv) $$\frac{\text{V–V rating}}{\Delta-\Delta \text{ rating}} = \frac{51.96}{3\times 30} = .5773 \text{ i.e, } 57.73\%$$

One transformer rating is 30 *KVA*

3 transformer rating for Δ – Δ connection is = 3 × 30 = 90 *KVA*

Ans. (i) 34.64 KVA; (ii) 115.4%;
(iii) 51.96 *KVA* rating is carried by *V – V* bank; (iv) 57.7%

EXERCISE

9. What should be the KVA rating of each transformer in a *V – V* bank when the 3-phase balanced load is 40 KVA. If a third similar transformer is connected for operation, what is the rated capacity? What percentage increase in rating is affected in this way? **Ans.** 72.5%

10. A Δ–Δ bank consisting of three 20 KVA, 2300/230*V* transformers supplies a load of 40 *KVA*. If one transformer is removed, find for the resulting *V–V* connection.
   1. *KVA* load carried by each transformer
   2. %age rated load carried by each transformer
   3. Total *KVA* rating of *V–V* bank
   4. Ratio of the *V–V* bank to Δ–Δ bank

Ans. (i) 23.1 KVA, (ii) 115.5%; (iii) 34.64 KV, (iv) 57.7%

1.7 SCOTT CONNECTION OR T–T CONNECTION

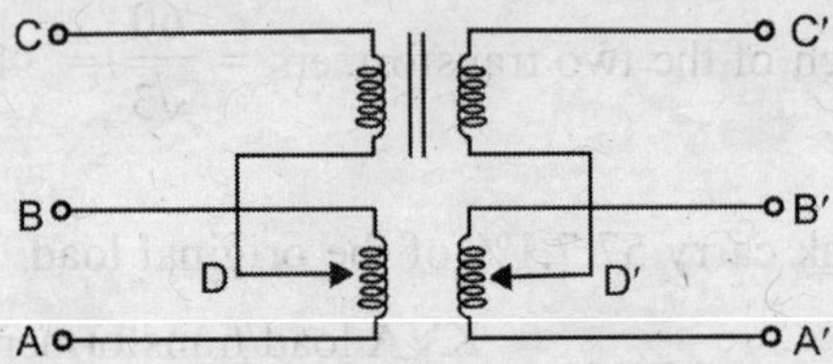

Fig. 1.11: Scott connection

This type of connections are formed when 2 single phase transformers are used to form a 3-phase transformer. This connection can be used for 3-phase to 2 phase transformation.

One of the transformer has centre taps on both the sides of primary and secondary which is known as main transformer. The other transformer is known as teaser transformer which has 0.866 tap. The one end of both the primary and secondary teaser transformer is connected to centre top of main transformer.

Assumptions Made .

1. Three phase supply line voltage is assumed to be 100*V*
2. Transformation ratio is unity.

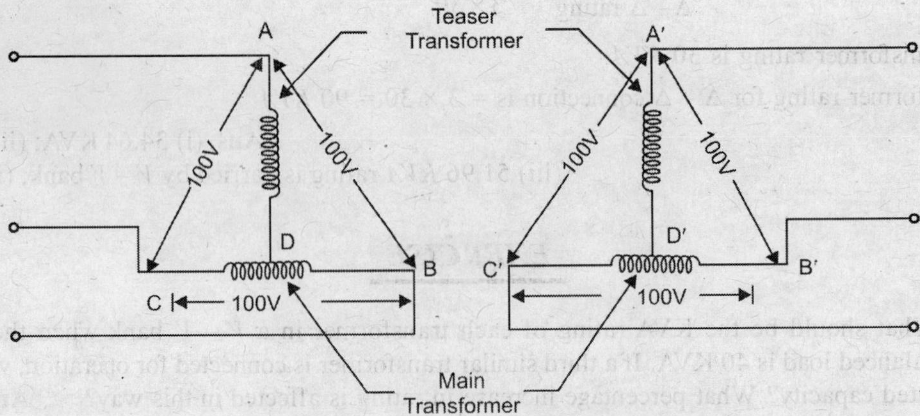

Fig. 1.12(a): Scott (or) T–T connection

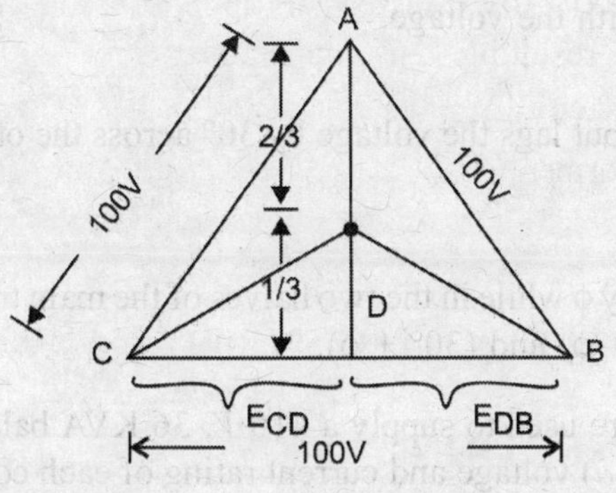

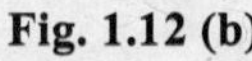

Fig. 1.12 (b)

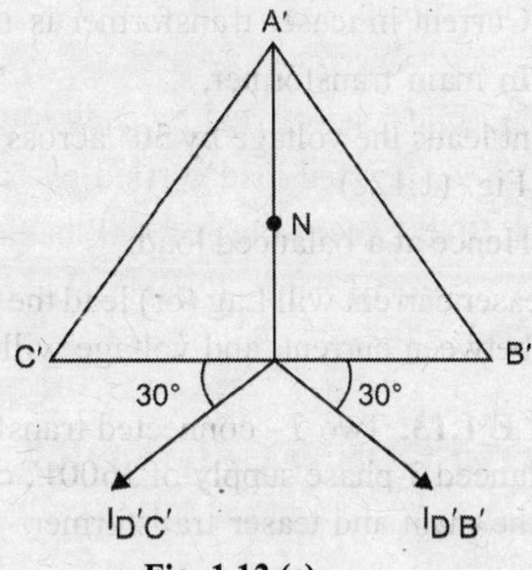

Fig. 1.12 (c)

From Fig. 1.12 (b)

E_{DC} and E_{DB} equal to $50V$ but differ in phase by 180°, because DB and DC are on the same magnetic circuit and are connected in opposition. Each side of the equilateral triangle represents $100V$.

Voltage E_{DA} is equal to $\left(\frac{\sqrt{3}}{2} \times 100\right)$ $86.6V$ and lags behind the voltage across the main by 90°

From Fig. 1.12 (c)

For unity power factor load,

$I_{D'B'}$ lags behind voltage $E_{D'B'}$ by = 30° and $I_{D'C'}$ lead $E_{D'C'}$ by 30°. (or) we can say that the teaser transformer and each half of the main transformer, all operate at different power factors.

The teaser transformer operates at only 0.866 of its rated voltages and the main transformer coil operates at cos 30° = 0.866 power factor, which is equivalent to the main transformer's coil working at 86.6 percent of this KVA rating. Hence the capacity of T–T connection is 86.6%. Both primary and secondary of teaser transformer are designed for 86.6 V only, then they will be operating at full rating.

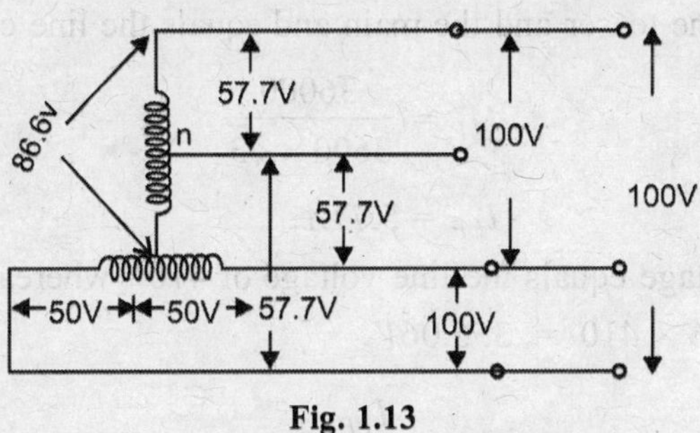

Fig. 1.13

From Fig. 1.13. The secondary of the T–T connection with its different voltages based on a nominal voltage of 100V.

If we draw vector diagram of secondary voltage and current at unity power factor (Load), the following points are noted,

- Current in teaser transformer is inphase with the voltage.
- In main transformer,

Current leads the voltage by 30° across one half but lags the voltage by 30° across the other half as shown in Fig. (1.12c)

- Hence at a balanced load,

The teaser current will Lag (or) lead the voltage by ϕ while in the two halves of the main transformer, the angle between current, and voltage will be $(30° - \phi)$ and $(30° + \phi)$.

EXAMPLE 1.13: Two l– connected transformers are used to supply a 410 V, 36 KVA balanced load from a balanced 3-phase supply of 3600V, calculate (a) voltage and current rating of each coil (b) KVA rating of the main and teaser transformer.

Solution:

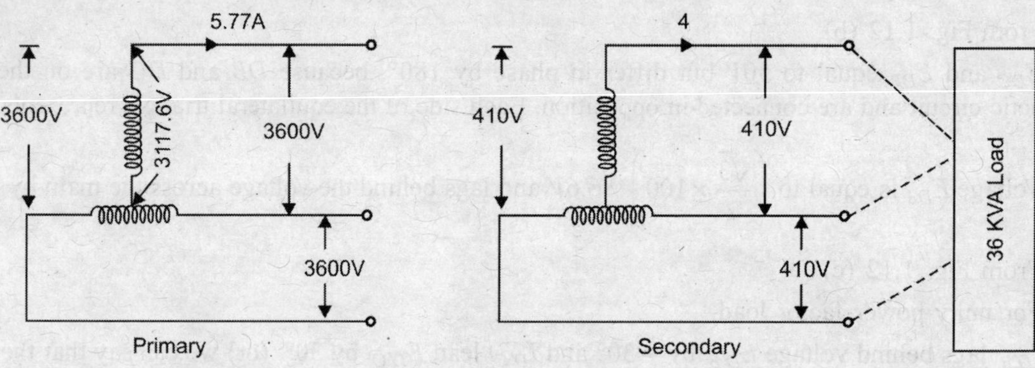

Fig. 1.13.1

(i) Voltage across primary (main transformer) = 3600V

Voltage across primary (teaser) = 3600 × .866

= 3117.6V

The current is same in the teaser and the main and equals the line current.

$$I_{LP} = \frac{36000}{3600 \times \sqrt{3}} \qquad \left[\because I = \frac{\text{Power}}{\text{Voltage}}\right]$$

(Primary line current) $I_{LP} = 5.77A$

The secondary main voltage equals the line voltage of 410V whereas teaser secondary voltage is

$$0.866 \times 410 = 355.06V$$

The secondary line current, $$I_{LS} = \frac{I_{LP}}{K} \qquad \left[\because K = \frac{\text{Secondary voltage}}{\text{Primary voltage}}\right]$$

$$= 5.7/(410/3600) = \frac{5.77}{.113}$$

$$I_{LS} = 51.06A$$

(ii) Main $KVA = 3600 \times 5.77 \times 10^{-3} = 20.77\ KVA$

Teaser $KVA = 0.866 \times$ main $KVA = 17.98\ KVA$

Ans. (a) Main transformer voltage $= 3600\ V.$

Teaser transformer voltage $= 3117.6V$

$I_{LP} = 5.77A$

$I_{LP} = 51.06\ A$

(b) Main $KVA = 20.77\ KVA$

Teaser $KVA = 17.98\ KVA$

EXERCISE

11. Two T connected transformers are used to supply a $440V$, 33 KVA balanced load from a balanced 3-phase supply of $3300V$. Calculate

(a) Voltage and current rating of each coil

(b) KVA rating of the main and teaser transformer.

1.8 THREE PHASE TO TWO PHASE CONVERSION

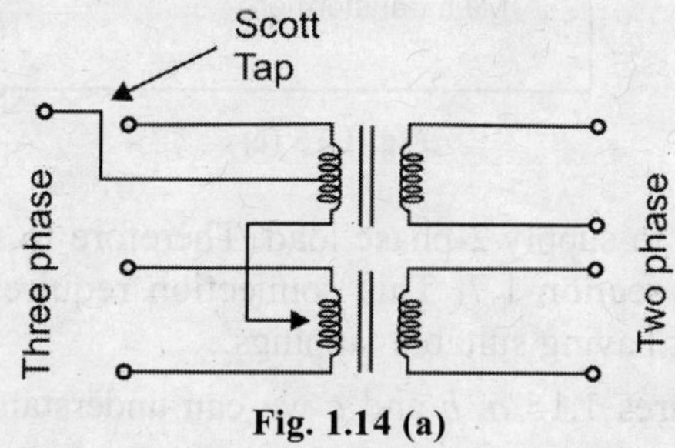

Fig. 1.14 (a)

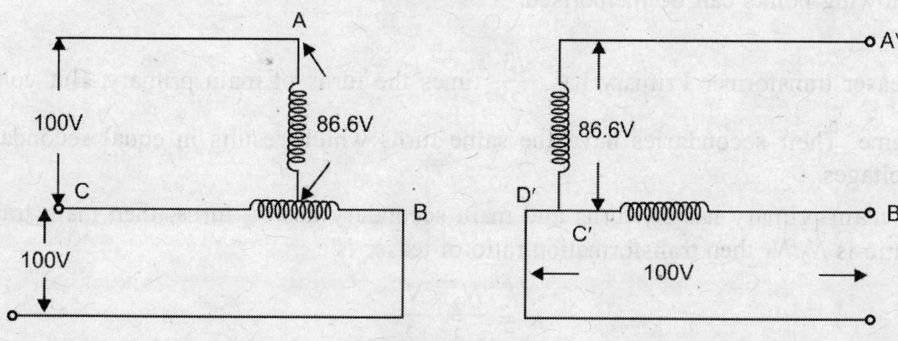

Fig. 1.14 (b)

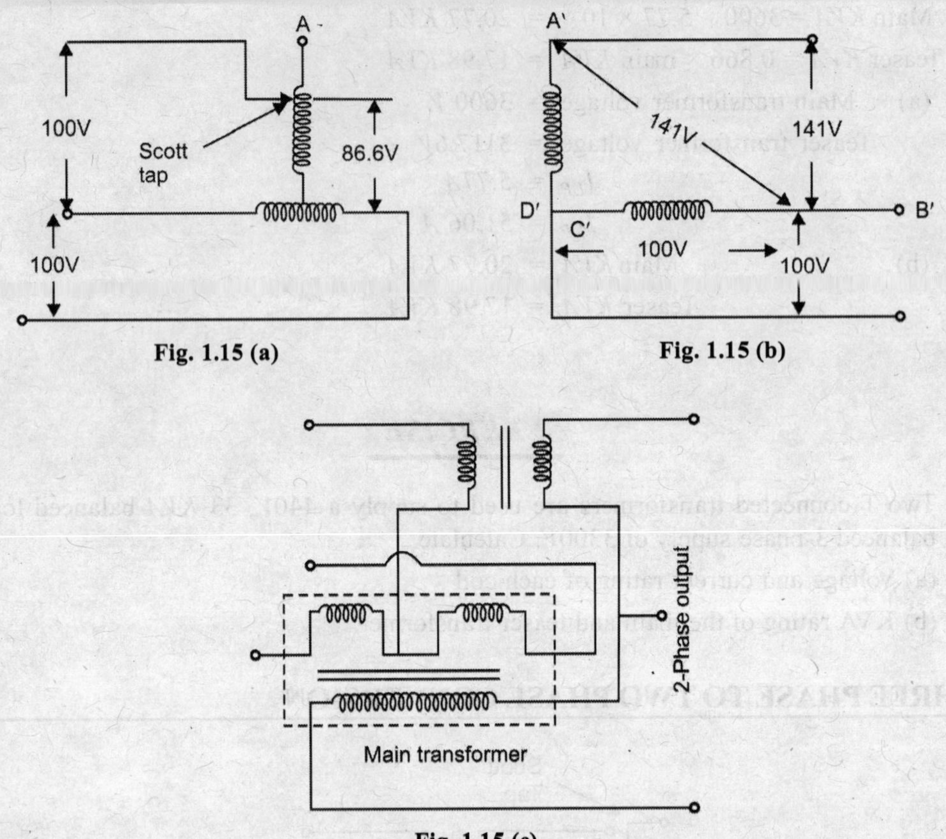

Fig. 1.15 (a)

Fig. 1.15 (b)

Fig. 1.15 (c)

This conversion is required to supply 2-phase load. Therefore to supply 2 phase load we may use SCOTT connection as seen in section 1.7. This connection requires two transformers of different ratings and may be identical but having suitable tappings.

By referring the above figures 1.15 *a*, *b* and *c* we can understand how to obtain 2-phase supply from 3-phase supply.

The following points can be memorised.

- Teaser transformer Primary has $\frac{\sqrt{3}}{2}$ times the turns of main primary. But volt/turn is the same. Their secondaries have the same turns which results in equal secondary terminal voltages.
- If main primary has N_1 turns and main secondary has N_2 turns, then main transformation ratio is N_2/N_1 then transformation ratio of teaser is

$$K = \frac{N_2}{\sqrt{3}} \cdot \frac{N_1}{2}$$

$$= 1.15 \frac{N_2}{N_1} = 1.15\ K$$

- If the load is balanced on one side, it is balanced on the other side as well.
- Under balanced load conditions, main transformer rating is 15% greater than that of the teaser.
- The current is either of the two halves of main primary or the vector sum of KI_{2m} and .58 KI_{2T} (or 1/2 I_{1T}).

where,

$I_{2m} \rightarrow$ Secondary current of main transformer

$I_{2T} \rightarrow$ Secondary current of teaser transformer

EXAMPLE 1.14: Two transformers are required for a scott connection operating from a 410V. 3-phase supply for supplying two single phase loads at 180 V on the two phase side. If the total output is 130 KVA, calculate the secondary to primary turns ratio and the winding currents of each transformer.

Solution: (i) Main transformer

$$\text{Primary volts} = 410\ V$$

$$\text{Secondary volts} = 180V$$

$$\therefore \quad K = \frac{N_2}{N_1}$$

$$\frac{180}{410} = .439$$

$$\text{Secondary current} = 130000/2 \times 180 = 361.114$$

$$\text{Primary current} = 361.11 \times K$$

$$= 361.11 \times .439 = 158.52\ A$$

***(ii)* Tease transformer**

$$\text{Primary volts} = \left(\frac{\sqrt{3}}{2} \times 410\right) = 355.06V$$

$$\text{Secondary volts} = 180V$$

$$\text{K} = \frac{\text{Secondary turns}}{\text{Primary turns}} = \frac{180}{355.06} = .506$$

Ans. (i) $K = .439$; (ii) $Py\ ct = 361.11A$, $Sy\ ct = 158.52A$

EXERCISE

12 Two transformers are required for scott connection operating from a 440V, 3-phase supply for supplying two-single phase loads at 200V on the two phase side. If the total out put is 150 KVA, calculate the secondary to primary turns ratio and the winding currents of each transformer.

[**Ans.** (i) $K = \frac{1}{2.2}$ and $K = \frac{1}{1.905}$

(ii) $Py\ ct = 197A$ and $Sy\ ct = 375\ A$]

1.9 SIX PHASE CONNECTION AND THEIR APPLICATIONS

By using single phase transformer not only 3-phase transformers are made but also six phase transformers are made. This kind of connections generally we use in rectifier where the conversion of *AC* to *DC* is taking place.

There are 4 types of basic connections in six phase transformers.

1. Double star connection
2. Double delta connection
3. Six phase star connection
4. Diametrical connection

Let us See One By One In Detail

1. Double Star Connection

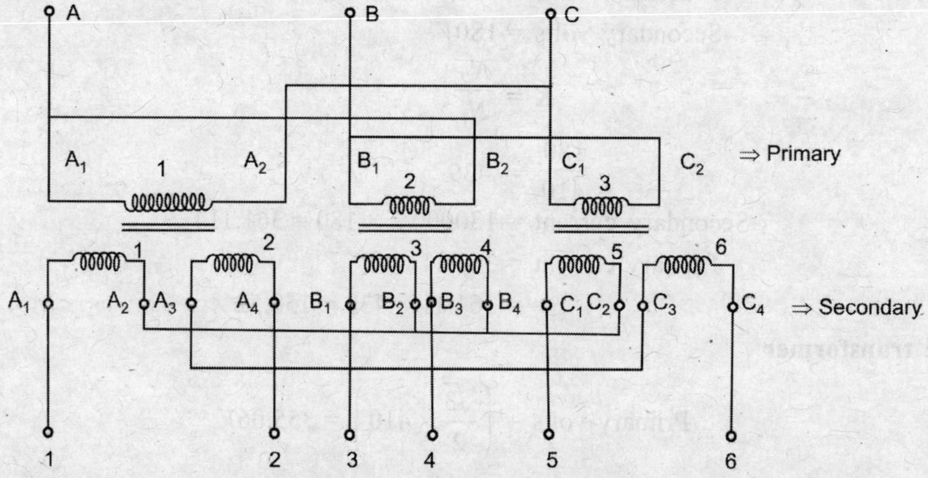

Fig. 1.16: Double star connection

- From Fig(1.16). we can see that there are 3 transformer in primary and six transformers in secondary. i.e., it is having 3-phase in primary and six phases in secondary.
- The primary should have delta connection
- Secondary can be splitted into two parts, where, 1 set is connected in reversed star and other is connected in star.
- It has two neutral points N_1 and N_2

Application

1. It can be used in *DC* link in the transmission line.
2. It can be used in rectifier where *AC–DC* conversion is taking place.
3. It also can be used in cyclo convertor.

2. Double Delta Connection

This is opposite to that of first case. This also has 3-phases in primary and six phases in secondary.

- Primary should have star connection.
- Secondary has two sections. One set is forming delta and other forms reversed delta.

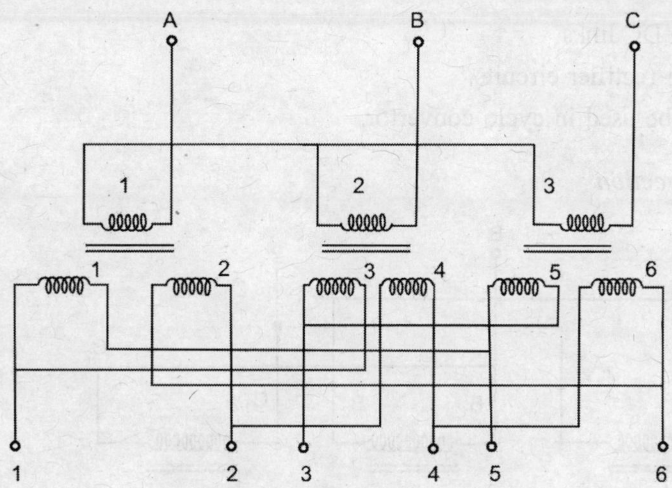

Fig. 1.17: Double delta connection

Application

- It can be used in DC link
- It can not be used in rectifier because it does not have neutral point.
- It can be used in cyclo converter.

3. Six Phase Star Connection

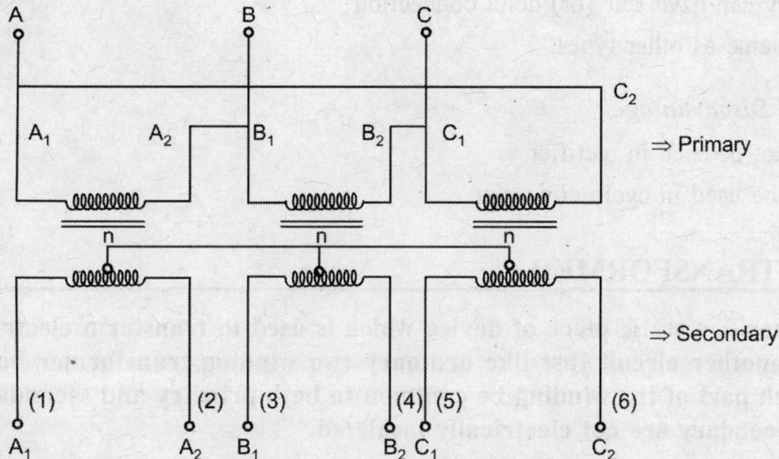

Fig. 1.18: Six phase star connection

- Secondary winding of each transformer is centrally tapped which are connected together to have common neutral point.
- Primary can be connected in delta (or) star.

Application

- It is used in DC links
- It is used in rectifier circuits.
- It also can be used in cyclo convertor.

4. Diametrical Connection

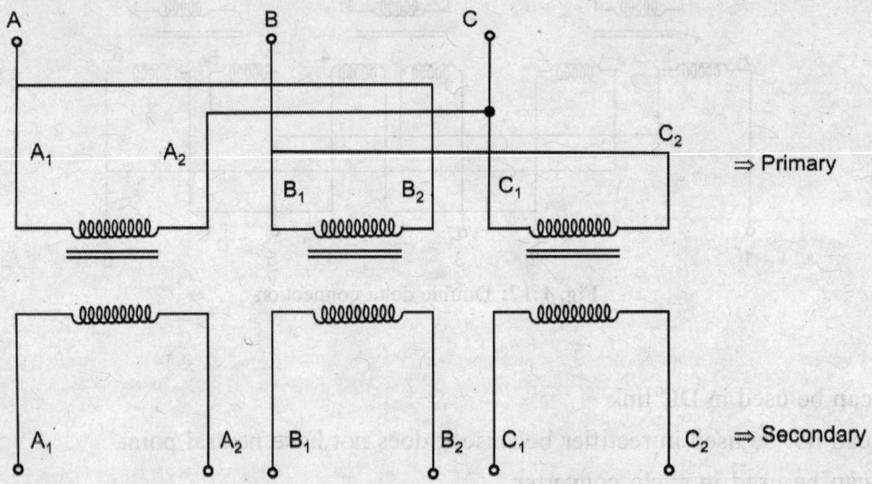

Fig. 1.19: Diametrical connection

- Primary can have star (or) delta connection

Other wise same as other types.

Application and Disadvantage

- It cannot be used in rectifier
- It can be used in cyclo converter.

1.10 AUTO TRANSFORMER

Auto transformer is a static piece of device which is used to transform electrical power from one circuit to another circuit just like ordinary two winding transformer but with only one winding in which part of the winding be common to both primary and secondary. In this type, primary and secondary are not electrically insulated.

Let us see the circuit diagram for auto transformer.

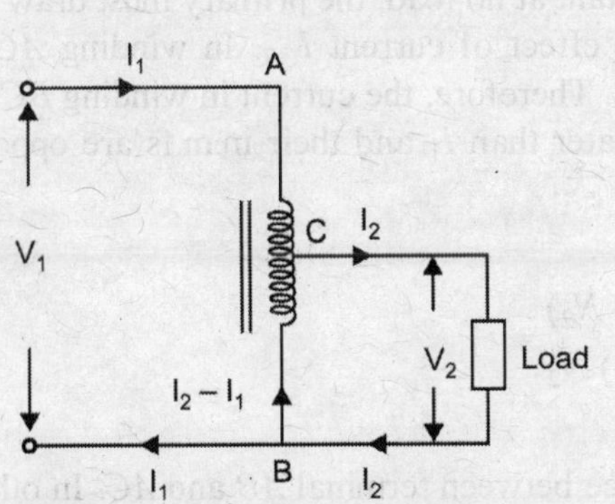

Fig. 1.20 (a): Step down auto transformer

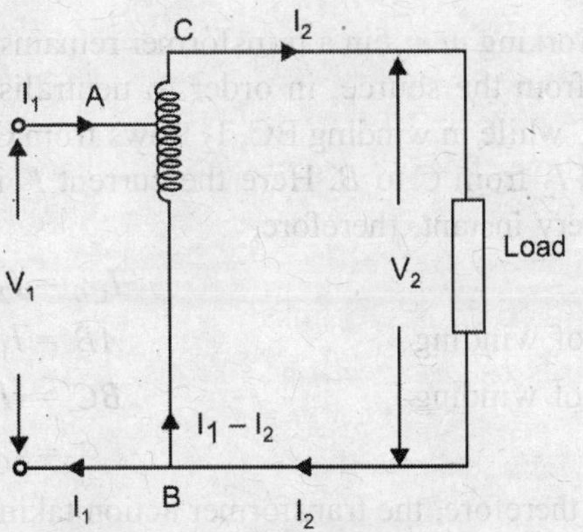

Fig. 1.20 (b) Step-up auto transformer

The above two figures show the basic circuit representation of auto transformer.

The operation is similar as two winding transformer but due to single winding the saving of copper is more in this type of transformer and copper loss also reduces by which high efficiency can be obtained.

In Fig. 1.20 (a) and (b)

$AB \rightarrow$ Primary winding with N_1 turns

$CB \rightarrow$ Secondary winding with N_2 turns

We know turns ratio, $$K = \frac{N_2}{N_1} = \frac{V_2}{V_1} = \frac{I_1}{I_2}$$

The current in section CB is the vector sum of I_2 and I_1.

$I_1 \rightarrow$ Primary current

$I_2 \rightarrow$ Secondary current

$V_1 \rightarrow$ Primary voltage

$V_2 \rightarrow$ Secondary voltage

It is same as two winding transformer but it has high efficiency and good and superior voltage regulation.

1.10.1 Operation of Auto Transformer

The circuit arrangement of the step down auto transformer is depicted in Fig. a and b. When voltage V_1 is applied to winding ABC, an exciting current starts flowing through the full winding AC. If the internal impedance drop is neglected, then the voltage per turn in winding AC is $\frac{V_1}{N_1}N_2$. In auto transformers a part of the power is conducted and the rest is transformed to load by transformer action. When the switch S is closed, a current I_2 starts flowing through the load.

Let terminal A in Fig. 1.21(a) be positive with respect to terminal C, at this instant the exciting current flows from A to C and it establishes a working *m.m.f* directed vertically downward in the core. When switch S is closed, the current in winding BC must flow from C to B, in order to create an *m.m.f.*

Since the working *m.m.f* in a transformer remains constant at no load, the primary must draw additional current I_1 from the source, in order to neutralise the effect of current I_{CB}. In winding AC, I_1 flows from A to C while in winding BC, I_2 flows from C to B. Therefore, the current in winding BC is I_1 from B to C and I_2 from C to B. Here the current I_2 is greater than I_1 and their m.m.fs are opposing each other at every instant, therefore,

$$I_{CB} = I_2 - I_1$$

m.m.f.of winding $$AB = I_1(N_1 - N_2)$$

m.m.f of winding $$BC = (I_2 - I_1)\, N_2 = I_{CB}\, N_2$$

It is in therefore, the transformer action taking place between terminal AB and AC. In other words, the volt amperes across winding AB are trasferred by transferred action to the load connected across winding BC.

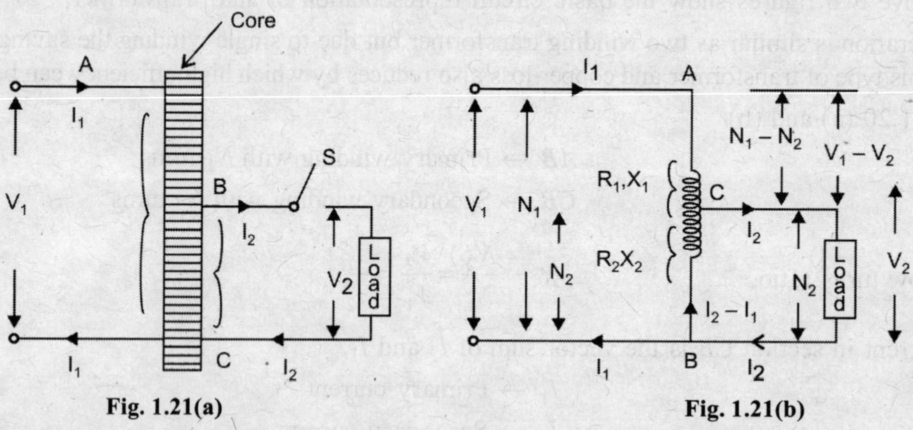

Fig. 1.21(a)

Fig. 1.21(b)

1.10.2 Volt Ampere Relation

(i) Step Down Transformer

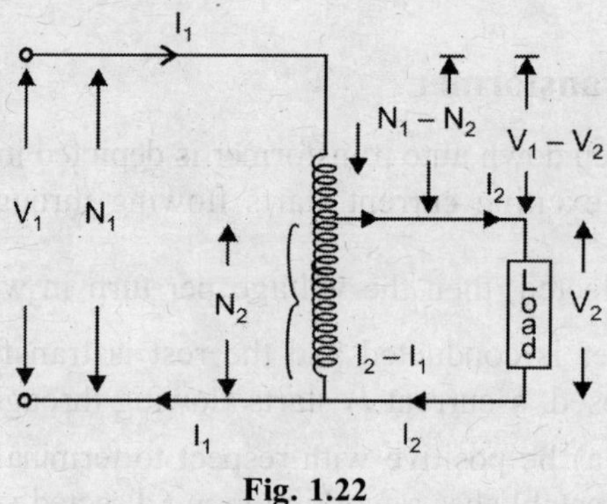

Fig. 1.22

In Fig.

$V_1 \rightarrow$ Primary voltage

$N_1 \rightarrow$ Primary number of turns

$I_1 \rightarrow$ Primary current

R_1 and $X_1 \rightarrow$ Resistance and reactance of primary

$V_2 \rightarrow$ Secondary voltage

$I_2 \rightarrow$ Secondary current

$N_2 \rightarrow$ Secondary number of turns

R_2 and $X_2 \rightarrow$ Secondary Resistance and reactance.

$N_1 - N_2 \rightarrow$ Number of turns between points AC

$V_1 - V_2 \rightarrow$ Voltage across points AC

$I_2 - I_1 \rightarrow$ Current through points BC

Voltage per turn in winding AB is $\frac{V_1}{N_1}$

i.e.,
$$\text{Voltage/turn} = \frac{\text{Total voltage}}{\text{Number of turns}}$$

$\therefore$ Voltage across winding CB is

$$V_{CB} = \left(\frac{V_1}{N_1}\right)N_2$$

Above equation is by voltage division rule.

In this Auto transformer we know,

$$\text{Input} = \text{output [Neglecting dosses]}$$

Where,
$$\text{Input power} = V_1 I_1 \cos\phi_1$$
$$\text{output power} = V_2 I_2 \cos\phi_2$$

If internal impedance drops and losses are neglected, then

$$\cos\phi_1 = \cos\phi_2$$

Then,
$$V_1 I_1 = V_2 I_2$$

or
$$\frac{V_1}{V_2} = \frac{I_2}{I_1} = \frac{N_1}{N_2} = \text{Turns ratio.}$$

The same calculation can be done for step up transformer also.)

1.10.3 Auto transformer efficiency

As we know the general formula for efficiency is ratio of output to input. The efficiency formula for auto transformer is also the ratio of output power to input power.

- Auto transformer has very high efficiency due to less copper loss.
- Due to less losses it has good voltage regulation.

1.10.4 Advantages

- Smaller in size
- Cheaper due to less amount of copper
- It has less copper loss
- High efficiency
- Variable output can be obtained
- It has good voltage regulation

1.10.5 Disadvantages

- There is some loss due to isolation problem between input and output circuit

1.10.6 Applications

1. It is used as a booster in distribution cable to correct voltage drop
2. It can be used as a starter to start induction type motors.
3. In can be used as control equipment such as electrical locomotives.

SHORT QUESTIONS AND ANSWERS

Q.1 Why *O.C* and *S.C* tests are conducted?

Ans. To determine the losses and thereby efficiency and regulation can be calculated.

Q.2 What are the information gained from *O.C* and *S.C* tests?

Ans. 1. Iron loss can be found from *O.C* test.

2. Copper loss can be found from *S.C* test.
3. Iron loss, $W_0 = V_0 I_0 \cos \phi_0$
4. Copper loss, $W_S = I_S^2 R_{01}$.

Q.3 Define efficiency.

Ans. It is defined as the ratio of output power to the input power.

Q.4 Define voltage regulation of transformer.

Ans. It is the ratio of total losses to the output power.

Q.5 What is the condition for maximum efficiency?

Ans. Variable loss = Constant loss

or Copper loss = Core loss

Q.6 What are the differant types of 3-phase transformer connections?

Ans. 1. $Y - Y$ connection

2. $\Delta - \Delta$ connection

3. $\Delta - Y$ connection

4. $Y - \Delta$ connection

5. $V - V$ connection

6. Scott connection

Q.7 What is the other name of $V - V$ connection and T connection

Ans. (i) Open delta connection

(ii) Scott connection

Q.8 What are the different types of six phase connections in transformer.

Ans. 1. Double delta connection

2. Double star connection

3. Six phase star connection

4. Diametrical connection

Q.9 What is auto transformer?

Ans. It is a static piece of device with one winding in which part of this being common to both primary and secondary winding.

Primary and secondary both are not electrically isolated from each other.

Q.10 What is the advantage of auto transformer over two winding transformer?

Ans. 1. Saving of copper due to only one winding.

2. High efficiency as composed to two winding transformer.

Q.11 Application of auto transformer.

Ans. This can be used as a booster in transmission system.

1.11 D. C MOTOR

DC motor is a machine which converts electrical power into mechanical power. This works on motor principle.

Wherever the current carrying conductor is placed in the magnetic field it experiences mechanical force.

Once the armature of the motor starts rotation in the magnetic fields then by Faraday's law of principle (just like generator) emf will be generated on it. The emf which is generated on armature opposes the applied voltage (Len'z law). Hence it is called as back emf (E_b)

The Eq. of E_b is

$$E_b = \frac{\phi ZNP}{60A} \quad (1.20)$$

Where, ϕ = Flux/pole

Z = Number of conductors

Where, Z = [Number of conductors/slot] × Number of slots

N = Speed of armature in RPM.

$P \rightarrow$ Number of poles

$A \rightarrow$ Number of parallel paths

A = 2(for wave winding)

A = P(for lap winding)

There are 3 types of DC motor, they are

(i) Series motor

(ii) Shunt motor

(iii) Compound motor

(i) Series Motor

By applying KVL in fig. (1.23*b*)

$$V_T = I_{se}R_{se} + I_aR_a - E_b$$

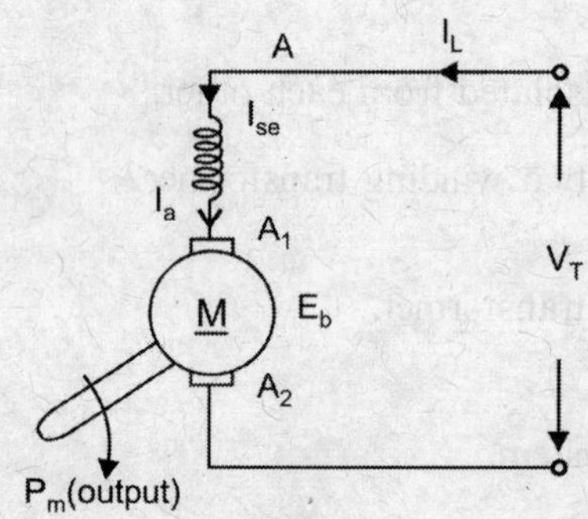

Fig. 1.23 (a)

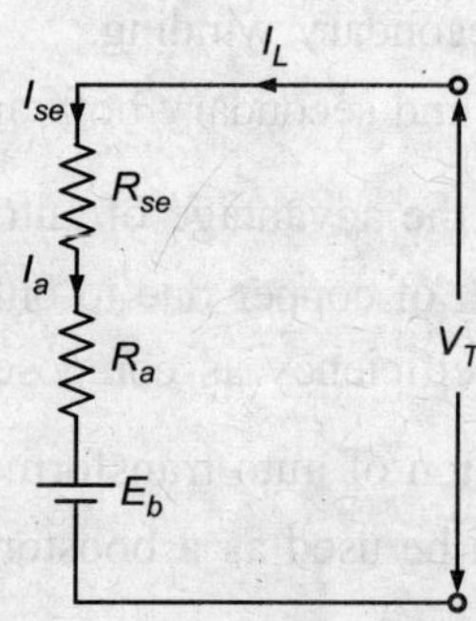

Fig 1.23 (b)

$\therefore$ $$E_b = V_T - I_a (R_a + R_{se}) \quad (1.21)$$

and $$I_a = I_{se} = I_L \quad (1.22)$$

Where, $E_b \rightarrow$ Back emf voltage

$V_T \rightarrow$ Terminal voltage

$I_a \rightarrow$ Armature current

$I_{se} \rightarrow$ Series field current

$R_{se} \rightarrow$ Series field resistance

$R_a \rightarrow$ Armature resistance.

$I_L \rightarrow$ Line current.

(ii) Shunt Motor

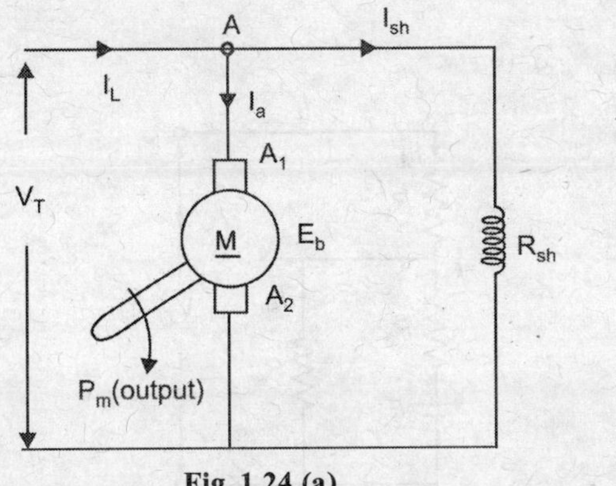

Fig. 1.24 (a)

⇒

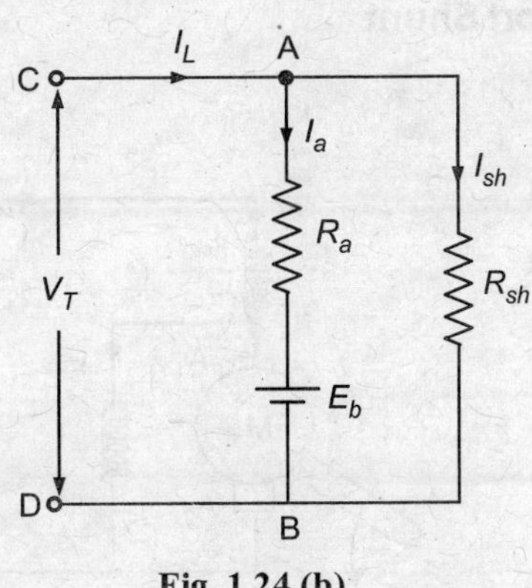

Fig. 1.24 (b)

At point A, in fig. 1.24(*b*) $I_L = I_a + I_{sh} \Rightarrow I_a = I_L - I_{sh}$ By KCL (1.23)

As we know the points in fig. 1.24(*b*) CD and AB are in parallel $V_T = E_b - I_aR_a$ by KVL

$$I_L = V_T/R_L \text{ By Ohm's Law} \quad (1.24)$$

$$I_{sh} = V_T/R_{sh} \text{ By Ohm's Law} \quad (1.25)$$

$$\boxed{E_b = V_T - I_aR_a}$$

(iii) Compound Motor

(i) Long Shunt

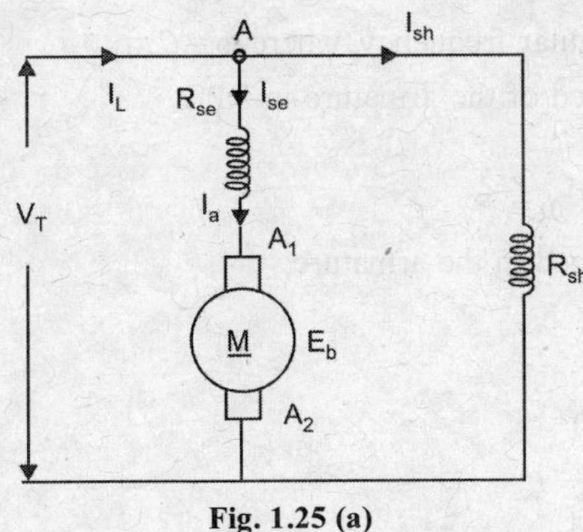

Fig. 1.25 (a)

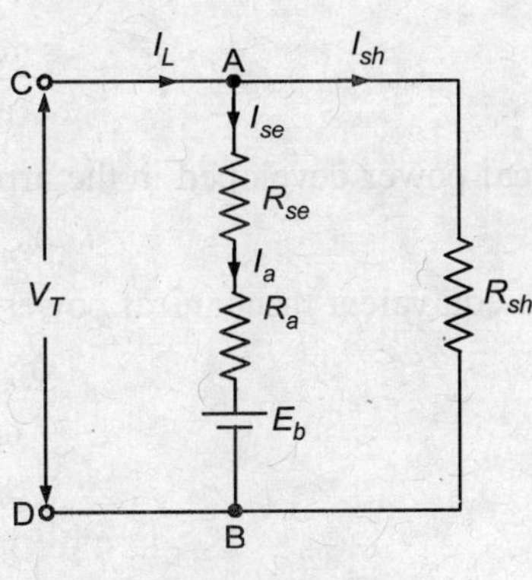

Fig. 1.25 (b)

From fig. 1.25*b*

The points *AB* and *CD* are in parallel, therefore,

$$V_T = I_{se}R_{se} + I_aR_a - E_b$$

Hence $E_b = V_T - I_a (R_a + R_{se})$ where $I_a = I_{se}$ (1.27)

By KCL, $$I_u = I_L - I_{sh} \tag{1.28}$$

(*ii*) Short Shunt

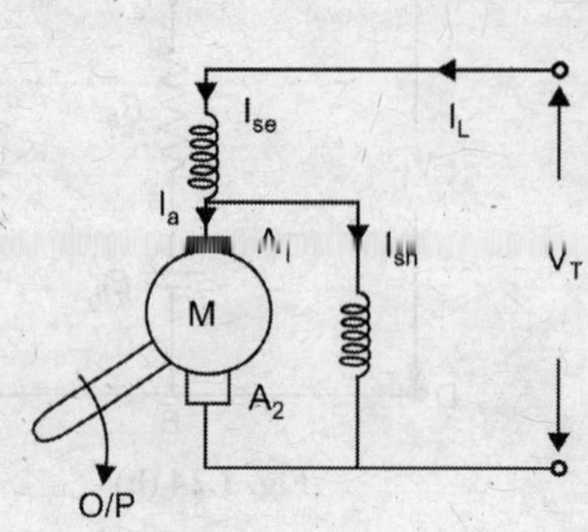

Fig. 1.26 (a)

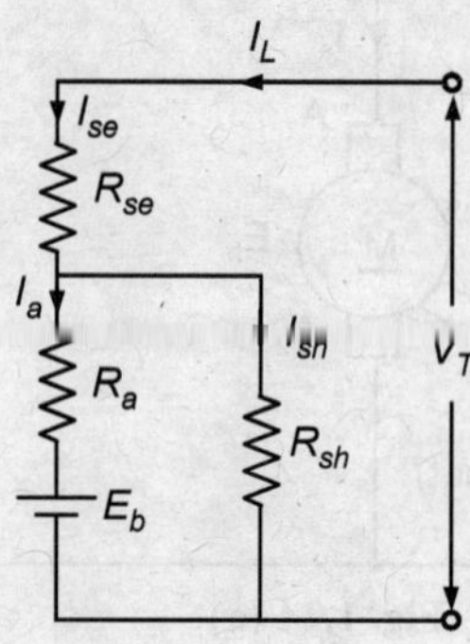

Fig. 1.26 (b)

From fig. (1.26(*b*)

$$I_L = I_{se} \tag{1.29}$$

By KCL, $$I_L = I_a + I_{sh} \Rightarrow I_a = I_L - I_{sh} \tag{1.30}$$

$$E_b = V_T - I_aR_a - I_{se}R_{se} \tag{1.31}$$

Torque Equation

T_a = Armature torque of dc motor

N = Speed of the armature in RPM

ω = Angular frequency, where $\omega = 2\pi\eta$

η = Speed of the armature in RPS

Mechanical power developed in the armature

$$\phi_m = T_a \times \omega = P_m$$

Electrical equivalent mechanical power developed in the armature,

$$\phi_{em} = E_b \times I_a$$

We know $$\omega = 2\pi n$$

$$n = N/60$$

$$\therefore \quad \omega = \frac{2\pi N}{60}$$

$$\therefore \quad P_m = \frac{2\pi N T_a}{60}$$

And also we know $$P_{em} = \frac{\phi \tau N P}{60 A} \times I_a$$

By equating P_m and P_{em}

$$\frac{2\pi N T_a}{60} = \frac{\phi \tau N P}{60 A} I_a$$

$$\boxed{T_a = \frac{\phi \tau P}{2\pi A} I_a \text{ N–m}} \quad ...(1.32)$$

where,

ϕ = Flux per pole

τ = Total No. of conductors

P = No. of Poles

A = Parallel Paths

In equation 1.32,

Z, p, 2π and A are constants

$$\therefore \quad \boxed{T_a \propto \phi I_a} \quad ...(1.33)$$

For Series Motor

$$\phi \propto I_{se} \propto I_a$$

$$\boxed{T \propto I_a^2} \quad ...(1.34)$$

As per equation (1.34) we conclude that series motor has high starting torque.

If the value of torque is varied from one stage to another stage then the equation can be written as the ratio of one stage to another stage,

$$\frac{Ta_1}{Ta_2} = \frac{Ia_1}{Ia_2} \times \frac{\phi_1}{\phi_2}$$

We know that flux is directly proportional to field current

$$\therefore \quad \frac{Ta_1}{Ta_2} = \frac{Ia_1}{Ia_2} \times \frac{If_1}{If_2} \quad ...(1.35)$$

For Shunt Motor

Flux ϕ is practically constant

$\therefore$ From equation 1.33

$$T_a \propto I_a$$

If the value of torque is varied from one stage to another stage then

$$\frac{Ta_1}{Ta_2} = \frac{Ia_1}{Ia_2} \quad \text{for shunt motor.}$$

$$\frac{Ta_1}{Ta_2} = \frac{Ia_1^2}{I_{a_2}^2} \quad \text{for series motor}$$

Speed Equation

As we know,
$$E_b = \frac{\phi \tau NP}{60A}$$
$$N = \frac{E_b}{\phi} \times \frac{60A}{\tau P}$$

The terms A, τ and P are constants

$$N \propto \frac{E_b}{\phi}$$

If the value of speed is varied from one range of speed into another speed then,

$$\frac{N_1}{N_2} = \frac{Eb_1}{Eb_2} \times \frac{\phi_2}{\phi_1}$$

For series motor, $\phi \propto I_{se}$

$$\therefore \quad \frac{N_1}{N_2} = \frac{Eb_1}{Eb_2} \times \frac{Ise_2}{Ise_1}$$

and also
$$I_{se} = I_a = \phi$$
$$\therefore \quad \frac{N_1}{N_2} = \frac{Eb_1}{Eb_2} \times \frac{Ia_2}{Ia_1} \quad ...(1.36a)$$

For shunt motor, ϕ is constant

$$\frac{N_1}{N_2} = \frac{Eb_1}{Eb_2} \quad ...(1.36b)$$

EXAMPLE 1.15: A 500 V dc shunt motor has an armature resistance of 0.6 Ω and shunt field resistance of 250 Ω. Evaluate the back emf when the motor output is 6 KW with efficiency of 80%.

Given data:
$$V_t = 500 \text{ V}$$
$$R_a = 0.6\ \Omega$$
$$R_{sh} = 250\ \Omega$$
$$P = 6 \text{ KW} = 6000 \text{ W}$$
$$\%\eta = 80\%$$

To find :
$$E_b = ?$$

Solution: We know
$$E_b = V_T - I_{ala}$$
$$I_a = I_L - I_{sh}$$
$$I_{sh} = \frac{V_T}{R_{sh}} = \frac{500}{250} = 2A$$

$$P = V_T I_L$$

$$6000 = 500 \times I_L \quad \Rightarrow \quad \boxed{I_L = 12\text{A}}$$

$$I_a = 12 - 2 = I_L - I_f = 10 \text{ A}$$

$$\therefore \quad E_b = 500 - 10(0.6) = 404\text{V}$$

Ans. $E_b = 404\text{V}$

EXAMPLE 1.16: A dc motor connected to a 400 V dc supply has an armature resistance of 1 Ω. Calculate the value of back emf, when the armature current is 100 A and what will be the armature current if back emf of dc motor is 380 V.

Given data:

$$V_t = 400 \text{ V}$$

$$R_a = 0.1 \text{ } \Omega$$

$$I_a = 100 \text{ A}$$

To find : $E_b = ?$

Solution:

$$E_b = V_T - I_a R_a$$

$$E_b = 400 - 100(0.1)$$

$$E_b = 400 - 10$$

$$\boxed{E_b = 390\text{V}}$$

(ii)

$$380 = 400 - I_a(0.1) - 20 = -0.1 I_a$$

$$I_a = 200 \text{ A}$$

Ans. $E_b = 390$ V, $I_a = 200$ A

EXAMPLE 1.17: A 300 V dc shunt motor runs at 700 rpm when the armature current is 40 A. What will be the new speed if the torque is doubled? Take $R_a = 0.3 \ \Omega$

Given data:

$$V_T = 300 \text{ V}$$

$$N_1 = 700 \text{ rpm}$$

$$I_a = 40 \text{ A}$$

$$R_a = 0.3 \ \Omega$$

To find : $N_2 = ?$

Solution:

$$\frac{Ta_1}{Ta_2} = \frac{40}{Ia_2}$$

$$\frac{Ta_1}{2Ta_1} = \frac{Ia_1}{Ia_2}$$

$$Ia_2 = 80\text{A}$$

We know,

$$E_b \propto N$$

$$\frac{Eb_1}{Eb_2} = \frac{N_1}{N_2}$$

$$\frac{V_T - Ia_1 R_a}{V_T - Ia_2 R_a} = \frac{300 - 40(.3)}{300 - 800(.3)} = \frac{700}{N_2}$$

$$\frac{288}{276} = \frac{700}{N_2}$$

$$1.043 = \frac{700}{N_2}$$

$$N_2 = \frac{700}{1.043} = 671 \text{ RPM}$$

Ans. $N_2 = 671$ RPM.

1.11.1 Concept of Starting

To understand the concept for starting methods of DC motor let us take one example of DC shunt motor.

We know,

$$E_b = V_T - I_a R_a$$

$$E_b - V_T = -I_a R_a$$

$$V_T - E_b = I_a R_a$$

$$\therefore \quad I_a = \frac{V_T - E_b}{R_a} \tag{1.37}$$

At the time of starting the value of E_b is Zero (*i.e.*, $E_b = 0$) substitute $E_b = 0$ in Eq. (1.37) we get

$$\therefore \quad I_a = \frac{V_T - 0}{R_a} \tag{1.38}$$

The Eq. (1.37) gives the value of I_a under running condition and the Eq. (1.38) gives the value of I_a under stand still condition (or) at the time of starting.

From Eq. (1.20), it is understood that at the time of starting the value of armature current is high. This high current may cause damages to the machine. So, it is important to limit the starting current. The starting current can be limited by using starters.

There are 3 types of *DC* starters. They are,

2 point starter

3 point starter

4 point starter

- 2 Point starter is used to start series motors.
- 3 Point starter is used to start shunt motors.
- 4 Point starter is used to start compound motors.

Now we will see all the types of starters in detail.

(i) Two point starter

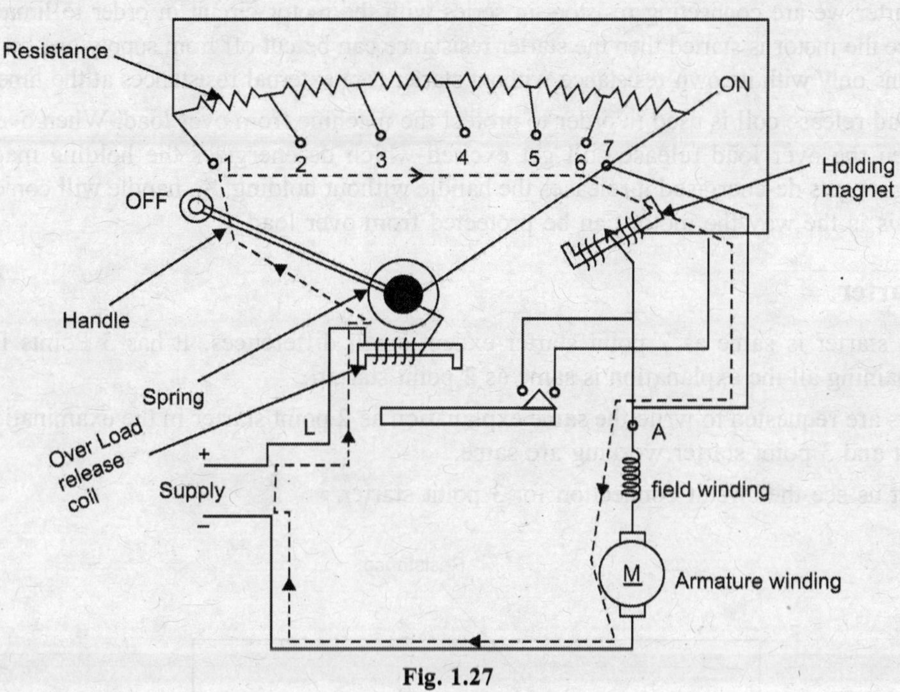

Fig. 1.27

Note: It has only two points. They are *L* and *A*.

where, $L \rightarrow$ Line
$A \rightarrow$ Armature

The Fig. (1.27) shows the circuit connection and internal connection of 2-point starter. It has two positions namely OFF position and *ON* position. It has number of resistors connected in series which is shown in Fig.(1.27) as 1, 2, 3, 7,

When supply is given to motor, first the handle is moved from left to right *i.e*, moved from OFF position to 1st position. When the handle is in position 1 then the circuit gets closed from positive terminal of supply to -ve terminal of supply through resistors which is shown in Fig. (1.27) by dotted line.

At the time of starting, the entire resistances of starter is connected in series with the motor by which the resistance value is increased. By Ohm's law we can understand that, when resistance value increases then the current values get reduced, by this way starting current is limited. As soon as motor is started then generated, back emf will be generated since back emf is generated we can cut the starter resistance one by one slowly by moving the handle from position 1 to 2, 3, 4 ..till last point. When the handle reaches last point then the handle gets held by holding magnet by magnetic field attraction. In Fig. 1.27 when handle is in last position there is no starter resistance included.

Conclusion

By using starter we are connecting resistors in series with the motor circuit in order to limit the starting current. Once the motor is started then the starter resistance can be cut off from supply one by one. Finally the motor runs only with its own resistance without starter (or) external resistances at the time of running.

Over load release coil is used in order to protect the machine from over load. When over voltage is supplied then the over load release coil get excited which de-energises the holding magnet. When holding magnet gets de-energised it releases the handle without holding. So handle will come to its *OFF* position. This is the way the motor can be protected from over load.

3 Point Starter

Three point starter is same as 2 point starter except small differences. It has 3 Points instead of 2 points. Remaining all the explanation is same as 2 point starter.

Students are requested to write the same explanation as 2 point starter in the examination. Because both 2 point and 3 point starter working are same.

Now let us see the circuit connection for 3 point starter.

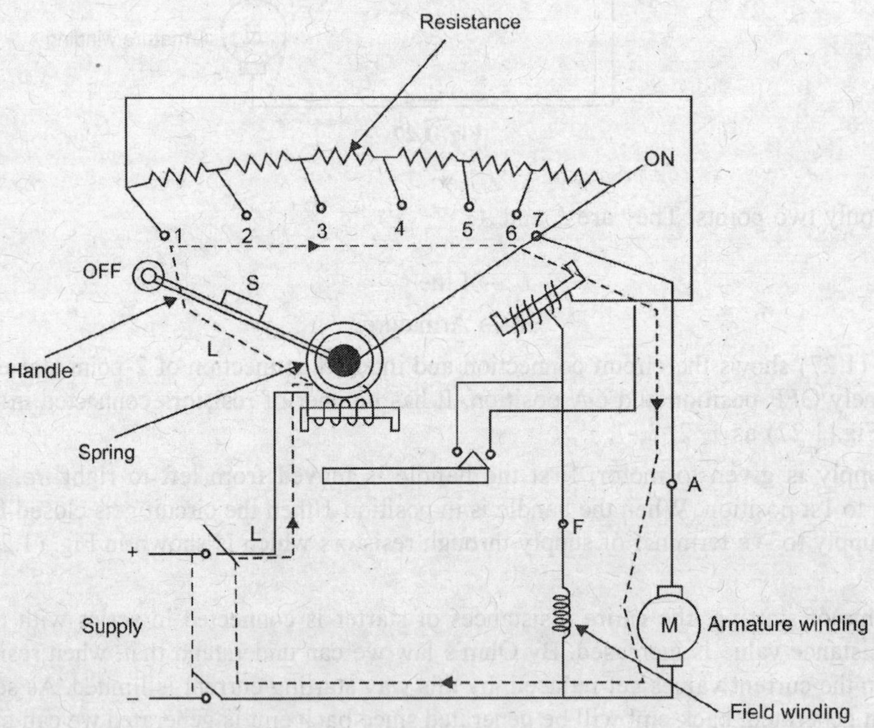

Fig. 1.28

From Fig. (1.28) we can see that there are three points namely *L*, *F* and *A*.

$L \rightarrow$ Line
$F \rightarrow$ Field
$A \rightarrow$ Armature

In case of series motor, series field and armature are connected in series so, it has only one point. So it has only two points namely,

$$L \rightarrow \text{Line}$$
$$A \rightarrow \text{Armature}$$

4–Point starter

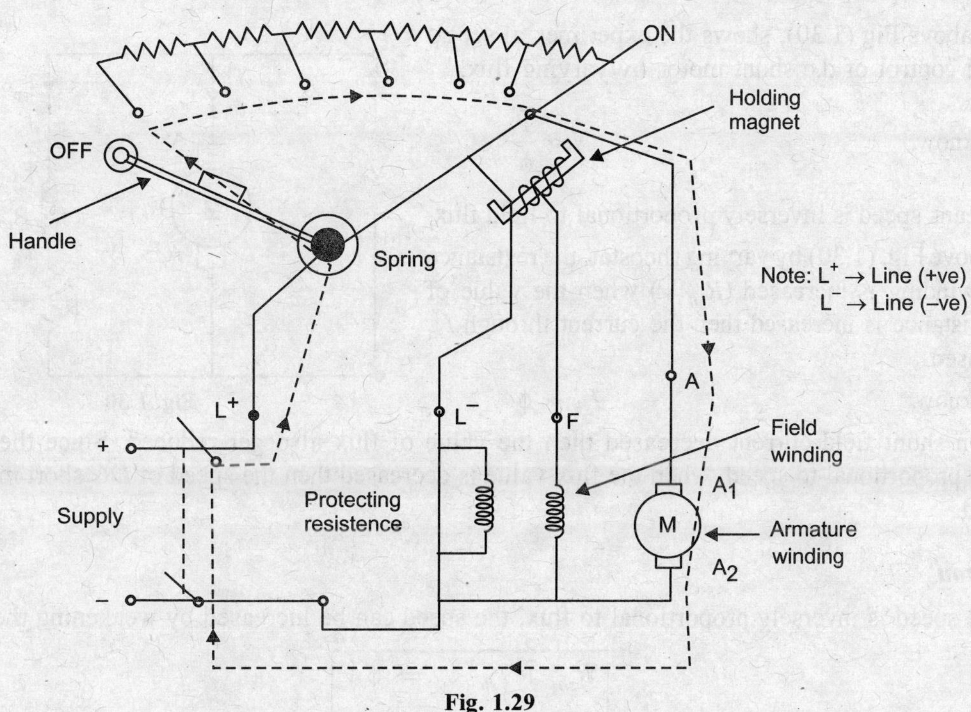

Fig. 1.29

This pointer is different from others. It has four points L^{+}, L^{-}, F and A compared to 3 points starter. The holding coil has been taken out of the Shunt field circuit and has been connected directly across the line through a protecting resistance. The remaining working principle is same as other pointers.

1.11.2 Speed control of DC motor

What is speed control?

Obtaining various speed ranges from a single machine is called speed control, eg: ceiling fan with regulator.

we know,

$$N = \frac{E_b \times 60 \times A}{\phi ZP} \tag{1.39}$$

Where, A, Z and P are constants.

$$\therefore \quad N \propto \frac{E_b}{\phi} \tag{1.40}$$

From above Eq. it is understood that speed can be varied by varying back emf (E_b) or by flux (ϕ)

Now we will see the speed control of *D-C* motor.

(I) Speed Control of* D.C *Shunt Motor

(i) Flux Control Method (or) Field Control (or) Variation flux Method

The above Fig (1.30), shows the experimental set up for speed control of d.c shunt motor by varying flux.

We know, $$N \propto \frac{1}{\phi}$$

It means speed is inversely proportional to field flux.

In above Fig. (1.30) by varying rheostat, the resistance of field winding is increased ($R_{sh} \uparrow$) when the value of shunt resistance is increased then the current through I_{sh} is decreased.

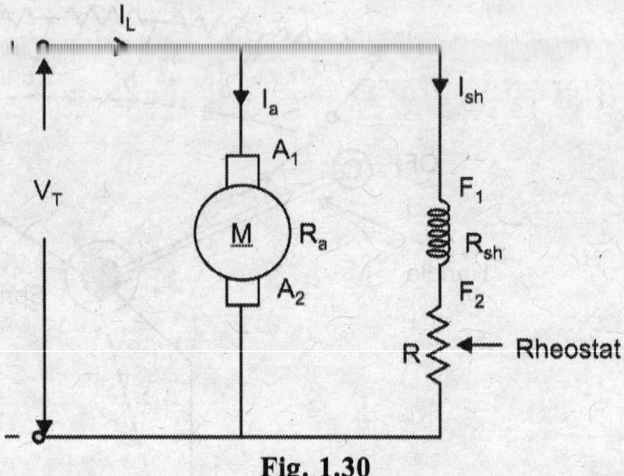

Fig. 1.30

We know, $$I_{sh} \propto \phi$$

When shunt field current decreased then the value of flux also get reduced. Since the flux is inversely proportional to speed, when the flux value is decreased then the speed of *DC* short motor get increased.

Conclusion

Since the speed is inversely proportional to flux, the speed can be increased by weakening the field.

$$R_{sh} \uparrow I_{sh} \downarrow \Rightarrow \phi \downarrow$$
$$\text{Since N} \propto \frac{1}{\phi}$$
$$\text{If } \phi \downarrow \Rightarrow \text{N} \uparrow$$

(ii) Armature Control

By varying the armature rheostat, the value of armature resitance can be increased.

When the value of armature resistance is increased then the voltage drop across armature resistance also increased. As we know that,

$$E_b = V_T - I_a R_a$$

When $I_a R_a$ drop increased then the value of back emf get reduced.

We know,

$$N \propto E_b$$

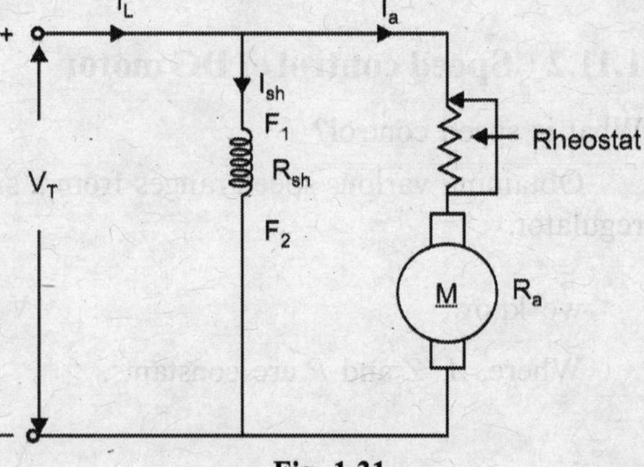

Fig. 1.31

When the value of back emf get reduced then the speed is also reduced.

Conclusion

Since the speed is directly proportional to back emf, the speed can be decreased by decreasing the value of back emf.

$$R_a \uparrow I_a R_a \uparrow \Rightarrow E_b \downarrow$$
$$\text{Since N} \propto E_b$$
$$\text{If } E_b \downarrow \Rightarrow N \downarrow$$

(iii) Voltage Control method (or) ward-leonard system

This kind of speed control is used in colliery winders, electric excavators, elevators etc.,

The main motor whose speed has to be controlled, the field of this motor is permanently connected to DC supply. By applying a variable voltage across armature, any required speed can be obtained. This variable voltage is supplied by a motor generator set which can be of *D.C* (or) *A.C.*

This method has following disadvantages.

1. Low efficiency at light loads
2. Costly

(II) Speed Control of DC Series Motor

(i) Flux Control Method

There are 4 types in flux control method. They are,

1. Field diverter method
2. Armature diverter method
3. Tapped field control
4. Parallelling field

(1) Field Diverter Method

The above Fig. (1.32) shows the experimental set up of field diverter method of speed control. In this generally the series winding is shunted by a variable resistance which is known as field diverter. Any designed amount of current can be passed through the diverter by adjusting its resistance value. Hence the value of flux can be varied.

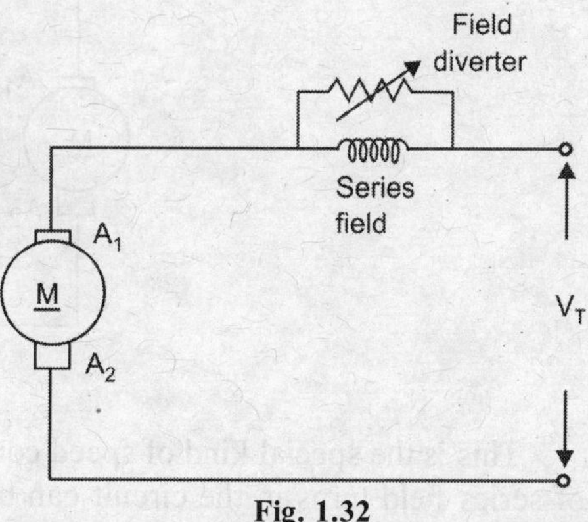

Fig. 1.32

By diverting some amount of current through diverter, the current flowing through series field winding is reduced.

We know, $I_{se} \propto \phi$

Therefore, the amount of flux value is also decreased.

Again we know, $N \propto \dfrac{1}{\phi}$

Therefore, the speed can be increased.

$$\text{By using diverter } I_{se} \downarrow \Rightarrow \phi \downarrow \Rightarrow N \uparrow$$

(2) Armature Diverter

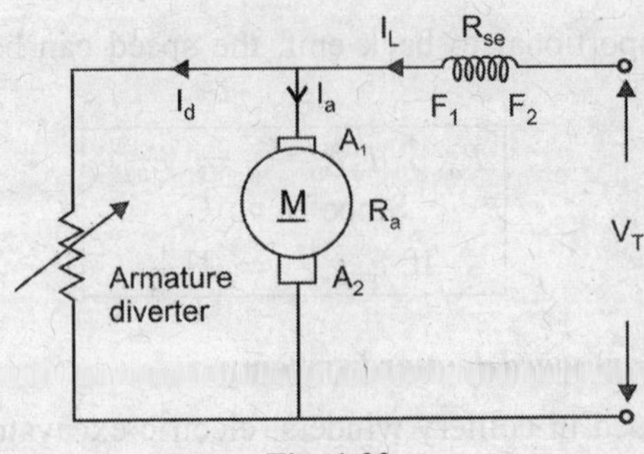

Fig. 1.33

where, $I_d \rightarrow$ Current in Armature diverter

By connecting diverter (variable rheostat) Parallel to armature, the value of armature current is reduced.

In series motor we know $I_{se} \propto I_a \propto \phi$

$\therefore$ The value of flux get decreased.

We know, $N \propto \dfrac{1}{\phi}$

Therefore, speed can be increased.

By diverter,
$I_a \downarrow \Rightarrow \phi \downarrow \Rightarrow N \uparrow$

(3) Tapped Field Control

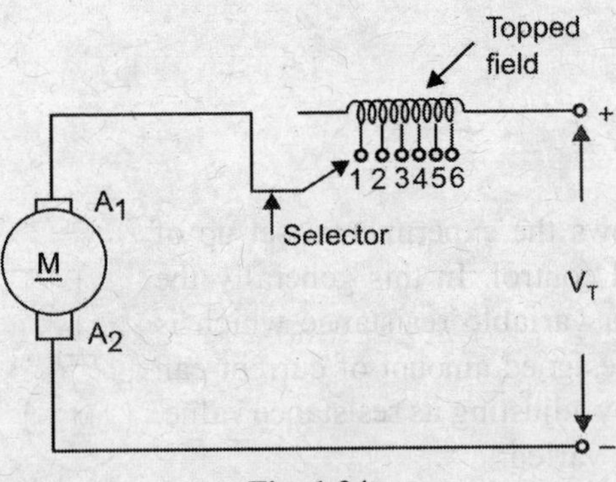

Fig. 1.34

This is the special kind of speed control. This method is often used in electric traction. The number of series field turns in the circuit can be changed by selector.

When the selector is in position 1, it means the full winding is selected (*i.e* high resistance is selected). Under this condition the value of field current will be therefore flux produced by field current also low and, ultimately speed will be very high.

When the selector is in position 6, it means the minimum part of the winding is selected, i.e, low resistance is selected. Under this condition the value of field current will be high so, the flux produced also high therefore, speed will be very low. From above points we can understand that when the selector moves from position 1 to 6 the speed start decreasing.

Position 1: $R \rightarrow$ high
$I \rightarrow$ low
$\phi \rightarrow$ low
$N \rightarrow$ high

Position 6: $R \rightarrow$ low
$I \rightarrow$ high
$\phi \rightarrow$ ligh
$N \rightarrow$ low

Speed start decreasing from maximum speed to minimum speed when selector moves from position 1 to 6.

(4) Paralleling Field Method

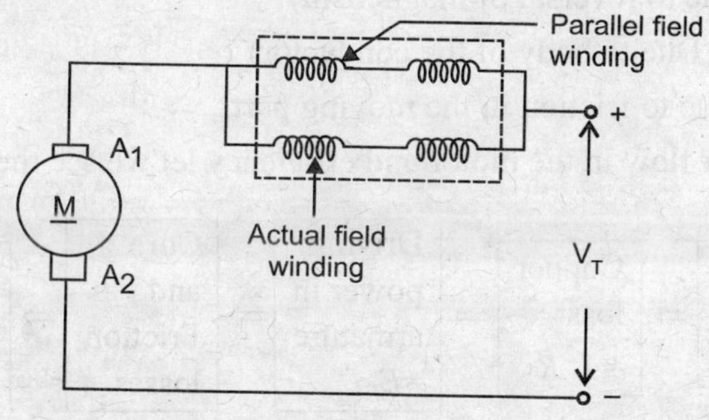

Fig. 1.35

In this method, as the name indicates the paralleling field, this has another field winding in parallel to actual field winding in order to obtain different range of speeds.

This method is used to obtain several speed ranges.

(ii) Armature Control

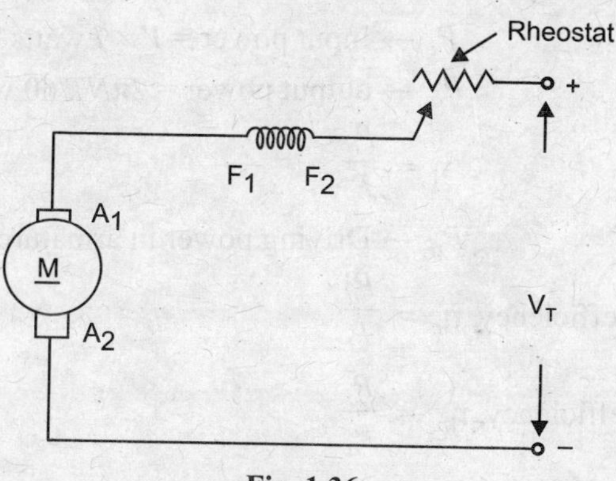

Fig. 1.36

This is same as shunt motor armature control. By varying the rheostat value, the value of armature resistance is increased and the value of back emf is reduced. as back emf is reduced, the speed also decreased.

$$R_a \uparrow \Rightarrow I_a R_a \uparrow \Rightarrow E_b \downarrow \Rightarrow N \downarrow$$

1.11.3 Losses and efficiency of a DC motor

There are three different losses in DC motor They are,

1. Copper loss
2. Magnetic (or) core loss
3. Mechanical loss

Copper loss is the loss which occurs in the motor due to current flow in the windings. magnetic losses are two types. They are

1. Hysterisis loss (Due to reversal of magnetism)
2. Eddy current loss (Due to body of the conductor)

Mechanical losses are due to friction in the moving part.

To understand the power flow in the motor and efficiency let we see the Fig. 1.37

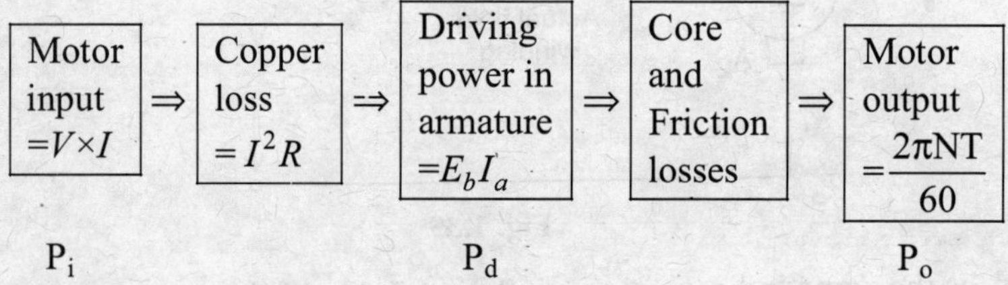

Fig. 1.37: Power stage of DC motor.

From above Fig. 1.37

Various formula's for calculating losses and efficiency can be formed.

Let, $P_i \rightarrow$ Input power $= V \times I$ watts

$P_o \rightarrow$ output power $= 2\pi NT/60$ watts

Over all efficiency, $$\eta = \frac{P_o}{P_i} \quad (1.41)$$

Let, $P_d \rightarrow$ Driving power in armature $= E_b I_a$

$\therefore$ Electrical efficiency, $$\eta_e = \frac{P_d}{P_i} \quad (1.42)$$

Mechanical efficiency, $$\eta_m = \frac{P_o}{P_d} \quad (1.43)$$

EXAMPLE 1.15: A 240 *V* shunt motor has an armature resistance of .8Ω and field resistance of 120Ω. The motor draws 6*A* at 1400 *r.p.m.* at no load. Calculate the speed when the motor draws 60*A* at rated voltage.

Given data:

$$R_a = .8\Omega$$
$$R_{sh} = 120\ \Omega$$
$$I_{L1} = 6A$$
$$N_1 = 1400\ r.p.m$$
$$I_{L2} = 60A$$
$$N_2 = ?$$

Solution:

$$I_{sh} = \frac{240}{120} = 2A$$
$$I_{a1} = 6 - 2 = 4A$$
$$I_{a2} = 60 - 2 = 58A \qquad [\therefore I_a = I_L - I_{sh}]$$
$$\therefore \quad E_{b1} = V_T - I_{a1}R_a = 240 - (4)(.8) = 263.8V$$
$$E_{b2} = V_T - I_{a2}R_a = 240 - 58(.8) = 193.6V$$

We know

$$N_2 = \frac{Eb_2}{Eb_1} \times N_1$$
$$N_2 = \frac{193.6}{236.8} \times 1400 = 1145\ R.P.M$$
$$\boxed{N_2 = 1145 R.P.M}$$

Ans. $N_2 = 1145$ *r.p.m*

EXAMPLE 1.16: A 240 V shunt motor on no load runs at 1200 r.p.m and takes 6A. The armature and shunt field resistances are .4Ω and 120Ω respectively. Calculate the speed when it draws 40A. Assume the armature reaction weakens the field by 5%

Given data:

$$V_T = 240V$$
$$N_1 = 1200\ r.p.m$$
$$I_{L1} = 6A$$
$$R_a = .4\Omega$$
$$R_{sh} = 120\Omega$$
$$I_{L2} = 40A$$
$$N_2 = ?$$

Solution:

$$I_{Sh} = \frac{V_T}{R_{sh}} = \frac{240}{120} = 2A$$

$$I_{a1} = I_{L1} - I_{sh} = 6 - 2 = 4A$$
$$I_{a2} = I_{L2} - I_{sh} = 40 - 2 = 38A$$
$$\therefore \quad E_{b1} = V_T - I_{a1}R_a = 240 - 4(.4) = 238.4\ V$$
$$E_{b2} = V_T - I_{a2}R_a = 240 - 38(.4) = 224.8V$$

Since the armature reaction weakens the field by 5%,

$$\phi_2 = .95\phi_1$$
$$\therefore \quad N_2 = \left(\frac{E_{b2}}{E_{b1}}\right) \times \left(\frac{\phi_1}{.95\phi_1}\right) \times 1200 \qquad \because N_2 = \frac{E_{b2}}{E_{b1}} \times \frac{\phi_1}{\phi_2} \times N_1$$
$$= \left(\frac{224.8}{238.4}\right)\left(\frac{1}{.95}\right) \times 1200$$
$$\boxed{N_2 = 1188 R.P.M}$$

Ans. $N_2 = 1188\ RPM$

EXAMPLE 1.17: A 400V d.c shunt motor takes a current of 7A on no load the resistance of the armature and field circuit are .4Ω and 200Ω respectively find (a) Efficiency when it draws 80A (b) percentage change in speed.

Given data:

$$V_T = 400V$$
$$I_{L1} = 7A$$
$$R_a = .4\Omega$$
$$R_{sh} = 200\Omega$$

Solution:

$$I_{sh} = \frac{V_T}{R_{sh}} = \frac{400}{200} = 2A$$
$$I_{a1} = 7 - 2 = 5A \qquad \because I_{a1} = I_{L1} - I_{sh}$$
$$E_{b1} = V_T - I_{a1}R_a = 400 - 5(.4)$$
$$E_{b1} = 398V$$

Armature copper loss = $I_{a1}^2 \times R_a = (5)^2(.4)$

$$\text{Armature cu loss} = 10 \text{ watts}$$
$$\text{Motor input} = V_T \times I_L = 400 \times 7$$
$$P_i = 2800 \text{ watts}$$
$$\therefore \quad \text{Constant losses} = 2800 - 10 = 2790 \text{ watts}$$

We know,

$$I_{L2} = 80A$$
$$I_{a2} = I_{L2} - I_{sh}$$
$$= 80 - 2 = 78A$$
$$E_{b2} = V_T - I_{a2}R_a$$
$$= 400 - (78)(.4)$$
$$\boxed{E_{b2} = 368.81}$$

$$\text{Armature copper loss} = (78)^2(.4)$$
$$= 2433.6 \text{ 'watts'}$$
$$\text{Total losses} = 2433.6 + 2790$$
$$= 5223.6 \text{ 'watts'}$$
$$\text{Motor input} = V_T \times I_L = 400 \times 80$$
$$= 32000 \text{ 'watts'}$$
$$\text{Motor output} = 32000 - 5223.6$$
$$= 26776.4 \text{ 'watts'}$$
$$\therefore \quad \text{Efficiency, } \eta = \frac{\text{output}}{\text{input}} = \frac{26776.4}{32000} = .8367$$
$$\%\eta = .8367 \times 100$$
$$\boxed{\eta = 83.67\%}$$
$$\text{Percentage change in speed} = \frac{N_2 - N_1}{N_1}$$
$$\text{Which is also equal to} = \frac{E_{b2} - E_{b1}}{E_{b1}} = \frac{368.81 - 398}{398} \times 100 = 7.3\%$$
$$\text{Percentage change in speed} = 7.3\%$$

Ans. (a) $\eta = 83.67\%$; (b) Change in speed = 7.3%

EXAMPLE 1.18: The armature winding of a 4 pole, 250V d.c shunt motor is lap connected. There are 120 slots, each slot containing 8 conductors. The flux per pole is 20 mwb and current taken by the motor is 25A. the resistance of armature and field circuit are 0.1 and 125Ω respectively. If the rotational losses amount to be 810W find efficiency.

Given data:

$$P = 4$$
$$V_T = 250V$$
$$Z = 120 \times 8 = 960 \text{ conductors}$$
$$\phi = 20 \times 10^{-3}\, wb$$
$$I_L = 25A$$
$$R_a = .1\Omega$$
$$R_{sh} = 125\Omega$$
$$\text{Rotational losses} = 810W$$
$$\eta = ?$$

Solution:

$$I_{sh} = \frac{V_T}{R_{sh}} = \frac{250}{125} = 2A$$

$$I_a = I_L - I_{sh}$$

$$I_a = 25 - 2 = 23A$$

$$\therefore \quad E_b = V_T - I_a R_a = 250 - (23 \times 0.1) = 247.7 \text{ V}$$

$$E_b = \frac{\phi ZNP}{60 \times A}$$

$$= \frac{20 \times 10^{-3} \times 9.60 \times N \times 4}{60 \times 4} = 247.7V$$

$$\therefore \quad N = 773 \ r.p.m$$

Armature Cu loss $= (23)^2 \times 0.1 = 53W$

Shunt Cu loss $= 250 \times 2 = 500W$

Rotational losses $= 810W$

Total motor losses $= 810 + 500 + 53$

$= 1363 \ W$

Motor input $= 250 \times 25 = 6250W$

Motor output $= 6250 - 1363 = 4887W$

$$\therefore \quad \eta = \frac{4887}{6250} = .782$$

$$\therefore \quad \eta = 78.2\%$$

Ans. $\% \ \eta = 78.2\%$

EXERCISE

13. A 220 V shunt motor has an armature resistance of 0.2Ω and field resistance of 110Ω. The motor draws 5A 1500 $r.p.m$ at no-load. Calculate the speed and shaft torque if the motor draws 52 A at rated voltage.

[**Ans.** N = 1426 $r.p.m$; T_{sh} = 65.5 N–M]

14. A 250V shunt motor on no loads runs at 100 $r.p.m$ and takes 5A. Armature and shunt field resistances are .2 and 250W respectively calculate the speed when Load is taking a current of 50A. The armature reaction weakens the field by 3%.

[**Ans.** N = 944 $r.p.m$]

15. A 500 V D C shunt motor takes current of 5A on no-load the resistances of the armature and field circuit are .22W and 125W respectively. Find (a) the efficiency when load is and taking a current of 100A (b) The percentage change of speed.

[**Ans.** (i) η = 90.8%;
(ii) % change in speed = 4.18%]

SHORT QUESTION AND ANSWERS

Q.12 What is motor?

Ans. Motor is a machine which converts electrical energy into mechanical energy.

Q.13 Why starter is needed to start *DC* motor?

Ans. *DC* motor is having high starting current at the time of starting so in order to start them we should limit the starting current. So by using starters we are limiting the starting current so as to start *DC* motors safely.

Q.14 What are the types of *DC* starts?

Ans. 1. 2 Point starter
2. 3 Point starter
3. 4 Point starter

Q.15 What is speed control?

Ans. Speed control is a technique which is used to obtain variable speed in single motor.

Q.16 What are the types of speed control in *DC* shunt motor?

Ans. 1. Voltage control (ward leonard)
2. Flux control
3. Armature control

Q.17 Types of speed control in *DC* series motor?

Ans. 1. Field diverter method
2. Armature diverter method
3. Paralleling field method

Q.18 What are the types of losses in *DC* motor.

Ans. 1. Mechanical losses
2. Core loss
3. Copper loss

CHAPTER 2

Three Phase Induction Motor

2.1 CONSTRUCTION OF THREE INDUCTION MOTOR

It has two main parts

1. Stator
2. Rotor

Stator

It consist of two main parts slots and three phase winding.

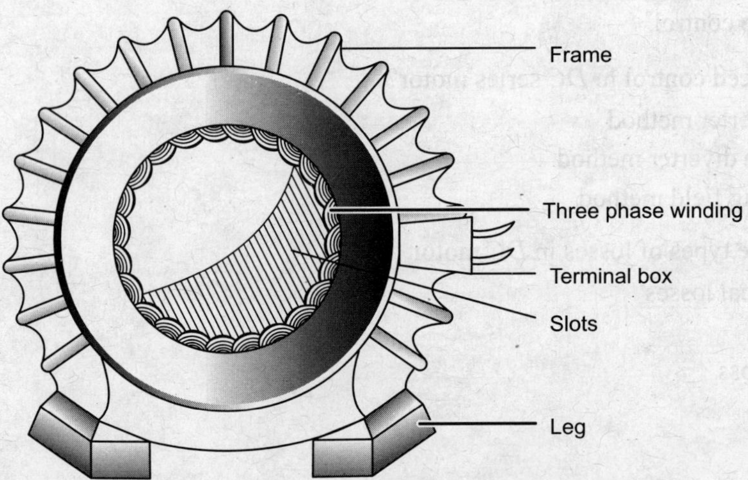

Fig. 2.1: Stator and frame construction of three phase induction motor.

- The stator of an induction motor principle is same as that of synchronous motor.
- It is made up of a number of stampings, which are slotted to receive the windings.
- The stator carries a three phase winding.
- Three phase winding fed from 3-phase supply it is wound for a definite number of poles.
- The exact number of poles being determined by the requirement of speed.

Rotor

The types of induction motor is classified by rotor construction

- Squirrel cage rotor type induction motor
- Slipring rotor type induction motor

(i) Squirrel cage rotor

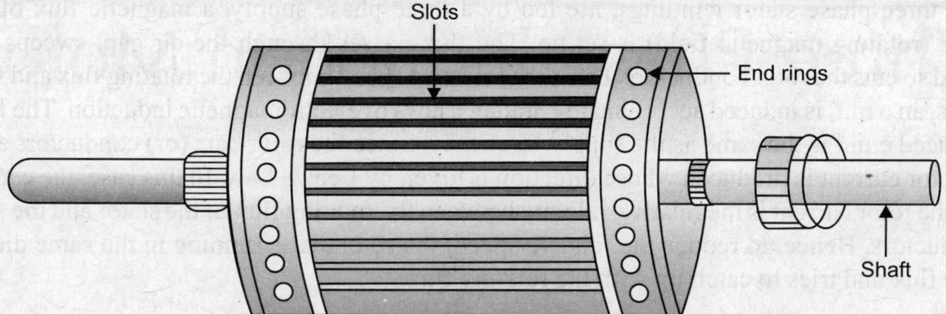

Fig. 2.2: Constructional view of slipring rotor.

- Almost 90 percent of induction motors are squirrel-cage type, because this type of rotor has the simplest and most rugged construction. The rotor consists of a cylindrical laminated core with parallel slots for carrying rotor conductors. It should be noted clearly, that they are not wires but consists of heavy bars of copper, aluminium (or) alloys. One bar is placed in each slot, rather the bars are inserted from the end when semi-closed slots are used.
- End rings are short circuited.

(ii) Slip ring type rotor

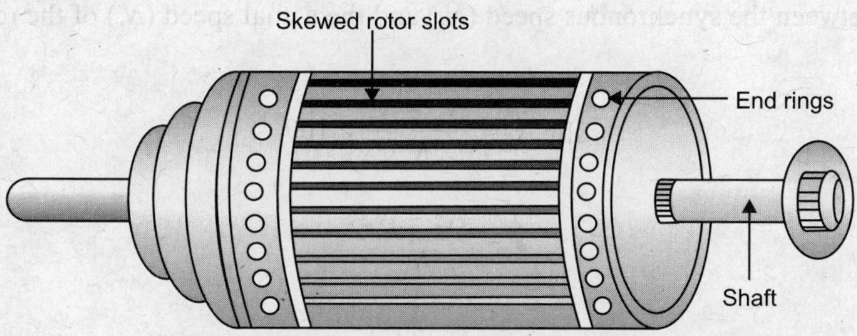

Fig. 2.3: Constructional view of slipring rotor.

- The rotor is wound for as many poles as the number of stator poles and is always wound for 3-phase.
- The winding terminals are brought out and connected to three insulated slip rings.
- External resistance can be connected.
- End rings are not short circuited.

Frame: Made up of grained alloy cast-iron.

Stator and rotor: Built from high-quality low loss silicon steel lamination.

- Ball and roller bearings are used.

2.2 PRINCIPLE OF OPERATION

When the three phase stator windings, are fed by a three phase supply, a magnetic flux of constant magnitude (rotating magnetic field) is set up. The flux passes through the air gap, sweeps the rotor surface and so cuts the rotor conductors. Due to the relative speed between the rotating flux and stationary conductors, an e.m.f. is induced according to Faraday's laws of electromagnetic induction. The frequency of the induced e.m.f is the same as the supply frequency. Since the rotor bars (or) conductors are closed circuits, rotor current is produced whose direction is given by Lenz's Law. In this case, the cause which produces the rotor current is the relative velocity between the rotating flux of the stator and the stationary rotor conductors. Hence, to reduce the relative speed, the rotor starts running in the same direction as that of the flux and tries to catch up with the rotating flux.

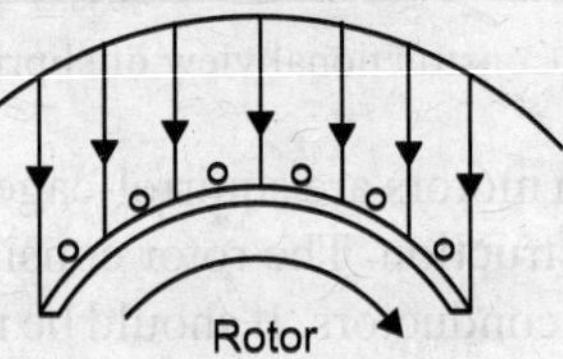

Fig. 2.4: Representation of principle of operation of three phase induction motor.

2.3 SLIP

The difference between the synchronous speed (N_s) and the actual speed (N_r) of the rotor is known as slip.

$$\%\text{ slip, } s = \frac{N_s - N_r}{N_s} \times 100 \quad (2.1)$$

$$s = \frac{N_s - N_r}{N_s} \quad (2.2)$$

$$s = 1 - \frac{N_r}{N_s}$$

$$\frac{N_r}{N_s} = (1 - s)$$

$$\therefore \quad \boxed{N_r = N_s(1 - s)} \quad (2.3)$$

Note: The flux which is produced by the stator winding is rotating in nature. The speed of the stator field depends on the frequency of supply and the number of poles. i.e.

$$N_s = \frac{120f}{P}$$

Frequency of rotor current is,

$$f' = s_f \quad (2.4)$$

2.4 SLIP TORQUE CHARACTERISTICS

Let we see the relation between slip and torque. The value of slip will be maximum when the speed of the rotor is decreased and we known that the speed and torque both are inversely proportional to each other.

From the above discussion.

$$s \propto \frac{1}{N}$$

$$N \propto \frac{1}{T}$$

$$\therefore \quad \boxed{s \propto T} \quad (2.5)$$

i.e. the slip is directly proportional to torque.

At the time of starting, $N_r = 0$

Substituting $N_r = 0$ in Eq. (2.2)

$$s = \frac{N_S - 0}{N_S}$$

i.e $s = 1$

At the time of starting the value of slip value will be equal to one.

The value of torque of a three phase induction motor depends on not only slip but also depends on rotor resistor (R_2) i.e. torque is directly proportional to slip and inversely proportional to rotor resistor (R_2).

i.e $T \propto \frac{s}{R_2}$; If R_2 is constant $T \propto s$.

For low values of slip, the torque/slip curve is approximately a straight line. As slip increase the torque also increases and becomes maximum when $s = \frac{R_2}{X_2}$. This torque is known as pull-out torque (or) breakdown torque. As the slip increases further with increase in load, then R_2 becomes negligible as compared to $s\,X_2$.

$\therefore$ For large values of slip

$$T \propto \frac{1}{s}$$

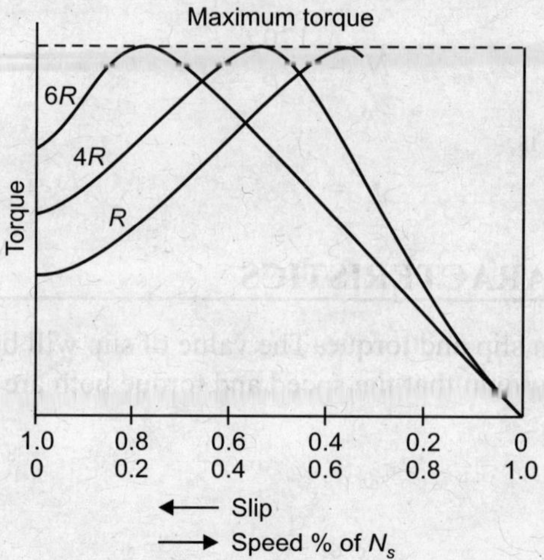

Fig. 2.5: Torque-slip characteristic of 3 phase induction motor.

2.5 APPLICATIONS

(i) Squirrel cage induction motor

1. Fans
2. Blowers
3. Centrifugal pumps
4. Conveyor belt system
5. Compressors
6. Crushers
7. Reciprocating pumps
8. Shearing machine
9. Punch pres and stamping

(ii) Slip ring induction motor

1. Hoist
2. Cranes
3. Compressors
4. Conveyors

EXAMPLE 2.1: Calculate the synchronous speed of a 6 pole, three phase induction motor which is getting 440 V, 50 Hz supply. Calculate the slip if motor is running at 960 r.p.m.

Given data:

$$P = 6$$
$$V = 440\ V$$
$$f = 50\text{ Hz}$$
$$N_r = 960\text{ r.p.m}$$

To find:

$$N_S = ?$$
$$s = ?$$

Solution:

$$N_S = \frac{120 f}{P} = \frac{120 \times 50}{6}$$

$$N_S = 1000 \text{ r.p.m}$$

$$\%s = \frac{N_S - N_r}{N_S} \times 100 = \frac{1000 - 960}{1000} \times 100$$

$$\%s = 4\%$$

Result: $N_S = 1000$ r.p.m.

$\%S = 4\%$

EXAMPLE 2.2: A 3-phase 4 pole induction motor is running with 6% slip. If supply frequency is 50 Hz then find the speed of the induction motor?

Given data:

$$\%s = 6\%$$

$$f = 50 \text{ Hz}$$

To find:

$$N_r = ?$$

Solution:

$$N_S = \frac{120 f}{P} = \frac{120 \times 50}{4}$$

$$= 1500 \text{ r.p.m}$$

$$N_r = N_S(1 - s)$$

$$N_r = 1500\left(1 - \frac{6}{100}\right)$$

$$N_r = 1500\ (1 - 0.06)$$

$$N_r = 1410 \text{ r.p.m.}$$

Result: $N_r = 1410$ r.p.m

2.6 TORQUE EQUATION OF AN 3 PHASE INDUCTION MONITOR

Torque is a twisting moment for rotor. We can classify the torque as different ways, they are starting torque and running torque.

Starting Torque

The torque developed on the rotor at the time of starting is called starting torque. This starting torque may be greater than the normal running torque (or) some cases it can be Less also.

Let,

$E_2 \rightarrow$ Rotor emf per phase

$R_2 \rightarrow$ Rotor resistance per phase

$X_2 \rightarrow$ Rotor reactance per phase

$Z_2 \rightarrow$ Rotor impedances phase

$$Z_2 = \sqrt{R_2^2 + X_2^2}$$

Then, $$I_2 = \frac{E_2}{Z_2} = \frac{E_2}{\sqrt{R_2^2 + X_2^2}}$$

We know power factor $$\cos\phi = \frac{R}{Z} \quad (2.1)$$

∴ Rotor power factor, $$\cos\phi_2 = \frac{R_2}{Z_2} = \frac{R_2}{\sqrt{R_2^2 + X_2^2}} \quad (2.2)$$

at the time of starting, torque ∝ $VI\cos\phi$ i.e. torque at the time of starting depends on power.

where,

V is rotor voltage E_2

I is rotor current I_2

$\cos\phi$ is rotor power factor $\cos\phi_2$

∴ $$T_{st} = kE_2I_2\cos\phi_2 \quad (2.3)$$

Substitute (2.1) and (2.2) in Eq. (2.3).

We get,

$$T_{st} = \frac{kE_2 \cdot E_2}{\sqrt{R_2^2 + X_2^2}} \cdot \frac{R_2}{\sqrt{R_2^2 + X_2^2}}$$

$$\boxed{T_{st} = \frac{KE_2^2 R_2}{R_2^2 + X_2^2}} \quad (2.4)$$

where,

K is torque constant

$$K = 3/2\pi N_s$$

If supply Voltage is constant, then the *emf* E_2 will be constant.

$$T_{st} = \frac{K \cdot R_2}{R_2^2 + X_2^2} = \frac{K \cdot R_2}{Z_2^2} \quad (2.5)$$

Note: Generally squirrel cage type rotor has poor starting torque where as slipring type rotor has high starting torque.

Condition for Maximum Starting Torque

The condition to obtain maximum starting torque is that rotor resistance equals to rotor reactance, i.e.

$$R_2 = X_2$$

we know, $$T_{st} = \frac{KR_2}{R_2^2 + X_2^2}$$

$$\frac{dT_{st}}{dt} = K\left[\frac{1}{R_2^2 + X_2^2} - \frac{R_2(2R_2)}{R_2^2 + X_2^2}\right] = 0$$

(or) $$R_2^2 + X_2^2 = 2R_2^2$$

$\therefore$ $$\boxed{R_2 = X_2}$$

From above Eq. it is proved that the condition to obtain maximum stating torque is $R_2 = X_2$

Note: $T_{st} \propto V^2$

The torque is directly proportional to square of supply voltage.

Torque Under Running Condition

[**Note:** There Will be change in rotor emf (E_2) and reactance (X_2) under running condition with respect to starting condition.

Let,

$E_2 \rightarrow$ rotor emf at the time of starting.

$X_2 \rightarrow$ rotor resistance at the time of starting.

But under running condition, rotor emf.

$$E_r = sE_2$$

$$X_r = sX_2]$$

Under running condition the torque,

$$T_{run} \propto E_r I_r \cos \phi_r$$

$$T_{run} = kE_r I_r \cos \phi_r \tag{2.6}$$

where,

$$E_r = sE_2$$

$$I_r = \frac{E_r}{Z_2} = \frac{sE_2}{\sqrt{R_2^2 + (sX_2)^2}} \tag{2.7}$$

$$\cos \phi_r = \frac{R_2}{\sqrt{R_2^2 + (sX_2)^2}} \tag{2.8}$$

Substitute (2.7) and (2.8) in Eq. (2.6)

$$\boxed{T_{run} = \frac{s \cdot E_2^2 R_2}{\sqrt{R_2^2 + (sX_2)^2}}} \tag{2.9}$$

CHAPTER 2

Condition for Maximum Running torque

Torque under running condition is maximum when

$$s = \frac{R_2}{X_2} \tag{2.10}$$

(or) $$R_2 = sX_2$$

Substituting (2.10) in Eq. (2.9) we get

$$\boxed{T_{\underset{(\text{run})}{\max}} = K\frac{E_2^2}{2X_2}} \tag{2.11}$$

EXAMPLE 2.3: A three phase induction motor having a star connected rotor has an induced emf of 100 V between slip rings at stand still on open circuit. The rotor has a resistance and reactance per phase of 2Ω and 5Ω respectively. Calculate current per phase and power factor when (a) slip rings are short circuited. (b) connected rheostat of 2Ω per phase.

Given data:

$$E_2 = 100 \text{ V}$$
$$R_2 = 2\Omega$$
$$X_2 = 5\Omega$$

Solution:

(a) Rotor emf /phase $= \dfrac{100}{\sqrt{3}} = 57.736$ V

Rotor impedance, $Z_2 = \sqrt{R_2^2 + X_2^2}$

$$Z_2 = \sqrt{2^2 + 5^2} = 5.385\ \Omega$$

Rotor current/phase, $I_2 = \dfrac{E_2}{Z_2} = \dfrac{57.736}{5.385}$

$$\boxed{I_2 = 10.721}\ A$$

Power factor, $\cos\phi_2 = \dfrac{R_2}{Z_2} = \dfrac{2}{\sqrt{2^2+5^2}}$

$$\boxed{\cos\phi_2 = .371}$$

(b) Rotor resistance/phase $= 2 + 2 = 4\Omega$[∵ Rotor resistance + Rheostate resistance]

Rotor impedance $= \sqrt{4^2 + 5^2} = 6.403\ \Omega$

∴ Rotor current/phase $= \dfrac{57.736}{6.403} = 9.01\text{ A}$

rotor power factor $= \dfrac{4}{6.403} = .624.$

$$\boxed{\cos\phi_2 = .624}$$

Ans.

(a) $I_2 = 10.721\text{ A}$

$\cos\phi_2 = .371$

(b) $I_2 = 9.01\text{ A}$

$\cos\phi_2 = .624$

EXAMPLE 2.4: A 1100 V, 50 Hz, delta connected induction motor has a star-connected slip-ring rotor with a phase transformation ratio of 3.8. The rotor resistance and stand still leakage reactance are 0.012 Ω and .25 Ω per phase respectively. Neglecting star impedance determine.

(a) Rotor current at start with slip ring shorted.

(b) Rotor power factor at start with slip ring shorted.

(c) Rotor current and power factor with 4 % slip.

Given data:

$$V = 1100\text{ V}$$
$$f = 50\text{ HZ}$$
$$k = 3.8$$
$$R_2 = .012\ \Omega$$
$$X_2 = .25\ \Omega$$

Solution:

Rotor phase voltage at stand still $= 1000 \times \dfrac{1}{3.8}$

$= 289.5\text{ V}$

Rotor impedances/phase $= \sqrt{0.012^2 + .25^2} = .2503\ \Omega$

(i) Rotor current at start $= \dfrac{289.5}{.2503} = 1156.6\,A$

(ii) Power factor $= \dfrac{R_2}{Z_2} = \dfrac{.012}{.2503} = .048(\text{lag})$

(iii) reactance at 4% slip $X_r = sX_2 = 0.04 \times .25 = .01\ \Omega$

CHAPTER 2

$$Z_r = \sqrt{0.012^2 + .01^2} = .0156\Omega$$

$$E_r = sE_2 = .04 \times 289.5 = 11.58 \text{ V}$$

$$I_2 = 11.58/.0156 = 742.3 \text{ A.}$$

$$\text{Power factor} = \frac{0.12}{.0156} = .77(\text{lag})$$

Ans.

(i) $I_2 = 1157 \text{ A}$

(ii) $\cos\phi_2 = .048 \text{ (lag)}$

(iii) $I_2 = 742.3\text{A} - \text{at } 4\% \text{ slip}$

$\cos\phi_2 = .77 \text{ (lag) at } 4\% \text{ slip.}$

EXAMPLE 2.5: A 300 kW, 2500 V, 50 Hz, 8 pole star connected induction motor has a star connected slip ring rotor with a transformation ratio of 3.8. The rotor resistance is 0.3 Ω/phase and its per phase leakage reactance is 3.8 mH. The stator impedances may be neglected. Find the starting current and starting torque on rated voltage with short circuited slip rings.

Given data:

$$P = 300 \text{ kW (power)}$$

$$V = 2500 \text{ V}$$

$$f = 50 \text{ Hz}$$

$$P = 8$$

$$1/K = 3.8$$

$$R_2 = .3\Omega/\text{phase}$$

$$L = 3.8 \text{ mH} = 3.8 \times 10^{-3} \text{ H}$$

$$X_2 = 2\pi f L = 2 \times 3.14 \times 50 \times 3.8 \times 10^{-3}$$

$$\boxed{X_2 = 1.193\Omega}$$

Solution:

$$K = \frac{1}{3.8}$$

$$\therefore \quad R_2' = \frac{R_2}{K^2} = .3 \times (3.8)^2 = 4.332\Omega$$

$$\boxed{R_2' = 4.332\Omega}$$

Similarly

$$X_2' = \frac{X_2}{K^2} = 1.193 \times (3.8)^2$$

$$\boxed{X_2' = 17.22\ \Omega}$$

$$I_{st} = \frac{V}{\sqrt{(R_2')^2 + (X_2')^2}} = \frac{2500}{\sqrt{(4.332)^2 + (17.22)^2}}$$

$$\boxed{I_{st} = 140.797\ \text{A}}$$

Now, Synchronous speed,

$$N_s = \frac{120f}{P} = \frac{120 \times 50}{8}$$

$$N_s = 750 \text{ r.p.m}$$

$$T_{st} = \frac{3}{2\pi N_s} \cdot \frac{V^2 R_2'}{(R_2')^2 + (X_2')^2} = \frac{3}{2 \times 3.14 \times 750} \cdot \frac{(2500/\sqrt{3})^2 (4.332)}{(4.332)^2 + (17.22)^2}$$

$$\boxed{T_{st} = 18.23\ \text{N-m}}$$

Solution:

$$I_{st} = 140.797 \text{ A}$$

$$T_{st} = 18.23 \text{ N-m}$$

EXAMPLE 2.6: The star connected rotor of an induction motor has a standstill impedances of $(0.4 + j4)\Omega$ per phase and the rheostat impedances per phase is $(6 + j2)\ \Omega$. The motor has an induced emf the of 80 V between slip rings at stand still when connected to its normal supply voltage. Find

(a) Rotor currents and p.f at standstill with the rheostat is in the circuit.

(b) When the slip rings are short circuited and motor is running with a slip of 3%.

Given data:

$$Z_2 = .4 + j4\ \Omega$$

$$\text{Rheostat, } Z = 6 + j2\ \Omega$$

$$E_2 = 80 \text{ V}$$

Solution:

(a) at standstill condition:

$$E_2 = \frac{80}{\sqrt{3}} = 46.2 \text{ V.}$$

Rotor plus rheostat impedances $= (6.4 + j6)\ \Omega$ $\quad [\because 0.44 + j4 + 6 + j2]$

$$Z_2 = 8.77 \angle 43.15^\circ$$

$$\text{Rotor current/phase} = \frac{46.2}{8.77} = 5.26 \text{ A} \qquad \left[\because I_2 = \frac{E_2}{Z_2}\right]$$

Power factor $= \cos[43.15^\circ]$ $\quad [\because \text{from } Z_2,\ \phi_2 = 43.15^\circ]$

$$\boxed{\cos\phi_2 = .729}$$

(b) at running condition:

$$E_r = sE_2 = 0.03 \times 46.2 = 1.386 \text{ V}$$

$$X_r = 0.03 \times 4 = .12\ \Omega$$

$$\therefore \quad Z_r = 0.4 + j0.12\Omega = .4176\angle 16.7^\circ$$

$$\text{rotor current} = 1.386/.4176 = 3.32 \text{ A}$$

$$\cos\phi_2 = \cos[16.7^\circ] = .96$$

Ans. (a) $I_2 = 5.26$ A

$\cos\phi_2 = .729$

(b) $I_2 = 3.32$ A

$\cos\phi_2 = .96.$

EXERCISE

1. A $3-\phi$ induction motor having a star-connected rotor has an included emf of 80 V between slip-rings at standstill on open circuit. The rotor has a resistance and reactance per phase of 1Ω and 4Ω respectively. Calculate currents/phase and power factor when

(a) slip rings are short circuited (b) slip rings are connected to a star connected rheostat of 3 Ω per phase.

[**Ans.** (a) 11.2A

.243

(b) 8.16A

.707]

2. A 150 kW, 3000 V, 50 Hz, 6 pole star connected induction motor has a star connected slip ring rotor with a transformation ratio of 3.6 (stator/ratio), The rotor resistance is 0.1 Ω/phase and its per phase Leakage reaction is 3.61 m.H. The stator impedances may be neglected. Find the starting current and starting torque on rated voltage with short-circuited slips rings.

[**Ans.** I_{st} = 117.4 A

T_{st} = 513. N-m]

3. A three phase, slip-ring induction rotor with star connected rotor has included emf of 120 V between slip rings at stand still with normal voltage applied to the stator. The rotor winding has a resistance per phase of 0.3 Ω and standstill Leakage reactance per phase of 1.5 Ω. Calculate.

(i) rotor current per phase when running short circuited with 4% slip and

(ii) the slip and rotor current per phase when the rotor is developing maximum torque.

[**Ans.** I_2 = 9 A

R_2 = 0.2 Ω

I_2 = 33 A]

2.7 EQUIVALENT CIRCUIT OF 3ϕ INDUCTION MOTOR

As we know that both transformer and induction motor works on induction principle, we can call 3ϕ induction motor as rotating transformer. Where stator is refereed as Primary and rotor is referred as secondary.

Let,

$R_1 \rightarrow$ Primary resistance (stator)

$X_1 \rightarrow$ Primary reactance (stator)

$R_2 \rightarrow$ Secondary resistance (rotor)

$X_2 \rightarrow$ Secondary reactance (rotor)

$R_0 \rightarrow$ Loss component (stator)

$X_0 \rightarrow$ Magnetising reactance (stator)

$I_m \rightarrow$ Magnetising current

$I_w \rightarrow$ Working current

$I_1 \rightarrow$ Primary current

$I_2 \rightarrow$ Secondary current

$I_0 \rightarrow$ No-load primary current

$I_2' \rightarrow$ Additional primary current.

$V_1 \rightarrow$ Input voltage (supply voltage)

$V_2 \rightarrow$ Secondary output voltage (load voltage)

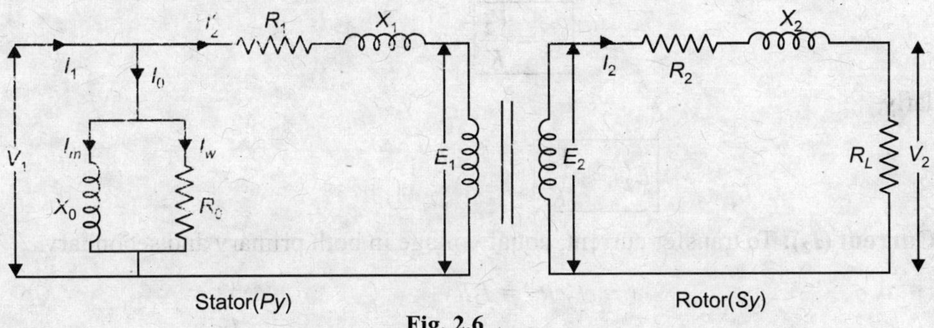

Fig. 2.6

What is the equivalent Circuit of an 3-ϕ induction motor?

The circuit which is used to reduce two different circuit into single and simple circuit is called equivalent circuit. The above circuit (Fig. 2.6) can be reduced into single circuit by transferring the elements from primary to secondary (or) from secondary to primary. Since secondary has less number of parameters it is easy to transfer. So now we will see how to transfer secondary parameters to primary side.

$$\text{Thus ratio, } K = \frac{N_2}{N_1} = \frac{V_2}{V_1} = \frac{E_2}{E_1} = \frac{R_2}{R_1} = \frac{X_2}{X_1} = \frac{Z_2}{Z_1} = \frac{I_1}{I_2}$$

First Let we transfer the elements R_2, X_2 and R_L.

To transfer the elements equate power in both primary and secondary.

Elements (R_2, X_2 and R_L)

(i) R_2:

$$I_1^2 R_2' = I_2^2 R_2$$

$$R_2' = \left(\frac{I_2}{I_1}\right)^2 R_2$$

$$\boxed{R_2' = \frac{R_2}{K^2}}$$

Similarly

(ii) X_2: $$\boxed{X_2' = \frac{X_2}{K^2}}$$

(iii) R_L: $$\boxed{R_L' = \frac{R_L}{K^2}}$$

(ii) Voltage (E_2 and V_2): To transfer voltage, equate current in both primary and secondary.

(i) V_2:

$$\frac{V_2'}{R_1} = \frac{V_2}{R_2}$$

$$V_2' = \left(\frac{R_1}{R_2}\right) V_2$$

$$\boxed{V_2' = \frac{V_2}{K}}$$

Similarly,

(ii) E_2: $$\boxed{E_2' = \frac{E_2}{K}}$$

(iii) Current (I_2): To transfer current, equal voltage in both primary and secondary.

(i) I_2:

$$I_2' R_1 = I_2 R_2$$

$$I_2' = I_2\left[\frac{R_2}{R_1}\right]$$

$$\boxed{I_2' = I_2 K}$$

Elements R_2, X_2 and R_2	Voltage V_2, E_2	Current I_2
$1/K^2$	$1/K$	K

Note:

$R_2 \rightarrow$ Secondary resistance

$R_2' \rightarrow$ Secondary resistance transferred to primary

$\therefore$ Power dissipation in primary due to R_2' is $I_1^2 R_2'$

and Power dissipation in secondary due to R_2 is $I_2^2 R_2$

Now equate the power dissipation due to secondary resistance, when it is in secondary side and when it is transferred to primary side i.e. $I_1^2 R_2' = I_2^2 R_2$.

Similarly,

$\frac{V_2}{R_2}$ is current in secondary due to secondary voltage (V_2)

$\frac{V_2'}{R_1}$ is current in primary due to secondary voltage transferred to primary (V_2').

Now Fig. (2.6) becomes.

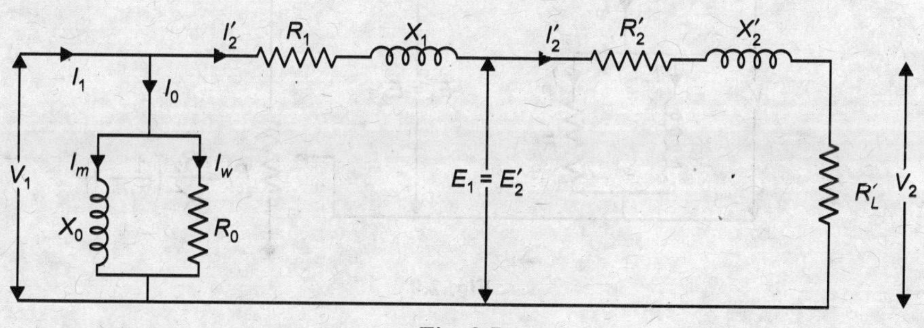

Fig. 2.7

From Fig. (2.7),

R_2 and R_2' are in series

$\therefore$
$$R_{0_1} = R_1 + R_2'$$

Similarly X_1 and X_2' are in series

$\therefore$
$$X_{0_1} = X_1 + X_2'$$

The circuit of Fig. (2.8) can be further reduced into, I_2'

The load resistance R_L is the electrical equivalent of mechanical load. The load resistance depends on slip. Therefore the mechanical load on an induction motor can be represented by a non inductive resistance of the value $R_2\left(\frac{1}{s}-1\right)$ i.e. $R_L = R_2\left[\frac{1}{s}-1\right]$

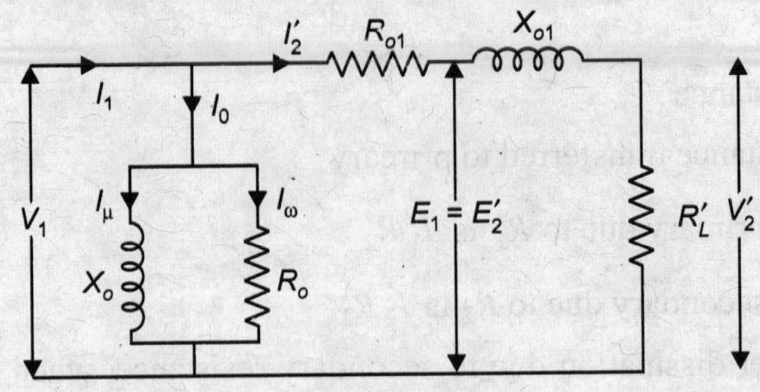

Fig. 2.8

$$\therefore \quad R_L' = \frac{R_L}{K^2} = \frac{R_2}{K^2}\left(\frac{1}{s} - 1\right)$$

Since slip (s) various with the respect to load condition, we can conclude that the load resistance also variable resistance. So, now we can redraw the Fig. (2.8) as.

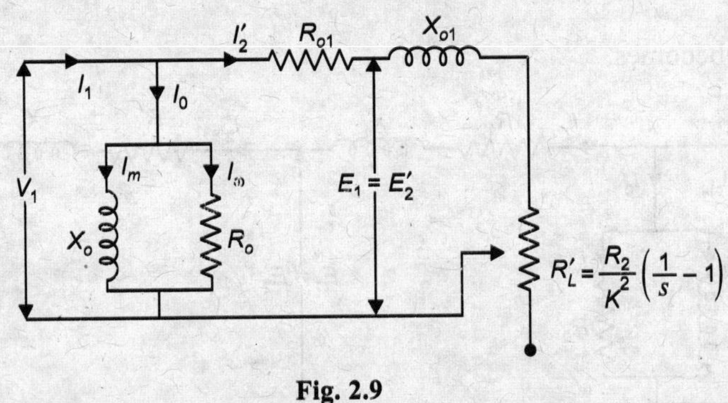

Fig. 2.9

Fig. 2.9 is the final equivalent circuit of 3ϕ induction motor.

EXAMPLE 2.7: The maximum torque of a 3-phase induction motor occur at a slip of 8%. The motor has an equivalent secondary resistance of 0.05Ω/phase. Calculate the equivalent load resistance R_L, the equivalent load voltage V_L and the current at this slip it the gross output power is 10000 watts.

Given data:

Slip, $s = 8\% = .08$

$R_2 = 0.05\ \Omega/\text{phase}$

output power $= 10000$ watts.

Solution:

$$R_L = R_2(1/s - 1)$$

$$= .05\left[\frac{1}{.08} - 1\right]$$

$$\boxed{R_L = .575\Omega/\text{phase}}$$

In rotor,

$$\text{Gross power, } P_g = 3I_2^2 R_L$$

(or)

$$P_g = 3\left[\frac{V^2}{R_L}\right]$$

$$\therefore \quad V = \sqrt{(P_g \times R_L / 3}$$

$$= \sqrt{10000 \times .575 / 3}$$

$$V = 43.77 \text{ V}$$

$\therefore$ Equivalent load current,

$$I_L = \frac{V}{R_L}$$

$$I_L = \frac{43.77}{.575}$$

$$\boxed{I_L = 76.13 \text{ A}}$$

Ans. R_L = .575Ω/phase
I_L = 76.13 A
V_L = 43.77 V.

EXAMPLE 2.8: A 3 phase, star connected 440 V, 50 Hz, 6 pole induction motor has the following per phase parameters in ohms, referred to the stator.

$$R_1 = 0.35\ \Omega,\ X_1 = .55\ \Omega$$

$$R_2' = .15\ \Omega,\ X_2' = .65\ \Omega \text{ and } X_m = 30.65$$

compute the stator current and power factor when the motor is operated at rated voltage and frequency with s = .06.

Given data:

V = 440 V
f = 50 Hz
P = 6
R_1 = 0.35 Ω
X_1 = .55 Ω
R_2' = .15 Ω
X_2' = .65 Ω
X_m = 30.65
s = .06

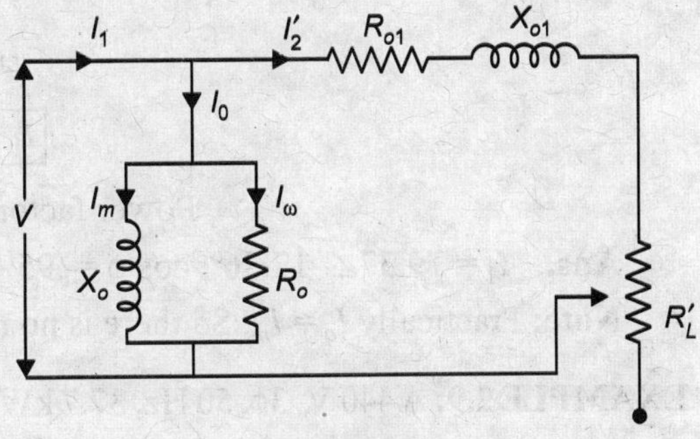

Fig. 2.8.1

Solution: The equivalent circuit with all values referred to stator is,

$$R'_L = R'_2\left(\frac{1}{s}-1\right)$$

$$= .15\left(\frac{1}{.06}-1\right)$$

$$\boxed{R'_L = 2.35\,\Omega}$$

From Fig. (2.8.1), $$I'_2 = \frac{V}{(R_{01}+R'_L)+jX_{01}}$$

where,

$$R_{0_1} = R_1 + R'_2 = .35 + .15 = .5\,\Omega$$

$$X_{0_1} = X_1 + X'_2 = .55 + .65 = 1.2\,\Omega$$

$$I'_2 = \frac{(440/\sqrt{3})}{(.5+2.35)+j1.2\,\Omega}$$

$$= \frac{254.04}{2.85+j1.2} = \frac{254.04}{3.09\angle 22.83^\circ}$$

$$I'_2 = 82.21\angle -22.83^\circ$$

$$I'_2 = 75.76 - j31.89\text{ A}$$

$$I_0 = \frac{440/\sqrt{3}}{X_m} = \frac{(440/\sqrt{3})}{j30.65} = j8.2$$

Stator current,

$$I_1 = I_0 + I'_2$$

$$= j8.2 + 75.76 - j31.89$$

$$I_1 = 75.76 - j23.69$$

$$\boxed{I_1 = 79.37\angle +17.36^\circ}$$

Power factor $= \cos\phi = \cos[+17.36] = 0.95$

Ans. $I_1 = 79.37\angle -17.36^\circ$; $\cos\phi = .95$

Note: Practically $I_o = I_m$. So there is no need to determine I_w.

EXAMPLE 2.9: A 440 V, 3ϕ, 50 Hz, 37.7 kW, *Y* connected induction motor has the following parameters: $R_1 = .1\,\Omega$, $X_1 = .4\,\Omega$, $R'_2 = .15\,\Omega$, $X'_2 = .44\,\Omega$. Motor has stator core loss of 1250 W and rotational loss of 1000 W. It draws a no-load line current of 20 A at p.f of .09 (lag). When motor operates at a slip of 3%

calculate (i) input line currents and power factor (ii) Electromagnetic torque developed in N-m (iii) output (iv) efficiency of the rotor.

Given data:

$$V = 440 \text{ V}$$
$$f = 50 \text{ Hz}$$
$$P = 37.7 \text{ kw} = 37700 \text{ watts.}$$
$$R_1 = .1\ \Omega$$
$$X_1 = .4\ \Omega$$
$$R_2' = .15\ \Omega$$
$$X_2' = .44\ \Omega$$
$$\text{Core loss} = 1250 \text{ W}$$
$$\text{Rotational loss} = 1000 \text{ W}$$
$$\left.\begin{aligned} I_0 &= 20 \text{ A} \\ \cos\phi &= .09 \text{ (Lag)} \end{aligned}\right\} \Rightarrow I_o = 20\angle -84.9^\circ = (1.78 - j19.92)$$
$$s = 3\% = .03$$

Solution:

$$V = \frac{440}{\sqrt{3}} = 254.04 \text{ V}$$
$$I_2' = \frac{V_1}{(R_1 + R_2'/s) + j(X_1 + X_2')}$$
$$= \frac{254.04\angle 0^\circ}{(0.1 + .15/.03) + j(0.4 + .44)}$$
$$= \frac{254.04\ \angle 0^\circ}{5.1 + j\,0.84}$$
$$= \frac{254.04\ \angle 0^\circ}{5.17\ \angle 9.3} = 49.13\angle -9.3^\circ$$
$$I_2' = 48.48 - j\,7.93$$

For all practical purposes, no-load motor current may be taken as equal to magnetising current i.e.

$$I_m = I_o$$

Hence,
$$I_1 = I_o + I_2' = 20\angle -84.9^\circ + 49.13\ \angle -9.3^\circ$$

(i)
$$I_1 = I_0 + I_2' = (48.4 - j\,7.9) + (1.78 - j\,19.92)$$
$$= 50.2 - j\,27.8$$
$$\boxed{I_1 = 57.4\angle -29^\circ}$$

$\therefore$ Power factor = cos 29° = .875 (Lag)

(ii) $$P_2 = 3I_2'^2\left(\frac{R_2}{s}\right) = 3 \times 49.13^2 \times (.15/.03)$$

$$P_2 = 36{,}160 \text{ W}$$

$$N_s = 1500 \text{ r.p.m}$$

$$T_g = 9.55 \times 36{,}160/1500 = 230 \text{ N-m}$$

(iii) $$P_m = (1 - s)\, P_2 = .097 \times 36{,}160 = 35075 \text{ W}$$

$$\text{Output power} = 35075 - 1000 = 34075$$

(iv) $$\text{Cone loss} = 1250 \text{ W}$$

$$\text{Stator copper loss} = 3I_1^2 R_1 = 3 \times 57.4^2 \times .1$$

$$= 988 \text{ W}$$

$$\text{Rotor Cu loss} = 3I_1^2 R_2' = sP_2$$

$$= .03 \times 36{,}100 = 1085 \text{ W}$$

$$\text{Frictional \& windage loss} = 1000 \text{ W}$$

$\therefore$ $$\text{Total loss} = 1250 + 988 + 1085 + 1000 = 4332 \text{ W}$$

$\therefore$ $$\%\text{ efficiency} = \frac{34075}{34075 + 4323} \times 100$$

$$\text{efficiency} = 88.7\ \%$$

Ans.

(i) $I_1 = 57.4\angle{-29^\circ}$ A

$\cos\phi = .875$ (Lag)

(ii) $T = 230$ N-m

(iii) $P_m = 34075$ W

(iv) $\%\eta = 88.7\ \%$

EXAMPLE 2.10: A 440 V, 3ϕ star connected induction motor has a stator impedances of $(.08 + j\,0.4)\Omega$ and an equivalent rotor impedances of $(.08 + j\,.44)\Omega$ (with respect to primary) Neglecting current, find the maximum gross power and the slip at which it occurs.

Given data:

$$V = 440 \text{ V}$$

$$Z_1 = 0.08 + \text{j}\ 0.04\ \Omega$$

$$Z_2 = 0.8 + \text{j}\ 0.44\ \Omega$$

Solution:

$$R_{01} = R_1 + R_2' = .08 + .08$$

$$R_{01} = .16\ \Omega$$

$$X_{01} = X_1 + X_2' = .04 + .44$$

$$X_{01} = .48\ \Omega$$

$$\mathbf{Z_{01}} = \sqrt{R_{01}^2 + X_{01}^2} = \sqrt{0.16^2 + .48^2}$$

$$= \sqrt{.0256 + .2304}$$

$$\boxed{Z_{01} = .0505\ \Omega}$$

slip, $$s = \frac{R_2}{R_2 + Z_1} = \frac{.08}{.08 + .505} = .13 = 13\%$$

voltage/phase, $$V_1 = \frac{440}{\sqrt{3}} = 254.04\ \text{V}$$

$$P_{g,\ max} = \frac{3V_1^2}{2(R_{01} + Z_{01})} = \frac{3(254.04)^2}{2(.16 \times .0505)}$$

$$= \frac{193608.96}{.421}$$

$$\boxed{P_{g(max)} = 459\ \text{KW}}$$

Ans.
(i) $s = 13\%$
(ii) 459 KW

EXERCISE

4 The maximum torque of a 3-phase induction motor occurs at slip of 12%. The motor has an equivalent secondary resistance of .08 Ω/phase. Calculate the equivalent load resistance R_L the equivalent voltage V_L the current at this slip if the gross power output is 9000 watts.

[**Ans.** $V = 42$ V, $I_L = 71.6$ A]

5 A 3-phase star connected 400 V, 50 Hz, 4 pole induction motor has the following per phase parameters in ohms, reformed to stator $R_1 = 015\ \Omega$, $X_1 = .45\ \Omega$, $R_2' = .12\ \Omega$, $X_2' = .45\ \Omega$, $X_m = 28.5\ \Omega$, compute the stator current and power factor when the motor is operated voltage and frequency with $s = .04$.

[**Ans.** $I_1 = 73.13\ \angle{-22^\circ}$, $\cos\phi = .927$ (lag)]

6 A 400 V, 3ϕ star-connected induction motor has a stator impedances of $(.06 + j\,0.2)\ \Omega$ and an equivalent rotor impedances of $(.06 + j\ .22\ \Omega)$ neglecting current, find the maximum gross power and the slip at which it occurs.

[**Ans.** $P_{g,\ max}$ =142,900 W]

2.8 SPEED CONTROL OF 3ϕ INDUCTION MOTOR

What is speed control?

Obtaining variable speed in a single machine is called as speed control. For example, we are changing the speed of a fan using regulator.

The speed control of 3ϕ induction motor can be done by two ways (i) from stator side (ii) from rotor side.

As we know the speed is directly proportional to voltage and frequency then the speed is inversely proportional to number of poles i.e) $N_s = \frac{120f}{P}$.

We can vary the speed by varying the above parameters. Let we see the method of speed control.

1. Control from stator side
   (i) By changing the applied voltage
   (ii) By changing the applied frequency
   (iii) By changing the number of poles
2. Control from the rotor side
   (iv) Rotor Rheostat control
   (v) Injection of emf the rotor circuit.

(i) By changing the applied voltage

As we know that the speed is directly proportional to voltage, we can change the speed by changing the applied voltage. The applied voltage can be varied either by rheostat (or) by auto transformer.

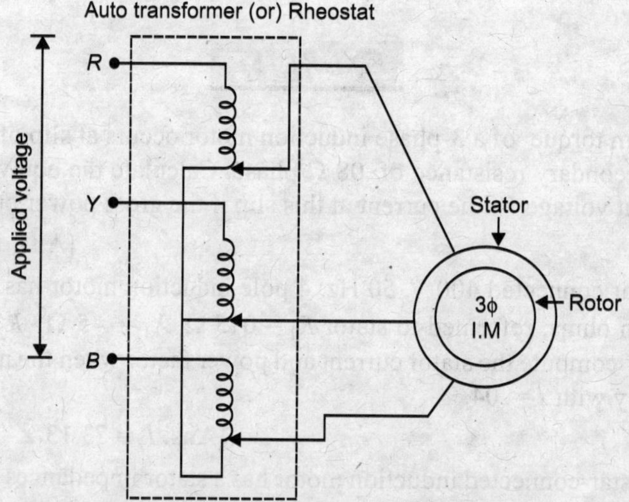

Fig. 2.10: Rheostat (or) auto transformer.

When supply is given to 3ϕ induction motor, It is given through either auto transformer (or) Rheostat where as the voltage can be varied. When the voltage is varied by autotransformer (or) rheostat and then the speed of 3ϕ induction motor also varied.

This method has following drawbacks.

(i) Large change in voltage is required for a relatively small change in speed.

(ii) Large change in voltage will result in a large change in the flux density there by the magnetic condition of motor get distributed.

(ii) By changing applied frequency

We know, $N_s = \frac{120f}{P}$.

The speed of 3ϕ induction motor directly proportional to frequency. There fore, by changing the applied frequency, the speed can be varied. The applied frequency can be varied by using cyclo converter. Cyclo converter is the device which gives variable frequency output from fixed input frequency.

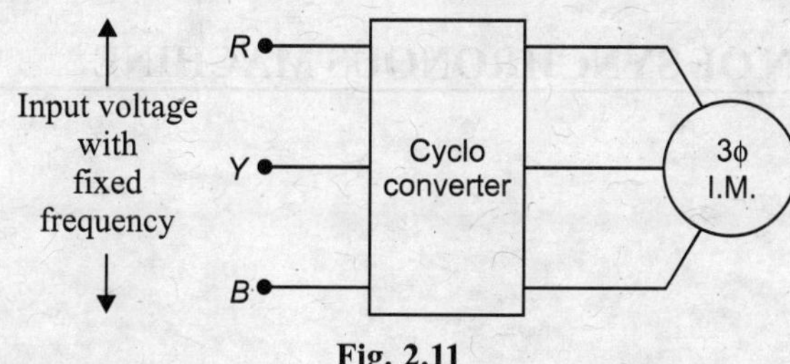

Fig. 2.11

The speed of the 3ϕ induction motor varies according to the output frequency of the cycloconverter.

(ii) By changing number of poles

The change of number of poles is achieved by having two or more independent stator windings in the same slots. Only one winding is used at a time, the other is entirely disconnected. It means if we are having 3 independent windings, then we can obtain 3 different speed ranges from single machine.

This method is adopted in Elevators, traction and small motors driving machine tools.

(iv) Rotor rheostat control

This method is applied to slip ring motors alone, The motor speed is reduced by introducing an external resistance in the rotor circuit. The speed can be decreases by increasing the rotor resistance R_2. We know torque (T) is directly proportional to rotor resistance (R_2). Therefore, torque value can be increased by increasing rotor resistance (R_2). We know the relation between speed and torque that they both are inversely proportional. Therefore when torque is increased speed get decreased thereby speed can be varied.

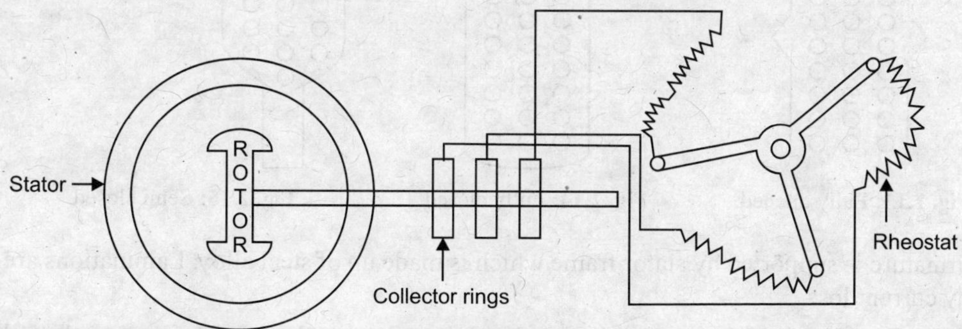

Fig. 2.12

This method has one disadvantage, this method has high I^2R loss which decreases the operating efficiency. So it is used where speed changes are needed for short period only.

(v) Injection of emf in rotor circuit

In this method, the speed is controlled by injecting an external emf to rotor circuit. It is necessary that the injected voltage should have the same frequency as the slip frequency.

When the emf is injected to the rotor circuit is phase opposition. Then the rotor resistance get increased. If the injected emf is in phase with rotor induced emf then the rotor resistance get decreased. Hence by changing the rotor resistance the speed can be controlled.

2.9 CONSTRUCTION OF SYNCHRONOUS MACHINE

It consist of two main parts

1. Stator
2. Rotor

Stator

It consists of frame, slots and conductors. The outer frame serves to carry the armature conductors and windings. Ventilation is made with the help of holes. The material used for frame is mild steel plates.

The slots are in three different shapes.

1. Fully closed
2. Fully opened
3. Semi closed

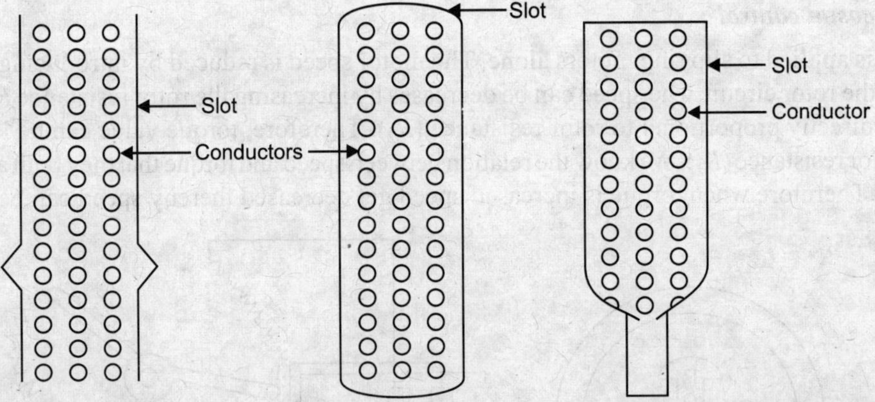

Fig. 2.13: Fully opened. **Fig. 2.14:** Fully closed. **Fig. 2.15:** Semi closed.

The armature is supported by stator frame which is made up of steel alloy. Laminations are used to avoid eddy current loss.

- Fully closed type of slots do not disturb the air-gap flux but they tend to increase the inductance of the windings.
  - For placing the conductor inside the slots will lead to increase in Labour cost.
  - Complicated for end ring connections.

- The fully opened type slot has the advantage of easy installation.
  – Removal is easy in case of repair.
  – But conductor flux will disturb the air gap flux.
- The semi-closed type slots are better in this respect, but do not allow the use of form-wound coils. It has more advantage compared to all types.

Rotor

There are two types of rotors are used

1. Salient pole type
2. Smooth cylindrical type

Salient (or) projecting pole type

It is used in Low- and medium speed machines. It has large number of projecting poles, having their cores bolted. It is made up of cast iron, (or) steel of good magnetic material. Such generators are characterised by their large diameters and short axial lengths. The pole and pole shoes are laminated to reduce eddy current loss.

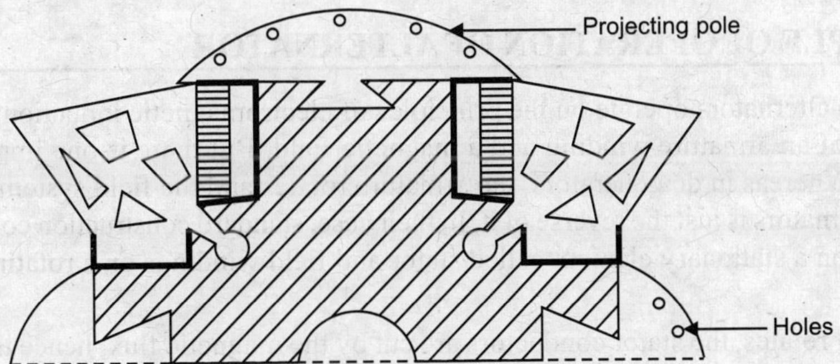

Fig. 2.16: Salient pole type construction.

Smooth cylindrical type

It is used for high speeds. The rotor consist of a smooth solid forged steel cylinder having a number of slots milled out at intervals along the outer periphery for accommodating field coils. These types of rotors are designed mostly for 2 pole. Poles are non salient i.e. they do not project out from the surface of the rotor. It is very small in diameter and large in axial length.

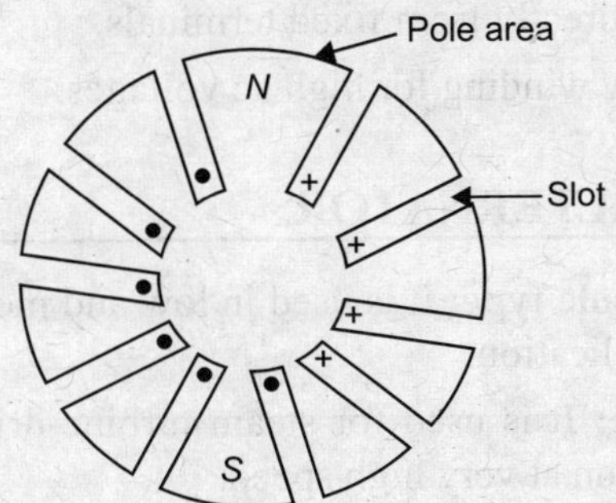

Fig. 2.17: Smooth cylindrical type construction.

Damper windings

Most of the alternators have damper winding in the rotor. It is made up of copper. These dampers are useful in preventing the hunting in generators and in motors. This also used to provide starting torque in case of motor. Under normal conditions, damper winding does not carry any current because rotor runs at synchronous speed.

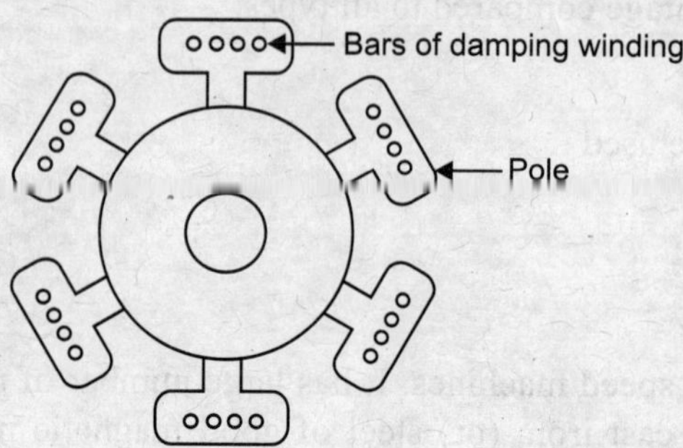

Fig. 2.18

2.10 PRINCIPLE OF OPERATION OF ALTERNATOR

AC generators (or) alternators operate on the principles of electromagnetic induction as dc generators. They also consist of an armature winding and a magnetic field. But there is one important difference between the two. Whereas in dc generators, the armature rotates and the field system is stationary, the arrangement in alternators is just the reverse of it. In their case, standard construction consists of armature winding mounted on a stationary element called stator and field windings on a rotating element called rotor.

When the rotor rotates, the stator conductors are cut by the magnetic flux, hence they have induced e.m.f produced in them. Because the magnetic poles are alternately N and S, they induce an e.m.f and hence current in armature conductors, which first flows in one direction and then in the other. Hence, an alternating e.m.f is produced in the stator conductors

(i) Whose frequency depends on the number of N and S poles.

(ii) Whose direction is given by Fleming's right hand rule.

There are several advantages of having stationary armature they are,

1. Output current can be led directly from fixed terminals.
2. Easier to insulate stationary winding for high ac voltages.

2.11 APPLICATIONS OF ALTERNATOR

(i) Salient (or) projecting pole type: It is used in low and medium speed applications. i.e. it is used in engine driven application.

(ii) Smooth cylindrical type: It is used for steam turbine-driven alternators i.e., it is used as turbo-alternators, which run at very high speed.

2.12 APPLICATIONS

1. It is used in power factor correction.
2. Constant speed application.
3. Constant load drives
4. Used to improve voltage regulation in transmission line
5. Reciprocating compressor drives

2.13 EMF EQUATION OF AN ALTERNATOR

$T \rightarrow$ Number of turn/phase

$Z \rightarrow$ Number of conductor/phase

$Z = 2T$ ($\therefore$ Each conductor has two turns ie, each coil has two sides).

$S \rightarrow$ Number of slots

$P \rightarrow$ Number of poles

$f \rightarrow$ Frequency of induced emf in Hz

$\phi \rightarrow$ Flux per pole in Wb.

$K_d \rightarrow$ Distribution factor $= \dfrac{\sin m\beta/2}{m \sin \beta/2}$

where,

$m \rightarrow$ Number of slots/phase/pole

$\beta \rightarrow \dfrac{180^\circ}{\text{number of slots/phase}}$

$\beta = \dfrac{180^\circ}{n}$

$k_P \rightarrow$ Pitch angle (or) span factor$= \cos \alpha/2$

where,

$\alpha \rightarrow$ Pitch angle ie, angle between full span and short pitch

$k_f \rightarrow$ From factor = 1.11

$N \rightarrow$ Rotor speed in R.P.M

According to faradays's second law,

$$e = \frac{d\phi}{dt}$$

In one revolution of the rotor, each stator conductor is cut by a flux of ϕP 'wb'

$$\therefore \quad d\phi = P\phi$$

$$dt = \frac{60}{N} \text{ sec}$$

$\therefore$ Average e.m.f per Conductor $= \dfrac{d\phi}{dt} = \dfrac{\phi P}{\dfrac{60}{N}}$

$$= \frac{\phi PN}{dt}$$

We know, $N = \dfrac{120f}{P}$

Average emf/conductor $= \dfrac{\psi P 120 f}{P \times 60}$

Average emf/conductor $= 2\phi f$ volts

The average emf for Z number of conductor $= 2\phi f(Z)$ volts

We know, $Z = 2T$

$\therefore$ average emf for number of conductor $= 2\phi f(2T)$

Average emf $= 4\phi f T$

R.M.S value of emf/phase $= 1.11 \times 4\phi f T$ i.e. $K_f \times$ Average emf

$$\boxed{\text{R.M.S emf/phase} = 4.44\phi f T} \text{ 'volts'}$$

Actual available voltage at the output depends on other two factors called distribution factor and pitch factor.

$\therefore$ emf/phase $= 4.44\phi f T k_d k_P$

(or) $$\boxed{\text{emf/phase} = 4\phi f T k_d k_P k_f} \text{ 'volts'}$$

EXAMPLE 2.11: The stator of a 3 phase 12 pole alternator has 156 slots and there are 6 conductors per slots connected in two layers and the conductors of each phase are connected in series. If the speed of the alternator is 400 r.p.m calculate the e.m.f included per phase. Resultant flux in the air gap is 6×10^{-2} wb per pole sinusoidally distributed. Assume the coil span as 160º electrical.

Given data:

$P = 12$

$S = 156$

$Z = 6$ conductors/slot

$N = 400$ r.p.m

$\phi = 6 \times 10^{-2}$

coil spans $= 160^\circ$ electrical

Solution:

$K_f = 1.11$

$\alpha = (180^\circ - 160^\circ) = 20^\circ$ (elect)

$k_P = \cos \alpha/2 = \cos 20^\circ/2 = 0.984$

$$n = \text{number of slots/pole}$$

$$n = \frac{156}{12} = 13$$

$$\beta = \frac{180^\circ}{3} = 13.84^\circ \approx 14^\circ \text{ i.e. } \beta = \frac{180^\circ}{n}$$

$$m = S/P/\text{Phase} = 156/12/3$$

$$\boxed{m = 4.33}$$

$$\therefore \quad k_d = \frac{\sin(4.33)(14^\circ)/2}{4.33 \sin 14^\circ/2} = \frac{.504}{.527}$$

$$\boxed{k_d = .956}$$

$$f = PN/120 = 12 \times 400/120$$

$$\boxed{f = 40 \text{ Hz}}$$

$$\text{Number of slots per/phase} = \frac{156}{3} = 52$$

$$\text{Number of conductors in series/phase} = 52 \times 6 = 312$$

$$\text{turns/phase} = \text{Conductors Per Phase}/2$$

$$T = 312/2 = 156$$

$$\text{emf/phase} = 4 k_f k_d f \phi T k_p$$

$$= 4 \times 1.11 \times .956 \times 40 \times 6 \times 10^{-2} \times 156 \times .984$$

$$\boxed{\text{emf/phase} = 1563.7 \text{ V}}$$

Ans. emf/phase = 1563.7 V

EXAMPLE 2.12: Find the no-load phase voltage of a star connected 3-phase 4 pole alternator which runs at 1400 rpm, having flux per pole of 0.3 wb. Its stator has 62 slots having double layer winding. Each coil has 6 turns and the pitch angle is 20º

Given data:

$$\alpha = 20^\circ$$

$$P = 4$$

$$N = 1400 \text{ rpm}$$

$$\phi = .3 \text{ wb}$$

$$\text{Slots} = 62$$

Solution:

$$k_p = \cos\alpha/2 = \cos 20^\circ/2 = 98^\circ$$

$$f = 4 \times 1400/120 = 47 \text{ Hz}$$

$$n = 62/4 = 15.5$$

$$\beta = 180°/15.5 = 11.61 \approx 12°$$

$$m = 62/4/3 = 5.16$$

$$\boxed{m = 5.16}$$

$$k_d = \frac{\sin(5.16)(12°/2)}{5.16 \sin 12°/2} = \frac{.514}{.539}$$

$$\boxed{k_d = .953}$$

$$Z = 62 \times 6/3 = 124$$

$$T = 124/2 = 62$$

$$\text{emf/phase} = 4.44 \times .953 \times .98 \times .3 \times 47 \times 62$$

$$\boxed{\text{emf/phase} = 3625 \text{ V}}$$

Ans. emf/phase = 3625 V

EXAMPLE 2.13: An alternator has 18 slots/pole and the first coil lies in slot 1 and 16. Calculate the pitch factor (k_P) (i) fundamental (ii) 3rd harmonic (iii) 5th harmonic

Given data:

$$\text{slots} = 18$$

Solution:

$$\text{Coil span} = (16 - 1) = 15\text{slots}$$

which has shortage of 3 slots in order to full fil 18 slots.

$\therefore$ $\alpha = 180° \times 3/18 = 30°$

(i) $k_p = \cos 30°/2 = .966$

(ii) $k_p = \cos 3 \times 30°/2 = .707$

(iii) $k_p = \cos 5 \times 30°/2 = .259.$

Ans.

(i) $k_p = .966$(Fundamental)

(ii) $k_p = .707$ (3rd harmonic)

(iii) $k_p = .259$ (5th harmonic)

EXAMPLE 2.14: A 3 Phase, 12 pole, alternator has a star-connected winding with 156 slots and 10 conductors per slot. The flux per pole is 0.03wb, Sinusoidally distributed and the speed is 375 r.p.m. Find the frequency and phase emf. Assume full-Pitched coil.

Given data:

$$f = \frac{PN}{120} = \frac{12 \times 375}{120}$$

$$f = 37.5 \text{ Hz}$$

$$k_p = 1[\therefore \text{ full pitched coil}]$$

$$n = \frac{156}{12} = 13$$

$$\beta = \frac{180^o}{13} = 14^o$$

$$m = \frac{156}{12 \times 3} = 4.33$$

$$k_d = \frac{\sin 4.33(14^o/2)}{4.33 \sin 14^o/2} = .956$$

$$Z = 156 \times 10/3 = 520$$

$$T = 520/2 = 260/\text{phase}$$

$$\text{emf/phase} = 4.44 \times 1 \times .956 \times 37.5 \times .03 \times 260$$

Ans. emf/phase = 1241.5 'volts'

EXERCISE

7 A phase, 16 pole alternator has a star connected winding with 144 slots and 10 conductors per slot. The flux per pole is 0.03 wb, Sinusoidally distributed and the speed is 375 r.p.m find the frequency and phase emf. (Assume full pitched coil).

[**Ans.** f = 50 Hz, E_{ph} = 2568 V]

8 Find the no-load phase and line voltage of a star connected 3-phase, 6 pole alternator which runs at 1200 rpm, having flux per pole of 0.1 wb sinusoidally distributed. Its stator has 54 slots having double layer winding. Each coil has 8 turns and the coil is chorded by 1 slot.

[**Ans.** E_{ph} = 1805 V, E_L = 3125 V]

9 The stator of a 3 phase, 16 pole alternator has 144 slots and there are 4 conductors per slot connected in two layers and the conductors of each phase are connected in series. If the speed of the alternator is 375 r.p.m, calculate the e.m.f induced per phase. Resultant flux in the air gap is 5×10^{-2} wb per pole sinusoidally distributed. Assume the coil. Span as 150° electrical.

[**Ans.** E_{ph} = 988 V]

2.14 VOLTAGE REGULATION

The voltage regulation of an alternative is defined as the "ratio of rise in voltage when full load is removed to the rated terminal voltage".

i e

$$\% \text{ regulation} = \frac{E_0 - V}{V} \times 100.$$

where,

E_0 → No-load induced emf

V → rated terminal voltage

$E_0 - V$ → amount of voltage rise when full load is removed

2.15 VOLTAGE REGULATION BY SYNCHRONOUS IMPEDANCES METHOD (OR) EMF METHOD

The voltage regulation of an alternator can be determined by many ways. One most important method among then is synchronous impedance method (or) E.M.F method.

We know,

$$\text{Regulation} = \frac{E_0 - V}{V}$$

- Terminal voltage V will be given.
- The voltage E_0 vary according to load. So we have to find E_0 for different type of loads such as lagging p.f load, Leading p.f load and unity p.f load.
- E_0 is the vector sum of terminal voltage (V), resistive drop (IR) and reactance drop (IX).

  We need to find the value of reactance (X). The value of reactance is determined using impedances (Z), i.e. $X = \sqrt{Z^2 - R^2}$ .That's why This method is called synchronous impedances method.

  where,

$$Z = \frac{\text{open circuit voltage}}{\text{short circuit current}} \Bigg| \text{ at same field current}$$

- After finding impedances value we can find E_0 value.

  Now we will see how to derive formula for E_0 for different types of loads.

(i) For Lagging Power factor Load

- Current I Lags terminal voltage (V) by an angle of ϕ.
- IR drop is a horizontal component and IX drop is a vertical component.

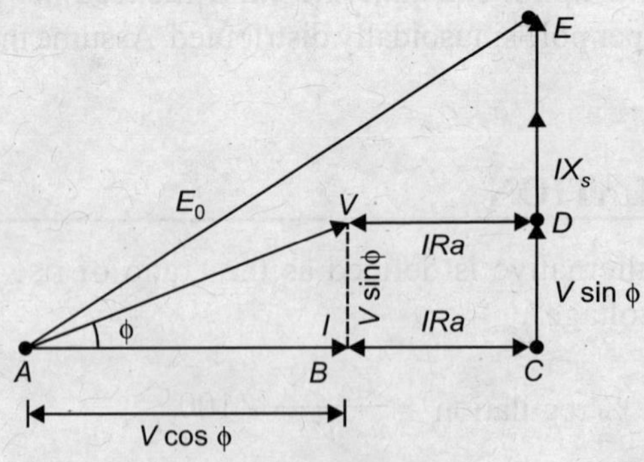

Fig. 2.19

In Fig. (2.19)

$$E_0 = AE$$

where

$$AE = \sqrt{(AC)^2 + (CE)^2}$$

and

$$AC = AB + BC = V\cos\phi + IR_a$$

$$CE = CD + DE = V\cos\phi + IX_s$$

$$\therefore \quad E_0 = \sqrt{(V\cos\phi + IR_a)^2 + (V\sin\phi + IX_s)^2}$$

After finding E_0 substitute in regulation formula.

i.e.,

$$\%\text{ Regulation} = \frac{E_0 - V}{V} \times 100.$$

(ii) For Leading Power factor Load

- Current I leads terminals voltage (V) by angle of ϕ
- IR drop is a horizontal and IX is vertical component.

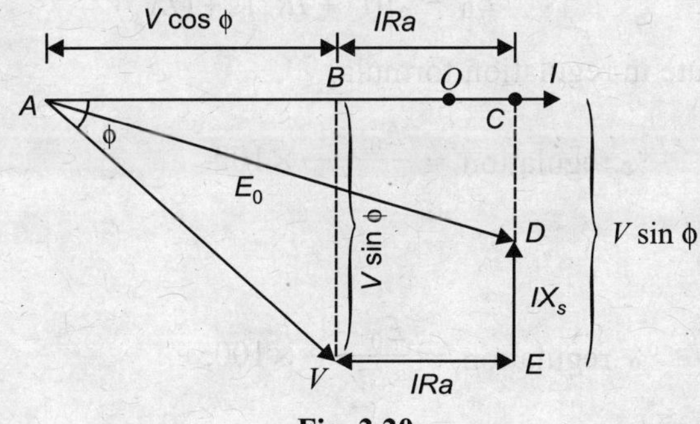

Fig. 2.20

From Fig. (2.20)

$$AD = E_0$$

where,

$$AD = \sqrt{(AC)^2 + (CD)^2}$$

and

$$CD = CE - DE = V\sin\phi - IX_s$$

$$AC = V\cos\phi + IR_a$$

$$\therefore \quad E_0 = \sqrt{(V\cos\phi + IR_a)^2 + (V\sin\phi - IX_s)^2}$$

After finding E_0 substitute in regulation formula.

i e,

$$\%\text{ regulation} = \frac{E_0 - V}{V} \times 100.$$

(iii) For Unity Power factor Load

- Current I in phase with the terminals voltage (V)

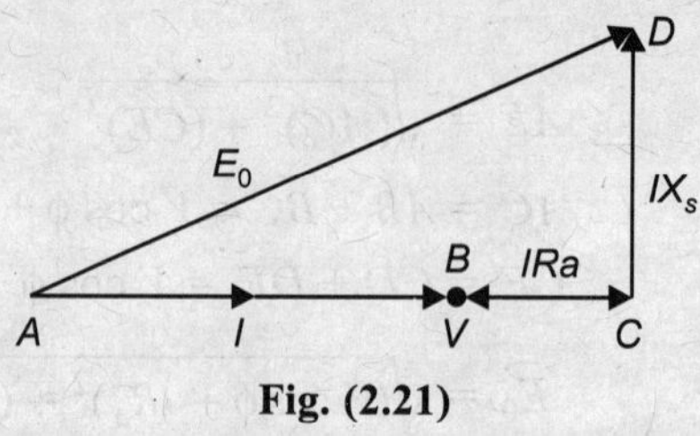

Fig. (2.21)

In Fig. (2.21),

$$E_0 = AD$$

$$AD = \sqrt{(AC)^2 + (CD)^2}$$

where,

$$AC = AB + BC = V + IR_a$$

$$CD = IX_s.$$

$\therefore$

$$E_0 = \sqrt{(V + IR_a)^2 + (IX_s)^2}$$

After finding E_0 substitute in regulation formula.

i e,

$$\%\ \text{regulation} = \frac{E_0 - V}{V} \times 100.$$

Conclusion

$$\%\ \text{regulation} = \frac{E_0 - V}{V} \times 100.$$

E_0 various according to load.

E_0 for Lagging p.f load,

$$E_0 = \sqrt{(V\cos\phi + IR_a)^2 + (V\sin\phi + IX_s)^2}$$

E_0 for leading p.f load,

$$E_0 = \sqrt{(V\cos\phi + IR_a)^2 + (V\sin\phi - IX_s)^2}$$

E_0 for unity p.f load,

$$E_0 = \sqrt{(V + IR_a)^2 + (IX_s)^2}$$

Note:

- This method is pessimistic method.
- This method is not accurate

EXAMPLE 2.15: From the following test result, determine the voltage regulation of a 1800 V, 1-phase alternator delivering a current of 80 A at (i) unity p.f (ii) 0.8 Leading p.f and (iii) .71 legging p.f.

Test results: Full load current of 80 A is produced on short circuit by a field excitation of 2 A. An emf of 500 V is produced on open circuit test by the same excitation. The armature resistance is .8 Ω

Given data:

Terminal voltage, $V = 1800$ V

$I = 80$ A

$V_{o.c} = 500$ V at 2 A of field current

$I_{sc} = 80$ A at 2 A of field current

$R_a = .8\ \Omega$

Solution:

we know,

$$\%\ \text{regulation} = \frac{E_0 - V}{V} \times 100\ \%$$

Case (i) Unity p.f:

we know E_0 at unity p.f

$$E_0 = \sqrt{(V + IR_a)^2 + (IX_s)^2}$$

Case (ii) at 0.8 Leading p.f

$$E_0 = \sqrt{(V\cos\phi + IR_a)^2 + (V\sin\phi - IX_s)^2}$$

Case (iii) at .71 Lagging p.f

$$E_0 = \sqrt{(V\cos\phi + IR_a)^2 + (V\sin\phi + IX_s)^2}$$

In above all 3-equations,

$V = 1800$ V [for all the cases]

$\cos\phi = 1$ (for 1st case)

$\cos\phi = .8$ (for 2nd case)

$\cos\phi = .71$ (for 3rd case)

$\therefore$ $\sin\phi = \sin[\cos^{-1}(1)] = 0$ [for 1st case]

$\sin\phi = \sin[\cos^{-1}(.8)] = 0.6$ [for 2nd case]

$\sin\phi = \sin[\cos^{-1}(.71)] = .70$ [for 3rd case]

$I = 80$ A [for all the cases]

$R_a = .8\ \Omega$ [for all the cases]

Now only unknown Parameter is X_s.

where,

$$X_s = \sqrt{Z_s^2 - R_a^2}$$

and $$Z_s = \frac{V_{o.c}}{I_{s.c}} = \frac{500}{80} = 6.25\Omega$$

$\therefore$ $$X_s = \sqrt{(6.25)^2 + (.8)^2}$$

$$\boxed{X_s = 2.36\Omega}$$

Now,

(i) at unity p.f, the % age voltage regulation

$$E_0 = \sqrt{(V + IR_a)^2 + (IX_s)^2}$$

$$= \sqrt{[1800 + (80)(.8)] + (80 \times 2.36)^2}$$

$$= \sqrt{(1864)^2 + (188.2)^2}$$

$$= \sqrt{3474496 + 35645.44}$$

$$= \sqrt{3510141.44}$$

$$E_0 = 1873.5 \text{ V}$$

$\therefore$ % regulation at unity P-f is $$= \frac{1873.5 - 1800}{1800} \times 100$$

$$\boxed{\% \text{ regulation} = 4.08\%}$$

(ii) % regulation at .8 Leading p.f

$$E_0 = \sqrt{(V\cos\phi + IR_a)^2 + (V\sin\phi - IX_s)^2}$$

$$= \sqrt{(1800 \times .8 + 80 \times .8)^2 + [1800 \times .6 - 80\,(2.36)]^2}$$

$$= \sqrt{(1440 + 64)^2 + [1080 - (188.8)]^2}$$

$$= \sqrt{(1504)^2 + (891.2)^2}$$

$$= \sqrt{2262016 + 794237.44}$$

$$= \sqrt{3056253.44}$$

$$\boxed{E_0 = 1748.21 \text{ V}}$$

$\therefore$ % regulation at unity P-f is $$= \frac{1748.5 - 1800}{1800} \times 100$$

$$\boxed{\% \text{ regulation} = -2.87\%}$$

(iii) regulation at .71 Lagging p.f

$$E_0 = \sqrt{(V\cos\phi + IR_a)^2 + (V\sin\phi + IX_s)^2}$$

$$= \sqrt{[1800\times(.71) + 80\times(.8)]^2 + [(1800\times.7) + (80\times 2.36)]^2}$$

$$= \sqrt{(1278+64)^2 + (1260-188.8)^2}$$

$$= \sqrt{(1342)^2 + (1448.8)^2}$$

$$= \sqrt{1800964 + 2099021.44}$$

$$= \sqrt{3899985.44}$$

$$\boxed{E_0 = 1974.83\text{ V}}$$

$$\therefore \quad \%\text{ regulation} = \frac{1984.83-1800}{1800}\times 100$$

$$\boxed{\%\text{ regulation} = +9.71\,\%}$$

Ans. (i) 4.08% (ii) – 2.87% (iii) 9.71%

EXAMPLE 2.16: A 3 phase, star connected alternator is rated at 2000 KVA, 15000 V. The armature resistance and synchronous reactance are 2Ω and 25Ω respectively per phase. Calculate the percentage regulation for a load of 1300 KW at 0.8 Leading power factor.

Given data:

Power = 2000 KVA

$V = 15000$ V

$R_a = 2\Omega$

$X_s = 25\Omega$

Load power = 13000 KW at 08 leading p.f

Solution:

$$\text{Load power} = \sqrt{3}\, V_L\, I_L \cos\phi$$

$$1300000 = \sqrt{3}\,(15000)\,(I)\,(.8)$$

$$\therefore \quad I = \frac{1300000}{\sqrt{3}\,(15000)\,(.8)} = \frac{1300000}{20784}$$

$$\boxed{I = 62.548\text{ A}}$$

$$\text{Voltage/phase} = \frac{15000}{\sqrt{3}} = 8660.50\text{ V}$$

$$E_0 = \sqrt{(V\cos\phi + IR_a)^2 + (V\sin\phi - IX_s)^2}$$

$$= \sqrt{[8660.50\times.8 + (62.548)(2)]^2 + [8660.50\times.6 - (62.548)(25)]^2}$$

$$= \sqrt{49751805.82 + 13195782.76}$$

$$= \sqrt{18170963.58}$$

$$\boxed{E_0 = 426.74\text{ V}}$$

$$\therefore \quad \%\text{ regulation} = \frac{1262.71 - 8660.50}{8660.50}\times 100$$

$$\boxed{\%\text{ regulation} = -50.77\,\%}$$

Ans. % regulation = – 50.77 %.

EXAMPLE 2.17: The following test result are obtained from a 3-phase 6000 KVA, 6600 star connected, 2 pole, 50 Hz turbo-alternator.

With a field current of 125 A the open circuit voltage is 8000 V at the rated speed; with the same field current and rated speed, the short circuit current is 800 A. The resistance drop is 3%. Find the regulation of the alternator on fall load and at .8 p.f lagging.

Given data:

Power = 6000 KVA

$V = 6{,}600$ V

Poles = 2

$f = 50$ Hz

$V_{o\cdot c} = 8000$ V at 125 A field current.

$I_{s\cdot c} = 800$ A at 125 A field current.

The resistance drop is 3%

$\cos\phi = .8$

Solution:

$$Z_s = \frac{V_{o\cdot c}}{I_{s\cdot c}} = \frac{8000/\sqrt{3}}{800} = 5.77\ \Omega$$

$$\text{Voltage/phase} = 6{,}600/\sqrt{3} = 3810\text{ V}$$

$$IR_a\text{ drop} = 3\%\text{ i.e } 3810\times .03$$

$$\boxed{IR_a = 114.3\text{ V}}$$

$$\text{Full load current} = 6000\times 10^2/\sqrt{3}\times 6{,}600$$

$$I = 525 \text{ A}$$

$$IR_a = 114.3$$

$$\therefore \quad R_a = 114.3/525 = .218\ \Omega$$

$$X_s = \sqrt{Z_s^2 - R_a^2} = \sqrt{5.77^2 - .218^2}$$

$$X_s = 5.74\ \Omega$$

$$E_0 = \sqrt{(3810 \times 0.8 + 114.3)^2 + (3810 \times 0.6 + 525 \times 5.74)^2}$$

$$E_0 = 6180 \text{ V}$$

$$\therefore \quad \%\text{ regulation} = \frac{(6180 - 3810)}{3810} \times 100 = 62.2\ \%$$

Ans. % regulation = 62.2 %

EXERCISE

10. From the following test results, determine the voltage regulation of a 2000 V, 1-phase alternator delivering a current of 100 A at (i) unity p.f (ii) 0.8 Leading p.f and (iii) 0.71 Lagging p.f.

Test results: Full load current 100 A is produced of short circuit by a field excitation of 2.5 A. An emf of 500 V is produced on open circuit by the same excitation. The armature resistance is 0.8 Ω.

[**Ans.** (i) 7 % (ii) –9 % (iii) 21.6 %]

11. A 100 kVA, 3000 V, 50 Hz, phase star connected alternator has effective armature resistance of .2Ω. The field current of 40 A produces short circuit current of 200 A and an open circuit emf of 1046 V (line value) calculate the full load voltage regulation at 0.8 p.f lagging ad 0.8 p.f Leading, Draw phasor diagram.

[**Ans.** (i) 2.2 % (ii) –1.8 %]

12. A3 phase, star connected alternator is rated at 1600 kVA, 13500 V. The armature and synchronous reactance are 1.5 Ω and 30 Ω respectively per phase. Calculate the percentage regulation for a load of 1280 kW at 0.8 Leading power factor.

[**Ans.** % regulation = – 14.11 %]

2.16 PRINCIPLES OF OPERATION OF SYNCHRONOUS MOTOR

When a 3 phase winding is fed by a 3 phase supply, then a magnetic flux is produced which is constant magnitude but rotating at synchronous speed. Consider a two pole stator of Fig. 2.19 in which two stator poles (marked N_S and S_S) rotating at synchronous speed in clockwise direction. With the rotor position as shown, suppose the stator poles are at that instant situated at points A and B. The two similar poles, N (of rotor) and N_S (of stator) as well as S and S_S will repel each other, with the result that the rotor tends to rotate in anticlockwise direction.

But half a period later, stator poles, having rotated around, interchange their positions i.e. N_S is at point B and S_S at point A. Under these conditions, N_S attracts S and S attracts N. Hence, rotor tends to rotate clockwise. Hence, we find that due to continuous and rapid rotation of stator poles, the rotor is subjected to a torque which is rapidly reversing i.e. in quick succession, the rotor is subjected to torque which tends to move it first in one direction and then in the opposite direction. Owing to its large inertia, the rotor cannot instantaneously respond to such quickly-reversing torque, with the result that it remains stationary. Now, consider the condition shown in Fig. (2.23). The stator and rotor poles are are attracting each other. Suppose that the rotor is not stationary, but is rotating clockwise, with such a speed that it turns through one pole-pitch by the time the stator poles interchange their positions, as shown in Fig. (2.24). In Fig. (2.24) also again the stator and rotor poles attract each other. If the rotor poles shift their positions along with the stator poles, then they will continuously experience a unidirectional torque in clockwise direction.

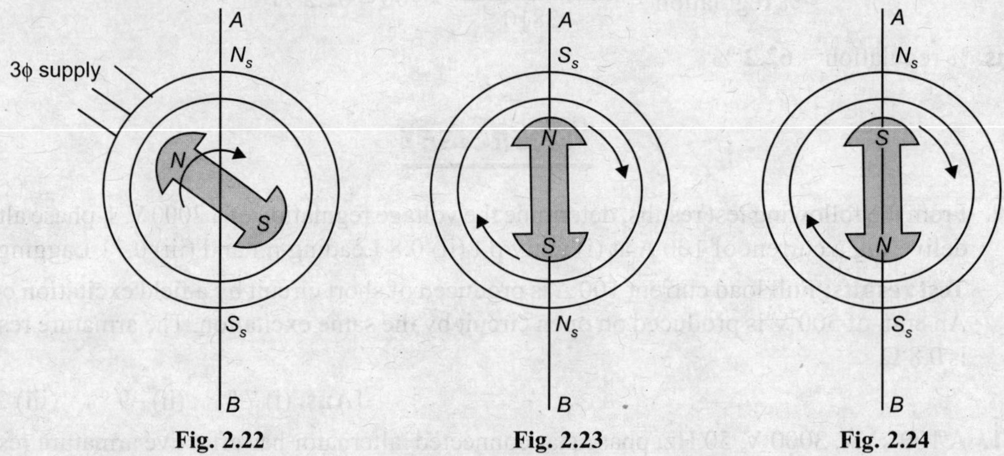

Fig. 2.22 Fig. 2.23 Fig. 2.24

2.17 STARTING METHODS OF SYNCHRONOUS MOTOR

Synchronous motors are not self starting motors. So in order to make it as self starting the following methods can be used.

1. Starting synchronous motor with dampers
2. By using pony motor.
3. By using Dc shunt motor (or) three phase induction motor.

Now we will see one by one.

1. Starting of synchronous motor with damper

The Synchronous motors are equipped with dampers (or) squirrel cage winding which consist of copper bars employed in the pole shoes and which are short circuited at both ends. Such motor will act as a induction motor and starts instantly.

Note: Operating principle will be same as 3 phase induction motor.

2. By using pony motor

As we know that synchronous motor is not self starting it needs some auxiliary motor which can help the synchronous motor to rotate at the time of starting. One method among the different types of starting method is starting synchronous motor using Pony (small) motor. Pony motor means small motor (which can be of self starting motor) initial torque is applied to rotor of synchronous motor by means of pony motor. As soon as the rotor of synchronous motor reaches its rated value then the small motor which is rotating the rotor of synchronous motor is being disconnected from the set up.

Note: Supply can be of common (or) it can be separate also.

From Fig. (2.25), the supply is given to both the motors as soon as supply is given, the pony motor starts rotate. Since the rotor of pony, the rotor of synchronous motor also starts rotate. Once the synchronous motor reaches its rated speed then the pony motor is disconnected (or) released from the operation.

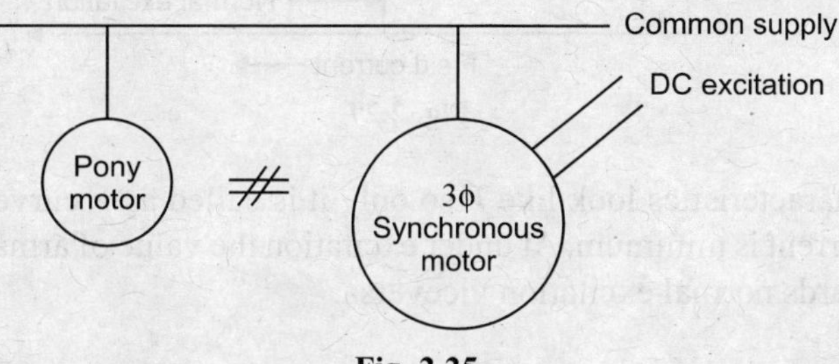

Fig. 2.25

3. By using three induction motor

This method is also same as 2nd case.

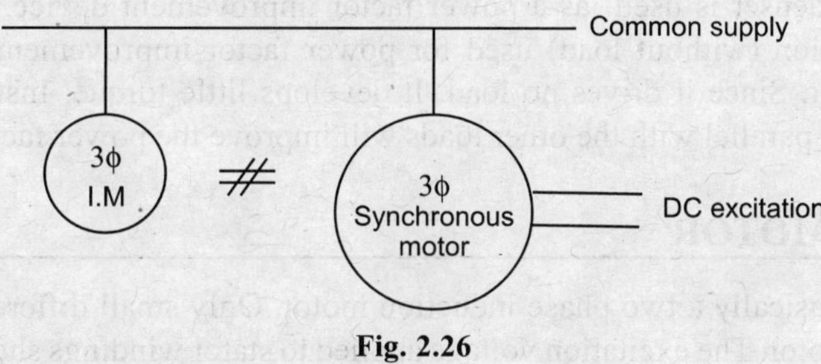

Fig. 2.26

Note 1: Explanation for this method is same as 2nd case. Just replace 3 phase induction motor in the place of pony motor.

Note 2: Instead of 3 phase induction motor DC shunt motor also can be used.

2.18 EFFECT OF EXCITATION ON LINE CURRENT (*V* CURVES)

As we know that, the synchronous motor is doubly excited motor. Due to doubly excitation, there are two currents are flowing in synchronous motor. One is stator current (or) Armature current (I_a) and another one is rotor current (or) field current (I_f). Under change of excitation (I_f) there is a effect on armature current. The excitation corresponding to unity power factor current drawn by the motor is called normal excitation. Excitation higher than the normal is called over excitation.

Excitation lower than the normal is called under excitation. Now we will see the relation between I_f and I_a for particular load in a synchronous motor.

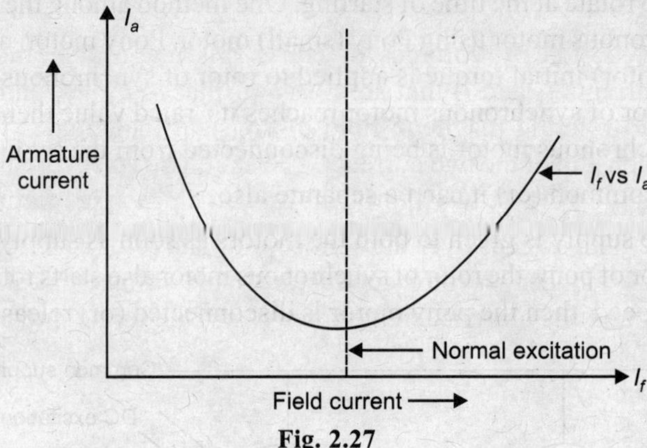

Fig. 2.27

The shape of the characteristics look like *V* so only it is called as *V* curve under normal excitation the value of armature current is minimum. At under excitation the value of armature current is maximum and gets decreased towards normal excitation viceversa.

2.19 SYNCHRONOUS CONDENSER

The synchronous motor is said to be synchronous condenser under over excitation condition. This synchronous condenser is used as a power factor improvement device. A synchronous motor without a shaft extension (without load) used for power factor improvement in a system, is called synchronous condenser. Since it drives no load. It develops little torque. Installation of overexcited synchronous motors in parallel with the other loads will improve the power factor of the line.

2.20 AC SERVOMOTOR

An ac servomotor is basically a two phase induction motor. Only small difference between AC servo motor and induction motor. The excitation voltage applied to stator windings should be displaced by 90º and the rotor is built with high resistance to obtain linear mechanical characteristics.

Construction

The AC servomotor is basically a two phase induction motor with some special design features. The stator consists of two pole-pairs (A–B and C–D) mounted on the inner periphery of the stator, such that their axes are at an angle of 90º in space. Each pole pair carries a winding. One winding is called reference winding and the other is called control winding. The supply used to drive the motor is single phase and so a phase capacitor is connected to one of phase to produce a phase difference of 90º.

The rotor construction is usually squirrel cage or drag-cup type. The squirrel cage rotor is made of laminations. The rotor bars are placed on the slots and short circuited at both ends by end rings. The diameter of the rotor is kept small in order to reduce inertia and to obtain good accelerating characteristics. The Drug-cup construction is employed for very low inertia applications. In this type of construction the rotor will be in the form of hollow cylinder made of aluminium.

Working

The stator windings are excited by voltages of equal rms magnitude and 90º phase difference. This results in exciting currents i_1 and i_2 that are phase displaced by 90º and have equal rms values. These currents give rise to rotating magnetic field of constant magnitude. The direction of rotation depends on the phase relationship of the two currents. The exciting current produces a clockwise rotating magnetic field and phase shift of 180º. It will produce an anti clockwise rotating magnetic field.

The rotating magnetic field sweeps over the conductors. The rotor conductors experience a change in flux and so voltages are induced in rotor conductors. This voltage circulates current in the short circuited rotor conductors and the currents create rotor flux. Due to the interaction of stator and rotor flux, a mechanical force is developed in the rotor and so the rotor starts moving in the same direction as that of rotating magnetic field.

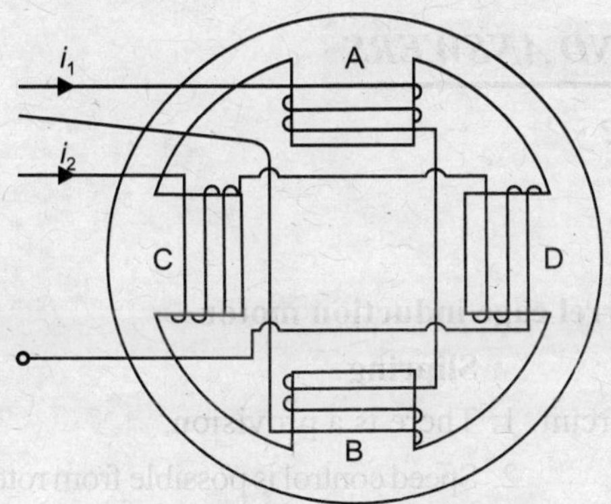

Fig. 2.28: Stator of AC servomotor.

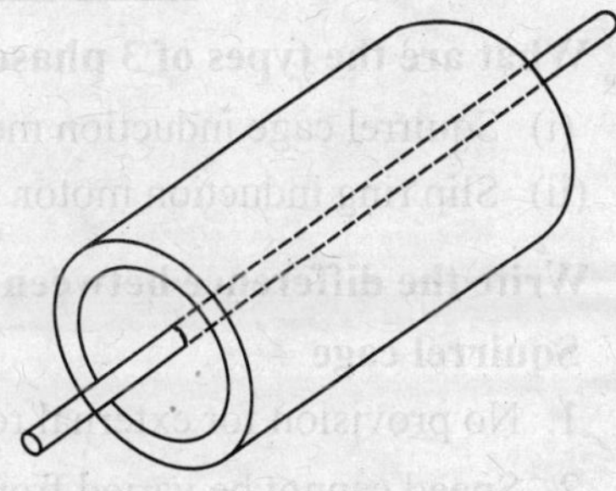

Fig. 2.29: Rotor of AC servomotor.

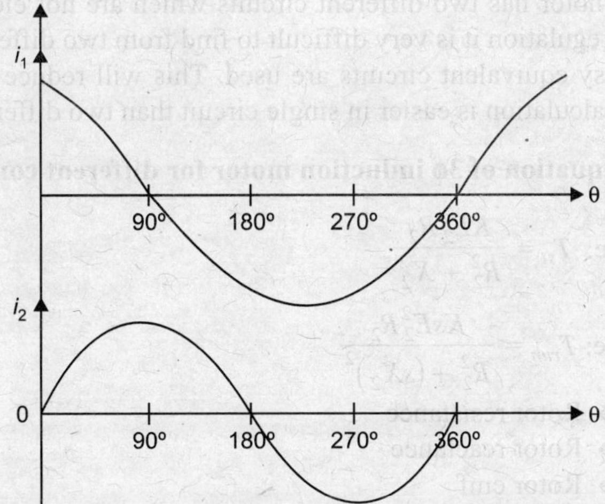

Fig. 2.30: Exciting currents of AC servomotor.

Advantages

1. Fast response
2. Steady state stability
3. Accurate position control
4. Easy construction.

Disadvantages

1. Lower power output compared to DC servomotor.
2. Linearity characteristics are not achieved easily as DC servomotor.
3. Starting torque is low as compared to DC servomotor.

Application

1. It is used in positioning system
2. It is used as a control the speed of other motor.

SHORT QUESTION AND ANSWERS

Q.1 What are the types of 3 phase induction motor?

Ans. (i) Squirrel cage induction motor
(ii) Slip ring induction motor.

Q.2 Write the difference between slipring and squirrel cage induction motor

Ans.

Squirrel cage	**Slipring**
1. No provision for external resistance in rotor circuit.	1. There is a provision.
2. Speed cannot be varied from rotor side.	2. Speed control is possible from rotor side.

Q.3 What is the use of equivalent circuit?

Ans. 3 phase induction motor has two different circuits which are not electrically connected. So to determine efficiency and regulation it is very difficult to find from two different circuits. So in order to make the procedure is easy equivalent circuits are used. This will reduce two different circuits into single circuit. Doing the calculation is easier in single circuit than two different circuits.

Q.4 Write the torque equation of 3ϕ induction motor for different conditions.

Ans. (i) Starting torque: $T_{st} = \dfrac{KE_2^2 R_2}{R_2^2 + X_2^2}$

(ii) Running torque: $T_{run} = \dfrac{KsE_2^2 R_2}{R_2^2 + (sX_2)^2}$

where, $R_2 \rightarrow$ Rotor resistance
$X_2 \rightarrow$ Rotor reactance
$E_2 \rightarrow$ Rotor emf
$s \rightarrow$ Slip
$K \rightarrow$ Rotor constant

Q.5 What is the condition for maximum torque under different conditions?

Ans. (i) Stating $R_2 = X_2$

(ii) Running $R_2 = sX_2$.

Q.6 What are the method of speed control for 3 phase induction motor?

Ans. ***From stator side***

1. By changing voltage
2. By changing frequency
3. By changing number of poles

From rotor side

4. Injecting emf in rotor circuit
5. Rotor Rheostat method.

Q.7 What is the emf equation of alternator?

Ans. emf $= 4.44 f\ \phi\ T k_d\, k_p$. 'volts'

f = frequency

ϕ = Flux per pole

k_d = Distribution factor

T = No. of turns

k_p = pitch factor

Q.8 What is voltage regulation of Alternator?

Ans. There is rise in voltage when full load is removed. The ratio of change in voltage from no load to full load to terminal voltage is defined as voltage regulation

$$\%\ \text{regulation} = \frac{E_0 - V}{V} \times 100.$$

Q.9 What are types of Synchronous machine?

Ans. (i) Salient pole type

(ii) Non-salient pole type.

Q.10 What are the starting methods of synchronous motor?

Ans. Starting as 3 f induction motor

Starting by using pony motor

Starting by using 3 induction motor (or) by using DC shunt motor.

Q.11 What is synchronous condenser?

Ans. A synchronous motor is said to be synchronous under over excitation.

Q.12 What is *V* curve?

Ans. When we draw the characteristics between Armature current and field current is gives the shape of *V* curve. So it is called as *V* curve.

The curve which is drawn between armature current (I_a) and field current (I_f) is known as '*V*' curve.

CHAPTER 3

Control System

I. CONTROL SYSTEM

3.1 INTRODUCTION

Control system theory evolved as an engineering discipline and due to universality of the principles involved, it is extended to various fields like economy, sociology, biology, medicine, etc. The automatic control has become an integral part of modern manufacturing industries, controlling pressure, temperature, humidity, viscosity and flow in process industry.

3.2 OPEN LOOP AND CLOSED LOOP CONTROLS

3.2.1 Control System Classification

When a number of elements (or) components are connected in a sequence to perform a specified function, the group thus formed is called a system.

In a system when the output quantity is controlled by varying the input quantity, the system is called control system. The output quantity is called controlled variable (or) response and input quantity is called command signal (or) excitation.

With respect to feedback signal, control system can be sub divided into two types (i) Open loop system and (ii) Closed loop system.

Control system can be classified in a number of ways. They are classified either as an open loop system (or) as a closed loop system. According to the differential equation describing the system, control systems are classified as linear or non linear, time varying or time invariant. control systems are classified as continuous data and discrete data systems according to the types of the signal found in the system. Control systems are also classified according to the components connected in the system. Electrical systems, mechanical systems, Hydraulic systems, Pneumatic systems, electromechanical systems etc. Depending upon the equations describing the behaviour, systems are classified as lumped or distributed system. System which have lumped parameters are described by differential equations whereas distributed system is described by partial differential equations. If the systems have fixed parameters and inputs they are called deterministic systems. On other hand systems which have randomness in parameters or inputs are called stochastic systems. Fig. (3.1) shows classification of control systems.

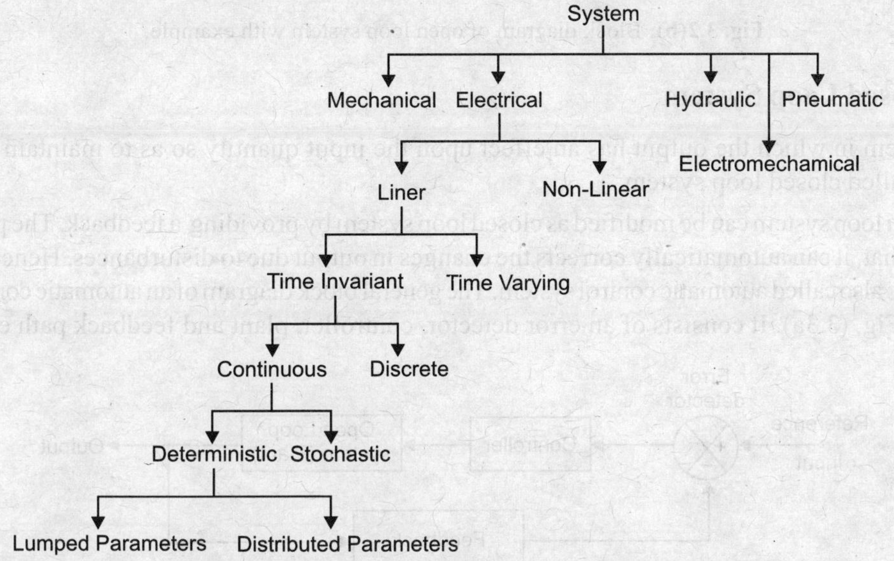

Fig. 3.1: Classification of control system.

Let we see about open loop and closed loop system in brief.

3.2.2 Open Loop System

Any physical system which does not automatically correct the variation in its output, is called an open loop system, or system in which the output quantity has no effect upon the input quantity is called open loop control system. This means that the output is not feedback to the input for correction.

In open loop control system the output can be varied by varying the input. But due to external disturbances the system output may change. When the output changes due to disturbances, it is not followed by changes in input to correct the output. The schematic block diagram of open loop control system shown in Fig. (3.2a).

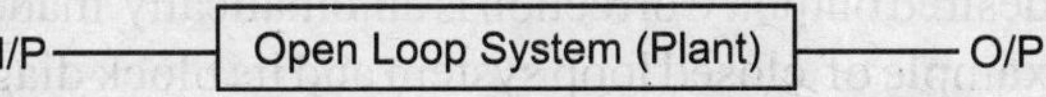

Fig. 3.2(a): Block diagram of open loop system.

To understand the principle of operation of open loop system, consider the action of a blind man who attempts to take a book from a table. The blind man gives command to his brain to take the book. The brain senses his command and gives out order to the arms to take the book from the table. The blind man stretches out his hands and tries to take the book from the table. In his single attempt to get the book, he may succeed (or) fail.

In case he fails he has to change the reference input. The blind man may or may not succeed in his attempt. The blind man's action is an example of open loop system. There is no automatic correction of the error. The block diagram of the above action is shown in Fig. (3.2b)

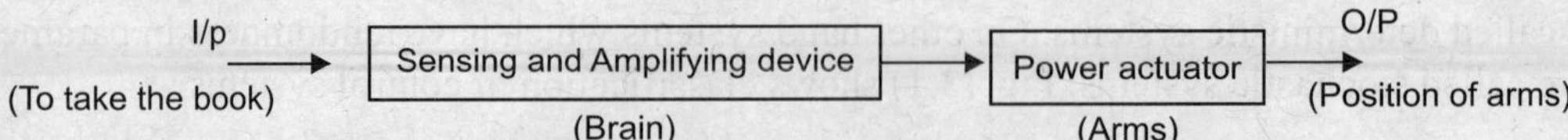

Fig. 3.2(b): Block diagram of open loop system with example.

3.2.3 Closed Loop System

Control system in which the output has an effect upon the input quantity so as to maintain the desired output are called closed loop system.

The open loop system can be modified as closed loop system by providing a feedback. The provision of feedback is that, it can automatically corrects the changes in output due to disturbances. Hence the closed loop system is also called automatic control system. The general block diagram of an automatic control system is shown in Fig. (3.3a). It consists of an error detector, controller, plant and feedback path elements.

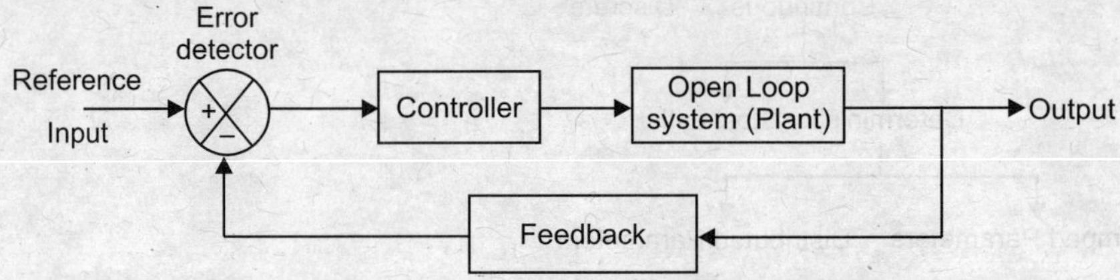

Fig. 3.3(a): Block diagram of closed loop system.

The feedback path elements samples the output and converts in to a signal of same types as that of reference signal. The feedback signal is proportional to output signal and it is fed to the error detector. The error signal which is generated by the error detector is the difference between reference signal and feedback signal. The controller modifies and amplifies the error signal to produce better control action. The modified error signal is fed to the plant to correct its output.

To understand the principle of operation of a closed-loop system, consider, the action of a normal man in picking up a book from the table. He gives command to the brain and the brain in turn commands and actuates the arms. The arms try to take the book If there is any error; suddenly the eyes detect that error and sends the signal to the brain. The brain gets aware of this error and moves the arms such that he pick up the book without miss. The eyes compare the input and output and tell the brain the amount of error between actual output and the desired output. Correction is automatically made by the brain until the desired result is achieved. This is an example of closed loop system and its block diagram is shown in Fig. (3.3b).

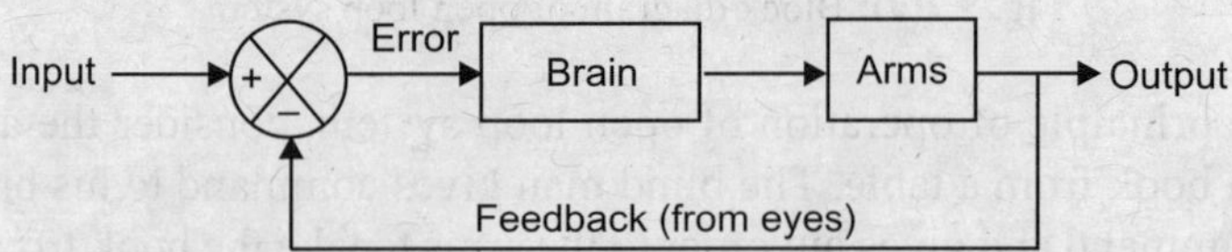

Fig. 3.3(b): Block diagram of closed loop system with example.

3.2.4 Advantages of Open Loop System

1. The open loop systems is simple and economical.

2. The open loop system is easier to construct.
3. Generally the open loop systems are stable.

3.2.5 Disadvantages of Open Loop Systems

1. The open loop systems are inaccurate and unreliable.
2. The changes in the output due to external disturbances are not corrected automatically.

3.2.6 Advantages of Closed Loop Systems

1. The closed loop system outputs are more accurate to desired output.
2. The output is accurate both for linear and non-linearity conditions.
3. Less noise.

3.2.7 Disadvantages of Closed Loop Systems

1. The closed loop systems are complex and costlier
2. The feedback in closed loop system may lead to oscillatory response.
3. The feedback reduces the overall gain of the system.
4. Stability is a major problem in closed loop system and more care is needed to design a stable closed loop system.

3.2.8 Examples of Control Systems

Example 1: Temperature Control

Open loop system

The electric furnace shown in Fig. (3.4a) is an example for open loop system. The output in the system is desired temperature. The temperature of the system is raised by heat generated by the heating element. The output temperature depends on the time during which the supply to heater remains ON.

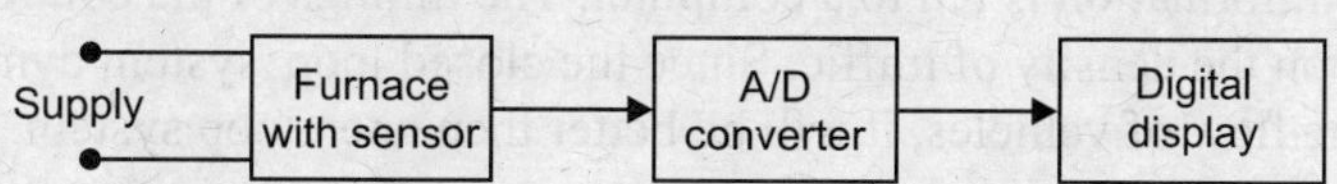

Fig. 3.4(a): Open loop temperature control system.

The ON and OFF of the supply is governed by the time setting. The temperature is measured by a sensor, which gives an analog voltage corresponding to the temperature of the furnace. The analog signal is converted to digital signal by an analog to digital converter (A/D converter). The digital signal is given to the digital display device to display the temperature.

Closed loop system

The electric furnace shown in Fig. (3.4b) is example for closed loop system. The output of the system is desired temperature and it depends on the time during which the supply remains ON.

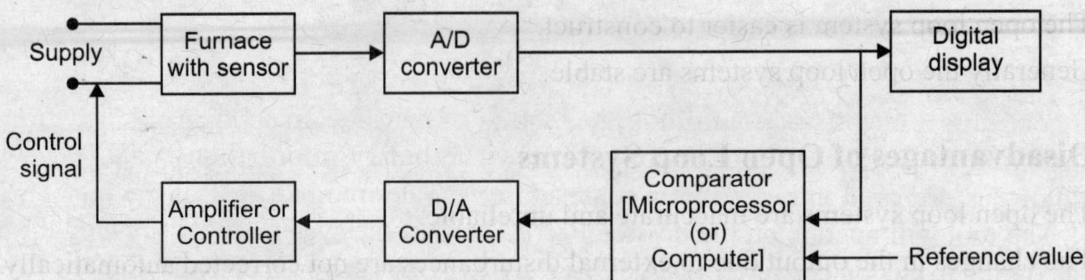

Fig. 3.4(b): Closed loop temperature control system.

The desired temperature is set to the system through keyboard or as a signal corresponding to desired temperature via ports. The actual temperature is sensed by sensor and converted to digital signal by the A/D converter. The comparator compares with desired temperature. If it finds any difference then it sends signals through D/A converter and Amplifier. Thus the system automatically corrects any change in output. Hence it is called closed loop system.

Example 2: Traffic Control

Open loop system

In traffic control, traffic signals which are operated on a time basis constitutes an open loop control system. The sequence of control signals are based on a time slot given for each signal. The time slots are decided based on a traffic study. The system will not measure the density of the traffic before giving the signal. Since the time slot does not change according to traffic density, the system is open loop system.

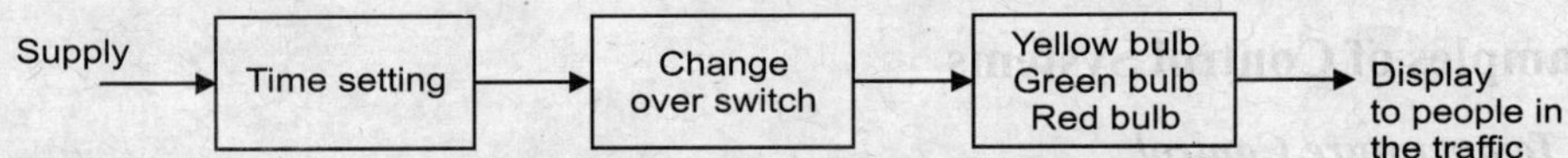

Fig. 3.5(a): Open loop traffic control system.

Closed loop system

Traffic control system can be made as a closed loop system if the time slots of the signals are decided based on the density of vehicles. In closed loop traffic control system, the density of the traffic is measured on all the sides and the information is fed to a computer. The timings of the control signals are decided by the computer based on the density of traffic. Since the closed loop system dynamically changes the timings according to the flow of vehicles, it will be better then open loop system.

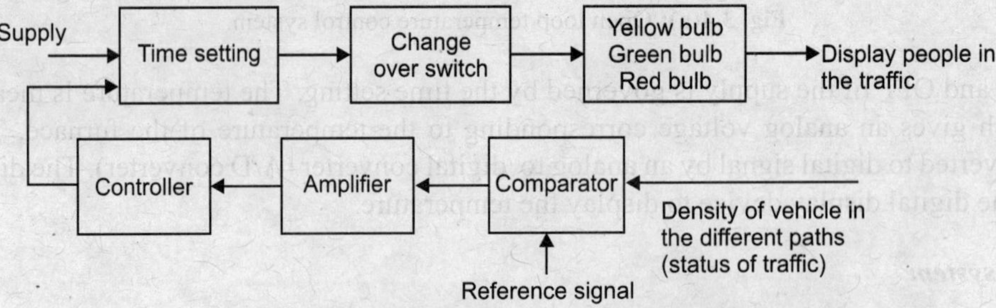

Fig. 3.5(b): Closed loop traffic control system.

Example 3: Numerical Control System

Open loop system

Numerical control is a method of controlling the motion of machine components by use of numbers. Here, the position of work head tool is controlled by the binary information contained on a disc. A magnetic disc is prepared in binary form representing the desired part P (P is the metal part to be machined). The tool will operate on the desired Part P. To start the system, the disc is fed through the reader to the digital to analog converter (D/A converter) The D/A converter converts the FM [frequency modulated] output of the reader to a analog signal. It is amplified and fed to servomotor which positions the cutter on the desired part P. The position of the cutter head is controlled by the angular motion of the servomotor. This is an open loop system since no feedback path exists between the output and input. The system positions the tool for a given input command. Any deviation in the desired position is not checked and corrected automatically.

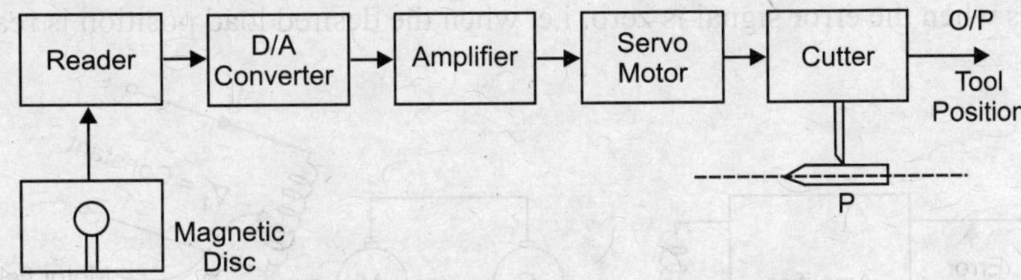

Fig. 3.6(a): Open loop numerical control system.

Closed loop system

A magnetic disc is prepared in binary form representing the desired part P (P is the metal part to be machined). To start the system, the disc is loaded in the reader. The controller compares the frequency modulated input pulse signal with the feedback pulse signal. The controller is a computer (or) micro-processor system. The controller carries out mathematical operations on the difference in the pulse signals and generates an error signal.

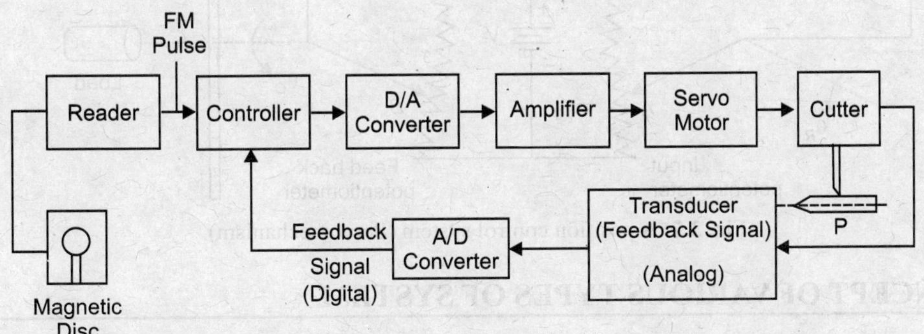

Fig. 3.6(b): Closed loop numerical control system.

The D/A converter converts the controller output pulse (error signal) into an analog signal. The amplified analog signal rotates the servomotor to position the tool on the job. The position of the cutter head is controlled according to the input of the servomotor. The transducer attached to the cutter head converts the motion into an electrical signal. The analog electrical signal is converted to the digital pulse signal by the A/D converter. Then this signal is compared with the input pulse signal. If there is any difference between these two, the controller sends a signal to the servomotor. Thus the system automatically corrects any deviation in the desired output position. An advantage of numerical control is that complex parts can be produced with uniform tolerances at the maximum milling speed.

Example 4: Position Control System using Servomotor

The position control system shown in Fig. 3.7 is a closed loop system. The system consists of a servomotor powered by a generator. The load whose position has to be controlled is connected to motor shaft through gear wheels. Potentiometers are used to convert the mechanical motion to electrical signals. The desired load position (θ_R) is set on the input potentiometer and the actual load position (θ_C) is fed to feedback potentiometer. The difference between the two angular positions generates an error signal, which is amplified and fed to generator field circuit. The induced emf of the generator drives the motor. The rotation of the motor stops when the error signal is zero. i.e. when the desired load position is reached.

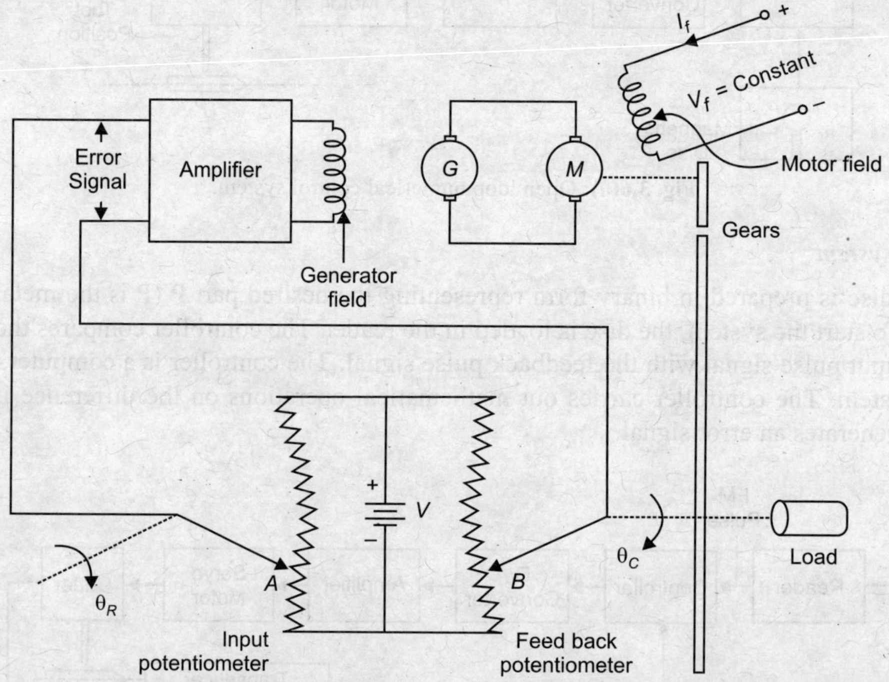

Fig. 3.7: A position control system (Servomechanism).

3.3 CONCEPT OF VARIOUS TYPES OF SYSTEM

3.3.1 Thermal System

List of symbols used in thermal system

$q \Rightarrow$ Heat flow rate, Kcal/sec

$\theta_1 \Rightarrow$ Absolute temperature of emitter, °K

$\theta_2 \Rightarrow$ Absolute temperature of receiver, °K

$\Delta\theta \Rightarrow$ Temperature difference, °C

$A \Rightarrow$ Area normal to heat flow, m^2

$K \Rightarrow$ Conduction coefficient, kcal/sec–°C

$K_f \Rightarrow$ Radiation coefficient, kcal/sec–°C

$K=HA \Rightarrow$ Convection coefficient, kcal/m^2–sec–°C

$K \Rightarrow$ Thermal conductivity, kcal/m–sec–°C

$\Delta x \Rightarrow$ Thickness of conductor, m

$R \Rightarrow$ Thermal resistance, °C–sec/kcal

$C \Rightarrow$ Thermal capacitance, kcal/°C

Heat flow rate

Thermal systems are those that involve the transfer of heat from one substance to another. There are different ways of heat flow from one substance to another. They are conduction, convection and radiation.

For conduction,

$$\text{Heat flow rate, } q = K\Delta\theta = KA/\Delta x$$

For Convection,

$$\text{Heat flow rate, } q = K\Delta\theta = HA\,\Delta\theta$$

For Radiation,

$$\text{Heat flow rate, } q = K_r\,(\theta^4{}_1 - \theta^4{}_2)$$

If $\theta_1 >> \theta_2$ then, $q = K_r\,\overline{\theta}^4$

where $$\overline{\theta}_4 = (\theta_1^4 - \theta_2^4)^{1/4}$$

Note: $\overline{\theta}_4$ is called effective temperature difference of the emitter and receiver.

The resistance and capacitance are basic elements of thermal system.

The thermal resistance, R (when heat is transfered between two substances) defined as the ratio of change in temperature and change in heat flow rate.

$$\text{Thermal resistance, } R = \frac{\text{Change in temperature, °C}}{\text{Change in heat flow rate, kcal/sec}} = \frac{d(\Delta\theta)}{dq}$$

For conduction or convection,

Heat flow rate, $q = K\Delta\theta$

On differentiating we get,

$$dq = Kd(\Delta\theta)$$

$$\therefore \quad \frac{d(\Delta\theta)}{dq} = \frac{1}{K}$$

But thermal resistance, $$R = \frac{d(\Delta\theta)}{dq}$$

CHAPTER 3

$\therefore$ Thermal resistance, $R = \frac{1}{K}$ for conduction

$$= \frac{1}{K} = \frac{1}{HA} \text{ for convection}$$

For radiation,

Heat flow rate, q = $k_r\ \overline{\theta}^4$

on differentiating we get $dq = K_r, 4\overline{\theta}^3\ d\overline{\theta}$

$$\therefore \quad \frac{d\overline{\theta}}{dq} = \frac{1}{K_r 4\overline{\theta}^3}$$

But thermal resistance, $R = \frac{d\theta}{dq}$

$\therefore$ Thermal resistance (for radiation)

$$R = \frac{1}{4K_r\overline{\theta}^3}$$

Thermal capacitance, C is defined as the ratio of change in heat stored and change in temperature.

Thermal capacitance, $C = \frac{\text{Change in heat stored, kcal}}{\text{Change in temperature, °C}}$

Let,

$M \rightarrow$ Mass of substance considered, kg

$C_p \rightarrow$ Specific heat of substance, kcal/ kg - °C

Now,

Thermal capacitance, $C = MC_p$

Consider a simple thermal system shown in Fig. (3.8). Let us assume that the tank is insulated to eliminate heat loss to the surrounding air, there is no heat storage in the insulation and liquid in the tank is kept at uniform temperature by perfect mixing. Let derive the transfer function for the Fig. (3.8).

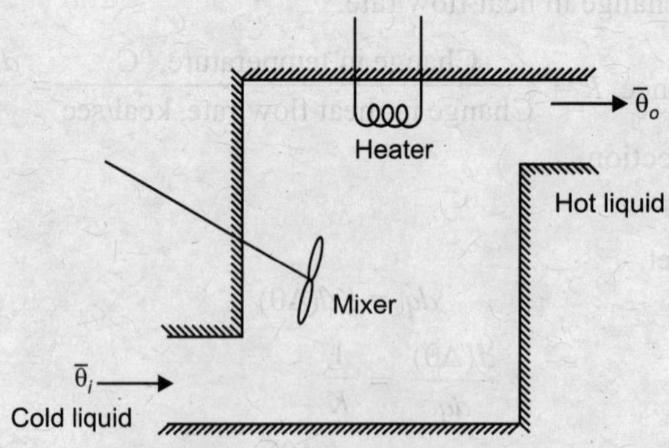

Fig. 3.8: Thermal system cold liquid.

Let,

$\overline{\theta}_i$ ⇒ Steady state temperature of inflowing liquid, °C.

$\overline{\theta}_0$ ⇒ Steady state temperature of outflowing liquid, °C.

G ⇒ Steady state liquid flow rate, kg/sec.

M ⇒ Mass of liquid in tank, kg.

c ⇒ Specific heat of liquid kcal/kg °C.

R ⇒ Thermal resistance, °C - sec/ kcal

C ⇒ Thermal capacitance, kcal/ °C

Q ⇒ Steady state heat input rate. kcal/sec.

Let us assume that the temperature of inflowing liquid is kept constant. Let the heat input rate to the system supplied by the heater is suddenly changed from $\overline{Q}$ to $\overline{Q} + q_i$.

Due to this, the heat output flow rate will gradually change from $\overline{Q}$ to $\overline{Q} + q_0$. The temperature of the outflowing liquid will also be changed from $\overline{Q}_0$ to $\overline{Q}_0 + Q$.

For this system the equation for q_0, C and R are obtained as follows,

Change in output heat flow rate q_0 = {Liquid flow rate, G} × {Specific heat of liquid, c} × {Change in temperature θ}

$$q_0 = Gc\theta \tag{3.1}$$

Thermal capacitance C = Mass, (M) × {Specific heat of liquid (c)}

$$C = Mc \tag{3.2}$$

$$\text{Thermal resistance, } R = \frac{\text{Change in temperature, } \theta}{\text{Change in heat flow rate, } q_0}$$

$$R = \frac{\theta}{q_0} \tag{3.3}$$

On substituting for q_0 from Eq. (3.1) in Eq. (3.3) we get,

$$R = \frac{\theta}{Gc\theta} = \frac{1}{Gc}$$

$$R = \frac{1}{Gc} \tag{3.4}$$

In this system, the rate of change of temperature is directly proportional to change in heat input rate.

$$\therefore \quad \frac{d\theta}{dt} \propto q_i - q_0 \tag{3.5}$$

Let the constant of proportionality is the capacitance C of the system.

$$\therefore \quad \frac{C\,d\theta}{dt} = q_i - q_o \tag{3.6}$$

Equation (3.6) is the differential equation governing the system. Since Eq. (3.6) is of first order equation, the system is first order system.

From Eq. (3.3), $R = \frac{\theta}{q_0}$;

$$\therefore \quad q_o = \frac{\theta}{R} \tag{3.7}$$

On substituting for q_0 from Eq. (3.7) in Eq. (3.6) we get

$$C\frac{d\theta}{dt} = q_i - \frac{\theta}{R}$$

$$C\frac{d\theta}{dt} = \frac{Rq_i - \theta}{R}$$

$$RC\frac{d\theta}{dt} = q_i - \theta$$

$$RC\frac{d\theta}{dt} + \theta = Rq_i - \theta \tag{3.8}$$

Let, $L(\theta) = \theta(s)$; $L\left[\frac{d\theta}{dt}\right] = s\theta(s)$ and $L[q_i] = Q_i(s)$

On taking Laplace transform of Eq. (3.8) we get,

$$RC\, s\theta(s) + \theta(s) = RQ_i(s)$$

$$\theta(s)\,[sRC + 1] = RQ_i(s)$$

$\frac{\theta(s)}{Q_i(s)}$ is the required transfer function of the system.

$$\frac{\theta(s)}{Q_i(s)} = \frac{R}{sRC+1} = \frac{R}{RC\left[s + \frac{1}{RC}\right]} = \frac{\frac{1}{C}}{s + \frac{1}{RC}}$$

$$\frac{\theta(s)}{Q_i(s)} = \frac{\frac{1}{C}}{s + \frac{1}{RC}} \tag{3.9}$$

The above Eq. (3.9) is the transfer function of thermal system of example which is taken for analysis.

The transfer function of thermal system of Fig. (3.8) is $\frac{\theta(s)}{Q_i(s)} = \frac{\frac{1}{C}}{s + \frac{1}{RC}}$. In Eq. the term RC is the time constant of the syste.

3.3.2 Hydraulic System

The hydraulic system may be classified into

1. Liquid level system
2. Hydraulic devices

The liquid level system consists of storage tanks and connecting pipes. The variables to be controlled are liquid height in tanks and flow rate in pipes. The driving force is the relative difference of the liquid heights in the tanks.

The hydraulic devices are using incompressible oil as their working medium. These devices are used for controlling the forces and motions. The driving force is the high pressure oil supplied by the hydraulic pumps.

The variables of hydraulic system are volumetric flow rate, q and pressure, P. The volumetric flow rate, q is through variable and it is analogous to current. The pressure, P is across variable and it is analogous to voltage.

Three basic elements of hydraulic systems are resistance, capacitance and Inertance. The liquid flowing out of a tank can meet the resistance in several ways. Liquid while flowing through a pipe meet with resistance due to the friction between pipe walls and liquid. Presence of valves, bends, coupling of pipe of different diameter also offer resistance to liquid flow.

The capacitance is an energy storage element and it represents storage in gravity field. The inertance represents fluid inertia and is derived from the inertia forces required to accelerate the fluid in a pipe. It is also an energy storage element. But the energy storage due to inertance element is negligible compared to that of capacitance element.

Consider the flow through a short pipe connecting two tanks. The resistance for liquid flow in such a pipe or restriction, is defined as the change in the level difference, necessary to cause a unit change in the flow rate.

$$R = \frac{\text{Change in level difference, m}}{\text{Change in flow rate, m}^3\text{/sec}}$$

Capacitance C of a tank is defined as the change in quantity of stored liquid, necessary to cause a unit change in the potential (head).

$$C = \frac{\text{Change in liquid stored, m}^3}{\text{Change in head, m}}$$

A simple liquid level system shown in Fig. 3.9(a) with the steady state flow rate, $\bar{Q}$ and steady state head, $\bar{H}$

Let,

$\bar{Q} \rightarrow$ Steady state flow rate (before any change occurred)

$q_i \rightarrow$ Small deviation of inflow rate from its steady state valve, m^3/sec

$q_o \rightarrow$ Small deviation of outflow rate from its steady state valve, m^3/sec

$\bar{H} \rightarrow$ Steady state head (before any change has occurred), m

$h \rightarrow$ Small deviation of head from its steady-state valve, m.

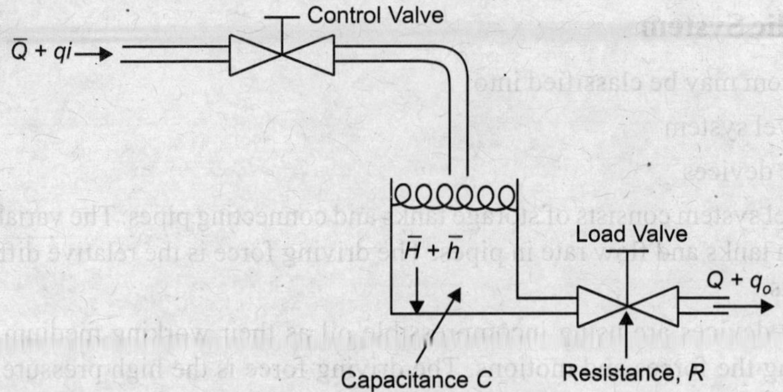

Fig. 3.9(a): Liquid level system.

Let the system be considered linear. The differential equation governing the system is obtained by equating, the change in flow rate to the amount stored in the tank. In a small time interval dt, Let the change in flow rate be $(q_i - q_o)$, and the change in height be dh.

Now, change in storage = change in flow rate

$$\therefore \quad C\,dh = (q_i - q_o)\,dt \tag{3.10}$$

The resistance,

$$R = \frac{\text{Change in head}}{\text{Change in outflow rate}} = \frac{h}{q_o} \tag{3.11}$$

$$\therefore \quad q_o = \frac{h}{R} \tag{3.12}$$

On substituting for q_0 from Eq. (3.12) in Eq. (3.10) we get,

$$C\,dh = \left(q_i - \frac{h}{R}\right)dt$$

$$C\,dh = \left(\frac{q_i R - h}{R}\right)dt$$

$$RC\frac{dh}{dt} = q_i R - h$$

$$RC\frac{dh}{dt} + h = q_i R \tag{3.13}$$

The Eq. (3.13) is the differential equation governing the system. The term RC is the time constant of the system. On taking Laplace transform of Eq. (3.13) we get,

$$RC\,sH(s) + H(s) = Q_i(s)\,R$$

$$(sRC + 1)\,H(s) = Q_i(s)\,R$$

$$\frac{H(s)}{Q_i(s)} = \frac{R}{(sRC+1)} = \frac{R}{RC\left(s + \frac{1}{RC}\right)} = \frac{\frac{1}{C}}{s + \frac{1}{RC}}$$

$$\frac{H(s)}{Q_i(s)} = \frac{\frac{1}{C}}{s + \frac{1}{RC}} \tag{3.14}$$

The Eq. (3.14) is the required transfer function of the system.

The hydraulic devices are used in hydraulic feedback systems and in combined electro-mechanical-hydraulic systems. In hydraulic devices, power is transmitted through the action of fluid flow under pressure and the fluid is in compressible. The fluid used are petroleum based oils or non-inflammable synthetic oils.

The most frequently used hydraulic device in control system is hydraulic motor-pump set. It consists of a variable stroke, hydraulic pump and a fixed stroke hydraulic motor as shown in Fig. (3.9b). The device accepts a linear displacement (stroke length) as input and delivers a large output torque.

The hydraulic motor is controlled by amount of oil delivered by the pump. By mechanically changing the pump stroke, the oil delivered by the pump is controlled. In a pump the input is mechanical power and output is hydraulic power.

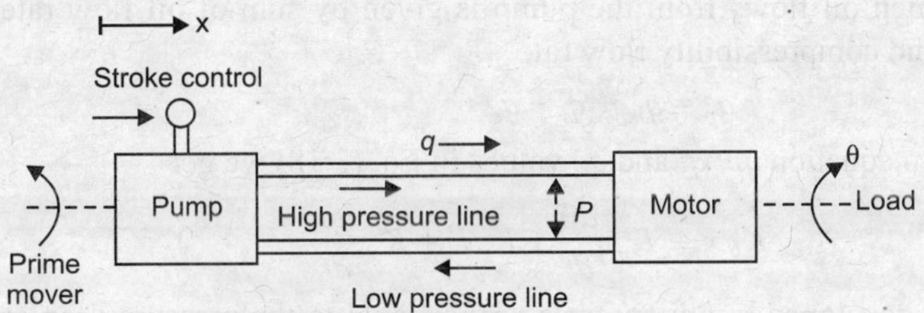

Fig. 3.9(b): Hydraulic motor-pump set.

Let,

$q_p \Rightarrow$ Rate at which the oil flows from the pump

$q_m \Rightarrow$ Oil flow rate through the motor

$q_i \Rightarrow$ Leakage flow rate

$q_c \Rightarrow$ Compressibility flow rate

$x \Rightarrow$ Input stroke length

$\theta \Rightarrow$ Output angular displacement of motor

$P \Rightarrow$ Pressure drop across motor.

The rate at which the oil flow from the pump is proportional to stroke angle, i.e., $q_p \propto x$.

$\therefore$ oil flow rate from the pump, $q_p = K_p x$ (3.15)

where K_p is a constant and is equal to the ratio of rate of oil flow to unit stroke angle.

The rate of oil flow through the motor is proportional to motor speed i.e., $q_m \propto \frac{d\theta}{dt}$

$\therefore$ oil flow rate through motor} $q_m = K_m \frac{d\theta}{dt}$ (3.16)

where $K_m \rightarrow$ motor displacement constant.

All the oil from the pump does not flow through the motor in the proper channels. Due to back pressure in the motor, a portion of the ideal flow from the pump leaks back in the pistons of motor and pump. The back pressure is the pressure that is built up by the hydraulic flow to overcome the resistance to free movement offered by load on motor shaft.

It is usually assumed that the leakage flow is proportional to motor pressure, i.e. $q_i \propto p$

∴ Leakage flow rate, $q_i = K_i P$

where, $K_i \rightarrow$ constant

The back pressure built up by the motor not only causes leakage flow in the motor and pump but also causes the oil in the lines to compress. Volume compressibility flow is essentially proportional to pressure and therefore the rate of flow is proportional to the rate of change of pressure, i.e., $q_c \propto \frac{dp}{dt}$

∴ Compressibility flow rate, $q_c = K_c \frac{dp}{dt}$ (3.17)

when $K_c \rightarrow$ coefficient of compressibility.

The rate at which oil flows from the pump is given by sum of oil flow rate through the motor, leakage flow rate and compressibility flow rate.

$$q_p = q_m + q_i + q_c \tag{3.18}$$

on substituting from equation *qn, qi* and *qc* values in Eq. (3.18) we get

$$K_p x = K_m \frac{d\theta}{dt} + K_i P + K_c \frac{dp}{dt} \tag{3.19}$$

The torque T_m developed by the motor's proportional to the pressure drop and balances the load torque.

∴ Hydraulic motor torque, $T_m \times P => T_m = K_t P$ (3.20)\

where K_t is motor torque constant.

If the load is consist of moment of inertia J and viscous friction with coefficient B,

Then, Load torque, $T_l = J \frac{d^2\theta}{dt^2} + B \frac{d\theta}{dt}$ (3.21)

Hydraulic power input = $q_m P$ (3.22)

Substitute for q_m, from Eq. (3.16) in Eq. (3.22)

$$\text{Hydraulic power input} = K_m \frac{d\theta}{dt} P \tag{3.23}$$

$$\text{Mechanical power output} = T_m \frac{d\theta}{dt} \tag{3.24}$$

Substitute for T_m from Eq. (3.20) in Eq. (3.24)

∴ $$\text{Mechanical power output} = K_t P \frac{d\theta}{dt} \tag{3.25}$$

It hydraulic motor losses are neglected then the hydraulic motor input is equal to mechanical power output of hydraulic motor.

$$\therefore \quad K_m \frac{d\theta}{dt} P = K_t P \frac{d\theta}{dt} \tag{3.26}$$

From Eq. (3.26), it is clear that $K_m = K_t$ Hence, Eq. (3.20) can be written as

$$T_m = K_t p = K_m P$$

Since the motor toque equals to load torque

$$T_m = T_l$$

$$\therefore \quad K_m P = J\frac{d^2\theta}{dt^2} + B\frac{d\theta}{dt}$$

$$P = \frac{J}{K_m}\frac{d^2\theta}{dt^2} + \frac{B}{K_m}\frac{d\theta}{dt} \tag{3.27}$$

On differentiating with respect to t, we get

$$\frac{dP}{dt} = \frac{J}{K_m}\frac{d^3\theta}{dt^3} + \frac{B}{K_m}\frac{d^2\theta}{dt} \tag{3.28}$$

On substituting for P and $\frac{dP}{dt}$ from Eq. (3.27) and Eq. (3.28) in Eq. (3.19) we get,

$$K_p x = K_m \frac{d\theta}{dt} + K_i\left[\frac{J}{K_m}\frac{d^2\theta}{dt^2} + \frac{B}{K_m}\frac{d\theta}{dt}\right] + K_c\left[\frac{J}{K_m}\frac{d^3\theta}{dt^3} + \frac{B}{K_m}\frac{d^2\theta}{dt^2}\right]$$

$$K_p x = \frac{K_c J}{K_m}\frac{d^3\theta}{dt^2} + \left[\frac{K_i J}{K_m} + \frac{K_c B}{K_m}\right]\frac{d^2\theta}{dt^2} + \left[K_m + \frac{K_i B}{K_m}\right]\frac{d\theta}{dt}$$

On taking Laplace transform with zero initial conditions, we get,

$$K_p X(s) = \frac{K_c J}{K_m}s^3\theta(s) + \left[\frac{K_i J}{K_m} + \frac{K_c B}{K_m}\right]s^2\theta(s) + \left[K_m + \frac{K_i B}{K_m}\right]s\theta(s)$$

$$\frac{\theta(s)}{X(s)} = \frac{K_p}{s\left[\frac{K_c J}{K_m}s^2 + \left[\frac{K_i J + K_c B}{K_m}\right]s + \frac{K_{m^2} + K_i B}{K_m}\right]} \tag{3.29}$$

In hydraulic systems, normally $K_c << K_m$,

therefore, Put $K_c = 0$ in Eq. (3.29)

$$\frac{\theta(s)}{X(s)} = \frac{K_p}{s\left[\frac{K_i J}{K_m}s + \frac{K_{m^2} + K_i B}{K_m}\right]} \tag{3.30}$$

The Eq. (3.30) is the required transfer function of the system

CHAPTER 3

3.3.3 Pneumatic System

Pneumatic system uses compressible fluid as working medium usually air. In pneumatic systems, compressibility effects of gas cannot be neglected and hence dynamic equations are obtained using conservation of mass. In pneumatic systems, change in fluid inertia energy and the fluid's internal thermal energy are assumed negligible.

The variables of pneumatic system are mass flow rate, q_m and pressure, P. The mass flow rate is through variable and it is analogous to current. The pressure is across variable and it is analogous to voltage. The two basic elements of pneumatic system are the resistance and capacitance. The restrictions in the pipes and valves offers resistance to gas flow.

The gas flow resistance, R is defined as the rate of change in gas pressure difference for a change in gas flow rate.

$$R = \frac{\text{Change in gas pressure difference, N/m}^2}{\text{Change in gas flow rate, kg/sec}}$$

The pneumatic capacitance is defined for a pressure vessel and depends on the type of expansion process involved. The capacitance of a pressure vessel may be defined as the ratio of change in gas stored for a change in gas pressure.

$$C = \frac{\text{Change in gas stored, kg}}{\text{Change in gas pressure, N/m}^2}$$

Pneumatic devices are employed in guided missiles, air craft systems, automation of production machinery and in many other fields as automatic controllers.

A simple pneumatic system shown in Fig. (3.10) and it consists of a pneumatic bellows in line with the restriction. The pneumatic bellows consists of a hollow chamber with thin pneumatic walls. The side walls of bellows are corrugated and the input and output surface are flat. An increase in pressure within the bellows results in an increase in separation between the input and output surfaces.

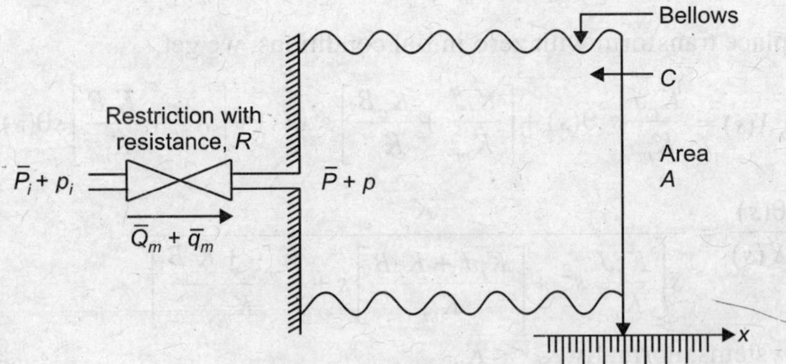

Fig. 3.10: Pneumatic system.

Let

$\bar{P}_i \Rightarrow$ Steady state value of input air pressure.

$p_i \Rightarrow$ Increase in the pressure of air-source.

$\bar{P} \Rightarrow$ Steady state value of pressure inside the bellows.

$p \Rightarrow$ Increase in pressure inside the bellow.

$\bar{Q}_m \Rightarrow$ Steady-state valve of air flow rate.

$q_m \Rightarrow$ Increase in air flow rate

$A \Rightarrow$ Area of each flat surface of the bellows

$R \Rightarrow$ Resistance of the restriction

$C \Rightarrow$ Capacitance of the bellows

$x \Rightarrow$ Displacement of the movable surface of the bellows.

Let the pressure of air source be increased from its steady state valve by an amount p_i. Thus results in an increase in air flow by q_m and increase in the pressure inside the bellows by p. Due to increase in pressure, there will be a displacement of the movable surface of the bellows, by an amount x. Here, the terms p_i, q_m, p and x are all functions of time t and therefore can be expressed as $p_i(t)$, $q_m(t)$, p(t) and $x(t)$.

The force exerted on the movable surface of the bellows is proportional to increase in pressure inside the bellows, i.e., $f_b \propto p(t)$

$\therefore$ Force exerted on the movable surface of the bellow,

$$f_b = Ap(t) \tag{3.31}$$

The force opposing the movement of the flat surface of bellow walls is proportional to the displacement i.e., $f_0 \propto x(t)$.

$\therefore$ Force opposing the motion,

$$f_0 = Kx(t) \tag{3.32}$$

where $K \rightarrow a$ constant representing the stiffness of the bellows.

At steady state the above two forces are balanced,

$$\therefore \quad f_b = f_0 \quad \therefore Ap(t) = Kx(t) \tag{3.33}$$

$$\text{The resistance, } R = \frac{\text{Difference between change in pressure}}{\text{Change in air flow rate}} = \frac{p_i(t) - p(t)}{q_m(t)}$$

$$\therefore \quad q_m(t) = \frac{p_i(t) - p(t)}{R} \tag{3.34}$$

$$\text{The capacitance,} \quad C = \frac{\text{Change in air flow rate}}{\text{Rate of change of pressure}} = \frac{q_m(t)}{dp(t)/dt}$$

$$\therefore \quad q_m(t) = C\frac{dp(t)}{dt} \tag{3.35}$$

On equating (3.34) and (3.35) we get

$$C\frac{dp(t)}{dt} = \frac{p_i(t) - p(t)}{R} \tag{3.36}$$

$$RC\frac{dp(t)}{dt} + p(t) = p_i(t) \tag{3.37}$$

CHAPTER 3

From the Eq. (3.33), we get

$$p(t) = \frac{K}{A}\,x(t) \tag{3.38}$$

On differentiating Eq. (3.38) with respect to t, we get

$$\frac{dp(t)}{dt} = \frac{K}{A}\frac{dx(t)}{dt} \tag{3.39}$$

On substituting for $p(t)$ and $\frac{dp(t)}{dt}$ from Eq. (3.38) and (3.39) in Eq. (3.37) we get,

$$RC\left[\frac{K}{A}\frac{dx\,(t)}{dt}\right] + \frac{K}{A}x(t) = p_i(t) \tag{3.40}$$

On taking Laplace transform with zero initial conditions we get

$$RC\left[\frac{K}{A}s\,X(s)\right] + \frac{K}{A}X(s) = P_i(s)$$

$$\frac{X(s)}{P_i(s)} = \frac{A/K}{RCs+1} = \frac{A/K}{\tau s+1}$$

where $\tau = RC \rightarrow$ time constant of the system

$$\frac{X(s)}{P_i(s)} = \frac{A/K}{\tau s+1} \tag{3.41}$$

The Eq. (3.41) is the required transfer function of the system.

3.3.4 Electrical System

The models of electrical systems can be obtained by using resistor, capacitor and inductor. The current-voltage relation of resistor, inductor and capacitor are given in table 1. For modelling electrical systems the electrical network or equivalent circuit is formed by using R, L and C and voltage or current source.

Table 1

Element	Voltage across the element	Current through the element
i(t) R, + V(t) − (resistor)	$V(t) = Ri\,(t)$	$i(t) = V(t)/R$
i(t) L, + V(t) − (inductor)	$V(t) = L\frac{d}{dt}i(t)$	$i(t) = \frac{1}{L}\int V(t)\,dt$
i(t) C, + − V(t) (capacitor)	$V(t) = \frac{1}{C}\int i(t)\,dt$	$i\,(t) = C\frac{dV(t)}{dt}$

The differential equations governing the electrical systems can be formed by writing Kirchoff's current law and Kirchoff's voltage Law equations. The transfer function can be obtained by taking Laplace transform of the differential equations and rearranging them as a ratio of output to input.

Let consider one electrical network and derive transfer function for the Fig. (3.11a).

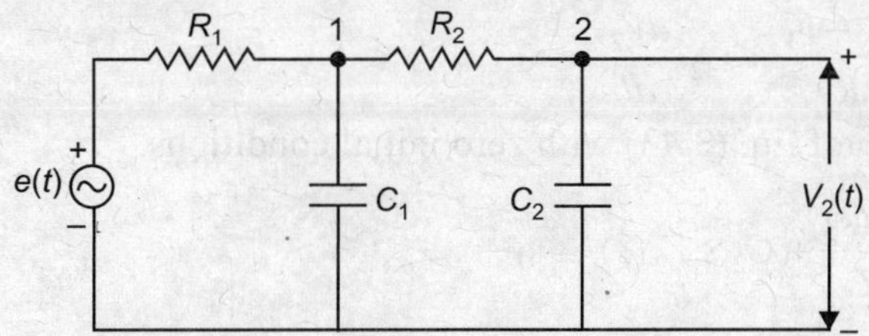

Fig. 3.11(a): Electrical network.

$e(t) \rightarrow$ Input voltage

$V_2(t) \rightarrow$ Output voltage

Laplace transform of $e(t) = L(e\,(t)) = E(s)$

Laplace transform of $V_2(t) = V_2(s)$

The transfer function of the network is $\dfrac{V_2(s)}{E(s)}$

Convert the voltage source which is in series with resistance R_1 into equivalent current source as shown in Fig. (3.11b). The network has two nodes. Let the node voltages be V_1 and V_2. The laplace transform of node voltages V_1 and V_2 are $V_1(s)$ and $V_2(s)$ respectively. The differential equations governing the network are given by the Kirchoff's Current Law equations at these nodes.

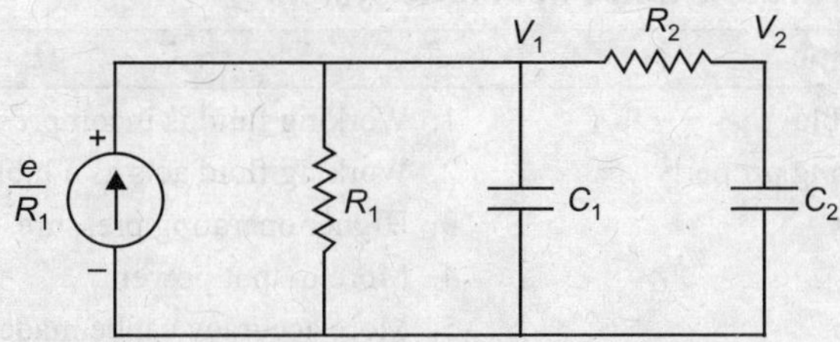

Fig. 3.11(b): Electrical network (After converting voltage source into current source).

At node 1, by Kirchoff's current law (Fig. 3.11b)

$$\frac{V_1}{R_1} + C_1\frac{dV_1}{dt} + \frac{V_1 - V_2}{R_2} = \frac{e}{R_1} \tag{3.42}$$

on taking laplace transform with zero initial conditions.

$$\frac{V_1(s)}{R_1} + C_1\,sV_1(s) + \frac{V_1(s)}{R_2} - \frac{V_2(s)}{R_2} = \frac{E(s)}{R_1} \tag{3.43}$$

$$V_1(s)\left[\frac{1}{R_1}+sC_1+\frac{1}{R_2}\right]-\frac{V_2(s)}{R_2}=\frac{E(s)}{R_1}$$

At node 2, by Kirchoff's current law Fig. (3.11b)

$$\frac{V_2-V_1}{R_2}+C_2\frac{dV_2}{dt}=0 \qquad (3.44)$$

on taking laplace transform of Eq. (3.43) with zero initial conditions.

$$\frac{V_2(s)}{R_2}-\frac{V_1(s)}{R_2}+C_2\,s\,V_2(s)=0$$

$$\frac{V_1(s)}{R_2}=\frac{V_2(s)}{R_2}+C_2\,s\,V_2(s)=\left[\frac{1}{R_2}+s\,C_2\right]V_2(s)$$

$$V_1(s)=[1+sC_2R_2]\;V_2(s) \qquad (3.45)$$

substituting for $V_1(s)$ from Eq. (3.45) in Eq. (3.43) we get,

$$(1+s\,R_2\,C_2)\,V_2(s)\left[\frac{1}{R_1}+s\,C_1+\frac{1}{R_2}\right]-\frac{V_2(s)}{R_2}=\frac{E(s)}{R_1}$$

$$\left[\frac{(1+s\,R_2\,C_2)\,(R_2+R_1+s\,C_1\,R_1\,R_2)-R_1}{R_1R_2}\right]V_2(s)=\frac{E(s)}{R_1}$$

$$\frac{V_2(s)}{E(s)}=\frac{R_2}{[(1+s\,R_2\,C_2)\,(R_1+R_2+s\,C_1\,R_1\,R_2)-R_1]} \qquad (3.46)$$

The Eq. (3.46) is the required transfer function of the system.

Difference between Hydraulic and Pneumatic system

Pneumatic	Hydraulic
1. Working fluid is compressible	1. Working fluid is incompressible
2. Working fluid lack lubricating property	2. Working fluid acts as a lubricant
3. Operating pressure is lower	3. Higher operating pressure
4. Output power is less	4. More output power
5. Accuracy of actuator is poor	5. More accuracy canbe made
6. External leakage is permissible but internal leakage must to avoided	6. Internal leakage is permissible and external leakage must be avoided
7. Non-return pipes are required	7. Return pipes are required
8. In sensitive to temperature changes	8. Sensitive to changes in temperature.
9. Fire and explosion proof	9. Not a fire and explosion proff.

3.3.5 Electromechanical System

Electro mechanical system consists of both electrical system and mechanical system. For example generator and motor both are called as electromechanical system. In case of generator input is mechanical and output is electrical then in case of motor input is electrical and output is mechanical.

The differential equation governing the electrical system can be formed by Kirchoff's laws and for mechanical system, the equation is formed by newton's second law.

Example 1: Transfer function of Armature Controlled DC motor

The speed of DC motor is directly proportional to armature voltage and inversely proportional to flux. In armature controlled DC motor, the desired speed is obtained by varying the armature voltage. This is an electromechanical control system. The electrical system consists of armature and field circuits but for analysis purpose, only the armature circuit is considered because the field is excited by constant voltage. The mechanical system consists of the rotating part and load connected to the shaft of the motor. The armature controlled DC motor system is shown in Fig. (3.12a).

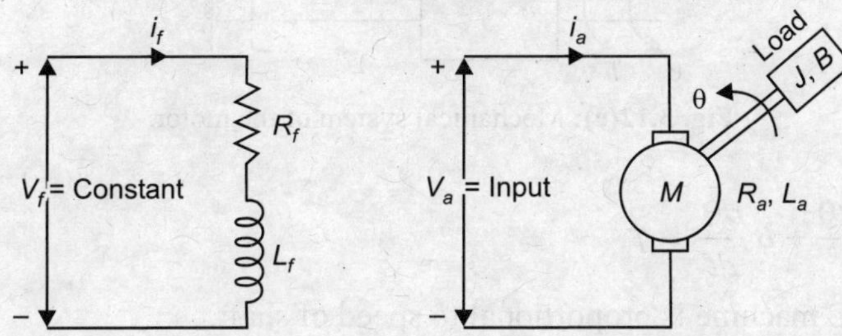

Fig. 3.12(a): Armature controlled DC motor.

Let,

$R_a \rightarrow$ Armature Resistance, Ω

$L_a \rightarrow$ Armature Inductance, H

$i_a \rightarrow$ Armature Current, A

$V_a \rightarrow$ Armature voltage, V

$e_b \rightarrow$ Back emf, V

$K_t \rightarrow$ Torque constant, $N-m/A$

$T \rightarrow$ Torque developed by motor, $N-m$

$\theta \rightarrow$ Angular displacement of shaft, Radian

$J \rightarrow$ Moment of inertia of motor and load, kgm^2/Rad

$B \rightarrow$ Frictional co-efficient of motor and load, $N-m/$ (rad - sec)

$K_b \rightarrow$ Back emf constant, $V/$(rad - sec)

The equivalent circuit of armature is shown in Fig. (3.12b)

By applying KVL to Fig. (3.12b), we get,

$$i_a R_a + L_a \frac{di_a}{dt} + e_b = V_a \tag{3.47}$$

Torque of DC motor is proportional to the product of flux and current. Since flux is constant in this system the torque is proportional current alone.

$$T \,\alpha\, i_a$$

$\therefore$ Torque $$T = K_t\, i_a \tag{3.48}$$

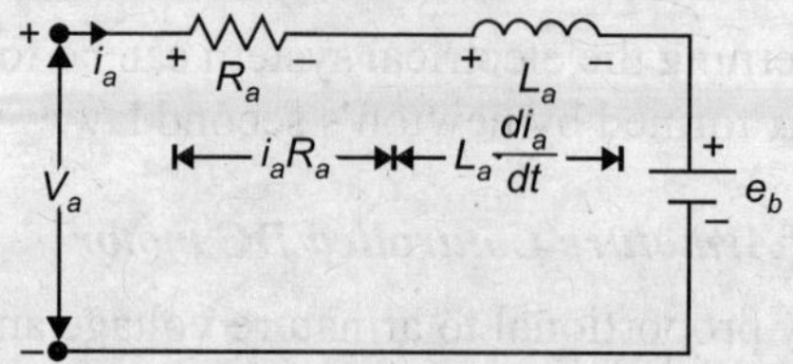

Fig. 3.12(b): Equivalent circuit of Armature.

The mechanical system of the motor is shown in Fig. (3.12c). the differential equation governing the mechanical system of motor is given by

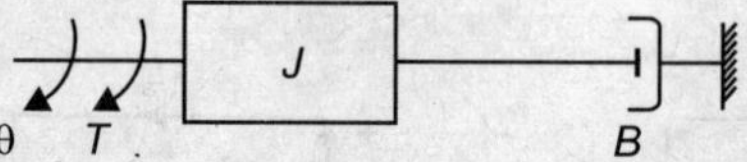

Fig. 3.12(c): Mechanical system of the motor.

$$J\frac{d^2\theta}{dt^2} + B\frac{d\theta}{dt} = T \qquad (3.49)$$

The back emf of DC machine is proportional to speed of shaft.

$$\therefore \qquad e_b \;\alpha \frac{d\theta}{dt};\; e_b = K_b\frac{d\theta}{dt} \qquad (3.50)$$

The laplace transform of various time domain signals involved in this system are shown below

$$L(V_a) = V_a(s); \quad L[e_b] = E_b(s)$$
$$L[i_a] = I_a(s); \quad L[\theta] = \theta(s)$$
$$L(T) = T(s)$$

The differential equations governing the armature controlled DC system are

$$i_aR_a + L_a\frac{di_a}{dt} + e_b = V_a; \qquad J\frac{d^2\theta}{dt^2} + B\frac{d\theta}{dt} = T$$

$$T = K_t\, i_a; \qquad e_b = K_b\frac{d\theta}{dt}$$

on taking laplace transform of the differential equations with zero initial conditions for above equations, we get,

$$I_a(s)\,R_a + L_a\, sI_a(s) + E_b(s) = V_a(s) \qquad (3.51)$$
$$T(s) = K_t\, I_a(s) \qquad (3.52)$$
$$Js^2\,\theta(s) + Bs\,\theta(s) = T(s) \qquad (3.53)$$
$$E_b(s) = K_b s \cdot \theta(s) \qquad (3.54)$$

on equating (3.52) and (3.53)

$$K_t\, I_a(s) = J^2 s\theta(s) + Bs\theta(s)$$
$$K_t\, I_a(s) = \theta(s)\,[Js^2 + Bs] \qquad (3.55)$$

$$I_a(s) = \theta(s)\left[\frac{Js^2 + Bs}{K_t}\right] \tag{3.56}$$

Equation (3.51) can be written as

$$(R_a + sL_a)\, I_a(s) + E_b(s) = V_a(s) \tag{3.57}$$

Substituting $E_b(s)$ and $I_a(s)$ from Eq. (3.54) and (3.56) in Eq. (3.57)

$$(R_a + sL_a)\,\frac{(Js^2 + Bs)}{K_t}\,\theta(s) + K_b\, s\theta(s) = V_a(s)$$

$$\left[\frac{(R_a + sL_a)(Js^2 + Bs) + K_b K_t s}{K_t}\right]\theta(s) = V_a(s) \tag{3.58}$$

The required transfer function is $\theta(s)/V_a(s)$

From Eq. (3.58)

$$\frac{\theta(s)}{V_a(s)} = \frac{K_t}{[(R_a + sL_a)(Js^2 + Bs) + K_b\, K_t s]} \tag{3.59}$$

$$= \frac{K_t}{R_a Js^2 + R_a Bs + L_a Js^3 + L_a Bs^2 + K_b K_t s}$$

$$= \frac{K_t}{s[JL_a s^2 + (JR_a + BL_a)s + (BR_a + K_b K_t)]}$$

$$\frac{\theta(s)}{V_a(s)} = \frac{K_t/JL_a}{s\left[s^2 + \left(\frac{JR_a + BL_a}{JL_a}\right)s + \left(\frac{BR_a + K_b K_t}{JL_a}\right)\right]} \tag{3.60}$$

This also can be written in another form i.e.

From Eq. (3.59)

$$\frac{\theta(s)}{V_a(s)} = \frac{K_t}{[(R_a + sL_a)(Js^2 + Bs) + (K_b K_t s)]}$$

$$\frac{\theta(s)}{V_a(s)} = \frac{K_t}{R_a\left[\frac{sL_a}{R_a} + 1\right] Bs\left[1 + \frac{Js^2}{Bs}\right] + K_b\, K_t s}$$

$$\frac{\theta(s)}{V_a(s)} = \frac{K_t/R_a B}{s\left[(1 + s\,T_a)(1 + s\,T_m) + \left(\frac{K_b K_t}{R_a B}\right)\right]} \tag{3.61}$$

where, $\frac{L_a}{R_a} = T_a \rightarrow$ Electrical time constant

CHAPTER 3

$$\frac{J}{B} = T_m \rightarrow \text{Mechanical time constant}$$

The Eq. (3.61) is the required transfer function.

Example 2: Transfer function of field controlled DC Motor

The speed of a DC motor is directly proportional to armature voltage and inversely proportional to flux. In field controlled DC motor the armature voltage is kept constant and the speed is varied by varying the flux of the field. Since flux is directly proportional to field current, the flux is varied by varying field current. The speed control is an electromechanical control system. The electrical system consists of armature and field excited by a constant voltage. The mechanical system consists of the rotating part and the load connected to the shaft of the motor. The field controlled DC motor shown in Fig. (3.12d).

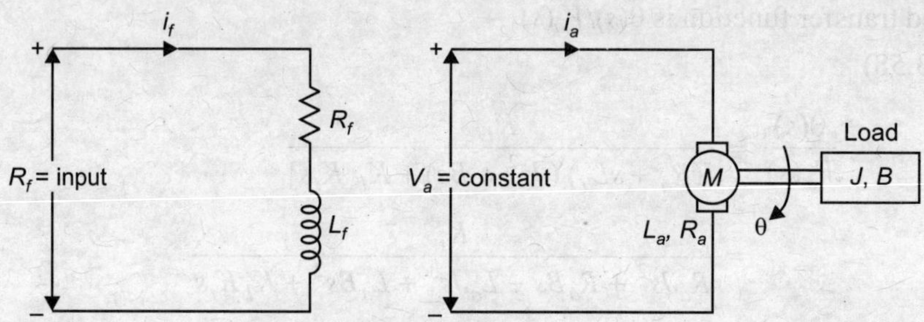

Fig. 3.12(d): Field controlled DC motor.

Let,

$R_f \rightarrow$ Field resistance, Ω

$L_f \rightarrow$ Field inductance, H

$i_f \rightarrow$ Field current, A

$V_f \rightarrow$ Field voltage, V

$T \rightarrow$ Torque developed by motor, N–m

$K_{tf} \rightarrow$ Torque constant, N–m/A

$J \rightarrow$ Moment of inertia of motor and load, kg-m²/rad

$B \rightarrow$ Frictional coefficient of motor and load, N–m/(rad/sec)

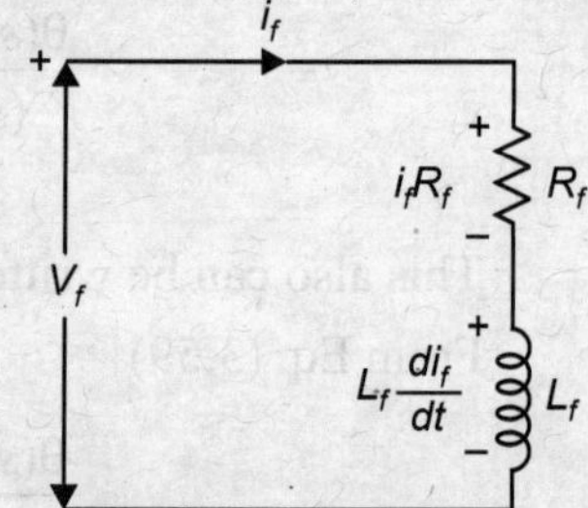

Fig. 3.12(e): Equivalent circuit of field.

The equivalent circuit of field in shown in Fig. (3.12e).

By Kirchoff's voltage law, we can write,

$$R_f\, i_f + L_f \frac{di_f}{dt} = V_f \qquad (3.62)$$

The Torque of DC motor is proportional to product of flux and armature current. Since armature current is constant in this system, the torque is proportional to flux alone, but flux is proportional to field current.

$$T \propto i_f \;\therefore\; \text{Torque, } T = Kt_f\, i_f \qquad (3.63)$$

The mechanical system of the motor is shown in Fig. (3.12f).

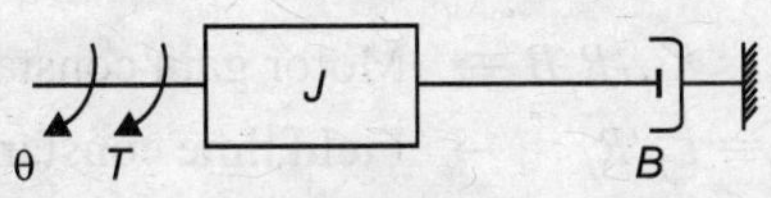

Fig. 3.12(f)

The differential Eq. governing the mechanical system of the motor is given by

$$J\frac{d^2\theta}{dt^2} + B\frac{d\theta}{dt} = T \tag{3.64}$$

The laplace transform of various time domain signals involved in this system are shown below.

$$L\,[i_f] = I_f\,(s);\ L(T) = T(s);\ L(V_f) = V_f(s)$$

$$L[\theta] = \theta(s)$$

The differential Eqs. governing the field controlled DC motor are

$$R_f i_f + L_f\frac{di_f}{dt} = V_f;\quad T_f = Kt_f\, i_f;\quad J\frac{d^2\theta}{dt^2} + B\frac{d\theta}{dt} = T$$

on taking laplace transform of the system differential Eq., we get

$$R_f I_f(s) + L_f\, sI_f(s) = V_f(s) \tag{3.65}$$

$$T(s) = K_{tf} I_f(s) \tag{3.66}$$

$$Js^2\,\theta(s) + Bs\,\theta(s) = T(s) \tag{3.67}$$

on equating (3.66) and (3.67)

$$K_{tf}\, I_f(s) = Js^2\theta\,(s) + Bs\,\theta(s)$$

$$K_{tf}\, I_f(s) = \theta(s)\,[Js^2 + Bs]$$

$$I_f(s) = s\,\frac{(Js + B)}{K_{tf}}\,\theta(s) \tag{3.68}$$

The Eq. (3.65) can be written as

$$(R_f + sL_f)\, I_f(s) = V_f(s) \tag{3.69}$$

on substituting for $I_f(s)$ from Eq. (3.68) in Eq. (3.69) we get,

$$(R_f + sL_f)\; s\,\frac{(Js + B)}{K_{tf}}\,\theta(s) = V_f(s)$$

$$\frac{\theta(s)}{V_f(s)} = \frac{K_{tf}}{s(R_f + sL_f)\,(B + s\,J)}$$

$$= \frac{K_{tf}}{s\,R_f\left(1 + \frac{s\,L_f}{R_f}\right)B\left(1 + \frac{s\,J}{B}\right)}$$

$$\frac{\theta(s)}{V_f(s)} = \frac{K_m}{s\,(1 + s\,T_f)\,(1 + s\,T_m)} \tag{3.70}$$

CHAPTER 3

where $K_m = K_{tf}/R_fB \rightarrow$ Motor gain constant

$T_f = L_f/R_f \rightarrow$ Field time constant

$T_m = J/B \rightarrow$ Mechanical time constant

II. MODELLING OF MECHANICAL SYSTEM

3.4 MECHANICAL TRANSLATION SYSTEMS (LINEAR MECHANICAL ELEMENTS)

The model of mechanical translational system can be obtained by using three basic elements like mass, spring and dash-pot. These three elements are essential for mechanical system.

The weight of the mechanical system is represented by the element mass and it is assumed to be concentrated at the centre of the body. The elastic deformation of the body can be represented by a spring. The friction existing in mechanical system can be represented by the dash-pot. The dash pot is a piston moving inside a cylinder filled with viscous fluid.

When a force is applied to a translational mechanical system, it is opposed by opposing forces due to mass, friction and elasticity of the system. The force acting on a mechanical body are governed by newton's second law of motion. For translational system it states that the sum of forces acting on a body is zero. (or Newton's second law states that the sum of applied forces is equal to the sum of opposing forces on a body.

List of symbols used in mechanical translational system.

$x \Rightarrow$ Displacement, m. (or) $y \Rightarrow$ Displacement, m.

$v \Rightarrow \frac{dx}{dt} \Rightarrow$ velocity, m/sec.

$a \Rightarrow \frac{d^2x}{dt^2} \Rightarrow$ Acceleration, m/sec^2.

$f \Rightarrow$ Applied force, N (Newtons)

$f_m \Rightarrow$ Opposing force offered by mass of the body, N

$f_k \Rightarrow$ Opposing force offered by the elasticity of the body (spring), N.

$f_b \Rightarrow$ Opposing force offered by the friction of the body (dash-pot), N.

$M \Rightarrow$ Mass, kg

$K \Rightarrow$ Stiffness of spring, N/m.

$B \Rightarrow$ Viscous friction co-efficient, N-sec/m

Note: Lower case letters are functions of time.

Force balance equations of Idealized elements

1. Consider an ideal mass elements shown in Fig. (3.16a) which has negligible friction and elasticity.

Let a force be applied on it. The mass will offer an opposing force which is proportional to acceleration of the body.

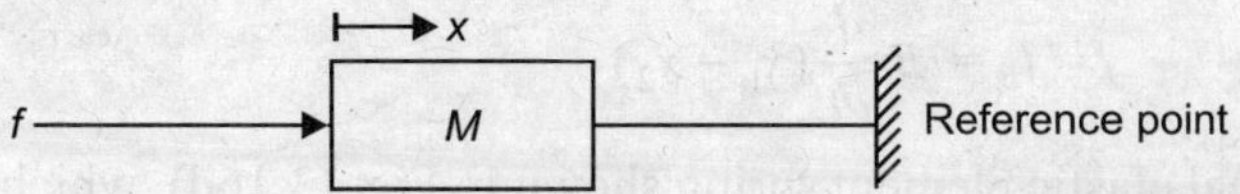

Fig. 3.16(a): Ideal mass element.

Let,

$f \rightarrow$ Applied force

$f_m \rightarrow$ Opposing force due to mass

Here $\quad f_m \, \alpha \, \dfrac{d^2x}{dt^2}$ (or) $f_m = \dfrac{Md^2x}{dt^2}$

By Newton's second law,

$$f = f_m = \frac{Md^2x}{dt^2} \tag{3.71}$$

2. Consider an ideal frictional element dashpot shown in Fig. (3.16b) which has negligible mass and elasticity. Let a force be applied on it. The dashpot will offer an opposing force which is proportional to velocity of the body.

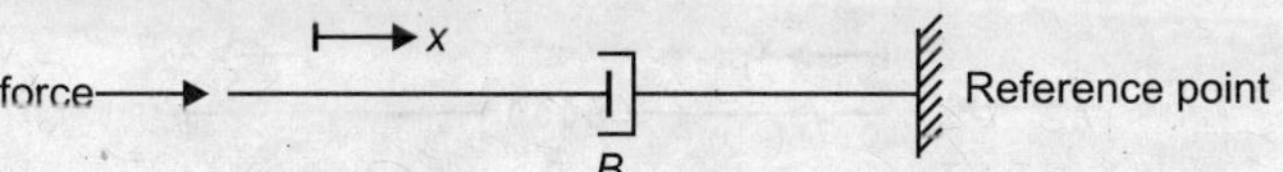

Fig. 3.16(b): Ideal dashpot element.

Let,

f = Applied force

f_b = Opposing force due to friction

Here, $\quad f_b \propto \dfrac{dx}{dt}$ or $f_b = B\dfrac{dx}{dt}$

By Newton's second law,

$$f = f_b = B\frac{dx}{dt} \tag{3.72}$$

3. When the dashpot has displacement at both ends as shown in Fig. (3.16c), the opposing force is proportional to differential velocity.

Fig. 3.16(c): Ideal dashpot with displacement at both ends.

$$f_b \propto \frac{d}{dt}(x_1 - x_2);$$

$$f_b = B\frac{d}{dt}(x_1 - x_2)$$

$$f = f_b = B\frac{d}{dt}(x_1 - x_2) \tag{3.73}$$

4. Consider an ideal elastic element spring shown in Fig. (3.16d), which has negligible mass and friction. Let a force applied on it. The spring will offer an opposing force which is proportional to displacement of the body

Fig. 3.16(d): Ideal spring element.

Let,

f = Applied force

f_k = Opposing force due to elasticity

Here, $f_k \propto x$ (or) $f_k = Kx$

$$f = f_k = Kx \tag{3.74}$$

5. When the spring has displacement at both ends as shown in Fig. (3.16e), the opposing force is proportional to differential displacement.

Fig. 3.16(e): Ideal spring with displacement at both ends.

$$f_k \propto (x_1 - x_2)$$

$$f_k = K(x_1 - x_2)$$

By Newton's second law.

$$f = f_k = K(x_1 - x_2) \tag{3.75}$$

6. Consider a liner arrangement is shown in Fig. [3.16f and 3.16g].

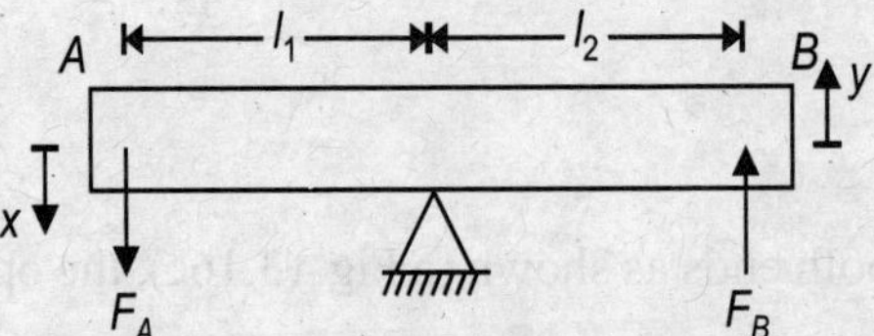

Fig. 3.16(f): A lever arrangement.

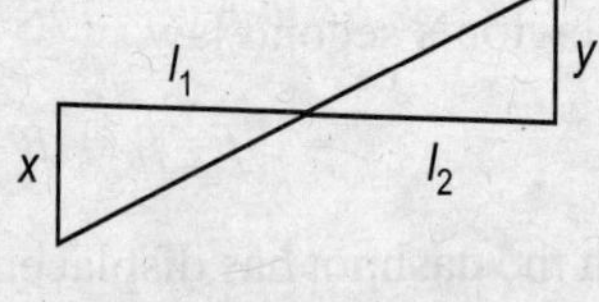

Fig. 3.16(g): A lever arrangement.

For Fig. (3.16f) a, the following equation can be written

$$F_A l_1 + F_B l_2 = 0 \tag{3.76}$$

From Fig. [3.16g], the following equation is written

$$\frac{x}{l_1} = \frac{y}{l_2} \tag{3.77}$$

7. Consider the gear arrangement shown in Fig. (3.16h) the primary gear has N_1 teeth and the secondary gear has N_2 teeth. the torque applied to the shaft of N_1 gear is $T_1(t)$ and the torque available at the shaft of N_2 gear will be $T_2(t)$.

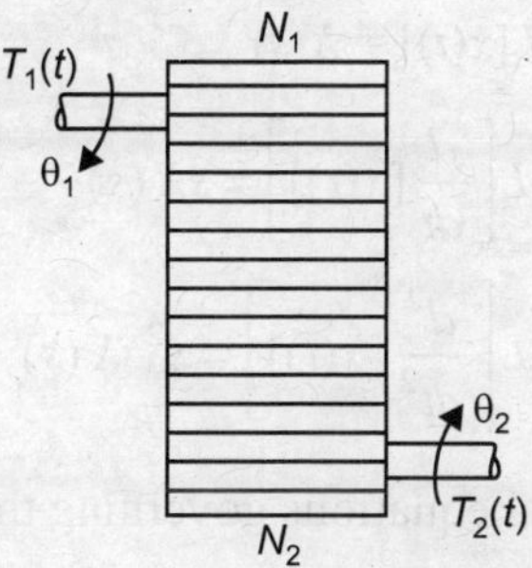

Fig. 3.16(h): Gear arrangement.

The following equations are written for Fig. (3.16h)

$$T_1\theta_1 = T_2\theta_2$$

$$N_1\theta_1 = N_2\theta_2$$

$$\therefore \quad \frac{T_1}{T_2} = \frac{\theta_2}{\theta_1} = \frac{N_1}{N_2} \tag{3.78}$$

The gear arrangement is similar to lever arrangement the dynamic equations for rotational systems are written on similar line for translational system. The linear displacement x meter is replaced by angular displacement θ radians, The force $f(t)$ is replaced by the torque $T(t)$ kg-m.

Guidelines to determine the transfer function of mechanical translational system

1. In mechanical translational system, the differential equations governing the system are obtained by writing force balance equations at every nodes in the system. The nodes are meeting point of elements. Generally the nodes are mass elements in the system. In some cases the nodes may be without mass element.
2. The linear displacement of the masses (nodes) are assumed x_1, x_2, x_3, etc. and assign a displacement to each mass (node). The first derivative of the displacement is velocity and the second derivative of the displacement is acceleration.
3. Draw the free body diagrams of the system. The free body diagram is obtained by drawing each mass separately and then marking all the forces acting on that mass (nodes). Always the opposing force acts in a direction opposite to applied force. The mass has to move in the direction of the applied force. Hence the displacement, velocity and acceleration of the mass will be in the direction of the applied force. If there is no applied force then the displacement, velocity and acceleration of the mass is in a direction opposite to that of opposing force.
4. For each free body diagram write one differential equation by equating the sum of applied forces to the sum of oppcsing forces.
5. Take laplace transform of differential equations to convert them to algebraic equations.

Then rearrange the *s*-domain equations to eliminate the unwanted variables and obtain the ratio between output variable to input variable. This ratio is the transfer function of the system.

Note:

Laplace transform of $x(t) = L[x(t)] = X(s)$

$$\text{Laplace transform of } \frac{dx(t)}{dt} = L\left[\frac{d}{dt}[x(t)]\right] = sX(s)$$

$$\text{Laplace transform of } \frac{d^2x(t)}{dt^2} = L\left[\frac{d^2}{dt^2}[x(t)]\right] = s^2X(s)$$

EXAMPLE 3.1: Write the differential equations governing the mechanical system shown in Fig. (3.1) and determine the transfer function.

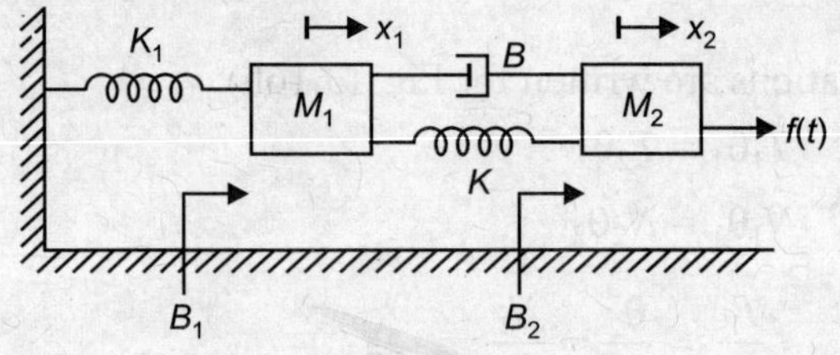

Fig. 3.1

Solution: In the given system, applied force $f(t)$ is the input and displacement x is the output.

Let laplace transform of $f(t) = L[f(t)] = F(s)$

and laplace transform of $x = L[x] = X(s)$

The required transfer function is $\dfrac{X_2(s)}{F(s)}$

The system has two nodes and they are mass M_1 and M_2. The differential equations governing the system are given by force balance equations at these nodes.

Let the displacement of mass M_1 be x_1. The free body diagram of mass M_1 is shown in Fig. (3.1.1). The opposing forces acting on mass M_1 are marked as $f_{m1}, f_{b1}, f_b, f_{k1}$ and f_k

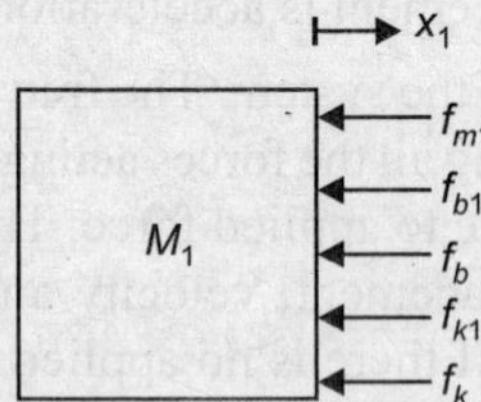

Fig. 3.1.1: Freebody diagram of mass M_1.

$$f_{m1} = M_1\frac{d^2x_1}{dt^2}; \quad f_{b1} = B_1\frac{dx_1}{dt}; \quad f_{k1} = K_1x_1$$

$$f_b = B\frac{d}{dt}(x_1 - x_2); \quad f_k = K(x_1 - x_2)$$

By newton's second law,

$$f_{m1} + f_{b1} + f_b + f_{k1} + f_k = 0$$

$$M_1\frac{d^2x_1}{dt^2} + B_1\frac{dx_1}{dt} + B\frac{d(x_1 - x_2)}{dt} + K_1x_1 + K(x_1 - x_2) = 0 \quad (1.1.1)$$

On taking laplace transform with zero initial conditions

$$M_1s^2X_1(s) + B_1sX_1(s) + Bs[X_1(s) - X_2(s)] + K_1X_1(s) + K[X_1(s) - X(s)] = 0$$

$$X_1(s)\,[M_1s^2 + (B_1 + B)\,s + (K_1 + K)] - X_2(s)\,[Bs + K] = 0$$

$$X_1(s)\,[M_1s^2 + (B_1 + B)\,s + (K_1 + K)] = X_2(s)\,[Bs + K]$$

$$X_1(s) = X_2(s)\frac{Bs + K}{M_1s^2 + (B_1 + B)s + (K_1 + K)} \quad (3.1.2)$$

Note: Laplace transform of $x_1 = L(x_1) = X_1(s)$

The freebody diagram of mass M_2 is shown in Fig. (3.1.2). The opposing forces acting on M_2 are marked as f_{m2}, f_{b2}, f_b and f_k.

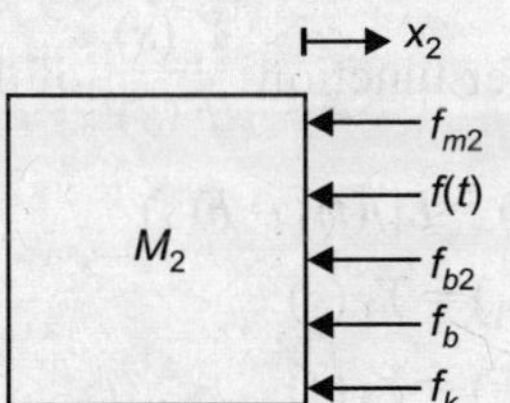

Fig. 3.1.2

$$f_{m2} = M_2\frac{d^2x_2}{dt^2}; \quad f_{b2} = B_2\frac{dx_2}{dt}$$

$$f_b = B\frac{d}{dt}(x_2 - x_1); \quad f_k = K(x_2 - x_1)$$

By newton's second law,

$$f_{m2} + f_{b2} + f_b + f_k = f(t)$$

$$M_2\frac{d^2x_2}{dt^2} + B_2\frac{dx_2}{dt} + B\frac{d}{dt}(x_2 - x_1) + K(x_2 - x_1) = f(t) \quad (3.1.3)$$

On taking laplace transform with zero initial conditions.

$$M_2s^2\,X_2(s) + B_2\,s\,X_2(s) + Bs[X_2(s) - X_1(s)] + K[X_2(s) - X_1(s)] = F(s)$$

$$X_2(s)\,[M_2s^2 + (B_2 + B)\,s + K] - X_1(s)\,[Bs + K] = F(s) \quad (3.1.4)$$

Substituting for $x_1(s)$ from Eq. (3.1.2) in (3.1.4) we get,

$$X_2(s)\,[M_2s^2 + (B_2 + B)s + K] - X_2(s)\frac{(Bs + K)^2}{M_1s^2 + (B_1 + B)s + (K_1 + K)} = F(s)$$

CHAPTER 3

$$X_2(s)\left[\frac{[M_2s^2+(B_1+B)\,s+(K_1+K)\,(M_1s^2+(B_2+B)s+K)-(Bs+K)^2]}{M_1s^2+(B_1+B)s+(K_1+K)}\right]=F(s)$$

$$\frac{X_2(s)}{F(s)}=\frac{M_1s^2+(B_1+B)s+(K_1+K)}{[M_1s^2+(B_1+B)s+(K_1+K)\,[M_2s^2+(B_2+B)s+K]-(Bs+K)^2]} \quad (3.1.5)$$

The Eq. (3.1.5) is the required transfer function.

Result: The differential equations governing the system are

1. $M_1\dfrac{d^2x_1}{dt^2}+B_1\dfrac{dx_1}{dt}+B\dfrac{d}{dt}\;(x_1-x_2)+K_1x_1+K(x_1-x_2)=0$

2. $M_2\dfrac{d^2x_2}{dt^2}+B_2\dfrac{dx_2}{dt}+B\dfrac{d}{dt}\;(x_2-x_1)+K(x_2-x_1)=f(t)$

The transfer function of the system is

$$\frac{X(s)}{F(s)}=\frac{M_1s^2+(B_1+B)s+(K_1+K)}{[M_1s^2+(B_1+B)s+(K_1+K)][M_2s^2+(B_2+B)s+K]\cdot(Bs+K)^2}$$

EXAMPLE 3.2: Determine the transfer function $\dfrac{Y_2(s)}{F(s)}$ of the system shown in Fig. (3.2).

Solution: Let, laplace transform of $f(t)=L[f(t)]=F(s)$

Laplace transform of $y_1=L(y_1)=Y_1(s)$

Laplace transform of $y_2=L(y_2)=Y_2(s)$

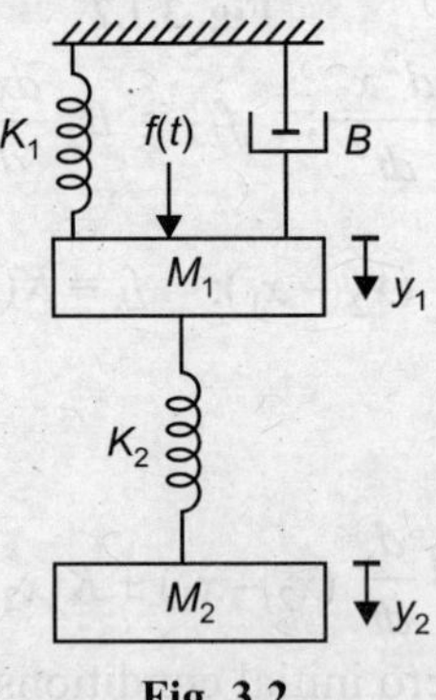

Fig. 3.2

The system has two nodes and they are mass M_1 and M_2, the differential equations governing the system and the force balance equations at these nodes are,

The freebody diagram of mass M_1 is shown in Fig. (3.2.1). The opposing forces are marked as f_{m1}, f_b, f_{k1} and f_{k2}.

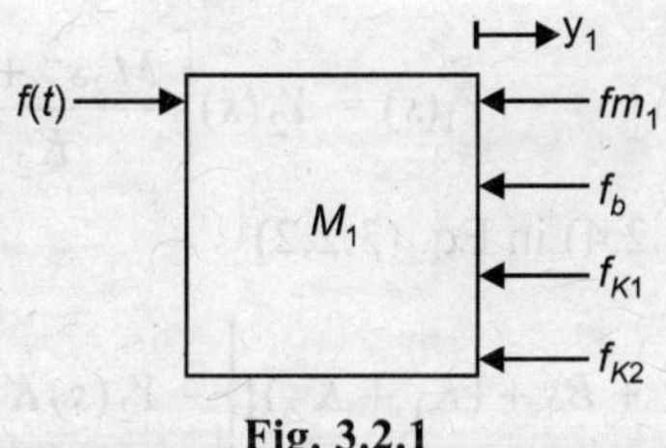

Fig. 3.2.1

$$f_{m1} = M_1 \frac{d^2 y_1}{dt^2}; \quad f_b = B \frac{dy_1}{dt}; \quad f_{k1} = K_1 y_1$$

$$f_{k2} = K_2 (y_1 - y_2)$$

By newton's second law,

$$f_{m1} + f_b + f_{k1} + f_{k2} = f(t)$$

$$M_1 \frac{d^2 y_1}{dt^2} + B \frac{dy_1}{dt} + K_1 y_1 + K_2 (y_1 - y_2) = f(t) \quad (3.2.1)$$

On taking laplace transform with zero initial condition

$$M_1 s^2 Y_1(s) + sBY_1(s) + K_1 Y_1(s) + K_2 [Y_1(s) - Y_2(s)] = F(s)$$

$$Y_1(s) [M_1 s^2 + Bs + K_1 + K_2] - Y_2(s) K_2 = F(s) \quad (3.2.2)$$

The free body diagram of mass M_2 is shown in Fig. (3.2.2). The opposing forces acting on M_2 are f_{m2} and f_{k2}

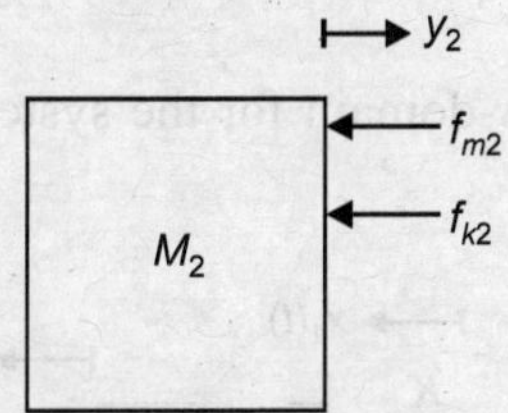

Fig. 3.2.2

$$f_{m2} = M_2 \frac{d^2 y_2}{dt^2}; \quad f_{k2} = K_2 (y_2 - y_1)$$

By newton's second law, $f_{m2} + f_{k2} = 0$

$$M_2 \frac{d^2 y_2}{dt^2} + K_2 (y_2 - y_1) = 0 \quad (3.2.3)$$

On taking laplace transform,

$$M_2 s^2 Y_2(s) + K_2 [Y_2(s) - Y_1(s)] = 0$$

$$Y_2(s) [M_2 s^2 + K_2] - Y_1(s) K_2 = 0$$

$$Y_1(s) = Y_2(s)\left[\frac{M_2 s^2 + K_2}{K_2}\right] \tag{3.2.4}$$

Substituting for $Y_1(s)$ from Eq. (3.2.4) in Eq. (3.2.2)

$$Y_2(s)\left\{\left[\frac{M_2 s^2 + K_2}{K_2}\right][M_1 s^2 + Bs + (K_1 + K_2)]\right\} - Y_2(s)K_2 = F(s)$$

$$Y_2(s)\left[\frac{(M_2 s^2 + K_2)(M_1 s^2 + Bs + (K_1 + K_2) - K_2^2}{K_2}\right] = F(s)$$

$$\therefore \qquad \frac{Y_2(s)}{F(s)} = \frac{K_2}{[M_1 s^2 + Bs + (K_1 + K_2)][M_2 s^2 + K_2] - K_2^2}$$

Result: The differential equations governing the systems are

3. $M_1 \dfrac{d^2 y_1}{dt^2} + B\dfrac{dy_1}{dt} + K_1 y_1 + K_2(y_1 - y_2) = f(t)$

2. $M_2 \dfrac{d^2 y_2}{dt^2} + K_2(y_2 - y_1) = 0$

The transfer function of the system is

$$\frac{Y_2(s)}{F(s)} = \frac{K_2}{[M_1 s^2 + Bs + (K_1 + K_2)][M_2 s^2 + K_2] - K_2^2}$$

EXAMPLE 3.3: Write the equations in s-domain for the system shown in Fig. (3.3). Determine the transfer function of the system.

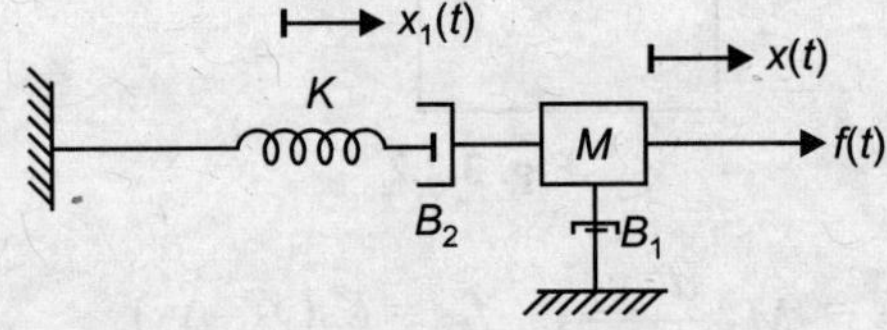

Fig. 3.3

Solution: Let,

Laplace transform of $x(t) = X(s)$

Lpalace transform of $f(t) = F(s)$

Let x_1 be the displacement at the meeting point of spring and dashpot. Laplace transform of x_1 is $X_1(s)$.

The system has two nodes and they are mass M and the meeting point of spring and dashpot. The differential equations governing the system are the force balance equations at these nodes.

The equations of motion in the s-domain are obtained by taking laplace transform of the differential equations

The freebody diagram of mass M is shown in Fig. (3.3.1). The opposing forces are marked as f_m, f_{b1} and f_{b2}

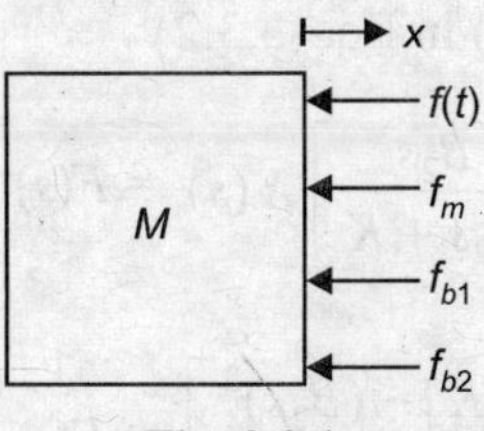

Fig. 3.3.1

$$f_m = M\frac{d^2x}{dt^2}; \quad f_{b1} = B_1\frac{dx}{dt}$$

$$f_{b2} = B_2\frac{d}{dt}(x - x_1)$$

By newton's second law the force balance equation is

$$f_m + f_{b1} + f_{b2} = f(t)$$

$$\therefore \qquad M\frac{d^2x}{dt^2} + B_1\frac{dx}{dt} + B_2\frac{d}{dt}(x - x_1) = f(t) \qquad (3.3.1)$$

on taking laplace transform

$$Ms^2 X(s) + B_1 s X(s) + B_2 s\,[X(s) - X_1(s)] = F(s)$$

$$[Ms^2 + (B_1 + B_2)\,s]\,X(s) - B_2 s\,X_1(s) = F(s) \qquad (3.3.2)$$

The free body diagram at the meeting point of spring and dashpot is shown in Fig. (3.3.2) the opposing forces are marked as f_k and f_{b2}.

$$f_{b2} = B_2\frac{d}{dt}(x_1 - x); \quad f_k = Kx_1$$

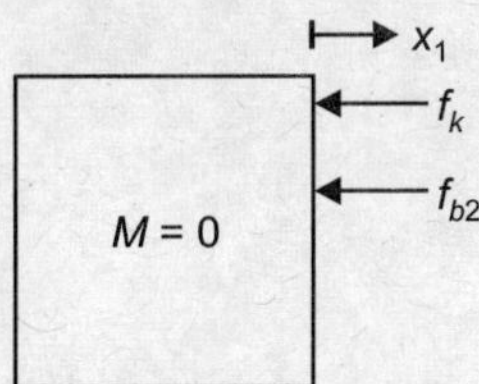

Fig. 3.3.2

By newton's second law, $f_{b2} + f_k = 0$

$$\therefore \qquad B_2\frac{d}{dt}(x_1 - x) + Kx_1 = 0 \qquad (3.3.3)$$

on taking laplace transform

$$B_2 s[X_1(s) - X(s)] + KX_1(s) = 0$$

$$(B_2 s + K) X_1(s) - B_2 sX(s) = 0 \quad (3.3.4)$$

$$X_1(s) = \frac{B_2 s}{(B_2 s + K)} X(s) \quad (3.3.5)$$

Substituting for X_1(s) from Eq. (3.3.5) in Eq. (3.3.2)

$$[Ms^2 + (B_1 + B)s] X(s) - B_2 s \left[\frac{B_2 s}{B_2 s + K}\right] X(s) = F(s)$$

$$X(s) \left[\frac{[Ms^2 + (B_1 + B_2)s][B_2 s + K] - (B_2 s)^2}{(B_2 s + K)}\right] = F(s)$$

$$\therefore \quad \frac{X(s)}{F(s)} = \frac{(B_2 s + K)}{[Ms^2 + (B_1 + B_2)s][B_2 s + K] - (B_2 s)^2} \quad (3.3.6)$$

Result: The differential Eqs. governing the system are,

1. $\dfrac{Md^2x}{dt^2} + B_1 \dfrac{dx}{dt} + B_2 \dfrac{d}{dt}(x - x_1) = f(t)$

2. $B_2 \dfrac{d}{dt}(x_1 - x) + Kx_1 = 0$

The transfer function of the system is

$$\frac{X(s)}{F(s)} = \frac{B_2 s + K}{[Ms^2 + (B_1 + B_2)s][B_2 s + K] - (B_2 s)^2}$$

EXAMPLE 3.4: Determine the transfer function $X_1(s)/F(s)$ and $\dfrac{X_2(s)}{F(s)}$, for the system shown in Fig. (3.4)

Solution: Let,

Laplace transform of $f(t) = F(s)$

Laplace transform of $x_1 = X_1(s)$

Laplace transform of $x_2 = X_2(s)$

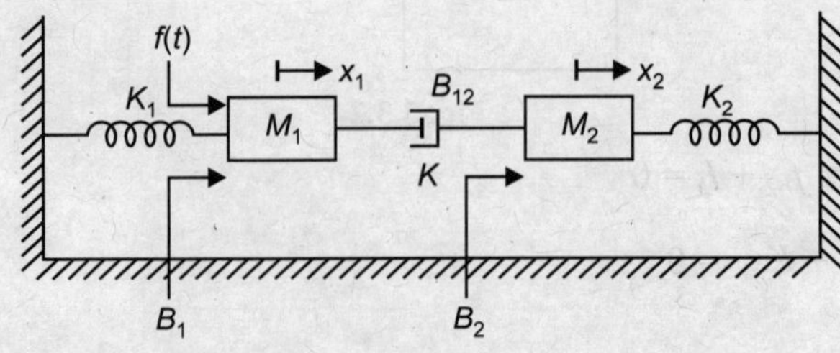

Fig. 3.4

The system has two nodes and they are mass M_1 and M_2. The differential equations governing the system are the force balance equations at these nodes. The free body diagram of mass M_1 is shown in Fig. (3.5.1). The opposing forces are marked as f_{m1}, f_{b1}, f_{b12} and f_{k3}.

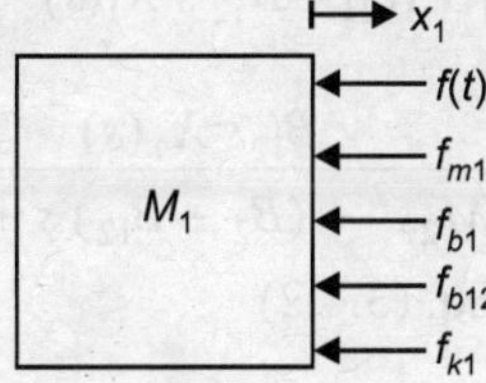

Fig. 3.4.1

$$f_{m1} = M_1 \frac{d^2x_1}{dt^2}; \quad f_{b1} = B_1 \frac{dx_1}{dt}; \quad f_{b12} = B_{12} \frac{d}{dt}(x_1 - x_2) \quad f_{k1} = K_1 x_1$$

By newton's second law, $f_{m1} + f_{b1} + f_{b12} + f_{k1} = f(t)$

$$M_1 \frac{d^2x_1}{dt^2} + B_1 \frac{dx_1}{dt} + B_{12} \frac{d}{dt}(x_1 - x_2) + K_1 x_1 = f(t) \tag{3.4.1}$$

on taking laplace transform with zero initial conditions

$$M_1 s^2 X_1(s) + B_1 s X_1(s) + B_{12} [X_1(s) - X_2(s)] + K_1 X_1(s) = F(s)$$

$$X_1(s) [M_1 s^2 + (B_1 + B_{12})s + K_1] - B_{12} sX_2(s) = F(s) \tag{3.4.2}$$

The free body diagram of mass M_2 is shown in Fig. (3.5.2) the opposing forces are marked as f_{m2}, f_{b2}, f_{b12} and f_{k2}

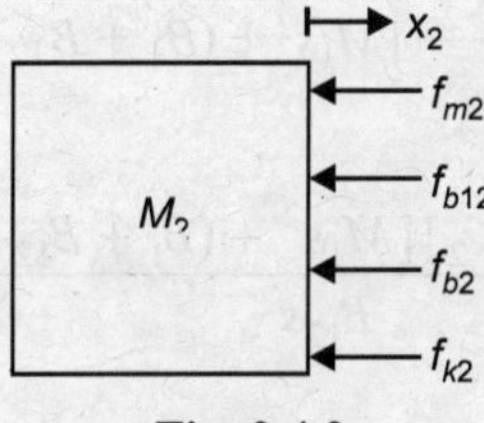

Fig. 3.4.2

$$f_{m2} = M_2 \frac{dx_2}{dt^2}; \; f_{b2} = B_2 \frac{dx_2}{dt}$$

$$f_{b12} = B_{12}[x_2 - x_1]; \; f_{k2} = K_2 x_2$$

By newton's second law,

$$f_{m2} + f_{b2} + f_{b12} + f_{k2} = 0$$

$$M_2 \frac{d^2x_2}{dt^2} + B_2 \frac{dx_2}{dt} + B_{12} \frac{d}{dt}(x_2 - x_1) + K_2 x_2 = 0 \tag{3.4.3}$$

on taking laplace transform

$$M_2s^2 X_2(s) + B_2s X_2(s) + B_{12} s[X_2(s) - X_1(s)] + K_2 X_2(s) = 0$$

$$X_2(s) [M_2s^2 + B_2s + B_{12}s + K_2] - X_1(s) B_{12}s = 0$$

$$X_2(s) [M_2s^2 + s[B_2 + B_{12}] + K_2] - B_{12} s X_1(s) = 0$$

$$X_2(s) = \frac{B_{12}s X_1(s)}{[M_2s^2 + (B_2 + B_{12}) s + K_2]} \quad (3.4.4)$$

Substituting $x_2(s)$ from Eq. (3.4.4) in Eq. (3.4.2)

$$X_1(s) [M_1s^2 + (B_1 + B_{12}) s + K_1] - \frac{(B_{12}s)^2 X_1(s)}{(M_2s^2 + (B_2 + B_{12}) s + K_2} = F(s)$$

$$\frac{X_1(s)\left[[M_1s^2 + (B_1 + B_{12})s + K_1][M_2s^2 + (B_2 + B_{12})s + K_2] - (B_1s)^2\right]}{M_2s^2 + (B_2 + B_{12}s) + K_2} = F(s)$$

$$\therefore \quad \frac{X_1(s)}{F(s)} = \frac{M_2s^2 + (B_2 + B_{12})s + K_2}{[M_1s^2 + (B_1 + B_{12})s + K_1][M_2s^2 + (B_2 + B_{12})s + K_2] - [B_1s]^2} \quad (3.4.5)$$

From Eq. (3.4.4),

$$X_1(s) = \frac{\left[[M_2s^2 + (B_2 + B_{12})s + K_2]\right] X_2(s)}{B_{12}s} \quad (3.4.6)$$

Substituting $X_1(s)$ from Eq. (3.4.6) in Eq. (3.4.2)

$$\frac{X_2(s)[M_2s^2 + (B_2 + B_{12})s + K_2}{B_{12}s} [M_1s^2 + (B_1 + B_{12})s + K_1] - B_{12}s X_2(s) = F(s)$$

$$X_2(s)\left[\frac{[M_2s^2 + (B_2 + B_{12}) s + K_2][M_1s^2 + (B_1 + B_{12} s + K_1) - (B_{12}s^2)}{B_{12}s}\right] = F(s)$$

$$\therefore \quad \frac{X_2(s)}{F(s)} = \frac{B_{12}s}{[M_2s^2 + (B_2 + B_{12}) s + K_2][M_1s^2 + (B_1 + B_{12})s + K_1] - [B_{12} s]^2}$$

Result: The differential Eq. governing the system are,

1. $M_1 \frac{d^2x_1}{dt^2} + B_1 \frac{dx_1}{dt} + B_{12} \frac{d}{dt} (x_1 - x_2) + K_1 x_1 = f(t)$

2. $M_2 \frac{d^2x_2}{dt^2} + B_2 \frac{dx_2}{dt} + B_{12} \frac{d}{dt} (x_2 - x_1) + K_2 x_2 = 0$

The transfer function of the system are,

1. $$\frac{X_1(s)}{F(s)} = \frac{M_2s^2 + (B_2 + B_{12})s + K_2}{[M_1s^2 + (B_1 + B_{12})s + K_1][M_2s^2 + (B_2 + B_{12})s + K_2] - (B_{12}s)^2}$$

2. $$\frac{X_2(s)}{F(s)} = \frac{B_{12}s}{[M_2s^2 + (B_2 + B_{12})s + K_2][M_1s^2 + (B_1 + B_{12})\,s + K_1 - (B_{12}s)^2]}$$

EXAMPLE 3.5: For the electro-mechanical system shown in Fig. (3.5.1) determine the transfer function $\frac{Y(s)}{V(s)}$.

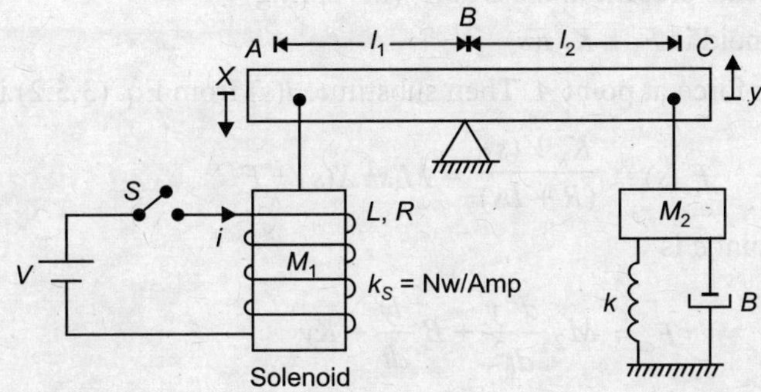

Fig. 3.5.1

Solution: In the given system, applied input is V and the output y is the linear displacement.

Let

Laplace transfer of $V = L(V) = V(s)$

Laplace transform of $y = L(y) = Y(s)$

Hence the required transfer function is $\frac{Y(s)}{V(s)}$

The system has two nodes and they are masses M_1 and M_2. The differential equations governing the system are given by force balance equations at these nodes.

Let we first write the loop equation for electrical circuit. The equivalent circuit of that electrical circuit as shown in Fig. (3.5.2).

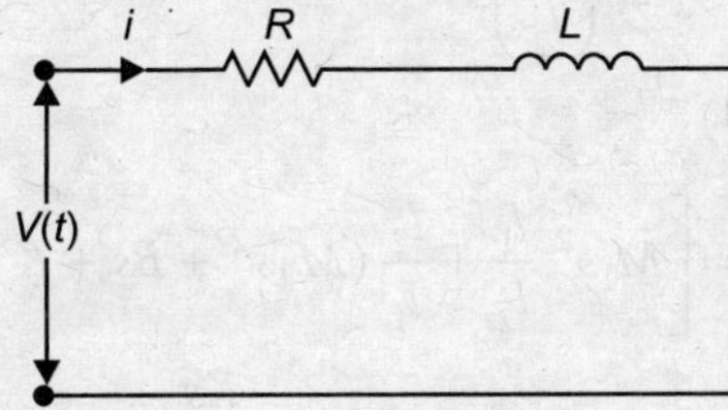

Fig. 3.5.2: Equivalent circuit of electrical circuit for given system.

By applying KVL, for Fig. (3.5.2).

$$V(t) = Ri + L\frac{di}{dt} \tag{3.5.1}$$

on taking Laplace transform of equation (3.5.1)

$$V(s) = R\,I(s) + Ls\,I(s)$$

$$V(s) = I(s)\,[R + Ls]$$

$$I(s) = \frac{V(s)}{[R + Ls]} \tag{3.5.2}$$

The mechanical elements present in mass is K_s (i.e. spring)

∴ The force on the solenoid is $F_s = K_s\,I(s)$ (3.5.3)

Let F_a be the reaction force at point A. Then substitute $I(s)$ from Eq. (3.5.2) in Eq. (3.5.3).

$$F_s(s) = \frac{K_s\,V(s)}{(R + Ls)} = M_1 s^2\,X(s) + F_a \tag{3.5.4}$$

The force acting at point c is

$$F_c = M_2\frac{d^2y}{dt^2} + B\frac{dy}{dt} + Ky$$

on taking laplace transform

$$F_c(s) = M_2 s^2\,Y(s) + Bs\,Y(s) + K\,Y(s)$$

$$Fc\,(s) = Y(s)\,[M_2 s^2 + Bs + K] \tag{3.5.5}$$

Taking moment at the point, the following equation is obtained

$$-F_a l_1 + F_c l_2 = 0$$

$$F_a = F_c\left(\frac{l_2}{l_1}\right) = \left(\frac{l_2}{l_1}\right)(M_2 s^2 + Bs + K)\,Y(s) \tag{3.5.6}$$

When the force is pulling the solenoid down, the following equation may be written,

$$\frac{x}{l_1} = \frac{y}{l_2}$$

$$\therefore \quad x = \left(\frac{l_1}{l_2}\right)y$$

Substituting (3.5.6) in Eq. (3.5.4)

$$\frac{K_s\,V(s)}{R + Ls} = \left[M_1 s^2\frac{l_1}{l_2} + \frac{l_2}{l_1}(M_2 s^2 + Bs + K)\right]Y(s) \qquad \left[\because X(s) = \frac{l_1}{l_2}Y(s)\right]$$

$$\frac{Y(s)}{V(s)} = \frac{Ks}{(R + Ls)\left(M_1 s^2\dfrac{l_1}{l_2} + \dfrac{l_2}{l_1}(M_2\,s^2 + Bs + K)\right)}$$

Results: The differential equations governing the system are,

1. $V(t) = R_i + L\dfrac{di}{dt}$

2. $F_s = \dfrac{Ks\,V(t)}{R + L\dfrac{di}{dt}} = M_1\dfrac{d^2x}{dt^2} + F_a$

3. $F_c = M_2\dfrac{d^2y}{dt^2} + B\dfrac{dy}{dt} + Ky$

4. $\dfrac{x}{l_1} = \dfrac{y}{l_2}$

The transfer function of the system is

$$\frac{Y(s)}{V(s)} = \frac{Ks}{[R + Ls]\left[M_1s^2\left(\frac{l_1}{l_2}\right) + \left(\frac{l_2}{l_1}\right)(M_1s^2 + Bs + K)\right]}$$

EXAMPLE 3.6: Consider the electro-mechanical system shown in Fig. (3.6.1). Assuming that the electrical force acting on the core of mass M is proportional to the current i in the inductor obtain the transfer function of the system.

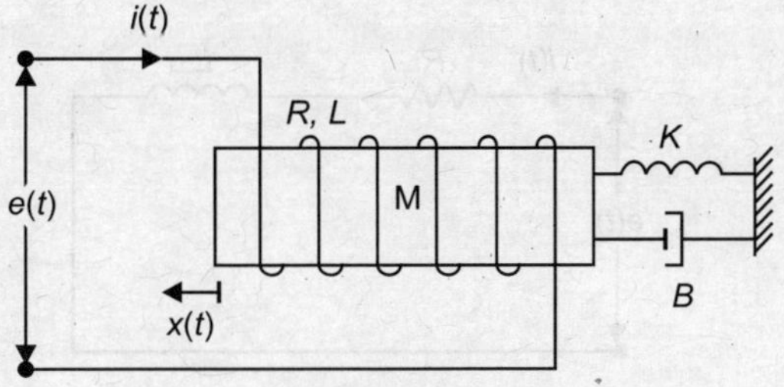

Fig. 3.6.1

Solution: In the given system the emf $e(t)$ is the input and the linear displacement $x(t)$ is the output.

Let,

Laplace transform of $x(t) = L[x(t)] = X(s)$

Laplace transform of $e(t) = L[e(t)] = E(s)$

Hence required transfer function in $\dfrac{X(s)}{E(s)}$.

The system has only one node, that is M. The differential equation are formed or mechanical system and also for electrical system.

First, Let write differential equation for mechanical system. The free body diagram of M is shown in Fig. (3.6.2). The opposing forces acting on M are marked as f_m, f_b and f_k.

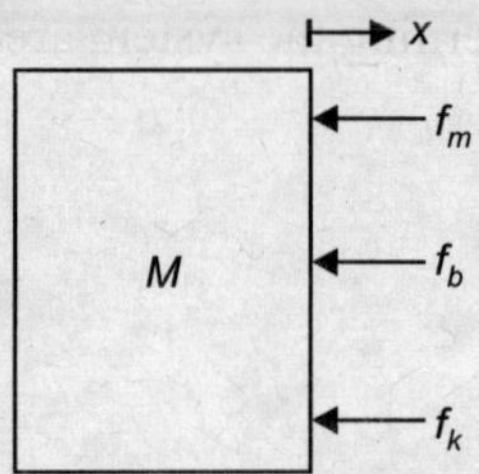

Fig. 3.6.2: Free body diagram of mass.

$$f_m = M\frac{d^2x}{dt^2};\quad f_b = B\frac{dx}{dt};\quad f_k = Kx$$

By newton's second law,

$$f(t) = f_m + f_b + f_k$$

$$f(t) = M\frac{d^2x}{dt^2} + B\frac{dx}{dt} + Kx \tag{3.6.1}$$

$f(t) = K_s i$ where $f(t)$ is the electrical force acting on the mass on taking Laplace transform of Eq. (3.6.1)

$$F(s) = Ms^2\,X(s) + Bs\,X(s) + K\,X(s)$$

$$F(s) = X(s)\,[Ms^2 + Bs + K] \tag{3.6.2}$$

The equivalent circuit for the electrical circuit as shown in Fig. (3.6.3)

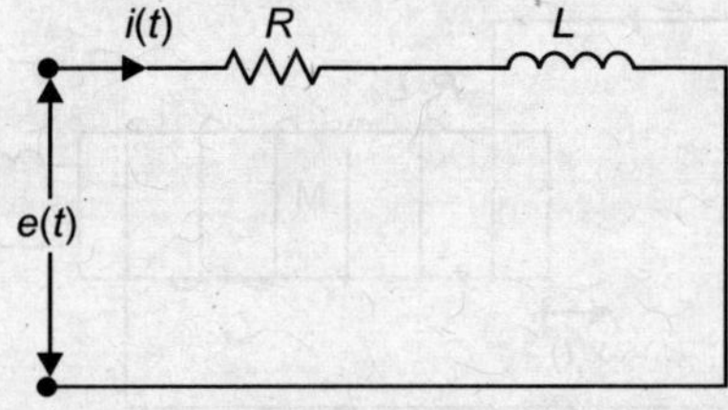

Fig. 3.6.3

By applying KVL,

$$e(t) = Ri\,(t) + L\frac{di(t)}{dt} \tag{3.6.3}$$

on taking laplace transform of Eq. (3.8.3)

$$E(s) = R\,I(s) + Ls\,I(s)$$

$$E(s) = I(s)\,[R + Ls] \Rightarrow I(s) = \frac{E(s)}{(R + Ls)} \tag{3.6.4}$$

and also,

$$f(t) \propto i(t) \tag{3.6.5}$$

$$f(t) = K_s\,i(t)$$

$\therefore$ $F(s) = K_s\,I(s)$ where, $K_S \rightarrow$ proportionality constent

Substitute $I(s)$ from Eq. (3.6.4) in Eq. (3.6.5)

$$F(s) = Ks\left[\frac{E(s)}{R+Ls}\right] \tag{3.6.6}$$

By equating Eqs. (3.6.6) and (3.6.2)

$$X(s)\,[Ms^2 + Bs + K] = K_s\left[\frac{E(s)}{R+Ls}\right]$$

$$\frac{X(s)}{E(s)} = \frac{K_s}{(R+Ls)\,(Ms^2+Bs+K)} \tag{3.6.7}$$

Result: The differential equations governing the system case,

1. $e(t) = iR + L\dfrac{di}{dt}$

2. $f(t) = M\dfrac{d^2x}{dt^2} + B\dfrac{dx}{dt} + Kx$

3. $f(t) = K_s\, i(t)$

The transfer function of the system is

$$\frac{X(s)}{E(s)} = \frac{K_s}{(R+Ls)\,(Ms^2+Bs+K)}$$

3.5 ELECTRICAL ANALOGOUS OF MECHANICAL TRANSLATIONAL SYSTEMS

Systems remain analogous as long as the differential equations governing the system or transfer functions are of identical form. The electric analogue of any other kind of system is of great importance since it is easier to construct electrical models and analyse them.

The three basic elements mass, dash-pot and spring that are used in modelling mechanical translational systems are analogous to resistance, inductance and capacitance of electrical system. The input force in mechanical system is analogous to either voltage source or current source in electrical systems.

The output velocity in mechanical system is analogous to either current (or) voltage in an element in electrical system. Since the electrical systems has two types of inputs either voltage or current source, there are two types of analogies: force-voltage analogy and force-current analogy.

Force-voltage Analogy

The force balance equations of mechanical elements and their analogous electrical elements in force-voltage analogy are shown in Table 1.

Table 1

Mechanical System	Electrical system
Input: Force Output: Velocity	Input: Voltage source Output: Current through the element.
$f = \frac{Bdx}{dt} = Bv$ ($v = dx/dt$)	$e = Ri$
$f = \frac{Md^2x}{dt^2} = \frac{Mdv}{dt}$ ($a = \frac{d^2x}{dt^2} = dv/dt$)	$e = \frac{Ldi}{dt}$
$f = Kx = \int v\, dt$	$e = 1/C \int idt$

Table 2

Item	Mechanical system	Electrical system
Independent variable (Input)	Force, f	Voltage, e
Dependent variable (Output)	Velocity, v	Current, i
Dissipative element	Frictional coefficient of dash-pot, B	Resistance, R
Storage element	Mass, M	Inductance, L
	Stiffness of spring, K	Inverse of capacitance, $\frac{1}{c}$
Physical law	Newton's second law $\Sigma f = 0$	KVL $\Sigma V = 0$
Changing the level of independent variable	Lever $\frac{f_1}{f_2} = \frac{l_1}{l_2}$	Transformer $\frac{e_1}{e_2} = \frac{N_1}{N_2}$

The following points serve as guide lines to obtain electrical analogous of mechanical systems based on force-voltage analogy.

1. In electrical system the elements which are in series will have same current, likewise in mechanical system, the elements having same velocity are said to be in series.

2. The elements having same velocity in mechanical system should have the same current in electrical analogous system.
3. Each node in the mechanical system corresponds to a closed loop in electrical system. A mass is considered as a node.
4. The number of meshes in electrical analogous is same as that of the number of nodes in mechanical system. Hence the number of mesh currents and system equations will be same as that of the number of velocities of nodes in mechanical system (number of masses = number of loops).
5. The mechanical driving sources and passive elements connected to the node in mechanical system should be represented as closed loop in electrical analogous system.
6. The elements connected between two masses in mechanical system is represented as a common elements between two meshes in electrical analogous system.

Note: Force voltage analogy is based on KVL.

Force Current Analogy

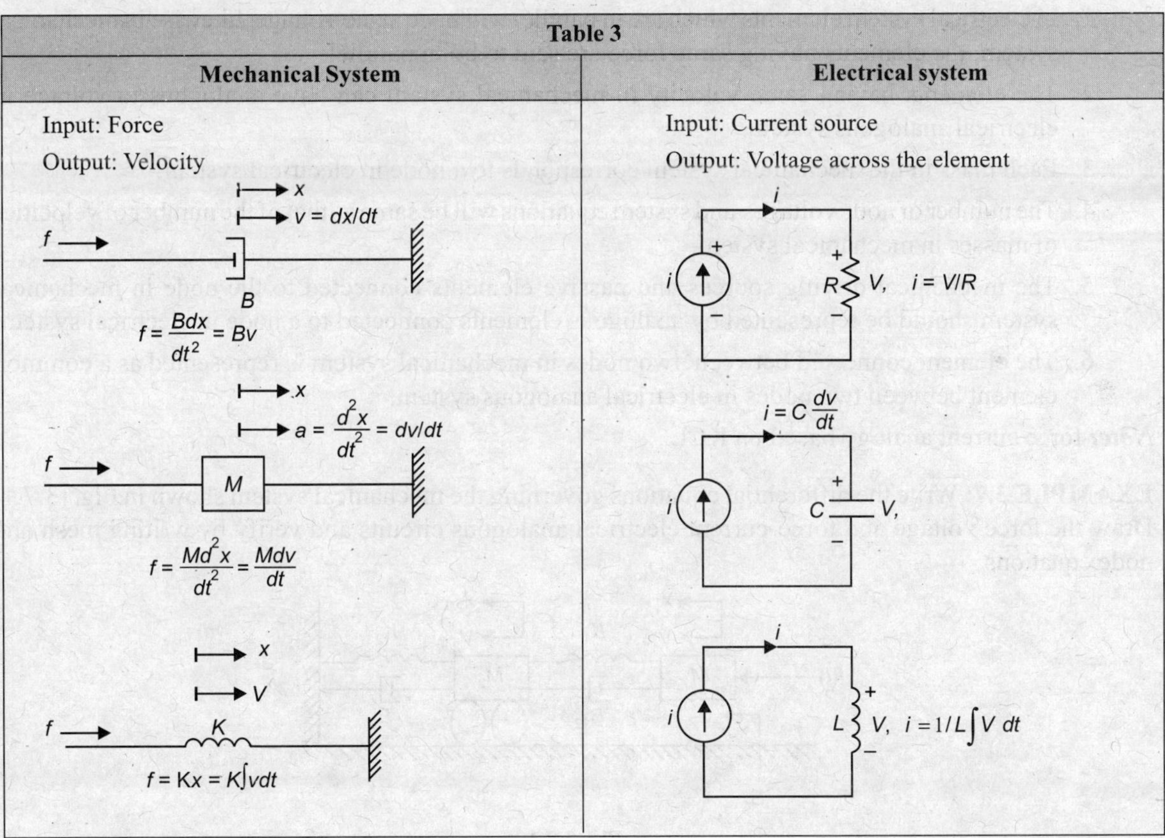

Table 3

Mechanical System	Electrical system
Input: Force Output: Velocity	Input: Current source Output: Voltage across the element
x, $v = dx/dt$, f, B $f = \frac{Bdx}{dt^2} = Bv$	i, R, V, $i = V/R$
x, $a = \frac{d^2x}{dt^2} = dv/dt$, f, M $f = \frac{Md^2x}{dt^2} = \frac{Mdv}{dt}$	$i = C\frac{dv}{dt}$, i, C, V
x, V, f, K $f = Kx = K\int vdt$	i, L, V, $i = 1/L\int V\,dt$

Table 4		
Item	**Mechanical system**	**Electrical system**
Independent variable (Input)	Force, f	Current, i
Dependent variable (Output)	Velocity, v	Voltage, V
Dissipative element	Frictional coefficient of dash-pot, B	Conductance, $G = 1/R$
Storage element	Mass, M Stiffness of spring, K	Capacitance, C Inverse of inductance, $1/L$
Physical law	Newton's second law $\Sigma f = 0$	KCL $\Sigma i = 0$
Changing the level of independent variable	Lever $\frac{f_1}{f_2} = \frac{l_1}{l_2}$	Transformer $\frac{i_1}{i_2} = \frac{N_1}{N_2}$

The following points serve as guidelines to obtain electrical analogous of mechanical system based on force-current analogy.

1. In electrical system elements which are in parallel will have same voltage, likewise in mechanical system, the elements having same force are said to be in parallel.
2. The elements having same velocity in mechanical system can have analogous to voltage in electrical analogous system.
3. Each mass in the mechanical system corresponds to a node in electrical system.
4. The number of node voltages and system equations will be same as that of the number of velocities of masses in mechanical system.
5. The mechanical driving sources and passive elements connected to the node in mechanical system should be represented by analogous elements connected to a node in electrical system.
6. The element connected between two nodes in mechanical system is represented as a common element between two nodes in electrical analogous system.

Note: force current analogy based on KCL.

EXAMPLE 3.7: Write the differential equations governing the mechanical system shown in Fig. (3.7.1) Draw the force voltage and force-current electrical analogous circuits and verify by writing mesh and node equations.

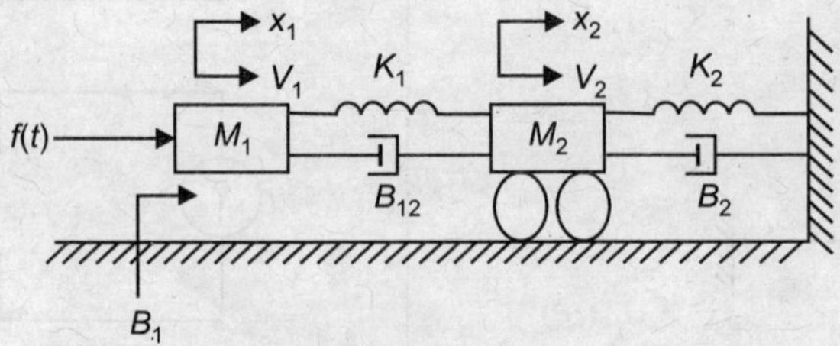

Fig. 3.7.1

Solution: Write force balance equation for given mechanical system.

Free body diagram of mass M_1 as shown in Fig. (3.7.2). Forces are marked as $f_{m1}, f_{b1}, f_{b12}, f_{k1}$ and $f(t)$

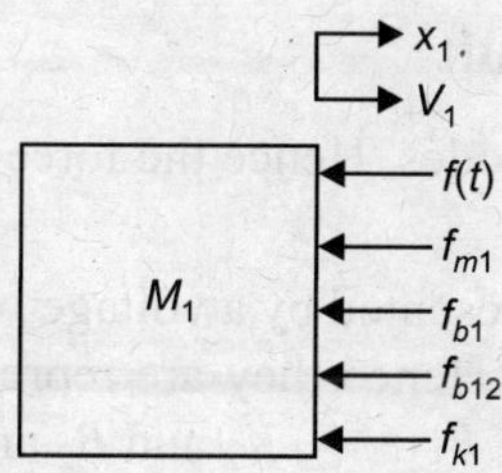

Fig. 3.7.2: Free body diagram of mass M_1.

By newton's second law,

$$f_{m1} + f_{b1} + f_{b12} + f_{k1} = f(t)$$

$$M_1 \frac{d^2 x_1}{dt^2} + B_1 \frac{dx_1}{dt} + B_{12} \frac{d}{dt}(x_1 - x_2) + K_1 (x_1 - x_2) = f(t) \tag{3.7.1}$$

The freebody diagram of mass M_2 as shown in Fig. (3.7.3). Forces are marked as, $f_{m2}, f_{b2}, f_{b12}, f_{k1}$ and f_{k2}.

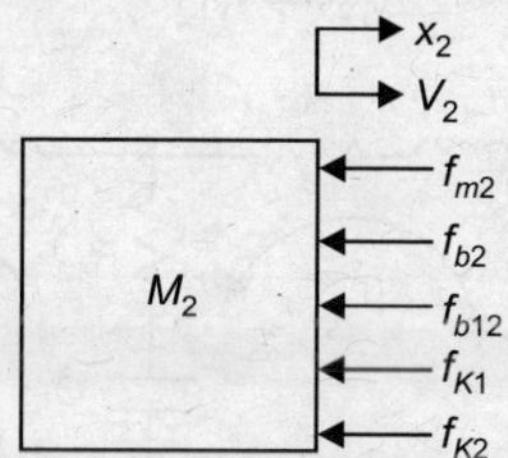

Fig. 3.7.3: Frebody diagram of mass M_2.

By newton's second law,

$$f_{m2} + f_{b2} + f_{b12} + f_{k1} + f_{k2} = 0$$

$$M_2 \frac{d^2 x_2}{dt^2} + B_2 \frac{dx_2}{dt} + K_2 x_2 + B_{12} \frac{d}{dt}(x_2 - x_1) + K_1 (x_2 - x_1) = 0 \tag{3.7.2}$$

on replacing the displacement by velocity in the differential Eqs. (3.7.1) and (3.7.2) of the mechanical system we get,

$$\left[\text{i.e., } \frac{d^2 x}{dt^2} = \frac{dv}{dt}; \quad \frac{dx}{dt} = v \quad \text{and} \quad x = \int v dt \right]$$

Eq. (3.7.1) become,

$$M_1 \frac{dv_1}{dt} + B_1 v_1 + B_{12} (v_1 - v_2) + K_1 \int (v_1 - v_2)\, dt = f(t) \tag{3.7.3}$$

Eq. (3.7.2) become,

$$M_2 \frac{dv_2}{dt} + B_2 v_2 + K_2 \int v_2 dt + B_{12}(v_2 - v_1) + K_1 \int (v_2 - v_2)\, dt = 0 \tag{3.7.4}$$

(i) Force-voltage Analogous Circuit

The given mechanical system has two nodes. Hence the force-voltage analogous electrical circuit will have two meshes.

The force applied to mass M_1 is represented by a voltage source in first mesh. The elements M_1, B_1, K_1 and B_{12} are connected to first node. Hence they are represented by analogous element in mesh 1 forming a closed path. The elements K_1, B_{12}, M_2, K_2 and B_2 are connected to second node. Hence they are represented by analogous element in mesh 2 forming a closed path.

The elements K_1 and B_{12} are common between node 1 and 2 and so they are represented by analogous element as common elements between two meshes. The force-voltage electrical analogous circuit is shown in Fig. (3.7.4). The electrical analogous elements for the elements of mechanical system are given below.

$f(t) \rightarrow e(t)$	$M_1 \rightarrow L_1$	$B_1 \rightarrow R_1$	$K_1 \rightarrow 1/C_1$
$x_1 \rightarrow i_1$	$M_2 \rightarrow L_2$	$B_2 \rightarrow R_2$	$K_2 \rightarrow 1/C_2$
$x_2 \rightarrow i_2$		$B_{12} \rightarrow R_{12}$	

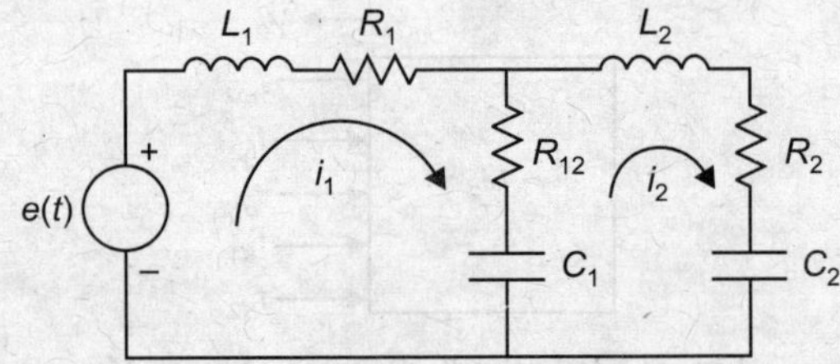

Fig. 3.7.4: Force voltage electrical analogous circuit.

The mesh Eqs. using KVL for the circuit shown in Fig. (3.7.4) are given below.

For first loop, apply Kirchoff's voltage law,

$$L_1 \frac{di_1}{dt} + R_1 i_1 + R_{12}(i_1 - i_2) + \frac{1}{C_1}\int (i_1 - i_2)\, dt = e(t) \tag{3.7.5}$$

For second loop, apply Kirchoff's voltage law,

$$L_2 \frac{di_2}{dt} + R_2 i_2 + \frac{1}{C_2}\int i_2\, dt + R_{12}(i_2 - i_1) + \frac{1}{C_1}\int (i_2 - i_1)\, dt = 0 \tag{3.7.6}$$

It is observed that the mesh basis Eqs. (3.7.5) and (3.7.6) are similar to the differential Eqs. (3.7.3) and (3.7.4) governing the mechanical system.

(ii) Force Current Analogous Circuit

The given mechanical system has two nodes. Hence the force current analogous electrical circuit will have two nodes.

The force applied to mass M_1 is represented as a current source connected to node 1 in analogous electrical circuit. The elements M_1, B_1, K_1 and B_{12} are connected to first node. Hence they are represented by analogous elements connected to node 1 in analogous electrical circuit. The elements K_1, B_{12} M_2, K_2 are B_2 are connected to second node. Hence they are represented by analogous electrical circuit.

The elements K_1 and B_{12} are common between 1 and 2 so they are represented as common element between two nodes in electrical analogous circuit. The force current electrical analogous circuit is shown in Fig. (3.7.4)

The electrical analogous elements for the elements of mechanical system are given below

$f(t) \rightarrow i(t)$	$M_1 \rightarrow C_1$	$B_1 \rightarrow 1/R_1$	$K_1 \rightarrow 1/L_1$
$x_1 \rightarrow V_1$	$M_2 \rightarrow C_2$	$B_2 \rightarrow 1/R_2$	$K_2 \rightarrow 1/L_2$
$x_2 \rightarrow V_2$			

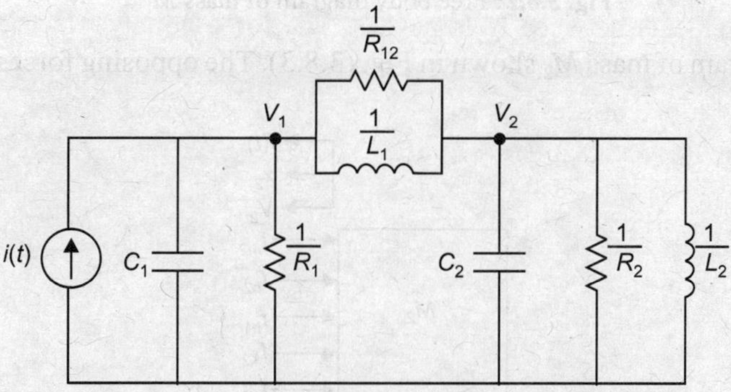

Fig. 3.7.5: Force-current electrical analogous circuit.

The node basis equations using Kirchoff's current Law for the circuit shown in Fig. (3.7.5) are given below.

For node 1,
$$C_1 \frac{dv_1}{dt} + \frac{1}{R_1} V_1 + \frac{1}{R_{12}} (V_1 - V_2) + \frac{1}{L_1} \int (V_1 - V_2)\, dt = i(t) \qquad (3.7.7)$$

For node 2,
$$C_2 \frac{dV_2}{dt} + \frac{1}{R_2} V_2 + \frac{1}{L_2} \int V_2 dt + \frac{1}{R_{12}} (V_2 - V_1) + \frac{1}{L_1} \int (V_2 - V_1)\, dt = 0 \qquad (3.7.8)$$

It is observed that the node basis Eqs. (3.7.7) and (3.7.8) are similar to the differential Eqs. (3.7.3) and (3.7.4) governing the mechanical system.

EXAMPLE 3.8: Write the differential equations governing the mechanical system shown in Fig. (3.8.1). Draw the force-voltage and force current electrical analogous circuits and verify by writing mesh node equations.

Solution: First let write force balance equation for given mechanical system.

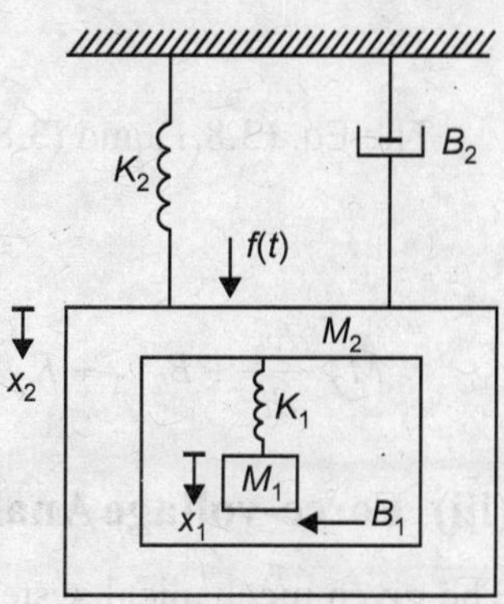

Fig. 3.8.1

Free body diagram of mass M_1 as shown in Fig. (3.8.2). Forces are marked as f_{m1}, f_{b1}, and f_{k3}.

By newton's second law

$$f_{m1} + f_{b1} + f_{k1} = 0$$

$$M_1 \frac{d^2 x_1}{dt^2} + B_1 \frac{d}{dt}(x_1 - x_2) + K_1 (x_1 - x_2) = 0 \quad (3.8.1)$$

Fig. 3.8.2: Free body diagram of mass M_1.

The free body diagram of mass M_2 shown in Fig. (3.8.3). The opposing forces are marked as $f_{m2}, f_{b2}, f_{b1}, f_{k2}$ and f_{k1}

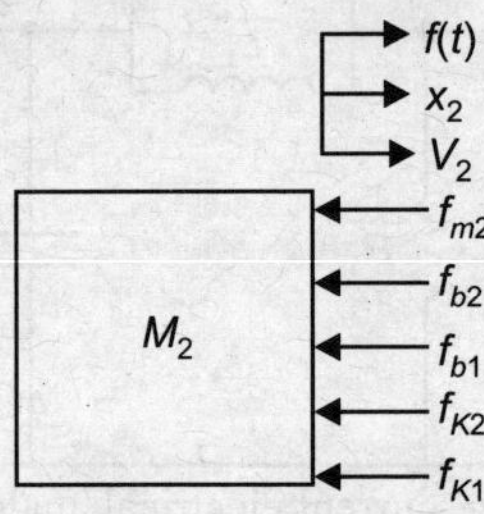

Fig. 3.8.3: Free body diagram of mass M_2.

By newton's second law

$$f_{m2} + f_{b2} + f_{b1} + f_{k2} + f_{k1} = f(t)$$

$$M_2 \frac{d^2 x_2}{dt^2} + B_2 \frac{dx_2}{dt} + K_2 x_2 + B_1 \frac{d}{dt}(x_2 - x_1) + K_1 (x_2 - x_1) = f(t) \quad (3.8.2)$$

on replacing the displacements by velocity in the differential Eq. (3.8.1) and (3.8.2) governing the mechanical system we get,

$$\left[\text{i.e.,} \frac{d^2 x}{dt^2} = \frac{dv}{dt}; \frac{dx}{dt} = v \text{ and } x = \int v dt \right]$$

The Eq. (3.8.1) and (3.8.2) become

$$M_1 \frac{dv_1}{dt} + B_1 (v_1 - v_2) + K_1 \int (v_1 - v_2)\, dt = 0 \quad (3.8.3)$$

$$M_2 \frac{dv_2}{dt} + B_2 v_2 + K_2 \int v_2 \, dt + B_1 (v_2 - v_1) + K_1 \int (v_2 - v_1)\, dt = f(t) \quad (3.8.4)$$

(iii) Force-voltage Analogous Circuit

The given mechanical system has two nodes. Hence the force voltage analogous electrical circuit will have two meshes. The force applied to mass, M_2 is represented by a voltage source is second mesh.

The elements M_1, K_1 and B_1 are connected to first node. Hence they are represented by analogous element in mesh 1 forming closed path. The elements M_2, K_2, B_2, B_1 and K_1 are connected to second node. Hence they are represented by analogous element in mesh 2 forming a closed path.

The element B_1 and K_1 are common between node 1 and 2 and so they are represented as common elements between mesh 1 and mesh 2. The force-voltage electrical analogous circuit shown in Fig. (3.8.4)

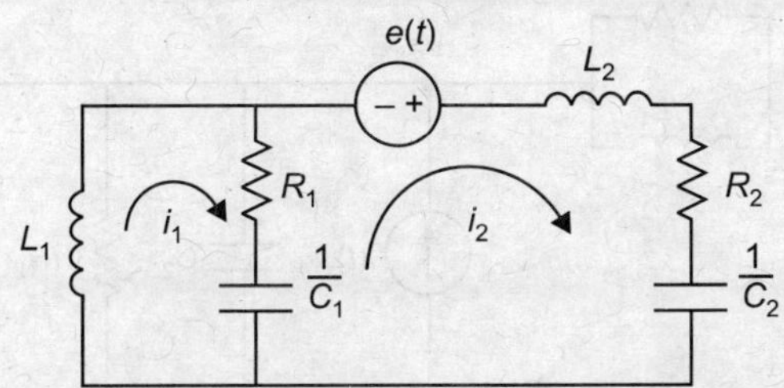

Fig. 3.8.4: Force-voltage electrical analogous circuit.

The electrical analogous elements for the elements of mechanical system are given below,

$f(t) \rightarrow e(t)$ $\quad x_1 \rightarrow i_1$ $\quad M_1 \rightarrow L_1$ $\quad K_1 \rightarrow 1/C_1$ $\quad B_1 \rightarrow R_1$

$x_2 \rightarrow i_2$ $\quad M_2 \rightarrow L_2$ $\quad K_2 \rightarrow 1/C_2$ $\quad B_2 \rightarrow R_2$

The mesh basis equations using KVL for the circuit shown in Fig. (3.8.4) are given below.

$$L_1 \frac{di_1}{dt} + R_1 (i_1 - i_2) + \frac{1}{C_1} \int (i_1 - i_2)\, dt = 0 \qquad (3.8.5)$$

$$L_2 \frac{di_2}{dt} + R_2 i_2 + \frac{1}{C_2} \int i_2 dt + \frac{1}{C_1} \int (i_2 - i_1)\, dt + R_1 (i_2 - i_1) = i(t) \qquad (3.8.6)$$

It is observed that the mesh basis Eqs. (3.8.5) and (3.8.6) are similar to the differential Eqs. (3.8.3) and (3.8.4) governing the mechanical system.

(iv) Force-Current Analogous Circuit

The given mechanical system has two nodes. Hence the force-current analogous electrical circuit will have two nodes. The force applied to mass M_2 is represented as a current source connected to node 2 in analogous electrical circuit.

The elements M_1 K_1 and B_1 are connected to first node. Hence they are represented by analogous elements as connected to node 1 in analogous electrical circuit. The elements M_2, K_2, B_2, B_1 and K_1 are connected to second node. Hence they are represented by analogous elements as elements connected to node 2 in analogous electrical circuits.

The elements K_1 and B_1 is common to node 1 and 2 and so they are represented by analogous elements as common elements between two nodes in analogous circuit. The force current electrical analogous circuit shown in Fig. (3.8.5)

The electrical analogous elements for the elements of mechanical system are given below

$$f(t) \rightarrow i(t) \quad x_1 \rightarrow V_1 \quad M_1 \rightarrow C_1 \quad B_1 \rightarrow 1/R_1 \quad K_1 \rightarrow 1/L_1$$

$$x_2 \rightarrow V_2 \quad M_2 \rightarrow C_2 \quad B_2 \rightarrow 1/R_2 \quad K_2 \rightarrow 1/L_2$$

The node basis equations using Kirchoff's current law for the circuit shown in Fig. (3.8.5)

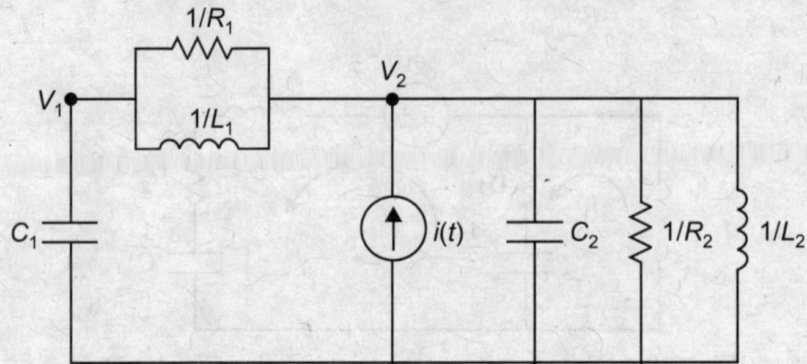

Fig. 3.8.5: Force current electrical analogous circuit.

$$C_1 \frac{dV_1}{dt} + \frac{1}{R_1}(V_1 - V_2) + \frac{1}{L_1}\int (V_1 - V_2)\, dt = 0 \tag{3.8.7}$$

$$C_2 \frac{dV_2}{dt} + \frac{1}{R_2} V_2 + \frac{1}{L_1}\int V_2\, dt + \frac{1}{R_1}(V_2 - V_1) + \frac{1}{L_1}\int (V_2 - V_1)\, dt = i(t) \tag{3.8.8}$$

It is observed that the node basis Eqs. (3.8.7) and (3.8.8) are similar to the differential Eqs. (3.8.3) and (3.8.4) governing the mechanical system.

3.6 BLOCK DIAGRAM

A control system may consist of a number of components. In control engineering to show the functions performed by each component, we commonly use a diagram called the block diagram. **A block diagram is a pictorial representation of the functions performed by each component and of the flow of signals. which shows interrelationships that exist among the various components. The elements of a block diagram are block, branch point and summing point.**

Block

In a block diagram all system variables are linked to each other through functional blocks. The functional block is a symbol for the mathematical operation on the input signal to the block that produces the output. The transfer functions of the components are usually entered in the corresponding blocks, which are connected by arrows to indicate the direction of the flow of signals. Fig. 3.18 (a) shows the functional block of the block diagram.

Fig. 3.18(a): Functional block.

The arrow pointing towards the block indicates the input, and the arrow leading away from the block represents the output. Such arrows are referred as signal flow. The output signal from the block is given by the product of input signal and transfer function in the block.

Summing Point

Summing points are used to add two (or) more signals in the system. Referring to Figure 3.18 (b). A circle with a cross is the symbol that indicates the summing operation.

The plus or minus sign at each arrow indicates whether the signal is to be added or subtracted.

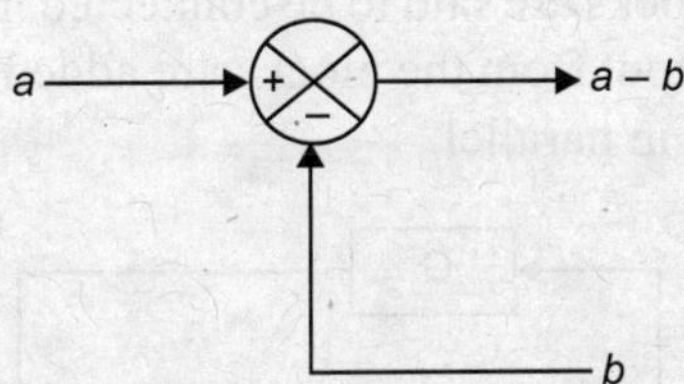

Fig. 3.18(b): Summing Point.

Branch Point

A branch point is a point from which the signal goes from one block to other blocks (or) to summing points.A branch point is shown in Fig. 3.18 (c)

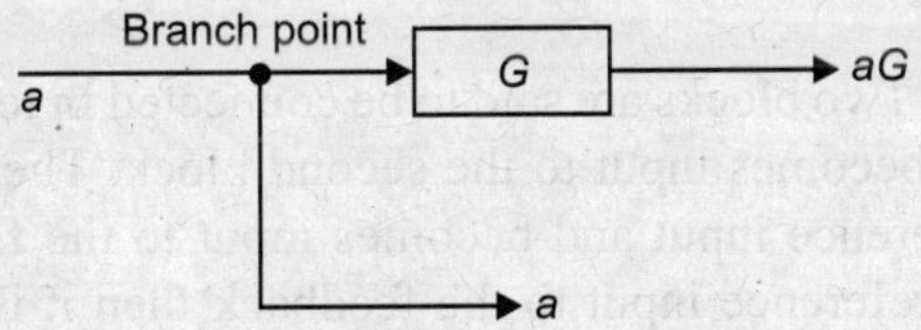

Fig. 3.18(c): Branch point.

Construction of blocK diagram for control systems

A control system can be represented diagrammatically by block diagram. The differential equations governing the system are used to construct the block diagram. By taking laplace transform the differential equations are converted to algebraic equations. The equations will have variables and constants. From the working knowledge of the system the input and output variables are identified and the block diagram for each equation can be drawn. Each equation gives one section of block diagram. The output of one section will be input for another section. The various sections are interconnected to obtain the overall block diagram of the system.

Basic connections for blocks

There are three basic connections namely (1) Cascade (2) Parallel (3) Feedback.

1. **Cascade connection:** Two blocks are said to be connected in cascade if the ouput of the first block becomes input to the second block. The cascade connected blocks are shown in Fig. 3.18 (d).

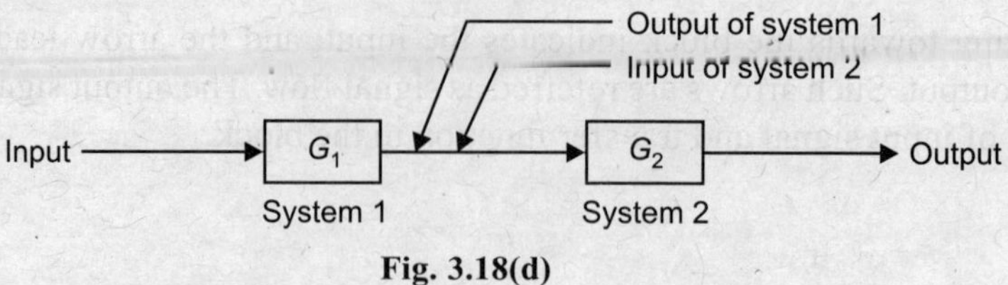

Fig. 3.18(d)

The same definition may be extended to n number of blocks.

Note: It also can be named as series connection.

2. **Parallel connection:** Two blocks are said to be connected in parallel if the same input is applied to both the blocks and the output from the blocks are added using summer. The same definition may be extended to n blocks in parallel.

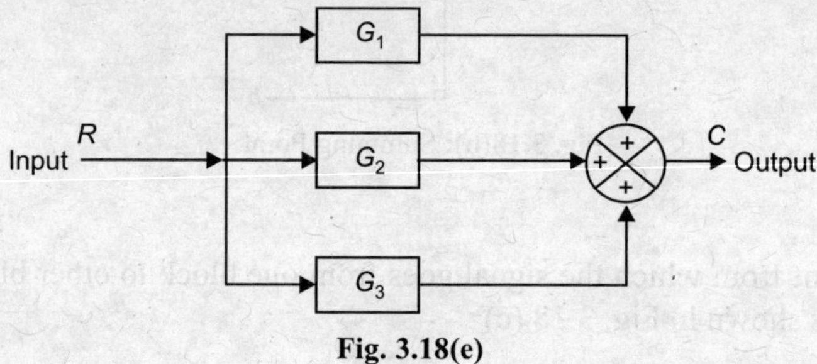

Fig. 3.18(e)

3. **Feedback connection:** Two blocks are said to be connected in feedback if the output of the first block is feedback and becomes input to the second block. The output of the second block is compared with the reference input and becomes input to the first block. If the output of the second block aids the reference input to the feedback then it is known as positive feedback. Usually the first block is called the forward path transfer function or open loop transfer function and is represented by G and feedback transfer function is usually represented by H.

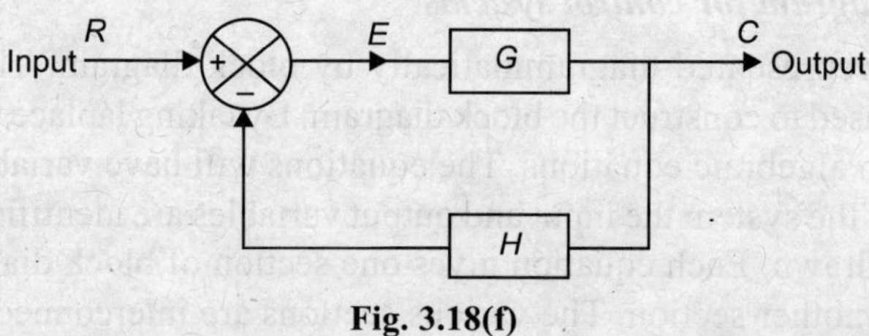

Fig. 3.18(f)

EXAMPLE 3.9: Consider the electromechanical system shown in Fig. (3.9.1). Assuming the magnetic pull produced by the solenoid is equal to K_0 times the current i, flowing through the solenoid, (a) draw a block diagram showing each component of the system. Indicate the input and output of each block clearly. (b) Determine the transfer function of each block. (c) Determine equivalent overall transfer function of the system. Take E as the input to the system and x_1 as the output of the system.

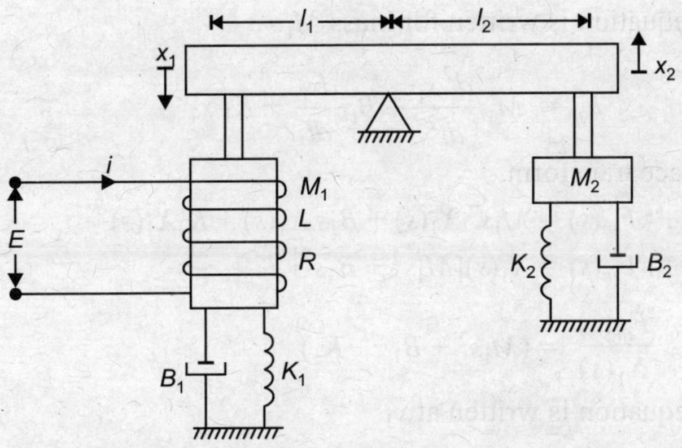

Fig. 3.9.1

Solution:

Step 1: Let F_1 is the force acting at x_1 and F_2 is the force acting at x_2, the following equation can be written

$$F_1 l_1 = F_2 l_2$$

$$\frac{F_2}{F_1} = \frac{l_1}{l_2} \tag{3.9.1}$$

Step 2: Let write equation for electrical circuit,

$$e(t) = Ri + L\frac{di}{dt}$$

$$E(s) = (R + sL)\, I(s)$$

$$I(s) = E(s)/(R + sL) \tag{3.9.2}$$

Step 3: $F(s) = K_0\, I(s)$

$$\frac{F(s)}{I(s)} = K_0 \tag{3.9.3}$$

Step 4: At point x_2, the following equation is written

$$F_2(t) = M_2\frac{d^2x_2}{dt^2} + B_2\frac{dx_2}{dt} + K_2 x_2$$

on taking laplace transform,

$$F_2(s) = (M_2 s^2 + B_2 s + K_2)\, X_2(s) \tag{3.9.4}$$

Step 5: $$\frac{x_1}{l_1} = \frac{x_2}{l_2}$$

$$\frac{x_1}{x_2} = \frac{l_1}{l_2} \tag{3.9.5}$$

Step 6: The following equation is written for mass M_1.

$$F_m = M_1 \frac{d^2 x_1}{dt^2} + B_1 \frac{dx_1}{dt} + K_1 x_1$$

on taking laplace transform,

$$F_m(s) = M_1 s^2 X_1(s) + B_1 s X_1(s) + K_1 X_1(s)$$

$$F_m(s) = X_1(s)[M_1 s^2 + B_1 s + K_1]$$

$$\frac{F_m(s)}{X_1(s)} = (M_1 s^2 + B_1 s + K_1) \tag{3.9.6}$$

Step 7: The following equation is written at x_1

$$F(s) - F_m(s) = F_1(s) \tag{3.9.7}$$

By combining all the equations, the block diagram can be drawn as shown in Fig. (3.9.2) and (3.9.3)

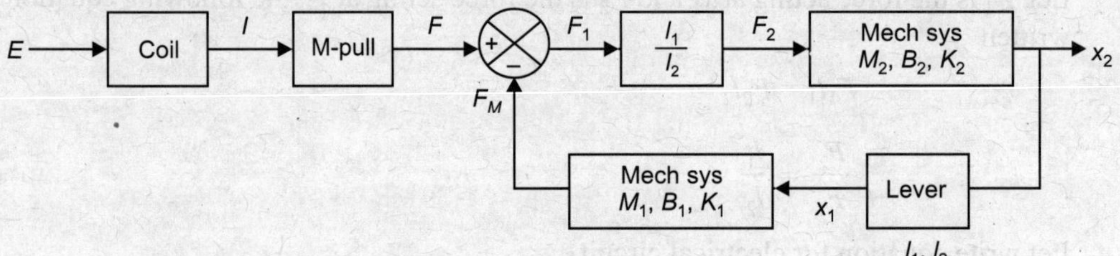

Fig. 3.9.2

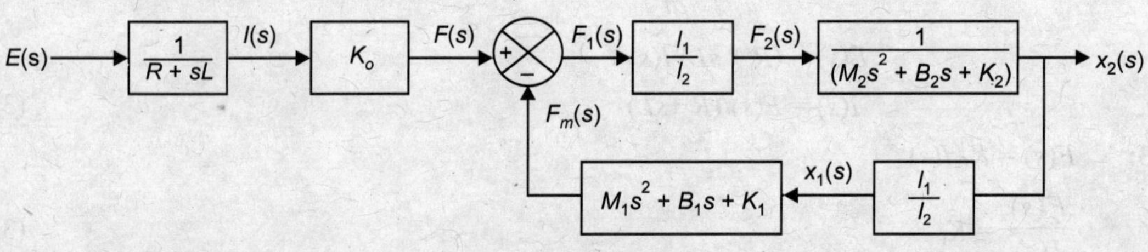

Fig. 3.9.3

using the block diagram reduction technique, we may obtain

$$\frac{X_1(s)}{E(s)} = \frac{K_0}{(R+sL)\left[\frac{l_1}{l_2}(M_1 s^2 + B_1 s + K_1) + (M_2 s^2 + B_2 s + K_2)\left(\frac{l_2}{l_1}\right)\right]}$$

EXAMPLE 3.10: Construct the block diagram of armature controlled dc motor.

Solution: The differential equations governing the armature controlled dc motor are (refer section 3.3.5)

$$V_a = i_a R_a + L_a \frac{di_a}{dt} + e_b \tag{3.10.1}$$

$$T = K_t \, i_a \tag{3.10.2}$$

$$T = J\frac{d\omega}{dt} + B\omega \tag{3.10.3}$$

$$e_b = K_b \omega \tag{3.10.4}$$

where, $$\omega = \frac{d\theta}{dt} \tag{3.10.5}$$

on taking laplace transform of Eq. (3.10.1) we get,

$$V_a(s) = I_a(s)\,R_a + L_a\,sI_a(s) + E_b(s)$$

$$V_a(s) - E_b(s) = I_a(s)\,[R_a + sL_a]$$

$$I_a(s) = \frac{[V_a(s) - E_b(s)]}{[R_a + sL_a]}$$

$$I_a(s) = \frac{1}{(R_a + sL_a)}\,[V_a(s) - E_b(s)] \tag{3.10.6}$$

from Eq. (3.10.6)

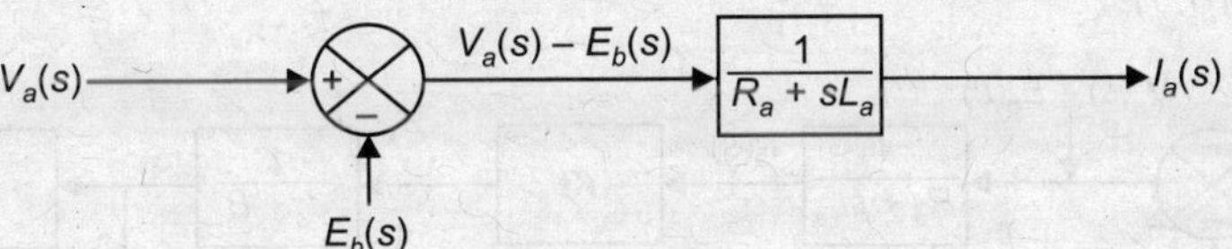

Fig. 3.10.1

on taking laplace transform of Eq. (3.10.2) we get

$$T(s) = K_t\,I_a(s)$$

from Eq. (3.10.7)

$I_a(s)$ → K_t → $T(s)$

Fig. 3.10.2

on taking laplace transform of Eq. (3.10.3) we get,

$$T(s) = \omega(s)\,[Js + B]$$

$$\omega(s) = \left(\frac{1}{Js + B}\right)T(s) \tag{3.10.8}$$

from Eq. (3.10.8)

$T(s)$ → $\frac{1}{Js + B}$ → $\omega(s)$

Fig. 3.10.3

CHAPTER 3

on taking laplace transform of Eq. (3.10.4) we get,

$$E_b(s) = K_b\, \omega(s) \tag{3.10.9}$$

ω(s) → K_b → $E_b(s)$

Fig. 3.10.4

on taking laplace transform of Eq. (3.10.5)

$$\omega(s) = s\theta(s)$$

$$\theta(s) = \frac{1}{s}\,\omega(s)$$

ω(s) → 1/s → θ(s)

Fig. 3.10.5

The overall block diagram of armature controlled dc motor is obtained by connecting various sections of Fig. (3.10.1 to 3.10.5).

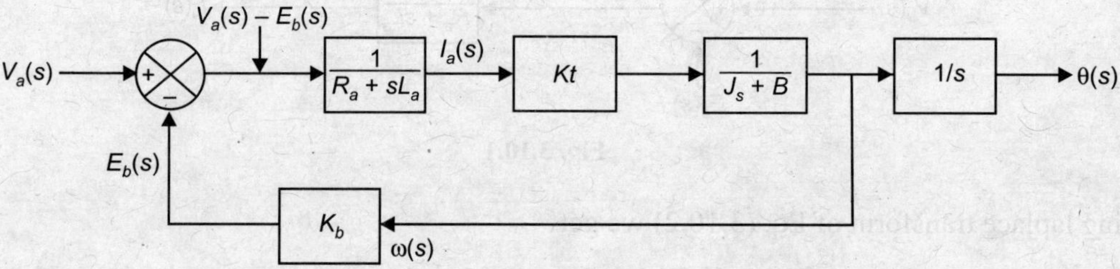

Fig. 3.10.6: Complete block diagram of armature controlled dc motor.

EXAMPLE 3.11: Construct the block diagram of field controlled dc motor.

Solution: The differential equations governing the field controlled dc motor are (refer section 3.3.5)

$$V_f = R_f i_f + L_f \frac{di_f}{dt} \tag{3.11.1}$$

$$T = Kt_f\, i_f \tag{3.11.2}$$

$$T = J\frac{d^2\theta}{dt} + B\frac{d\theta}{dt} \tag{3.11.3}$$

on taking laplace transform of Eq. (3.11.1)

$$V_f(s) = R_f I_f(s) + L_f s\, I_f(s)$$

$$V_f(s) = I_f(s)\,[R_f + s\, L_f]$$

$$I_f(s) = \frac{V_f(s)}{[R_f + s\,L_f]}$$

$$I_f(s) = \frac{1}{(R_f + s\,L_f)} V_f(s) \qquad (3.11.4)$$

From Eq. (3.11.4)

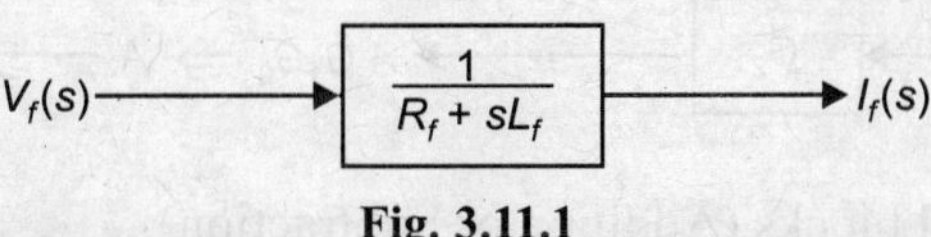

Fig. 3.11.1

on taking laplace transform of Eq. (3.11.2),

$$T(s) = K_{tf}\, I_f(s) \qquad (3.11.5)$$

From Eq. (3.11.5)

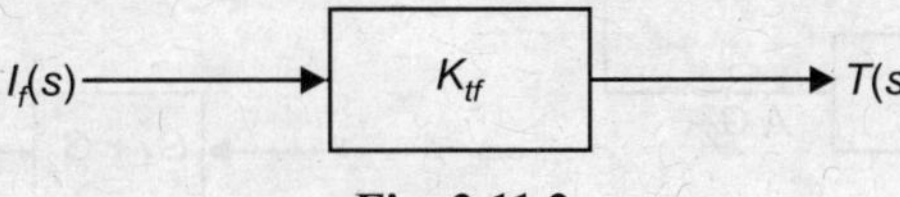

Fig. 3.11.2

on taking laplace transform of Eq. (3.11.3) we get,

$$T(s) = Js^2\,\theta(s) + Bs\,\theta(s)$$

$$T(s) = \theta(s)\,[Js^2 + Bs]$$

$$\therefore \qquad \theta(s) = \frac{1}{(Js^2 + Bs)} T(s) \qquad (3.11.6)$$

From Eq. (3.11.6)

T(s) → $\frac{1}{(Js^2 + Bs)}$ → θ(s)

Fig. 3.11.3

By combining Fig. (3.11.1), (3.11.2) and (3.11.3) the overall block diagram of field control DC motor can be obtained.

Fig. 3.11.4: Block diagram of field controlled DC motor.

Block diagram reduction

The block diagram can be reduced to find the overall transfer function of the system. The following rules can be used for block diagram reduction. The rules are framed such that any modification made on the diagram does not alter the input and output relation.

Rules of Block Diagram Algebra

Rule 1: Combining the blocks in Cascade (Multiplication)

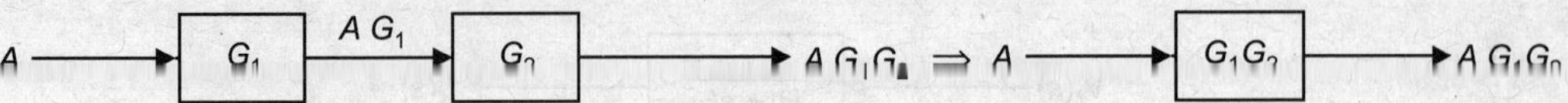

Rule 2: Combining parallel blocks (Addition or Subtraction)

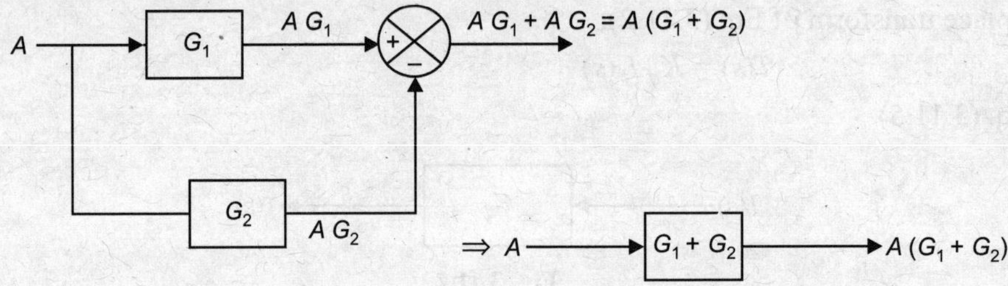

Rule 3: Moving the branch point ahead of the block [Branch point to right of gain (*BR* 1/*G*)]

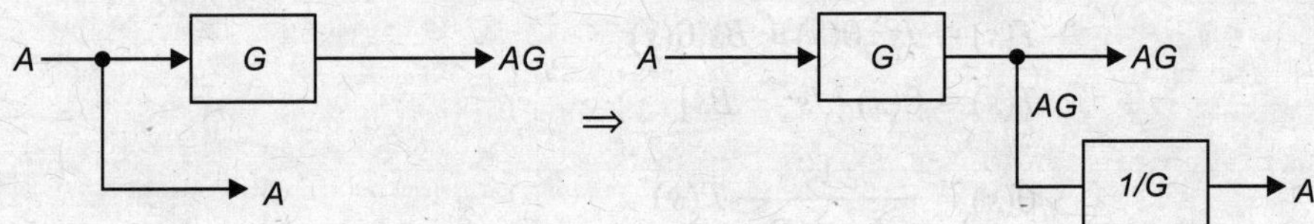

Note: When the branch point is moved to right then (YG) block should be included with the branch point.

Rule 4: Moving the branch point before the block [Branch point to left of gain (BLG)]

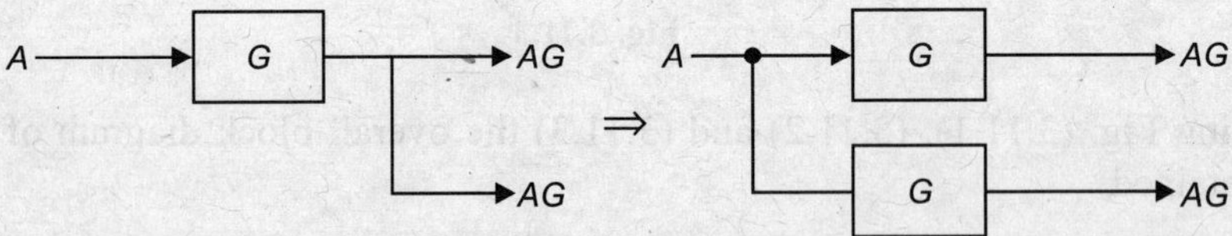

Note: When the brachn point is moved the left then (G) block should be included with the branch point.

(Contd...)

Rules of Block Diagram Algebra (*Contd...*)

Rule 5: Moving the summing point ahead of the block [summing point to right of gain (*SRG*)]

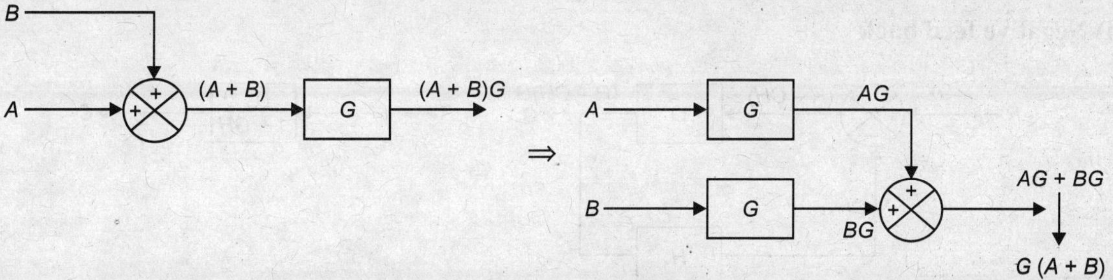

Rule 6: Moving the summing point before the block [Summing point to left of gain (*SL* 1/*G*)]

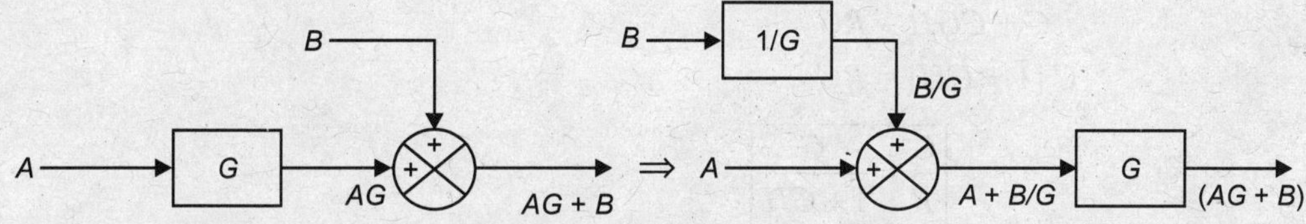

Rule 7: Interchanging summing point [No need to add any block]

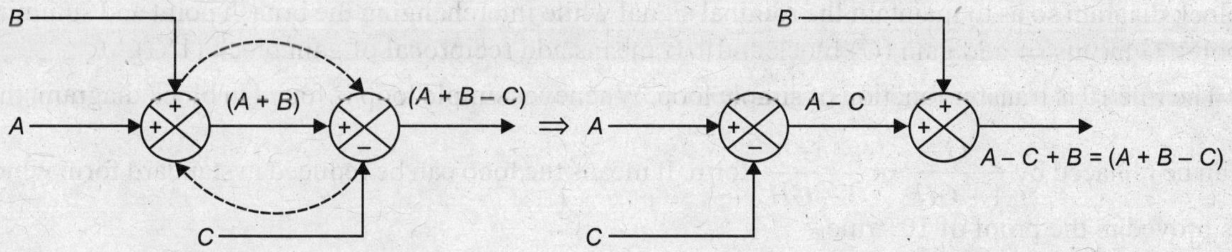

Rule 8: Splitting summing points

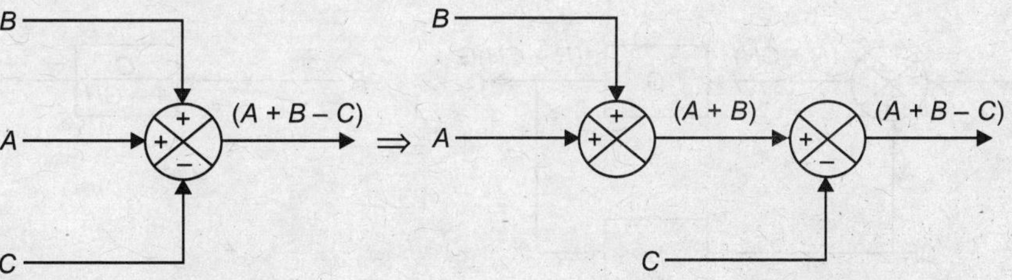

Rule 9: Combining summing points

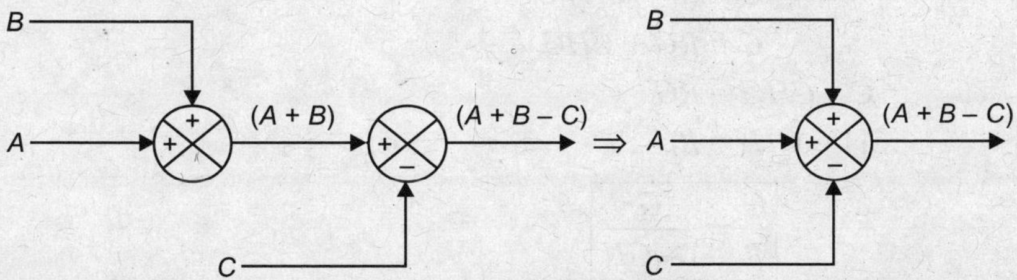

Rules of Block Diagram Algebra (*Contd...*)

Rule 10: Elimination of feedback loop [Finding transfer function for simple loop]

(i) Negative feed back

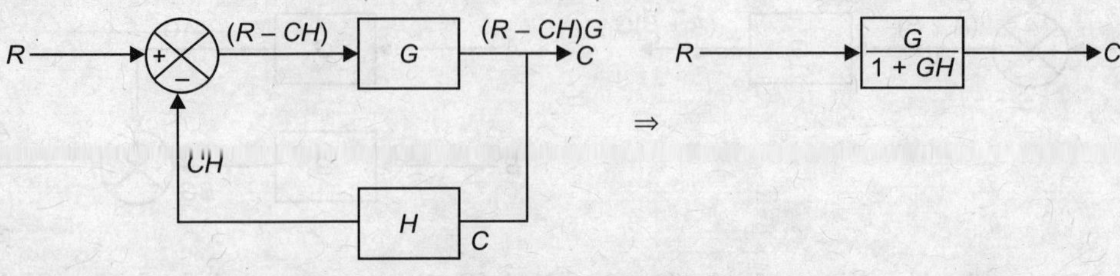

Proof:

$$C = (R - CH)\, G \Rightarrow C = RG - CGH$$

$$C + CGH = RG$$

$$C\,[1 + GH] = RG$$

$$\boxed{\frac{C}{R} = \frac{G}{1 + GH}}$$

Note: In rule 3, 4, 5 and 6, *G* and 1/*G* implies the compensation block which is to be a added to the block diagram so as to maintain the roiginal signal while interchanging the brnach point and summing point. *G* means ⇒ add gain (*G*) block and 1/*G* means add reciprocal of gain block (1/*G*).

* The rule 10 is transfer function of simple loop. Whenever simple loop is found in block diagram, that can be replaced by $\frac{G}{1+GH}$ or $\frac{G}{1-GH}$ form. It means the loop can be reduced in standard form which is proved in the proof of 10th rule.

(ii) Positive feedback

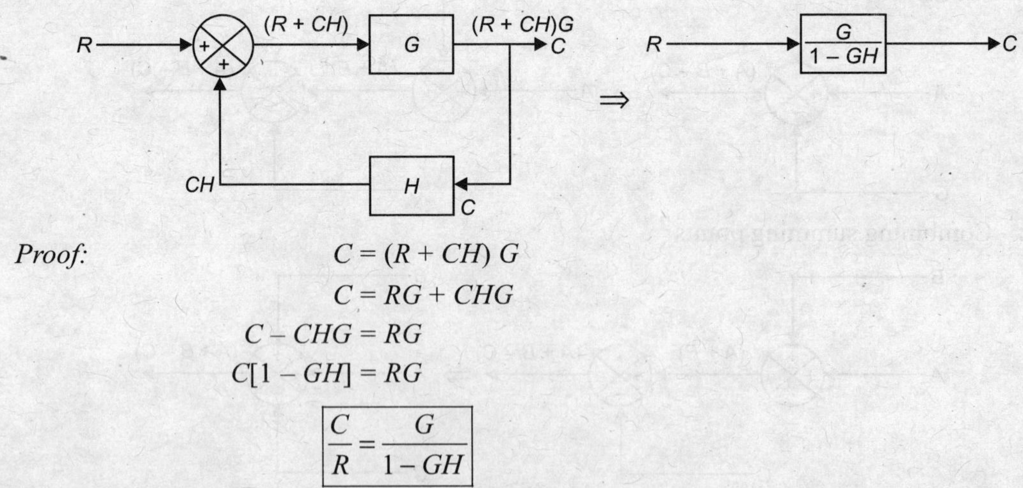

Proof:

$$C = (R + CH)\, G$$

$$C = RG + CHG$$

$$C - CHG = RG$$

$$C[1 - GH] = RG$$

$$\boxed{\frac{C}{R} = \frac{G}{1 - GH}}$$

EXAMPLE 3.12: Reduce the block diagram shown in Fig. (3.12.1) and find C/R.

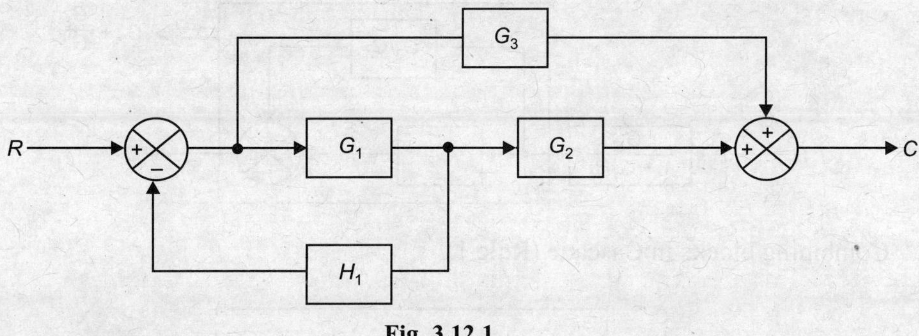

Fig. 3.12.1

Solution:

Step 1: Move the branch point after the block G_1 (Rule no. 3) [BR $1/G$]

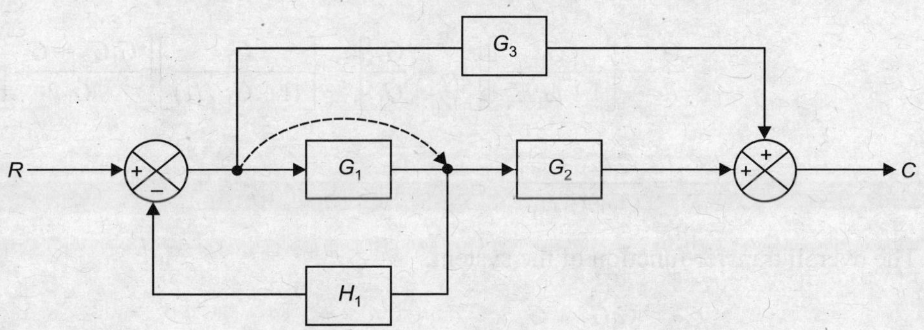

Step 2: Eliminate the feedback path and combine blocks in cascade. (Rule 1 and Rule 10)

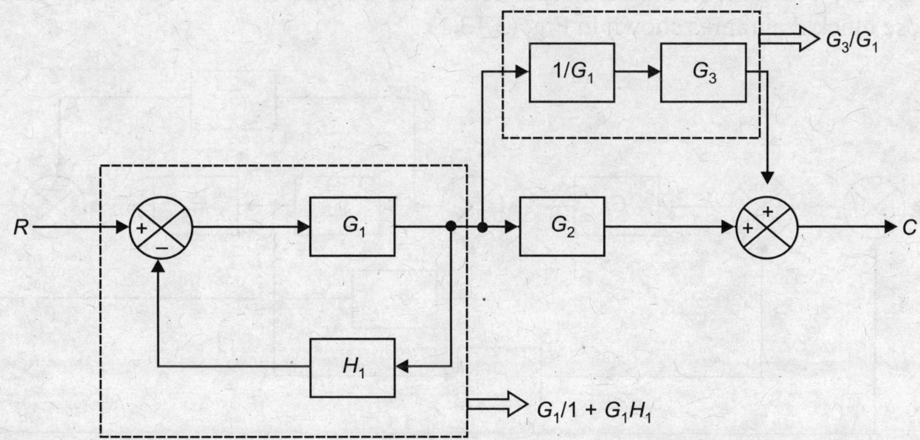

Step 3: Combining parallel blocks [Rule 2] [Addition]

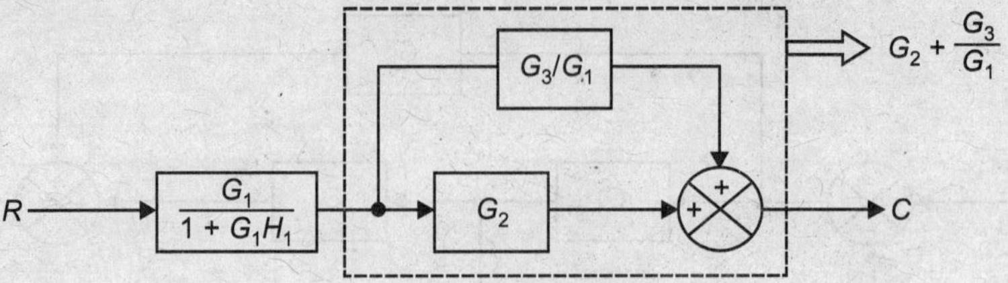

Step 4: Combining blocks in Cascade (Rule 1)

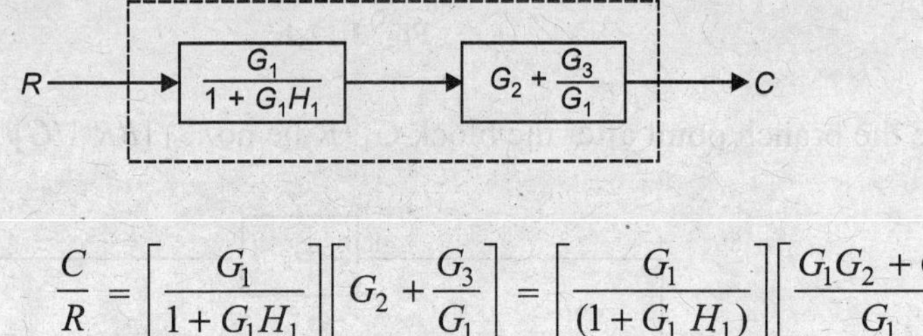

$$\frac{C}{R} = \left[\frac{G_1}{1+G_1H_1}\right]\left[G_2 + \frac{G_3}{G_1}\right] = \left[\frac{G_1}{(1+G_1\ H_1)}\right]\left[\frac{G_1G_2 + G_3}{G_1}\right]$$

$$\frac{C}{R} = \frac{G_1G_2 + G_3}{1+G_1H_1}$$

Results: The overall transfer function of the system,

$$\frac{C}{R} = \frac{G_1G_2 + G_3}{1+G_1H_1}$$

EXAMPLE 3.13: Using block diagram reduction technique find closed loop transfer function of the system whose block diagram is shown in Fig. (3.13.1).

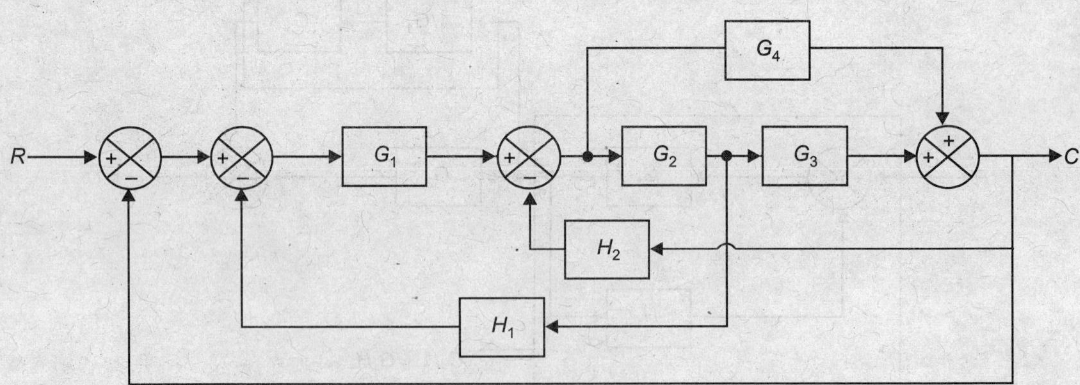

Fig. 3.13.1

Solution:

Step 1: Moving the branch point before the block G_2 (Rule 4) [*BLG*]

Note: According to rule 4, while changing the branch point before block G_2 [*BLG*], compensation block should added [G_2] with the corresponding branch poiint.

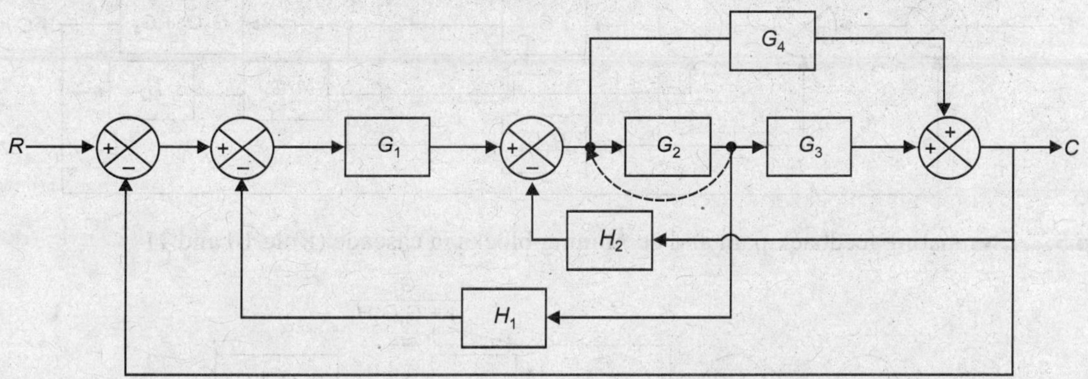

Step 2: Combining the blocks in cascade and eliminating parallel blocks [Rule 1 and 2]

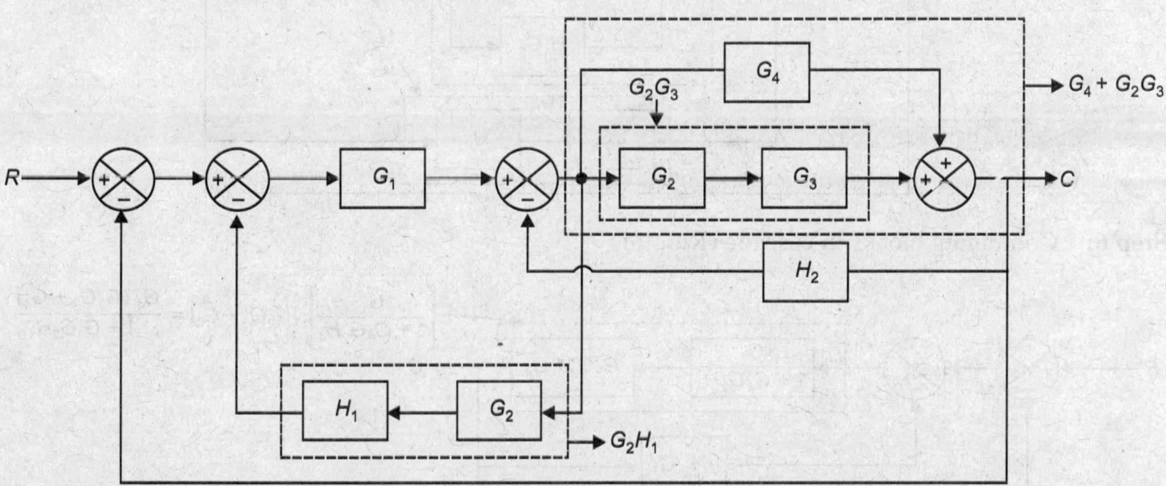

Step 3: Moving summing point before the block (Rule 6) [*SL* $1/G$]

Note: According to rule 6, while moving the summing point before block G_1, the compensation block should be added ($1/G_1$) with the summing point.

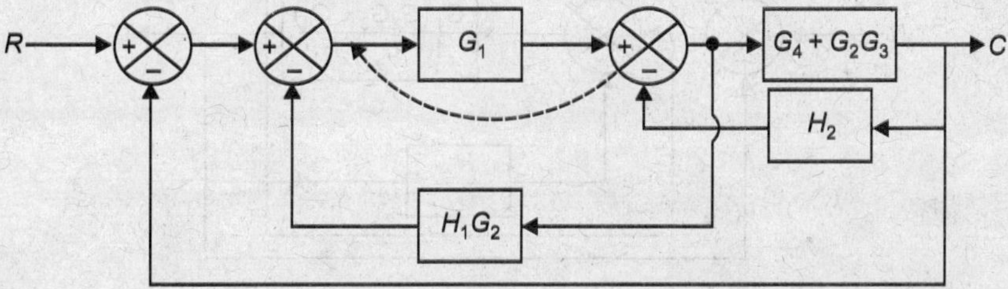

Step 4: Interchanging summing points (Rule 7) [According to rule 7, interchange summing point 2 and 3 without compensation)

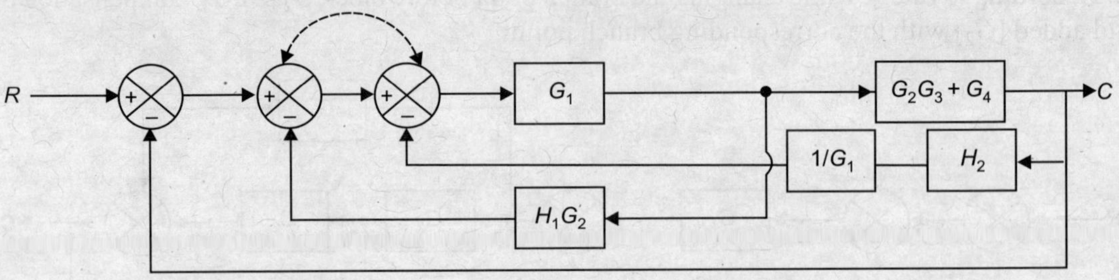

Step 5: Eliminating feedback path and combining blocks in cascade (Rule 10 and 1)

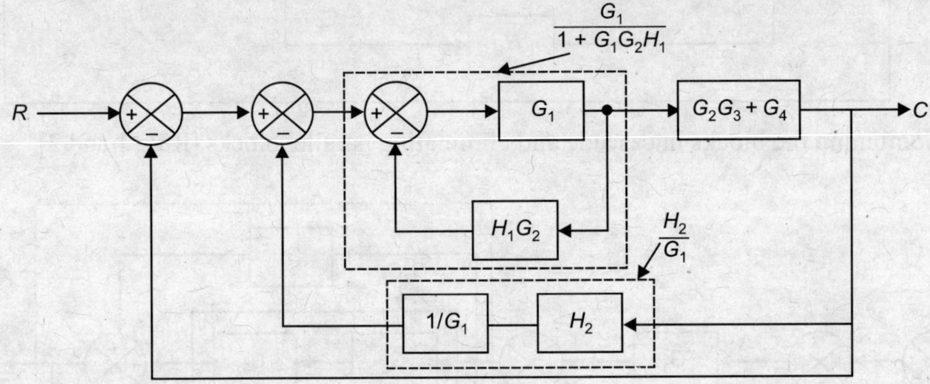

Step 6: Combining blocks in cascade (Rule 1)

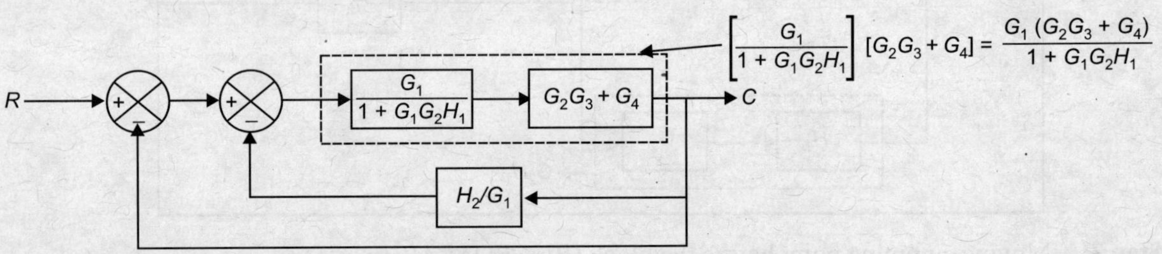

Step 7: Eliminating the feedback path (Rule 10)

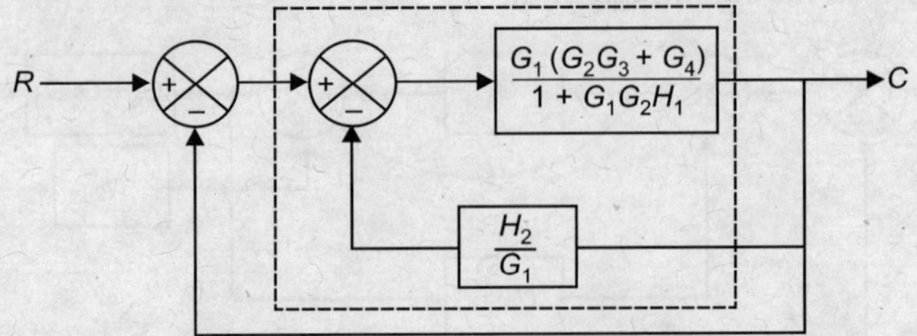

Step 8: Eliminating the feedback path

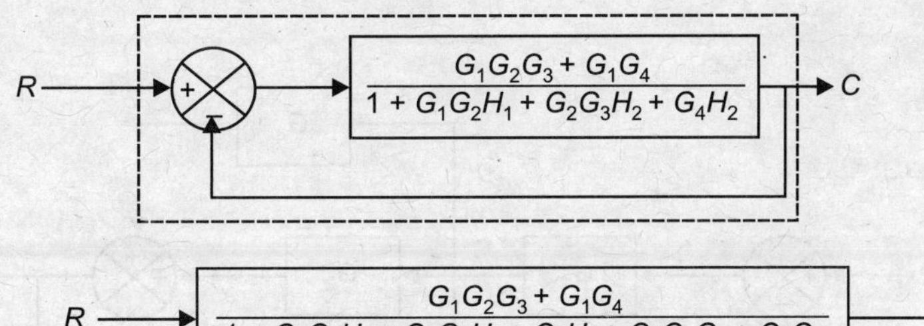

Step 9:

$$R \longrightarrow \boxed{\frac{G_1G_2G_3 + G_1G_4}{1 + G_1G_2H_1 + G_2G_3H_2 + G_4H_2 + G_1G_2G_3 + G_1G_4}} \longrightarrow C$$

Results: The overall transfer function is given by

$$\frac{C}{R} = \frac{G_1\, G_2\, G_3 + G_1\, G_4}{1 + G_1\, G_2\, H_1 + G_2\, G_3\, H_2 + G_4\, H_2 + G_1\, G_2\, G_3 + G_1\, G_4}$$

Proof for step 7

Rule 10 is

$$\frac{G}{1+GH} = \frac{\dfrac{G_1\,(G_2\, G_3 + G_4)}{1 + G_1\, G_2\, H_1}}{1 + \left(\dfrac{G_1\,(G_2\, G_3 + G_4)}{1 + G_1\, G_2\, H_1}\right)\left(\dfrac{H_2}{G_1}\right)}$$

$$\Rightarrow \quad \frac{G_1\,(G_2\, G_3 + G_4)}{1 + G_1\, G_2\, H_1 + G_2\, G_3\, H_2 + G_4\, H_2}$$

$$\Rightarrow \quad \frac{G_1\, G_2\, G_3 + G_1\, G_4}{1 + G_1\, G_2\, H_1 + G_2\, G_3\, H_2 + G_4\, H_2}$$

Proof for step 8: Here $H = 1$

$$\frac{G}{1+GH} \Rightarrow \frac{\dfrac{G_1\, G_2\, G_3 + G_1\, G_4}{1 + G_1\, G_2\, H_1 + G_2\, G_3\, H_2 + G_4\, H_2}}{1 + \dfrac{G_1\, G_2\, G_3 + G_1\, G_4}{1 + G_1\, G_2\, H_1 + G_2\, G_3\, H_2 + G_4\, H_2}}$$

$$\Rightarrow \quad \frac{\dfrac{G_1\, G_2\, G_3 + G_1\, G_4}{(1 + G_1\, G_2\, H_1 + G_2\, G_3\, H_2 + G_4\, H_2)}}{\dfrac{(1 + G_1\, G_2\, H_1 + G_2\, G_3\, H_2 + G_4\, H_2) + (G_1\, G_2\, G_3 + G_1\, G_4)}{(1 + G_1\, G_2\, H_1 + G_2\, G_3\, H_2 + G_4\, H_2)}}$$

$$\Rightarrow \quad \frac{G_1\, G_2\, G_3 + G_1\, G_4}{(1 + G_1\, G_2\, H_1 + G_2\, G_3\, H_2 + G_4\, H_2 + G_1\, G_2\, G_3 + G_1\, G_4)}$$

EXAMPLE 3.14: Consider the block diagram shown in Fig. (3.14.1) using the block diagram reduction method find C/R.

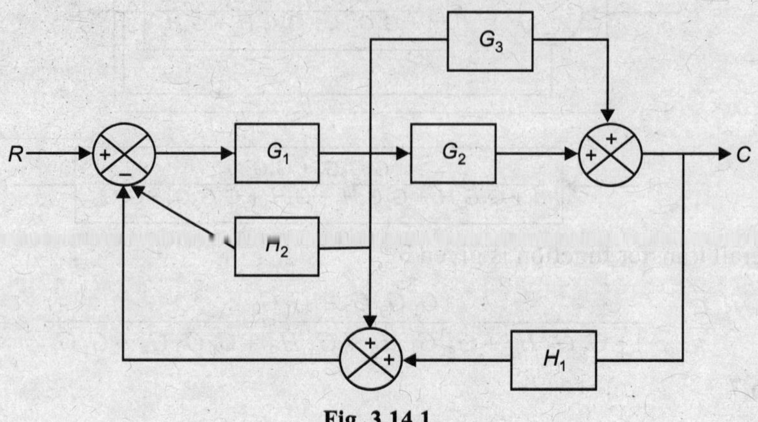

Fig. 3.14.1

Solution:

Step 1: Combining the blocks G_2 and G_3 [since two blocks are in parallel (Rule 2)]

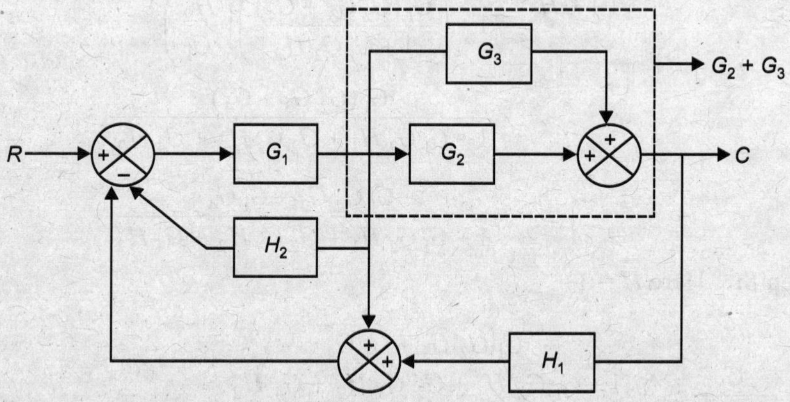

Step 2: Moving the branch point after the block $(G_2 + G_3)$ (Rule 3) $\left[BR\,\frac{1}{G}\right]$

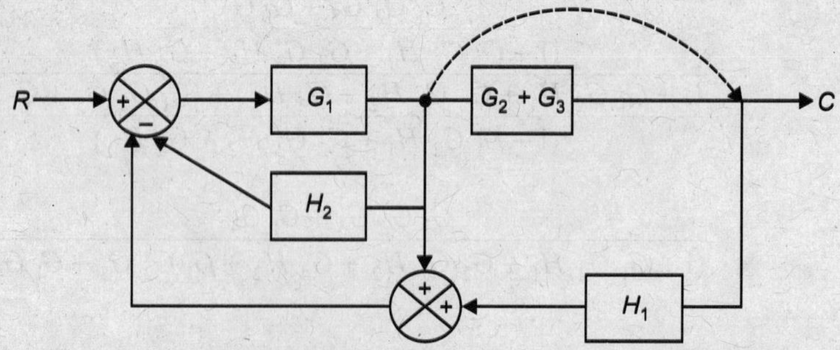

Step 3: Reducing parallel blocks and eliminating feedback path (Rule 2 and 10)

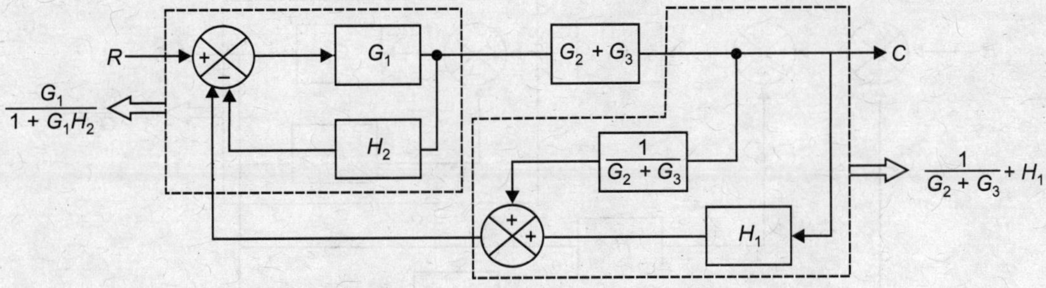

Step 4: Combining cascade blocks (Rule 1)

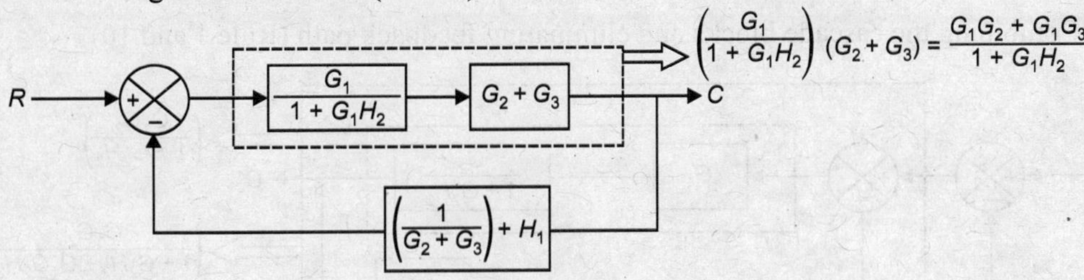

Step 5: Eliminating feedback path (Rule 10)

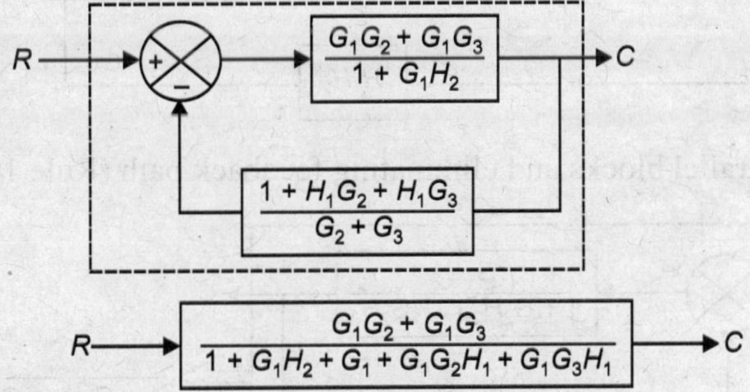

Step 6: R → $\frac{G_1G_2 + G_1G_3}{1 + G_1H_2 + G_1 + G_1G_2H_1 + G_1G_3H_1}$ → C

Result: The overall transfer function (*C*/*R*) is

$$C/R = \frac{G_1\,G_2 + G_1\,G_3}{1 + G_1\,H_2 + G_1 + G_1\,G_2\,H_1 + G_1\,G_3\,H_1}$$

EXAMPLE 3.15: Find *C*/*R* for the block shown in Fig. (3.15.1).

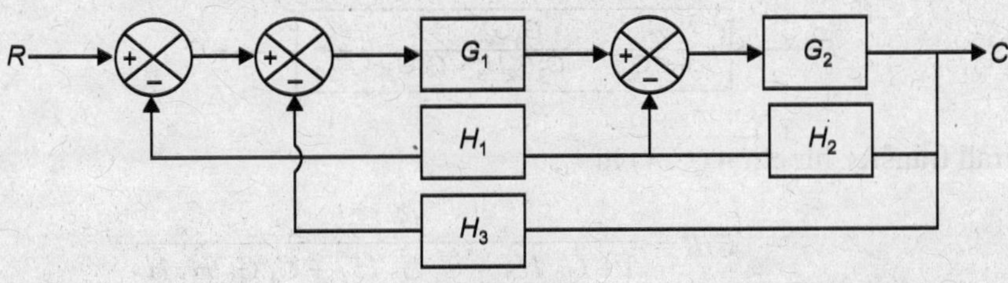

Fig. 3.15.1

CHAPTER 3

Step 1: Rearrange the feedback paths. Then eliminate feedback path (Rule 10)

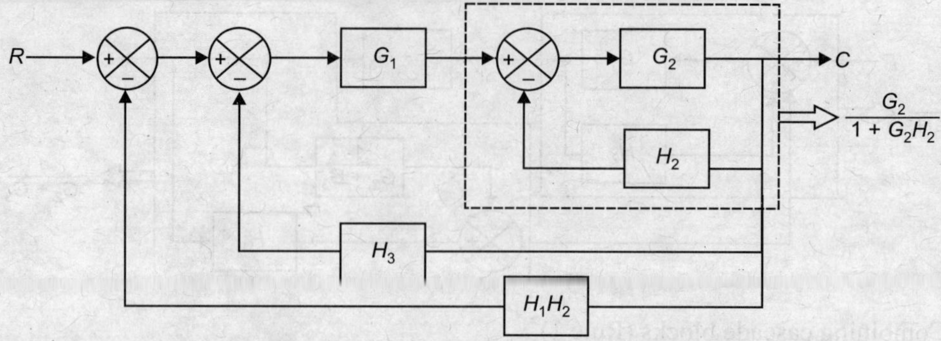

Step 2: Combining the cascade blocks and eliminating feedback path (Rule 1 and 10)

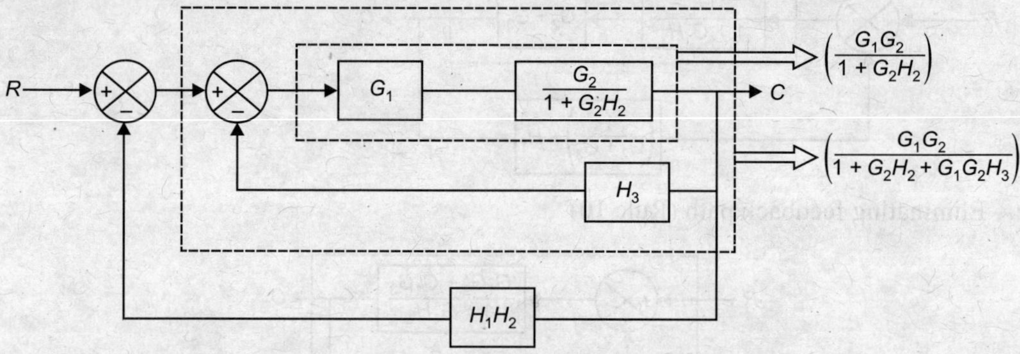

Step 3: Reducing parallel blocks and eliminating feedback path (Rule 10)

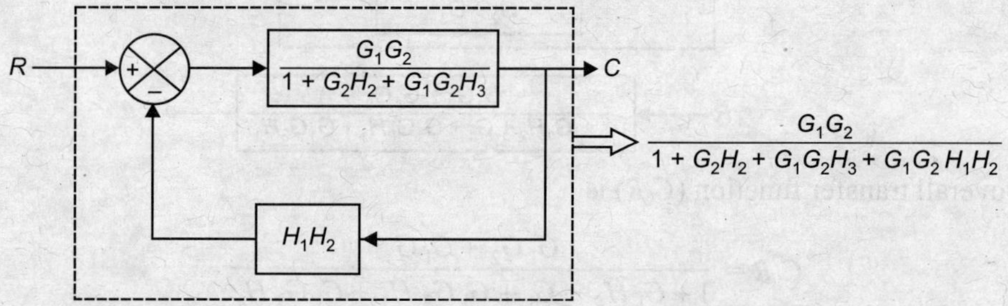

Step 4:

R → $\frac{G_1G_2}{1 + G_2H_2 + G_1G_2H_3 + G_1G_2H_1H_2}$ → C

Result: The overall transfer function (*C*/*R*) in

$$C/R = \frac{G_1\,G_2}{1 + G_2\,H_2 + G_1\,G_2\,H_3 + G_1\,G_2\,H_1\,H_2}$$

EXAMPLE 3.16: For the block diagram shown in Fig. (3.16.1) find C/R by block diagram reduction method.

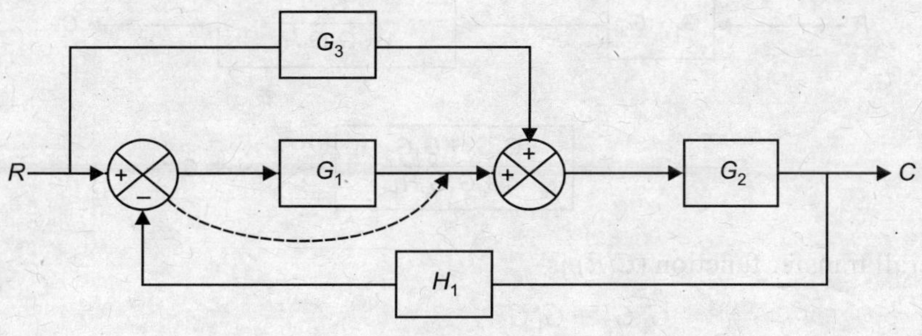

Fig. 3.16.1

Solution:

Step 1: Move the summing point after block G_1 (Rule 5) (*SRG*) and interchanging the summing point (Rule 7)

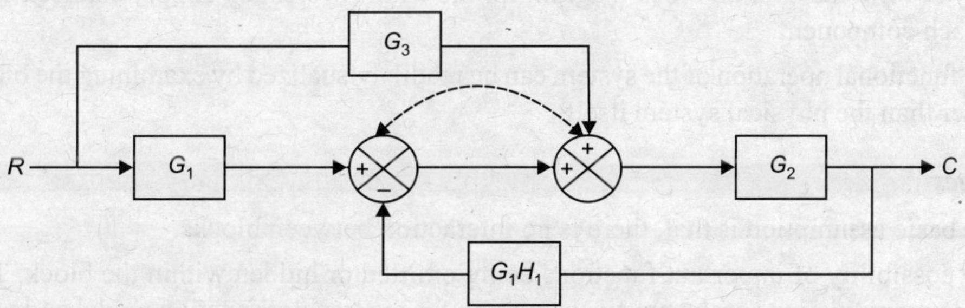

Step 2: Eliminate feedback path and parallel path (Rule 10 and 12)

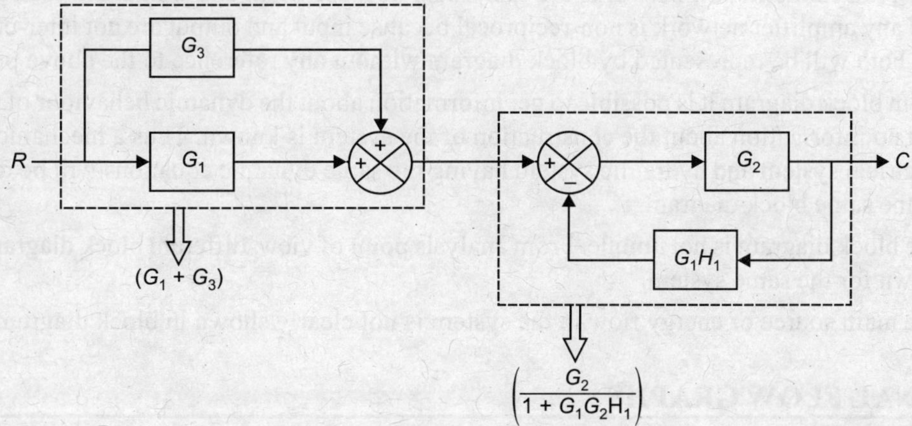

Step 3: Combine the blocks (Rule 1)

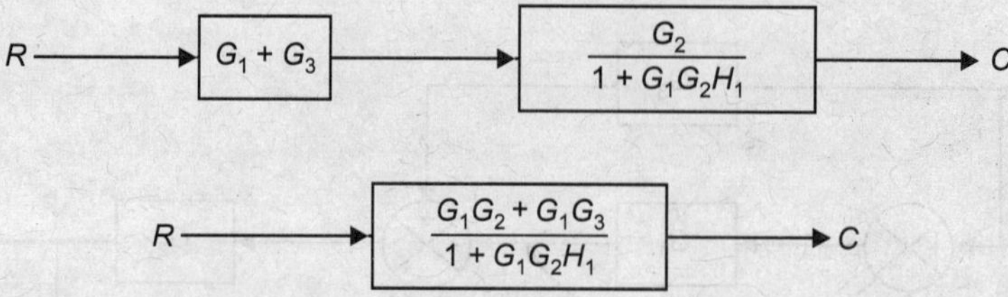

R → $\frac{G_1G_2 + G_1G_3}{1 + G_1G_2H_1}$ → C

Result: The overall transfer function (C/R) is

$$C/R = \frac{G_1 G_2 + G_1 G_3}{1 + G_1 G_2 H_1}$$

Advantages and Disadvantages of Block Diagram Representation

Advantages

1. Easy to form the over all block diagram for the whole system by simply connecting the blocks of each component.
2. The functional operation of the system can be readily visualized by examining the block diagram rather than the physical system itself.

Disadvantages

1. The basic assumption is that, there is no interaction between blocks.
2. The possibility of important functions being omitted or hidden within the block. This may be due to manipulations of the physical equation for the transfer function model and putting in the block diagram form.
3. Passive symmetrical π networks are reciprocal because input and output are interchangeable and any amplifier network is non-reciprocal because input and output are not inter-changeable, but both will be represented by block diagram without any reference to the above property.
4. From block diagram it is possible to get information about the dynamic behaviour of the system and no information about the construction of the system is known. Thus a mechanical system, electrical system and hydraulic system having the same dynamic equations will be represented by the same block diagram.
5. The block diagram is not unique. From analysis point of view, different block diagrams may be drawn for the same system.
6. The main source of energy flow in the system is not clearly shown in block diagram.

3.7 SIGNAL FLOW GRAPHS

The signal flow graph is used to represent the control system graphically and it was developed by S.J. Mason.

A signal flow graph is a diagram that represents a set of simultaneous linear algebraic equation by taking laplace transform, the time domain differential equations governing a control system can be transferred to a set of algebraic equations in s-domain. The signal flow graph of the system can be constructed using these equations.

The signal flow graph depicts the flow of signals from one point of a system to another and gives the relationships among the signals. A signal flow graph consists of a network in which nodes are connected by directed branches. Each node represents a system variable and each branch connected between two nodes acts as a signal multiplier. Each branch has a gain or transmittance when the signal pass through a branch, it gets multiplied by the gain of the branch. Note that the signal flow is only one direction. The direction of signal flow is directed by an arrow placed on the branch and the multiplication is indicated along the branch.

Definition of Terms Used in Signal Flow Graph

Node: A node is a point representing a variable or signal.

Example: ● a

Branch: A branch is directed line segment joining two nodes. The arrow on the branch indicates the direction of signal flow and the gain of a branch is known as transmittance.

Example: a ●——▶——● b (G_1)

Transmittance: The gain acquired by the signal when it travels from one node to another is called transmittance. The transmittance can be real (or) complex.

Example: a ●——▶——● b (G_1)

Here the transmittance is G_1

Input node (or) Source: It is a node that has only outgoing branches.

Example: a

Output node (or) Sink: It is a node that has only incoming branches.

Example:

Mixed node: It is a node that has both incoming and outgoing branches.

Path: A path is a traversal of connected branches in the directions of the branch arrows. The path should not cross a node more than once.

Open path: A open path starts at a node and ends at another node.

Example: a ●—▶—● b —▶—● c —▶—● d (G_1, G_2, G_3)

Closed path: Closed path starts and ends at same node.

Example: a G_1 b G_2

Forward path: It is a path from an input node to an output node that does not cross any node more than once.

Forward path gain: It is the product of the branch transmittances (gains) of a forward path.

Individual loop: It is a closed path starting from a node and after passing through a certain part of a graph arrives at the same node without crossing any node more than once.

Example:

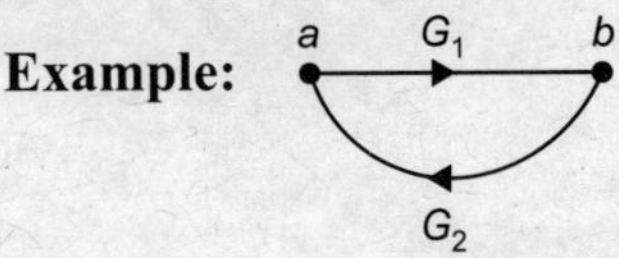

Loop gain: It is the product of the branch transmittances (gains) of a loop.

Non-touching loops: Number of loops does not have a common node, then they are said to be non-touching loops.

Example: a G_1 b G_2 c G_3 d H_1 H_2

Properties of Signal Flow Graph

The basic properties of signal flow graph are the following:

1. The algebraic equations which are used to construct signal flow graph must be in the form of cause and effect relationship.
2. It is applicable to linear system only.
3. A node represents the variable (or) signal.
4. A node adds the signals of all incoming branches and transmits the sum to all outgoing branches.
5. A mixed node which has both incoming and outgoing signals can be treated as output node by taking on outgoing branch of unity transmittance.
6. A branch indicates functional dependence of one signal on the other.
7. The signal travel along branches only in the marked direction and when it travels it gets multiplied by the gain (or) transmittance of the branch.
8. The signal flow graph of system is not unique. By rearranging the system equations different types of signal flow graphs can be drawn for a given system.

Signal Flow Graph Algebra

Signal flow graph for a system can be reduced to obtain the transfer function of the system using the following rules. Let see the rules to obtain transfer function from signal flow graph.

Rule 1: In coming signal to a node through a branch is given by the product of a signal at previous node and the gain of the branch.

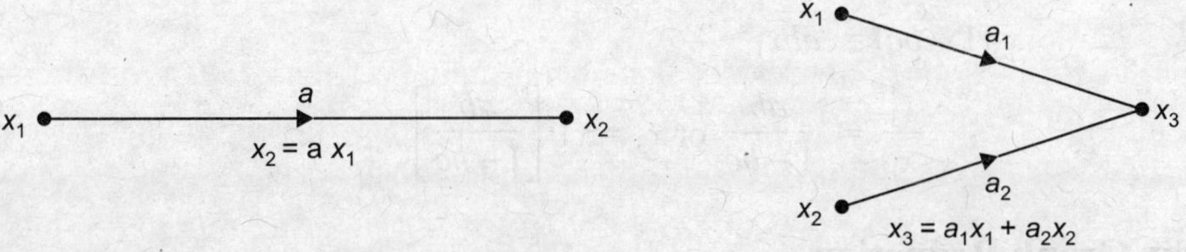

Rule 2: Cascaded branches can be combined to give a single branch whose transmittance is equal to the product of individual branch transmittance.

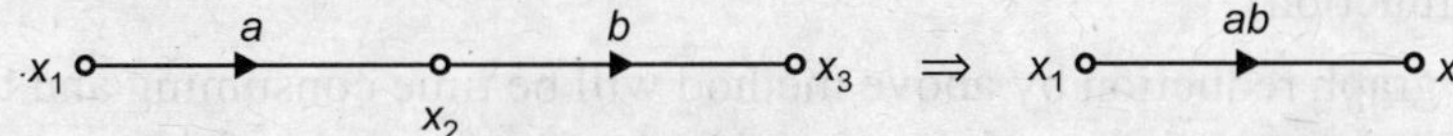

Rule 3: Parallel branches may be represented by single branch whose transmittance is the sum of individual branch transmittances.

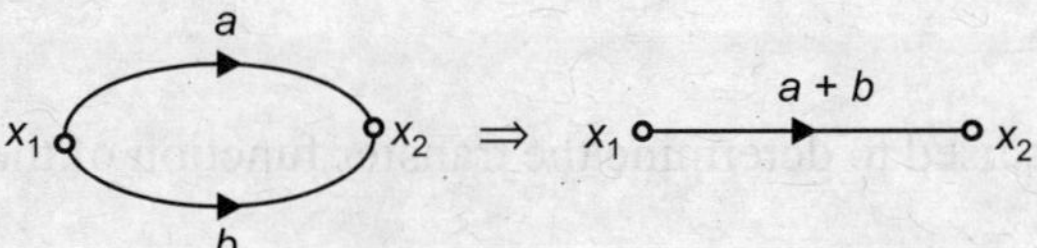

Rule 4: A mixed node can be eliminated by multiplying transmittance of outgoing branch (from the mixed node) to the transmittance of all incoming branches to the mixed node.

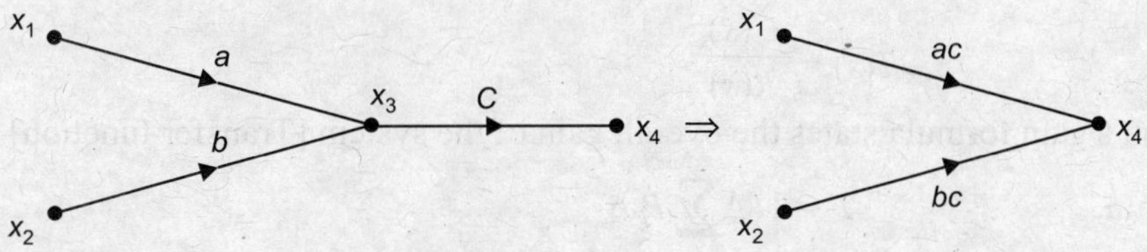

Rule 5: A loop may be eliminated by writing equations at the input and output node and rearranging the equations to find the ratio of output to input. This ratio gives the gain of resultant branch.

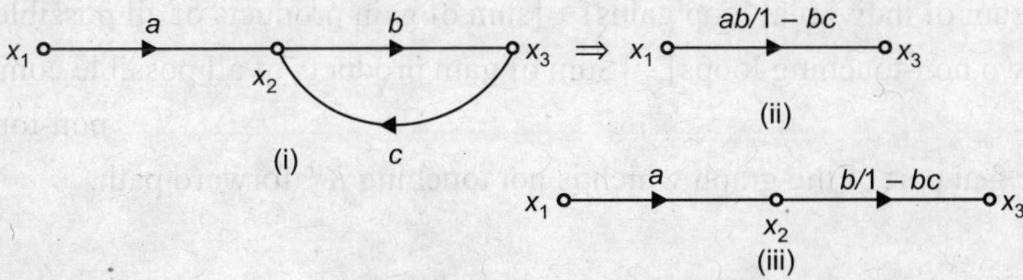

Proof:

$$x_2 = ax_1 + cx_3$$

$$x_3 = bx_2$$

Put *substitute* x_2 in equation for x_3

$$\therefore \quad x_3 = b\,[ax_1 + cx_3]$$

$$x_3 = abx_1 + bcx_3$$

$$x_3 - bcx_3 = abx_1$$

$$x_3[1 - bc] = abx_1$$

$$\frac{x_3}{x_1} = \frac{ab}{1-bc} \text{ or } x_3 = x1\left[\frac{ab}{1-bc}\right]$$

Signal Flow Graph Reduction

The signal flow graph of a system can be reduced either by using the rules of a signal flow graph algebra (i.e.) by writing equations at every node and then rearranging these equations to get the ratio of output and input (Transfer function).

The signal flow graph reduction by above method will be time consuming and tedious S.J. Mason have developed a simple procedure to determine the transfer function of the system represented as a signal flow graph. He has developed a formula called by his name Mason's gain formula which can be directly used to find the transfer function of the system.

Mason's Gain Formula

The Mason's gain formula is used to determine the transfer function of the system from the signal flow graph of the system.

Let, $R(s)$ = Input of the system

and $C(s)$ = Output of the system

Transfer function of the system

$$T(s) = \frac{C(s)}{R(s)}$$

The Mason's gain formula states the overall gain of the system [Transfer function] as follows

Overall gain, $$T = 1/\Delta \sum_k P_k \Delta_k$$

where, $T = T(s)$ = Transfer function of the system

P_K = Forward path gain of Kth forward path

Δ = 1 – (sum of individual loop gains) + [sum of gain products of all possible combinations of two non-touching loops] – [Sum of gain products of all possible combinations of three non-touching loops] + ...

Δ_K = Δ for that part of the graph which is not touching K^{th} forward path.

Construction of Signal Flow Graph for Control Systems

A control system can be represented diagrammatically by signal flow graph. The differential equations governing the system are used to construct the signal flow graph. By taking laplace transform the differential equations can be converted to algebraic equations. The constants and variables of the equations are identified. From the working knowledge of the system, the variables are identified as input, output and intermediate variables. For each variable a node is assigned in signal flow graph and constants are the gain or transmittance of the branches connecting the nodes. For each equation a signal flow graph and constants are the gain or transmittance of the branches connecting the nodes. For each equation a signal flow graph is drawn and then they are interconnected to give overall signal flow graph of the system

EXAMPLE 3.17: Construct a signal flow graph for armature controlled dc motor.

Solution: The differential equations governing the armature controlled dc motor (Refer section 3.3.5)

$$V_a = i_a R_a + L_a \frac{di_a}{dt} + e_b \tag{3.17.1}$$

$$T = K_t i_a \tag{3.17.2}$$

$$T = J\frac{d\omega}{dt} + B\omega \tag{3.17.3}$$

$$e_b = K_b\omega \tag{3.17.4}$$

whcre,

$$\omega = \frac{d\theta}{dt} \tag{3.17.5}$$

on taking laplace transform of Eq. (3.17.1 to 3.17. 5)

$$V_a(s) = I_a(s)\,[R_a + s\,La] + E_b\,(s) \tag{3.17.6}$$

$$T(s) = K_t\,I_a(s) \tag{3.17.7}$$

$$T(s) = J_s\,\omega(s) + B\omega(s) \tag{3.17.8}$$

$$E_b(s) = K_b\,\omega(s) \tag{3.17.9}$$

$$\omega(s) = s\,\theta(s) \tag{3.17.10}$$

The input and output variables of armature controlled dc motor are armature voltage $V_a(s)$ and angular displacement $\theta(s)$ respectively. The variables $I_a(s)$, $T(s)$, $E_b(s)$ and $\omega(s)$ are intermediate variables.

The Eq. (3.17.6) to (3.17.10) are rearranged and individual signal flow graph are shown in Fig. (3.17.4 to 3.17.5)

From Eq. (3.17.6)

$$V_a(s) = E_b(s) = I_a(s)\,[R_a + s\,L_a]$$

$$I_a(s) = \frac{1}{[R_a + s\,L_a]}[V_a(s) - E_b(s)]$$

CHAPTER 3

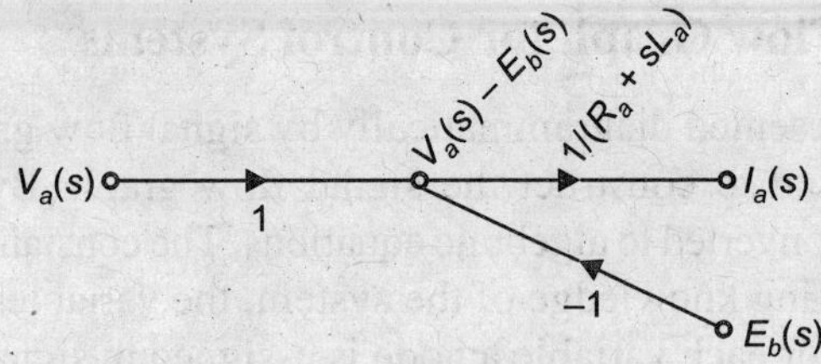

Fig. 3.17.1

from Eq. (3.17.7)

$$T(s) = K_t\, I_a(s)$$

$I_a(s)$ —— Kt ——→ $T(s)$

Fig. 3.17.2

from Eq. (3.17.8)

$$T(s) = \omega(s)\ [Js + B]$$

$$\omega(s) = \frac{1}{Js + B}\ T(s)$$

$T(s)$ —— $1/Js + B$ ——→ $\omega(s)$

Fig. 3.17.3

from Eq. (3.17.9)

$$E_b(s) = K_b\ \omega(s)$$

$\omega(s)$ —— K_b ——→ $E_b(s)$

Fig. 3.17.4

from Eq. (3.17.10)

$$\theta(s) = \frac{1}{s}\ \omega(s)$$

$\omega(s)$ —— $1/s$ ——→ $\theta(s)$

Fig. 3.17.5

The overall signal flow graph of armature controlled dc motor is obtained by interconnecting the individual signal flow graphs. The overall signal flow graph is shown in Fig. 3.17.6

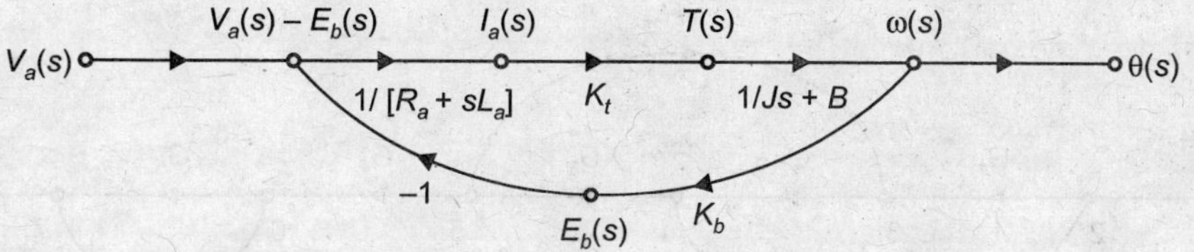

Fig. 3.17.6: Signal flow graph of armature controlled dc motor.

EXAMPLE 3.18: Construct a signal flow graph for field controlled dc motor

Solution: The differential equations governing the field controlled dc motor (Refer Section 3.3.5)

$$V_f = R_f i_f + L_f \frac{di_f}{dt} \tag{3.18.1}$$

$$T = K_{tf} i_f \tag{3.18.2}$$

$$T = J \frac{d^2\theta}{dt^2} + B \frac{d\theta}{dt} \tag{3.18.3}$$

on taking laplace transform of Eq. (3.18.1 to 3.18.3) we get,

from Eq. (3.18.1)

$$V_f(s) = R_f I_f(s) + L_f s\, I_f(s)$$

$$V_f(s) = V_f(s)/ [R_f + s\, L_f]$$

$$I_f(s) = V_f(s)/ (R_f + s\, L_f) \tag{3.18.4}$$

from Eq. (3.18.2)

$$T(s) = K_{tf} I_f(s) \tag{3.18.5}$$

from Eq. (3.18.3)

$$T(s) = Js^2\, \theta(s) + Bs\, \theta(s)$$

$$\theta(s) = \frac{1}{Js^2 + B} T(s) \tag{3.18.6}$$

From Eq. (3.18.4, 3.18.5 and 3.18.6)

$I_f(s)$ $T(s)$

$V_f(s)$ → $1/[R_f + sL_f]$ → K_{tf} → $1/Js^2 + B$ → $\theta(s)$

Fig. 3.18.1

The Fig. (3.18.1) shows the overall signal flow graph of field controlled dc motor.

EXAMPLE 3.19: Find the overall transfer function of the system whose signal flow graph is shown in Fig. (3.19.1)

CHAPTER 3

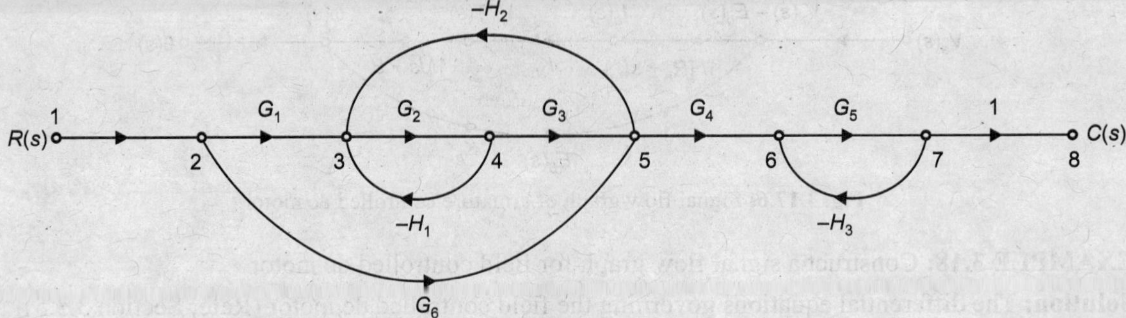

Fig. 3.19.1

Solution:

Step 1: Forward path gains

There are two forward paths. $\therefore K = 2$

Let forward path gains be P_1 and P_2

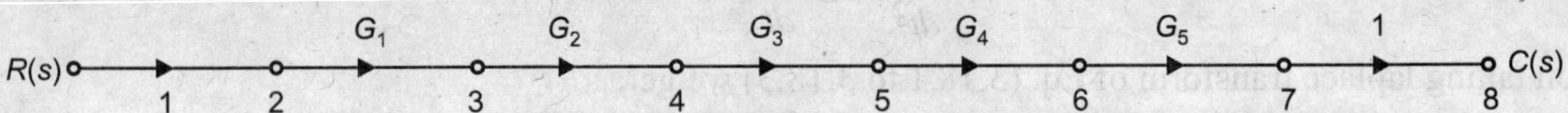

Fig. 3.19.2: Forward path-1

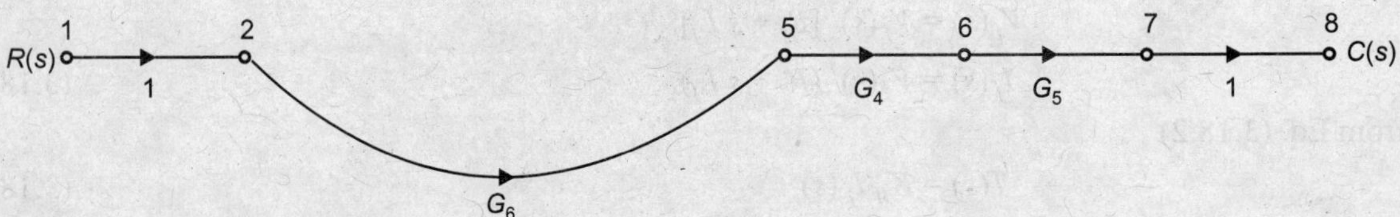

Fig. 3.19.3: Forward path-2.

Gain of forward path – 1, $P_1 = G_1\ G_2\ G_3\ G_4\ G_5$

Gain of forward path – 2, $P_2 = G_4\ G_5\ G_6$

Step 2: Individual Loop Gain

There are three individual loops. Let individual loop gains be P_{11}, P_{21} and P_{31}.

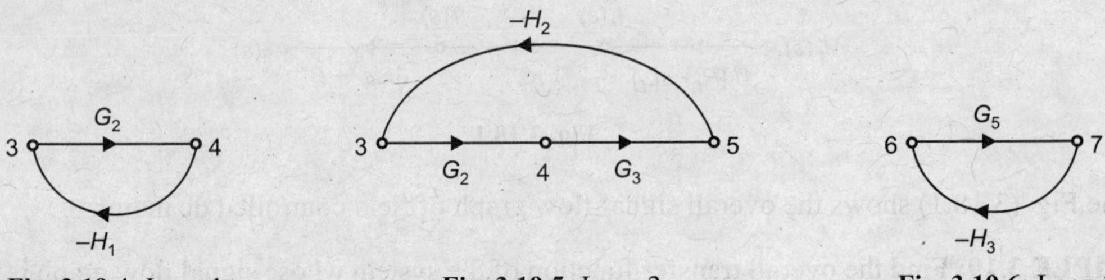

Fig. 3.19.4: Loop 1 **Fig. 3.19.5:** Loop 2 **Fig. 3.19.6:** Loop 3

Loop gain of individual loop – 1, $P_{11} = -\ G_2\ H_1$

Loop gain of individual loop – 2, $P_{21} = - G_2 G_3 H_2$

Loop gain of individual loop – 3, $P_{31} = - G_5 H_3$

Step 3: Gain Products of Two Non-Touching Loops

There are two combinations of two non-touching loops. Let the gain products of two non-touching loops be P_{12} and P_{22}

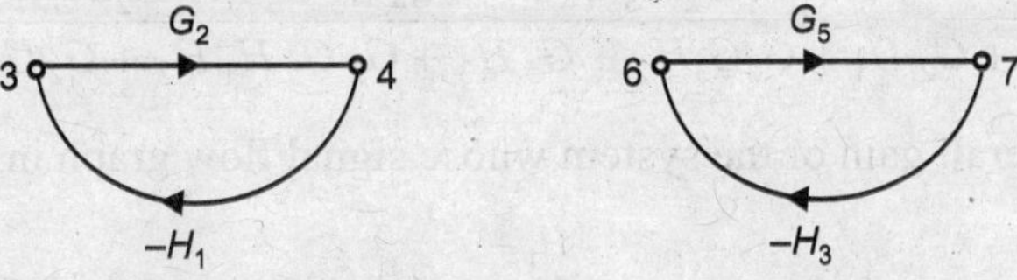

Fig. 3.19.7: First combination of two non-touching loops

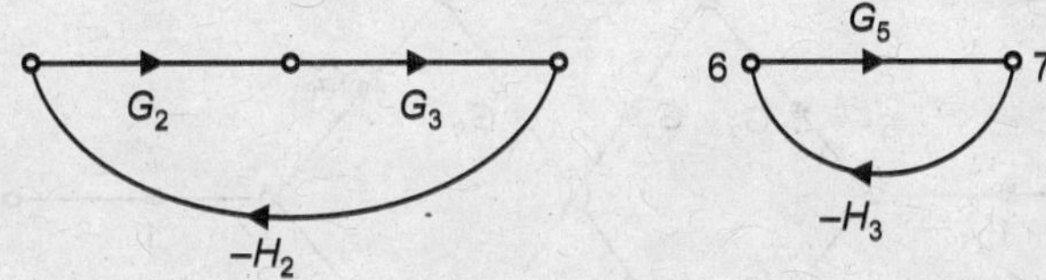

Fig. 3.19.8: Second combination of two non-touching loops

Gain product of first combination of two non touching loops

$$P_{12} = P_{11} P_{31} = (- G_2 H_1)(- G_5 H_3) = G_2 G_5 H_1 H_3$$

Gain product of second combination of two non touching loops

$$P_{22} = P_{21} P_{31} = (- G_2 G_3 H_2)(- G_5 H_5) = G_2 G_3 G_5 H_2 H_3$$

Calculation of Δ and ΔK

$$\Delta = 1 - (P_{11} + P_{21} + P_{31}) + (P_{12} + P_{22})$$

$$\Delta = 1 + G_2 H_1 + G_2 G_3 H_2 + G_5 H_3 + G_2 G_5 H_1 H_3 + G_2 G_3 G_5 H_2 H_3$$

$\Delta_1 = 1$ [Since there is no part of graph which is not touching with first forward path]

The part of the graph which is non touching with second forward path is shown in Fig. (3.22.8)

$$\Delta_2 = 1 - P_{11} = 1 - (- G_2 H_1)$$

$$\Delta_2 = 1 + G_2 H_1$$

G2
3
4
–H1

Fig. 3.19.9

Transfer Function, *T*

By Mason's gain formula the transfer function

$$T = \frac{1}{\Delta}\sum_k P_k \Delta_k$$

$$= \frac{1}{\Delta}[P_1\Delta_1 + P_2\Delta_2] \quad \text{[Here } K = 2\text{, since we have only two forward paths]}$$

$$\therefore \quad T = \frac{G_1 G_2 G_3 G_4 G_5 + G_4 G_5 G_6 (1 + G_2 H_1)}{1 + G_2 H_1 + G_2 G_3 H_2 + G_5 H_3 + G_2 G_5 H_1 H_3 + G_2 G_3 G_5 H_2 H_3}$$

EXAMPLE 3.20: Find overall gain of the system whose signal flow graph in shown in Fig. (3.20.1)

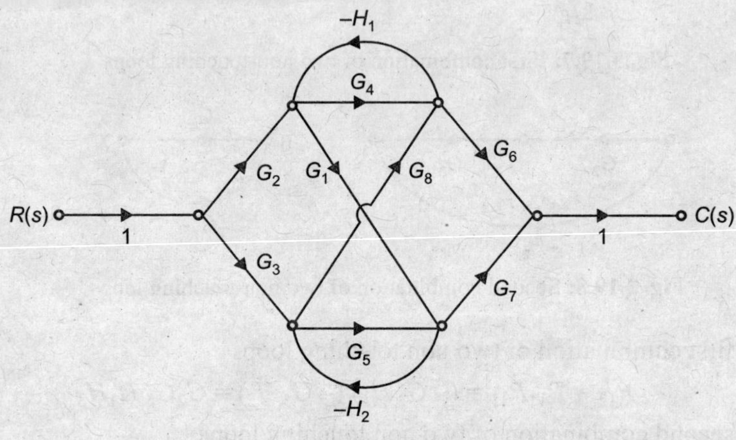

Fig. 3.20.1

Solution: Let us number the node as shown in Fig. (3.20.2)

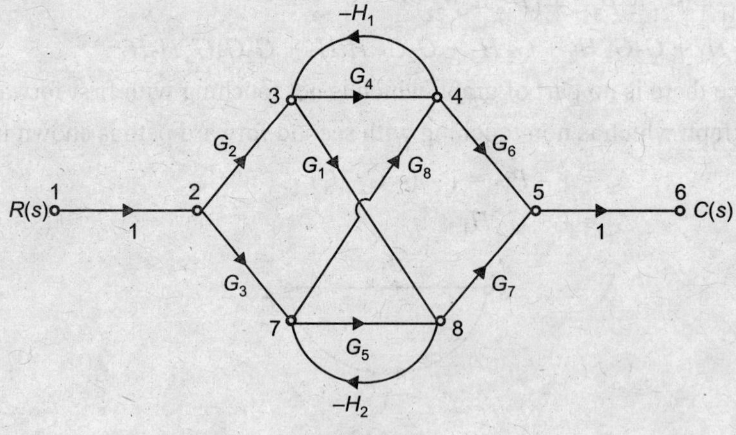

Fig. 3.20.2

Forward Path Gains

There are six forward path. $\therefore K = 6$.

Let the forward path gains be P_1, P_2, P_3, P_4, P_5 and P_6.

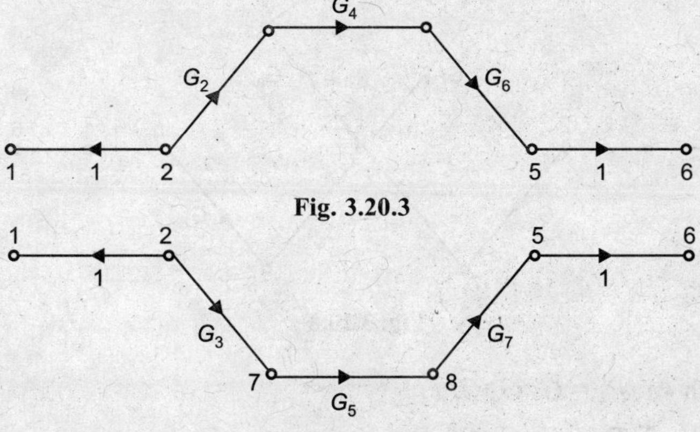

Fig. 3.20.3

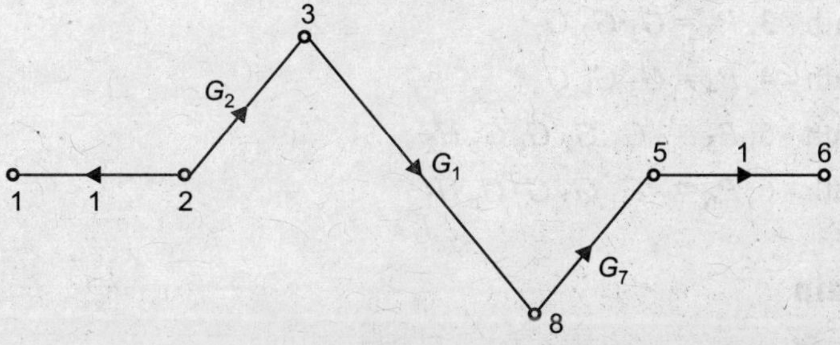

Fig. 3.20.4

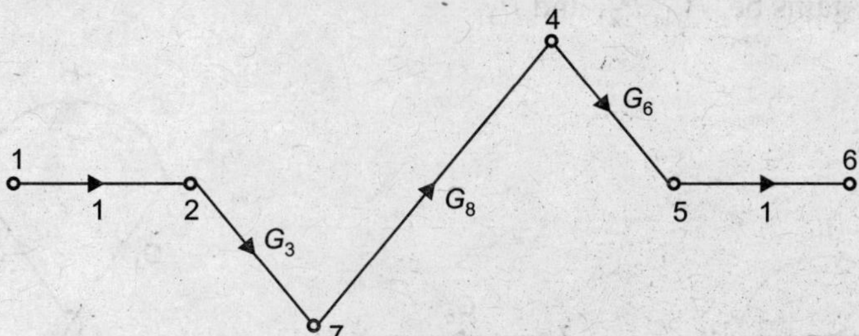

Fig. 3.20.5

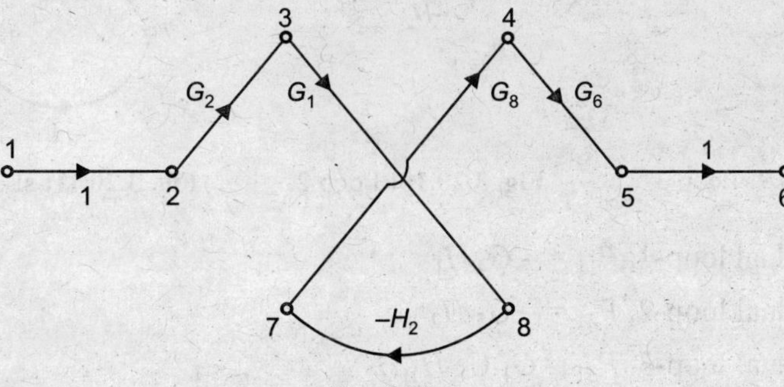

Fig. 3.20.6

Fig. 3.20.7

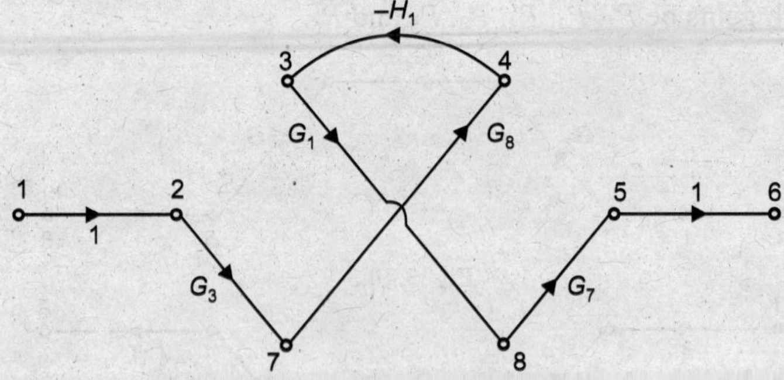

Fig. 3.20.8

Gain of forward path - 1, $P_1 = G_2\ G_4\ G_6$

Gain of forward path - 2, $P_2 = G_3\ G_5\ G_7$

Gain of forward path - 3, $P_3 = G_1\ G_2\ G_7$

Gain of forward path - 4, $P_4 = G_3\ G_8\ G_6$

Gain of forward path - 5, $P_5 = -G_1\ G_2\ G_6\ G_8\ H_2$

Gain of forward path - 6, $P_6 = -G_1\ G_3\ G_7\ G_8\ H_1$

Individual Loop Gain

There are three individual loops.

Let individual loop gains be, P_{11}, P_{21} and P_{31}

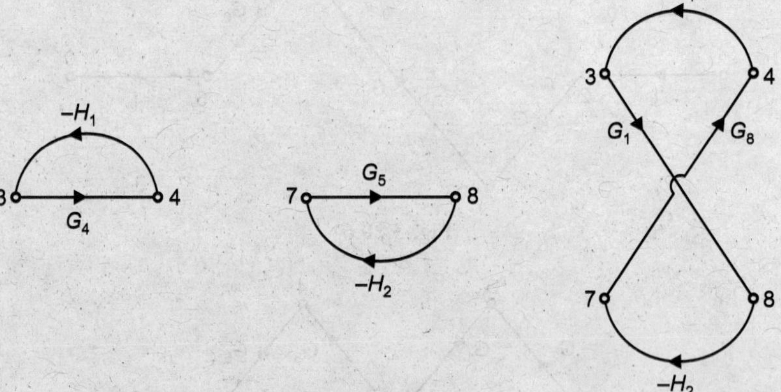

Fig. 3.20.9: Loop 3. **Fig. 3.20.10:** Loop 2. **Fig. 3.20.11:** sLoop 3.

Loop gain of individual loop-1, $P_{11} = -\ G_4\ H_1$

Loop gain of individual loop-2, $P_{21} = -\ G_5\ H_2$

Loop gain of individual loop-3, $P_{31} = G_1\ G_8\ H_1\ H_2$

Gain Products of Two non-Touching Loops

There is only one combination of two non-touching loops. Let the gain product of two non-touching loops be P_{12}.

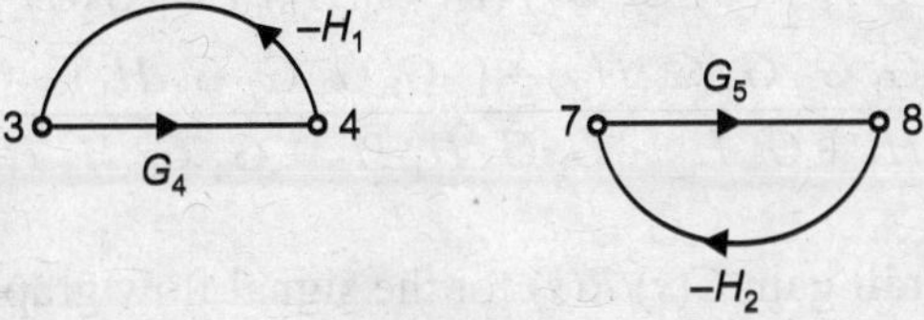

Fig. 3.20.12: First combination of two non-touching loops.

Gain product of first combination of two non-touching loops

$$P_{12} = P_{11}\, P_{21} = (-\,G_4\, H_1)\,(-\,G_5\, H_2) \Rightarrow G_4G_5H_1H_2$$

Calculation of Δ and ΔK

$$\begin{aligned}\Delta &= 1 - (P_{11} + P_{21} + P_{31}) + P_{12}\\ &= 1 - (-G_4\, H_1 - G_5\, H_2 + G_1\, G_8\, H_1\, H_2) + (G_4\, G_5\, H_1\, H_2)\\ &= 1 + G_4\, H_1 + G_5\, H_2 - G_1\, G_8\, H_1\, H_2 + G_4\, G_5\, H_1\, H_2\end{aligned}$$

The part of the graph non-touching forward path –1 is shown in Fig. (3.20.13)

$\therefore$ $$\Delta_1 = 1 - (-G_5\, H_2) = 1 + G_5\, H_2$$

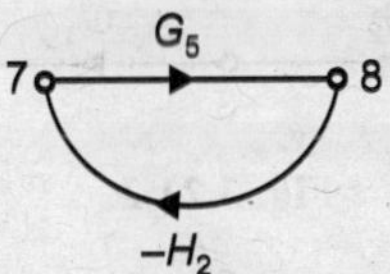

Fig. 3.20.13

The part of the graph non-touching forward path–2 is shown in Fig. (3.20.14)

$\therefore$ $$\Delta_2 = 1 - (-G_4\, H_1) = 1 + G_4\, H_1$$

Fig. 3.20.14

There is no part of the graph which is non-touching with forward paths 3, 4, 5, and 6

$\therefore$ $$\Delta_3 = \Delta_4 = \Delta_5 = \Delta_6 = 1$$

Transfer Function

By Mason's gain formula, $T = \dfrac{1}{\Delta}\left[\sum_k P_k\, \Delta_k\right]$

$$= \frac{1}{\Delta}\,[P_1\Delta_1 + P_2\Delta_2 + P_3\Delta_3 + P_4\Delta_4 + P_5\Delta_5 + P_6\Delta_6]$$

(Here $K = 6$, since we have six forward paths)

$$T = \frac{(G_2\,G_4\,G_6)\,(1 + G_5\,H_2) + (G_3\,G_5\,G_7)\,(1 + G_4\,H_1) + G_1\,G_2 G_7 + G_3\,G_8\,G_6 + (-G_1\,G_2\,G_6\,G_8\,H_2) + (-G_1\,G_3\,G_7\,G_8\,H_1)}{1 + G_4\,H_1 + G_5\,H_2 - G_1\,G_8\,H_1\,H_2 + G_4\,G_5\,H_1\,H_2}$$

EXAMPLE 3.21: Find the overall gain $C(s)/R(s)$ for the signal flow graph shown in Fig. (3.21.1)

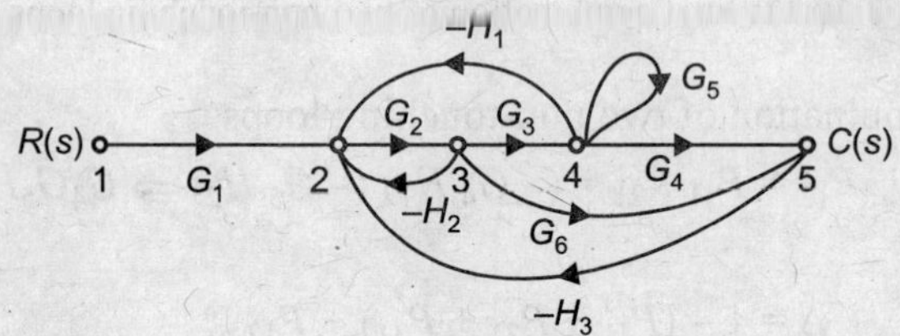

Fig. 3.21.1

Solution: Forward Path Gains

There are two forward paths. $\therefore\ K = 2$

Let the forward path gains be P_1 and P_2

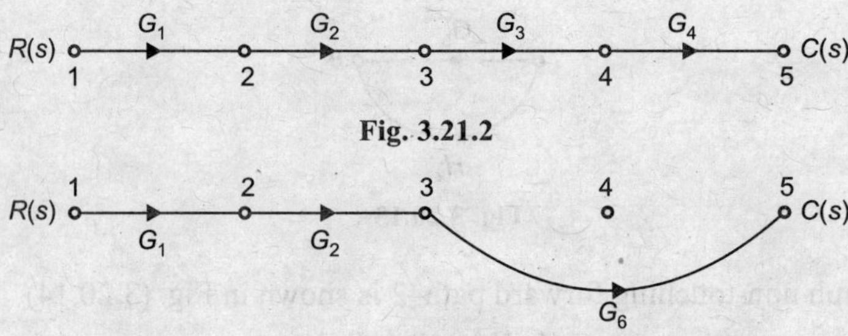

Fig. 3.21.2

Fig. 3.21.3

Gain of forward path-1 $P_1 = G_1\,G_2\,G_3\,G_4$

Gain of forward path-2 $P_2 = G_1\,G_2\,G_6$

Individual Loop Gain

There are five individual loops. Let the individual loop gains be P_{11}, P_{21}, P_{31}, P_{41} and P_{51}

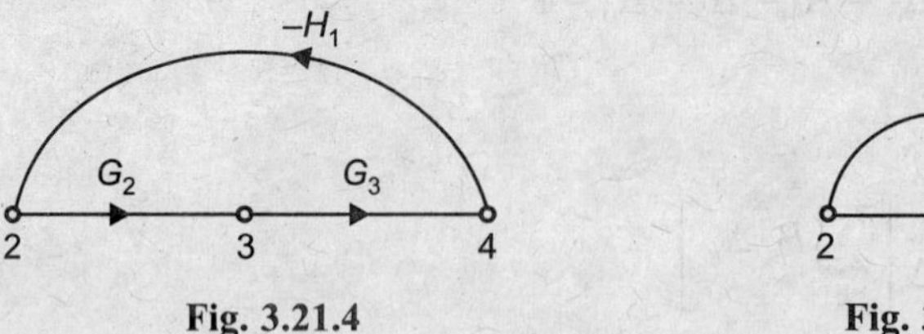

Fig. 3.21.4

Fig. 3.21.5

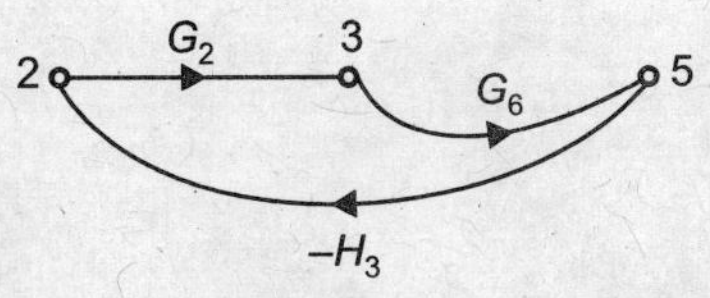

Fig. 3.21.6

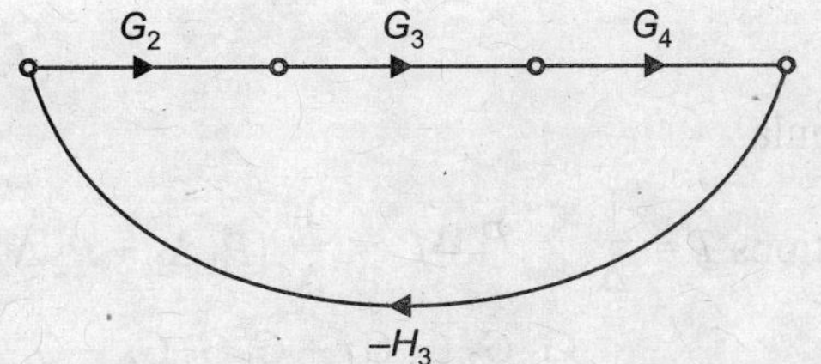

Fig. 3.21.7

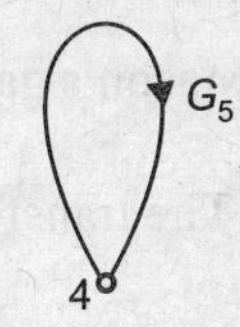

Fig. 3.21.8

Loop gain of individual loop-1, $P_{11} = -\, G_2\, G_3\, H_1$

Loop gain of individual loop-2, $P_{21} = -\, H_2\, G_2$

Loop gain of individual loop-3, $P_{31} = -\, G_2\, G_6\, H_3$

Loop gain of individual loop-4, $P_{41} = -\, G_2\, G_3\, G_4\, H_3$

Loop gain of individual loop-5, $P_{51} = G_5$

Gain Products of Two Non-touching Loops

There are two combinations of two non-touching loops.

Let the gain products of two non-touching loops be P_{12} and P_{22}.

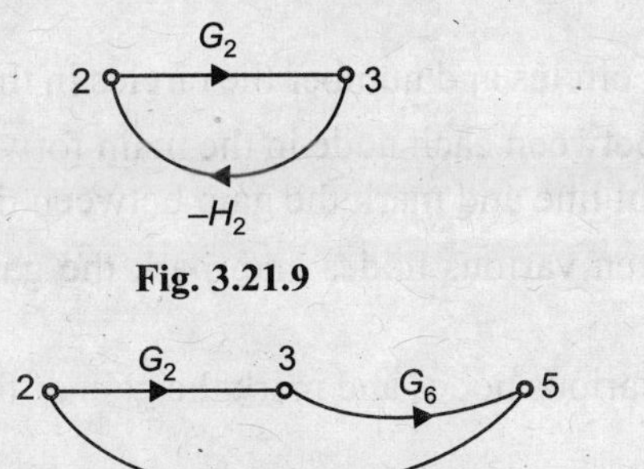

Fig. 3.21.9

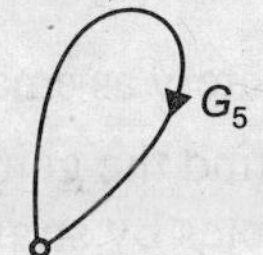

Fig. 3.21.10

Fig. 3.21.11

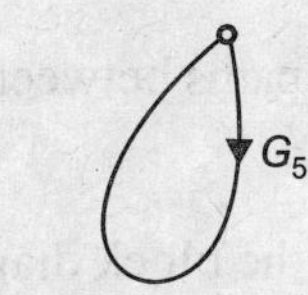

Fig. 3.20.12

Gain product of first combination of two non-touching loops

$$P_{12} = P_{21}\, P_{51} = (-\, G_2 H_2)\,(G_5) = -\, G_2\, G_5\, H_2$$

Gain product of second combination of two non-touching loops

$$P_{22} = P_{31}\, P_{51} = (-\, G_2\, G_6\, H_3) = -\, G_2\, G_5\, G_6\, H_3$$

Calculation of Δ and ΔK

$$\Delta = 1 - (P_{11} + P_{21} + P_{31} + P_{41} + P_{51}) + (P_{12} + P_{22})$$

$$= 1 - (-G_2\, G_3\, H_1 - H_2\, G_2 - G_2\, G_3\, G_4\, H_3 + G_5 - G_2\, G_6\, H_3) + \ (-\, G_2\, H_2\, G_5 - G_2\, G_5\, G_6\, H_3)$$

Since there is no part of graph which is not touching forward Path-1.

$\therefore$
$$\Delta = 1$$

$$\Delta_2 = 1 - G_5$$

Transfer Function

By Mason's gain formula

$$\text{The transfer functions } T = \frac{1}{\Delta}\sum_k P_k\,\Delta_k = \frac{1}{\Delta}\,[P_1\,\Delta_1 + P_2\,\Delta_2]$$

$$T = \frac{G_1\,G_2\,G_3\,G_4 + G_1\,G_2\,G_6 - G_1\,G_2\,G_5\,G_6}{1 + G_2\,G_3\,H_1 + H_2\,G_2 + G_2\,G_3\,G_4\,H_3 - G_5 - G_2\,H_2\,G_5 + G_2\,G_6\,H_3 - G_2\,G_5\,G_6\,H_3}$$

3.8 PROCEDURE FOR CONVERTING BLOCK DIAGRAM TO SIGNAL FLOW GRAPH

The signal flow graph and block diagram of a system provides the same information but there is no standard procedure for reducing the block diagram to find the transfer function of the system. Also the block diagram reduction technique will be tedious and it is difficult to choose the rule to be applied for simplification. Hence it will be easier if the block diagram is converted to signal flow graph and Mason's gain formula is applied to find the transfer function. The following procedure can be used to convert block diagram to signal flow graph.

1. Assume nodes at input, output, at every summing point, at every branch point and in between cascade blocks.
2. Draw the nodes separately as small circles and number the circles in the order, 1, 2, 3, 4 … etc.
3. From block diagram find the gain between each node in the main forward path and connect all the corresponding circles by straight line and mark the gain between the nodes.
4. Draw the feed forward path between various nodes and mark the gain of feed forward path along with sign.
5. Draw the feedback paths between various nodes and mark the gain of feedback path along with sign.

EXAMPLE 3.22: Convert the block diagram to signal flow graph and determine the transfer function using Mason's gain formula.

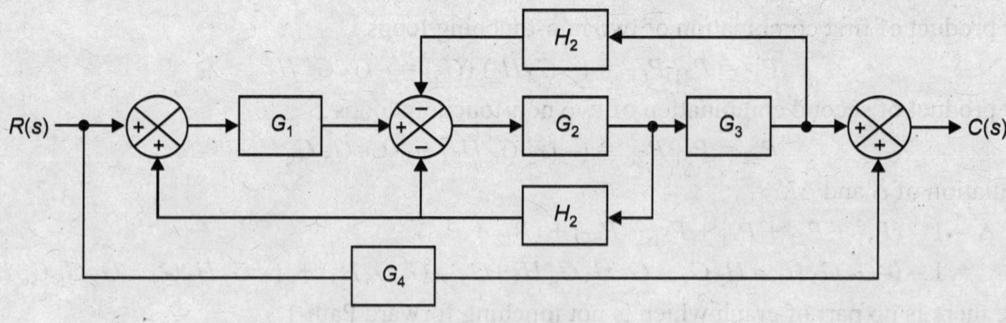

Fig. 3.22.1

Solution: The nodes are assigned at input/output, at every summing point and branch point as shown in Fig. 3.22.2.

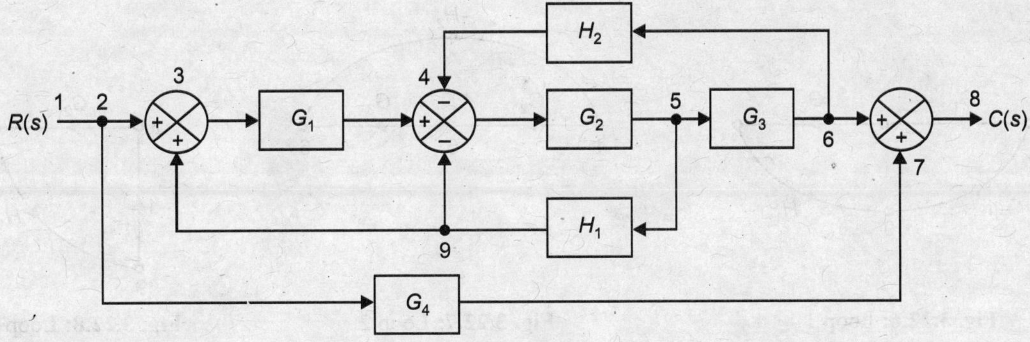

Fig. 3.22.2

The signal flow graph of above block diagram shown in Fig. (3.22.3)

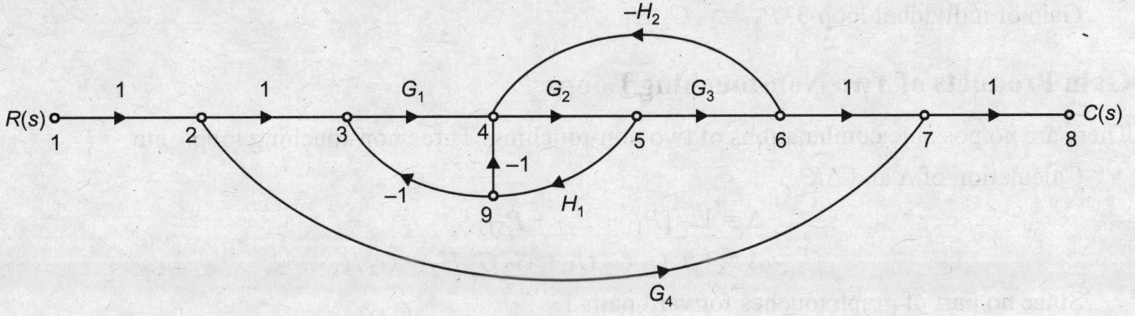

Fig. 3.22.3

Forward Path Gain

There are two forward path, $\therefore K = 2$

Let the forward path gains be P_1 and P_2.

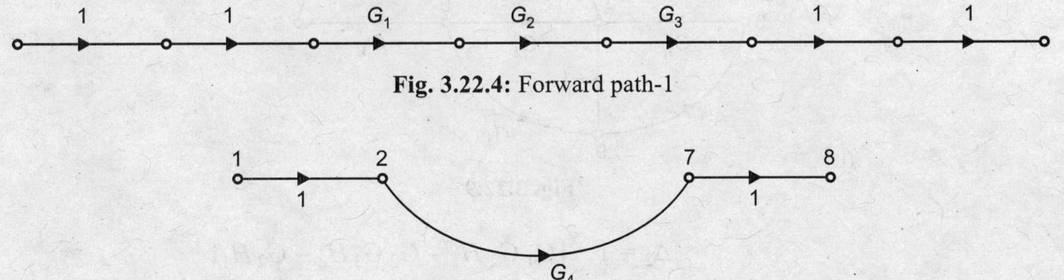

Fig. 3.22.4: Forward path-1

Fig. 3.22.5: Forward path-2

Gain of forward path-1, $P_1 = G_1 G_2 G_3$

Gain of forward path-2, $P_2 = G_4$

Individual Loop Gain

There are three individual loops with gains P_{11}, P_{21} and P_{31}

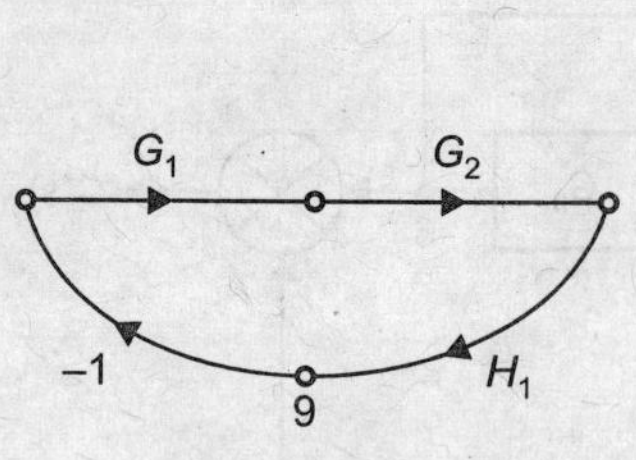

Fig. 3.22.6: Loop 1

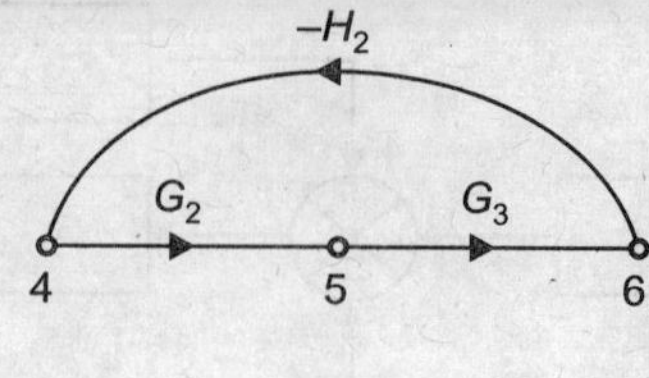

Fig. 3.22.7: Loop 2

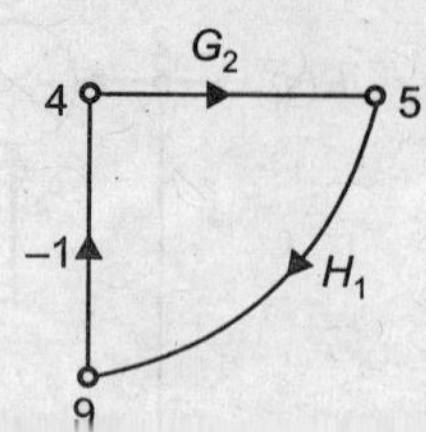

Fig. 3.22.8: Loop 3

Gain of individual loop-1, $P_{11} = - G_1 G_2 H_1$

Gain of individual loop-2, $P_{21} = - G_2 G_3 H_2$

Gain of individual loop-3, $P_{31} = - G_2 H_1$

Gain Products of Two Non-touching Loops

There are no possible combinations of two non-touching. Three non-touching loops, etc.

Calculation of Δ and ΔK

$$\Delta = 1 - [P_{11} + P_{21} + P_{31}]$$

$$\Delta = 1 + G_1 G_2 H_1 + G_2 G_3 H_2 + G_2 H_1$$

Since no part of graph touches forward path-1

$$\Delta_1 = 1$$

The part of graph non-touching forward Path-2 is shown in Fig. (3.22.9)

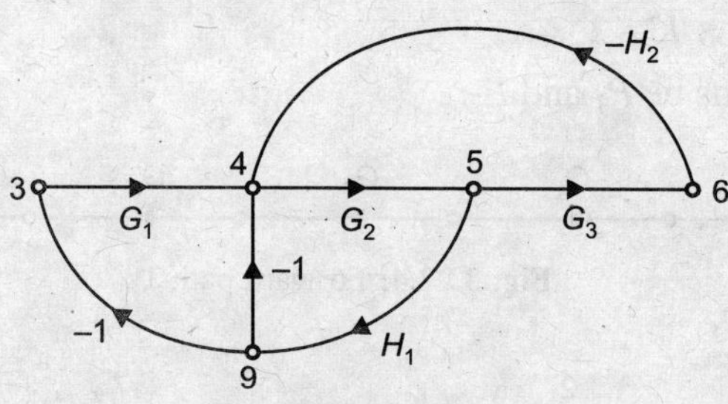

Fig. 3.22.9

$$\Delta_2 = 1 - [G_1 G_2 H_1 - G_2 G_3 H_2 - G_2 H_1]$$

$$\Delta_2 = 1 + G_1 G_2 H_1 + G_2 G_3 H_2 + G_2 H_1$$

Transfer Function, T

$$T = \frac{1}{\Delta} \sum_k P_k \Delta_k$$

$$= \frac{1}{\Delta} [P_1 \Delta_1 + P_2 \Delta_2]$$

$$T = \frac{G_1 G_2 G_3 + G_4 + G_1 G_2 G_4 H_1 + G_2 G_3 G_4 H_2 + G_2 G_4 H_1}{1 + G_1 G_2 H_1 + G_2 G_3 H_2 + G_2 H_1}$$

EXAMPLE 3.23: Convert the block diagram to signal flow graph and determine the transfer function using Mason's gain formula.

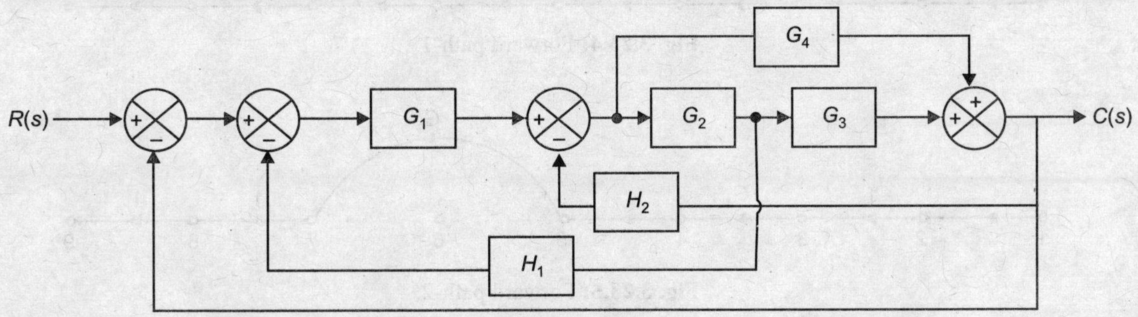

Fig. 3.23.1

Solution: The nodes are assigned at input, output, at every summing point and branch point as shown in Fig. (3.23.2)

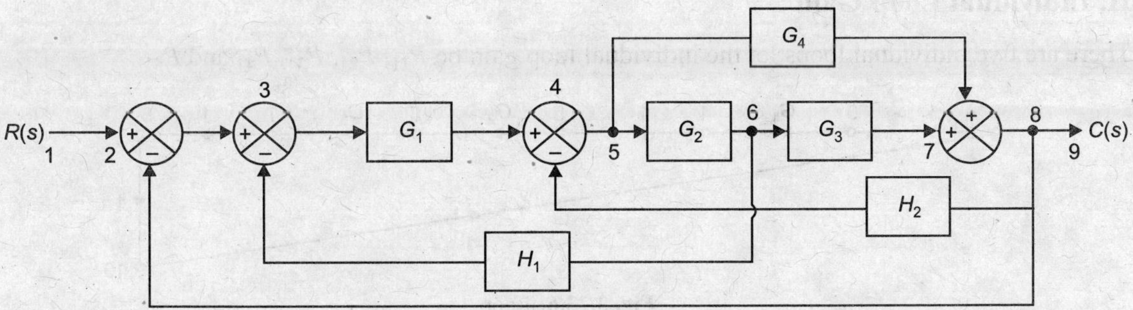

Fig. 3.23.2

The signal flow graph for the above block diagram shown in Fig. (3.23.3)

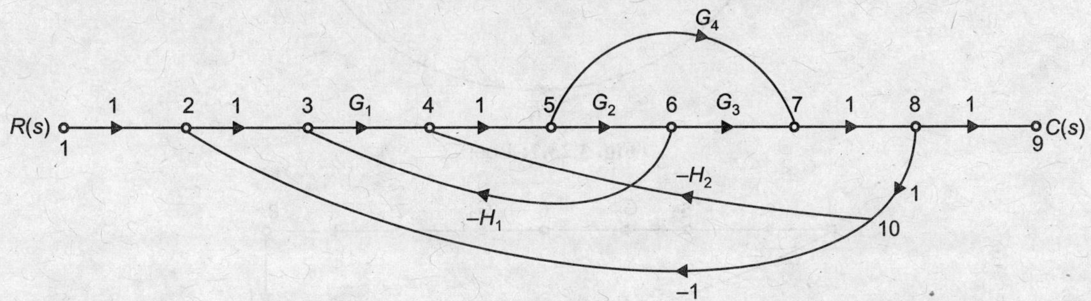

Fig. 3.23.3

CHAPTER 3

I. Forward Path Gain

There are two forward paths ∴ $K = 2$

Let the gain of the forward paths be P_1 and P_2

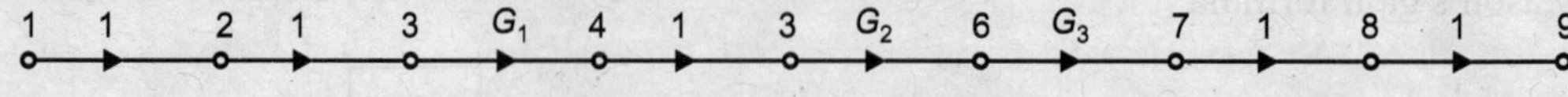

Fig. 3.23.4: Forward path-1.

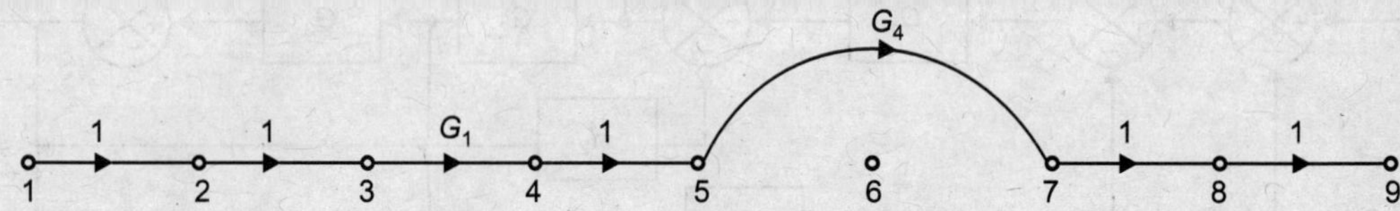

Fig. 3.23.5: Forward path-2.

Gain of forward path-1, $P_1 = G_1\, G_2\, G_3$

Gain of forward path-2 $P_2 = G_1\, G_4$

II. Individual Loop Gain

There are five individual loops let the individual loop gain be P_{11}, P_{21}, P_{31}, P_{41} and P_{51}.

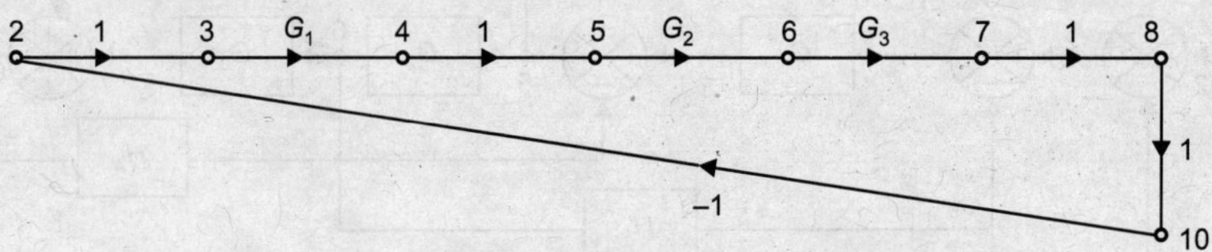

Fig. 3.23.6: Loop-1.

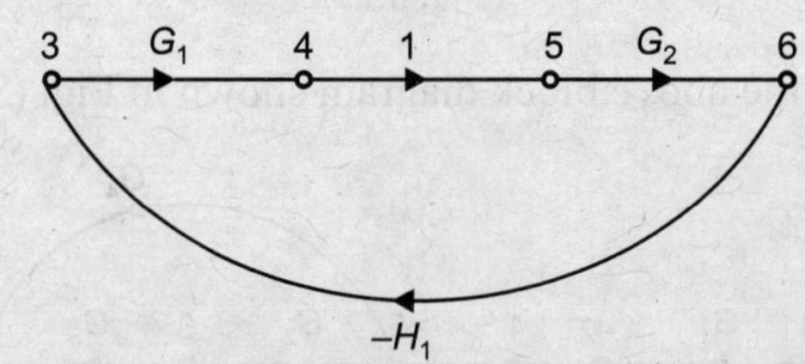

Fig. 3.23.7: Loop-2.

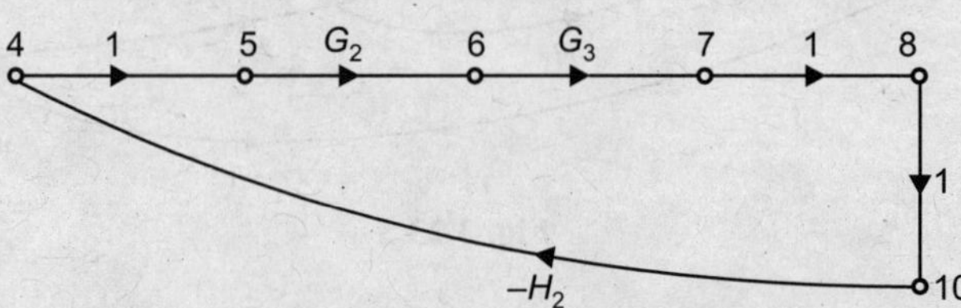

Fig. 3.23.8: Loop-3.

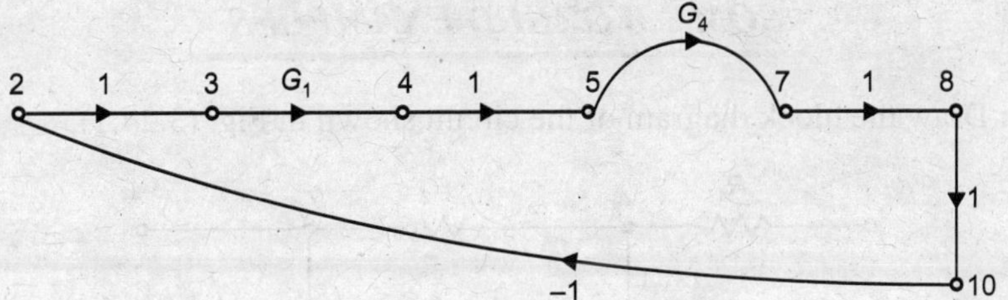

Fig. 3.23.9: Loop-4.

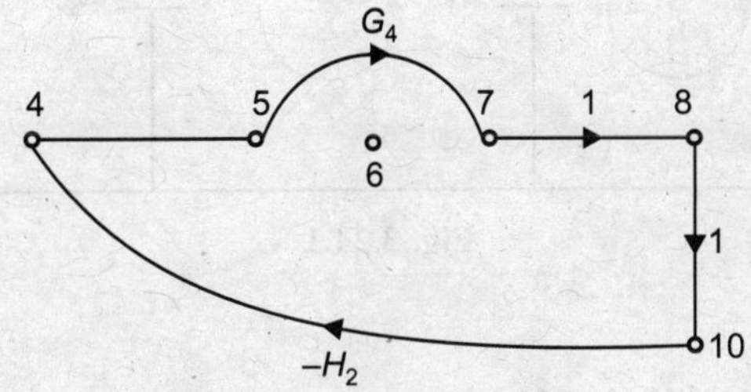

Fig. 3.23.10: Loop-5.

Loop gain of individual loop-1, $P_{11} = -\ G_1\,G_2\,G_3$

Loop gain of individual loop-2, $P_{21} = -\ G_2\,G_1\,H_1 = -G_1\ G_2\,H_1$

Loop gain of individual loop-3, $P_{31} = -\ G_2\ G_3\ H_2$

Loop gain of individual loop-4, $P_{41} = -\ G_1\ G_4$

Loop gain of individual loop-5, $P_{51} = -\ G_4 H_2$

III. Gain Products of Two Non-touching Loops

There are no possible combinations of two non-touching loops, threen-touching loops, etc.

IV. Calculation of Δ and ΔK

$$\Delta = 1 - [P_{11} + P_{21} + P_{31} + P_{41} + P_{51}]$$
$$= 1 + G_1\ G_2\ G_3 + G_1\ G_2\ H_1 + G_2\ G_3\ H_2 + G_1 G_4 + G_4 H_2$$

Since no part of graph is non-touching with forward paths 1 and 2, $\Delta_1 = \Delta_2 = 1$

V. Transfer Function, *T*

By Mason's gain formula

$$T = \frac{1}{\Delta}\sum_K P_K \Delta_K = \frac{1}{\Delta}\ [P_1\Delta_1 + P_2\Delta_2]$$

$$T = \frac{G_1\,G_2\,G_3 + G_1\,G_4}{1 + G_1\,G_2\,G_3 + G_1\,G_2\,H_1 + G_2\,G_3\,H_2 + G_1\,G_4 + G_4\,H_2}$$

SOME WORKED EXAMPLES

EXAMPLE 3.24: Draw the block diagram of the circuit shown in Fig. (3.24.1).

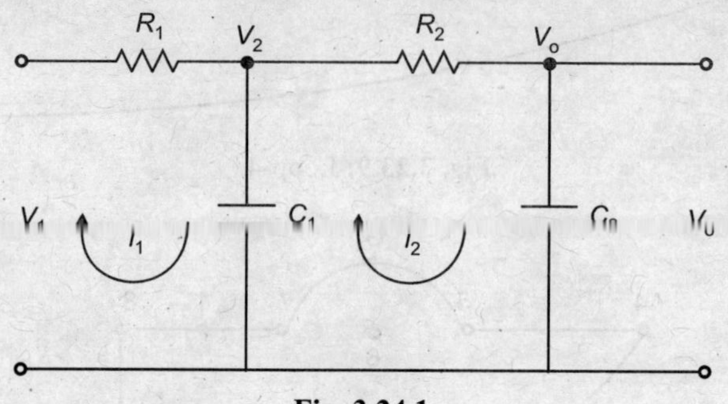

Fig. 3.24.1

Solution:

$$i_1(t) = V_1(t) - V_2(t)/R_1 \quad (3.24.1)$$

$$V_2(t) = \frac{1}{C_1}\int [i_1(t) - i_2(t)]\, dt \quad (3.24.2)$$

$$i_2(t) = \frac{1}{R_2}[V_2(t) - V_0(t)] \quad (3.24.3)$$

$$V_0(t) = \frac{1}{C_2}\int i_2\, dt \quad (3.24.4)$$

By taking laplace transform of Eq. (3.24.1 to 3.24.4)

$$I_1(s) = \frac{1}{R_1}[V_1(s) - V_2(s)] \quad (3.24.5)$$

$$V_2(s) = \frac{1}{sC_1}[I_1(s) - I_2(s)] \quad (3.24.6)$$

$$I_2(s) = \frac{1}{R_2}[V_2(s) - V_0(s)] \quad (3.24.7)$$

$$V_0(s) = \frac{1}{sC_2} I_2(s) \quad (3.24.8)$$

from Eq. (3.24.5)

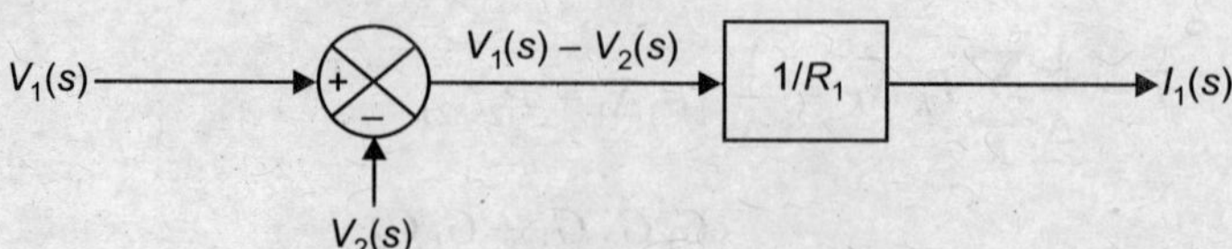

Fig. 3.24.2

from Eq. (3.24.6)

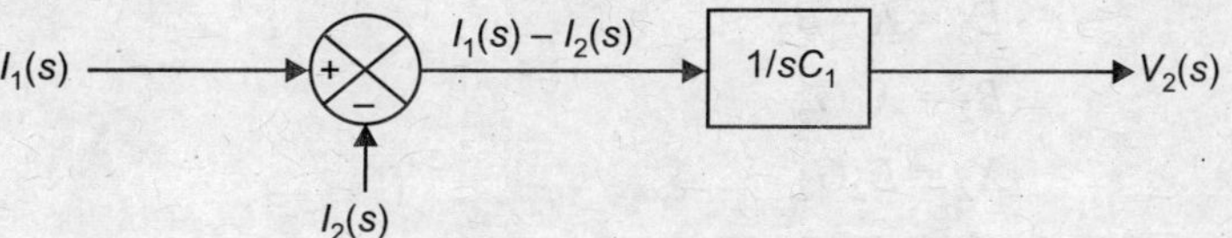

Fig. 3.24.3

from Eq. (3.27.7)

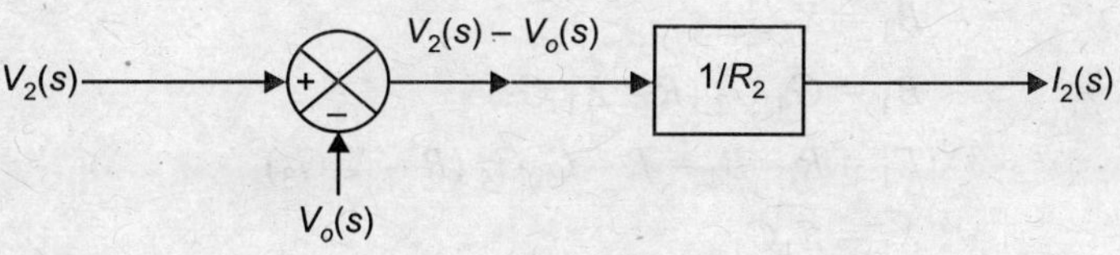

Fig. 3.24.4

from Eq. (3.28.8)

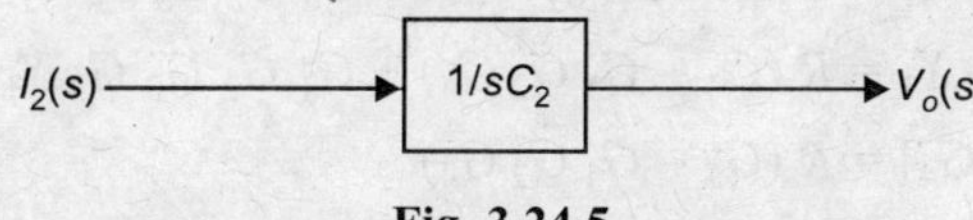

Fig. 3.24.5

Combining all the Fig. (3.24.2) to (3.24.5)

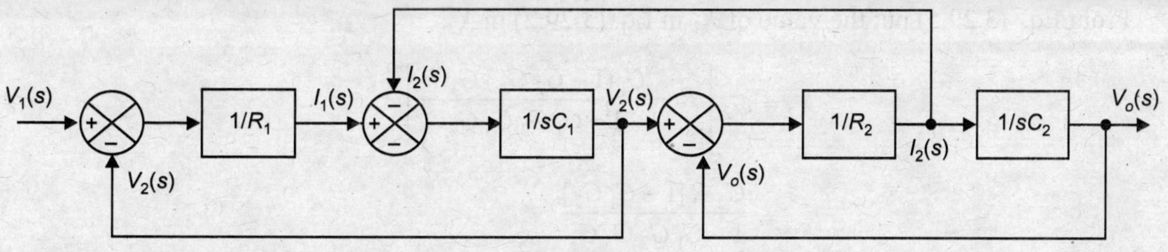

Fig. 3.24.6

EXAMPLE 3.25: Determine the ratio *C*/*R*

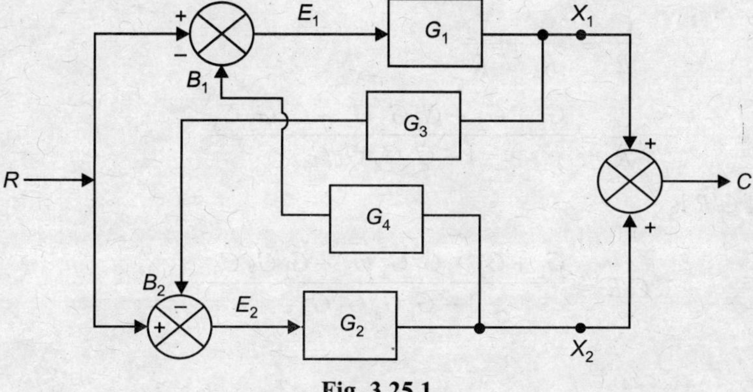

Fig. 3.25.1

Solution: From above block diagram

$$E_2 = R - B_2$$

$$B_2 = X_1G_3 \quad (3.25.1)$$

$$X_2 = E_2G_2$$

$$X_2 = G_2(R - B_2)$$

$$X_2 = G_2\,(R - X_1G_3) \quad (3.25.2)$$

$$B_1 = X_2G_4$$

$$B_1 = G_4\,G_2\,(R - X_1\,G_3)$$

$$E_1 = R - B_1 = R - G_2G_4\,(R - X_1G_3) \quad (3.25.3)$$

$$X_1 = E_1G_1$$

$$X_1 = G_1\,[R - G_2\,G_4\,(R - X_1\,G_3)]$$

$$X_1 = G_1R - G_1\,G_2\,G_4\,R + G_1\,G_2\,G_3\,G_4X_1$$

$$X_1 = R\,(G_1 - G_1\,G_2\,G_4) + G_1\,G_2\,G_3\,G_4X_1 \quad (3.25.4)$$

$$X_1\,[1 - G_1\,G_2\,G_3\,G_4] = R\,(G_1 - G_1\,G_2\,G_4)$$

$$\frac{X_1}{R} = \frac{G_1 - G_1\,G_2\,G_4}{1 - G_1\,G_2\,G_3\,G_4} \quad (3.25.5)$$

From Eq. (3.29.5) put the value of X_1 in Eq. (3.29.2) in X_2

$$X_2 = G_2\left[R - \frac{G_1\,(1 - G_2\,G_4)\,G_3\,R}{1 - G_1\,G_2\,G_3\,G_4}\right]$$

$$= \frac{G_2\,R\,[1 - G_1\,G_3]}{1 - G_1\,G_2\,G_3\,G_3}$$

$$\frac{X_2}{R} = \frac{G_2\,(1 - G_1\,G_3)}{1 - G_1\,G_2\,G_3\,G_4}$$

$$\frac{C}{R} = \frac{X_1}{R} + \frac{X_2}{R}$$

$$\frac{C}{R} = \frac{G_1 + G_2 - G_1\,G_2\,G_4 - G_1\,G_2\,G_3}{1 - G_1\,G_2\,G_3\,G_4}$$

Result: The ratio of C/R is

$$C/R = \frac{G_1 + G_2 - G_1\,G_2\,G_4 - G_1\,G_2\,G_3}{1 - G_1\,G_2\,G_3\,G_4}$$

EXAMPLE 3.26: Determine the transfer function C_1/R_1, C_2/R_2, C_1/R_2 and C_2/R_1 from the block diagram shown in Fig. (3.26.1).

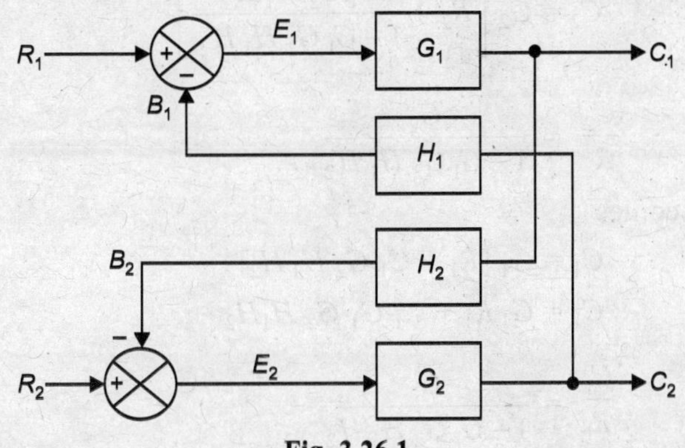

Fig. 3.26.1

Solution: From above block diagram

$$E_2 = R_2 - B_2 \tag{3.26.1}$$

$$B_2 = C_1 H_2 \tag{3.26.2}$$

$$C_2 = E_2 G_2 \tag{3.26.3}$$

Put the value of E_2 from Eq. (3.26.1) in (3.26.3)

$$C_2 = G_2 [R_2 - B_2] \tag{3.26.4}$$

from Eq. (3.26.2) put the value of B_2 in Eq. (3.26.4)

$$C_2 = G_2 (R_2 - C_1 H_2) \tag{3.26.5}$$

$$B_1 = C_2 H_1 \tag{3.26.6}$$

Put the value of C_2 from (3.26.5) in (3.26.6)

$$B_1 = G_2 H_1 (R_2 - C_1 H_2) \tag{3.26.7}$$

$$E_1 = R_1 - B_1 \tag{3.26.8}$$

$$E_1 = R_1 - G_2 H_1 (R_2 - C_1 H_2) \tag{3.26.9}$$

$$C_1 = E_1 G_1 \tag{3.26.10}$$

Also, from (3.26.8) put the value of E_1 in (3.26.9)

$$C_1 = G_1[R_1 - G_2 H_1 [R_2 - C_1 H_2]] \tag{3.26.11}$$

$$C_1 [1 - G_1 G_2 H_1 H_2] = G_1 R_1 - G_1 G_2 H_1 R_2 \tag{3.26.12}$$

when $R_1 = 0$ from Eq. (3.26.12)

$$\frac{C_1}{R_2} = \frac{-G_1 G_2 H_1}{1 - G_1 G_2 H_1 H_2} \tag{3.26.13}$$

CHAPTER 3

from Eq. (3.26.13) put the value of C_1 in Eq. (3.26.5)

$$C_2 = G_2\left[R_2 + \frac{G_1 G_2 H_1 H_2}{1 - G_1 G_2 H_1 H_2}\right]$$

$$\frac{C_2}{R_2} = \frac{G_2}{1 - G_1 G_2 H_1 H_2} \tag{3.26.14}$$

when $R_2 = 0$, (3.26.12) becomes

$$C_1 = G_1 [R_1 + C_1 G_2 H_1 H_2]$$

$$C_1 = G_1 R_1 + C_1 G_1 G_2 H_1 H_2$$

$$\frac{C_1}{R_1} = \frac{G_1}{1 - G_1 G_2 H_1 H_2} \tag{3.26.15}$$

when $R_2 = 0$, (3.26.5) becomes

$$C_2 = -G_2 C_1 H_2 \tag{3.26.16}$$

from Eq. (3.26.15) put the value of C_1 in (3.26.16)

$$C_2 = \frac{-G_1 G_2 R_1 H_2}{1 - G_1 G_2 H_1 H_2}$$

$$\frac{C_2}{R_1} = \frac{-G_1 G_2 H_2}{1 - G_1 G_2 H_1 H_2} \tag{3.26.17}$$

Result:

3. $\dfrac{C_1}{R_1} = \dfrac{G_1}{1 - G_1 G_2 H_1 H_2}$ 2. $\dfrac{C_2}{R_2} = \dfrac{G_2}{1 - G_1 G_2 H_1 H_2}$

3. $\dfrac{C_1}{R_2} = \dfrac{-G_1 G_2 H_1}{1 - G_1 G_2 H_1 H_2}$ 4. $\dfrac{C_2}{R_1} = \dfrac{-G_1 G_2 H_2}{1 - G_1 G_2 H_1 H_2}$

EXAMPLE 3.27: Draw the signal flow graph for the network shown in Fig. (3.27.1) take V_3 as output node.

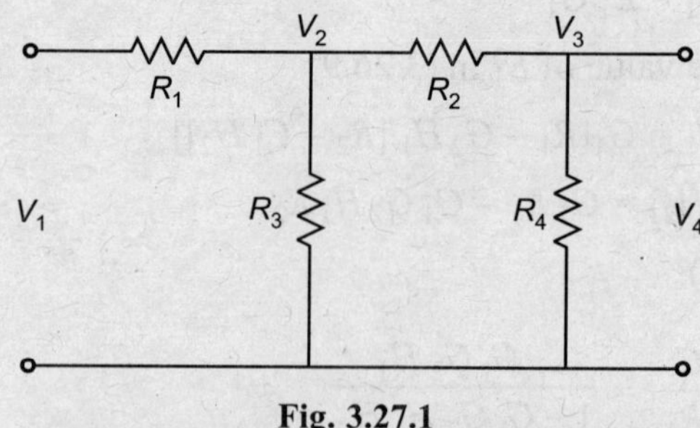

Fig. 3.27.1

Solution: The circuit can be redrawn with current directions.

From Fig. (3.27.2)

$$i_1 = \frac{V_1 - V_2}{R_1} = \frac{V_1}{R_1} - \frac{V_2}{R_1}$$

$$V_2 = I_3 R_3 = R_3 (i_1 - i_2) = R_3 i_1 - R_3 i_2$$

$$i_2 = \frac{V_2 - V_3}{R_2} = \frac{V_2}{R_2} - \frac{V_3}{R_2}$$

$$V_3 = R_4 i_4 = R_4 i_2$$

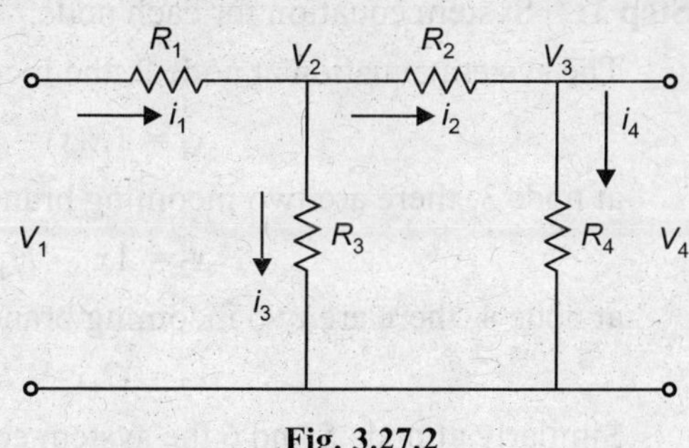

Fig. 3.27.2

Note: $i_2 = i_4$

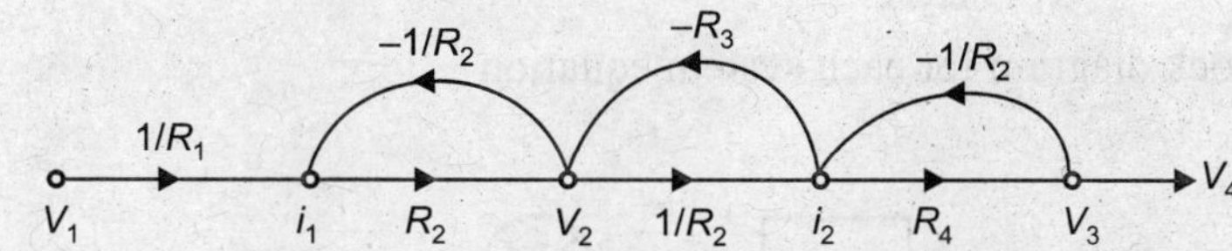

Fig. 3.27.3: Signal flow graph

3.9 PROCEDURE FOR CONVERTING SIGNAL FLOW GRAPH IN TO BLOCK DIAGRAM

Step 1: For given signal flow graph, write the system equations.

Step 2: At each node consider the incoming branches only.

Step 3: Add all incoming signals algebraically at a node.

Step 4: For +ve (or) – sign in system equations use a summing point

Step 5: For the gain of each branch of signal flow graph draw the block having the same transfer function as the gain of the branch.

EXAMPLE 3.28: Draw the block diagram from the given signal flow graph.

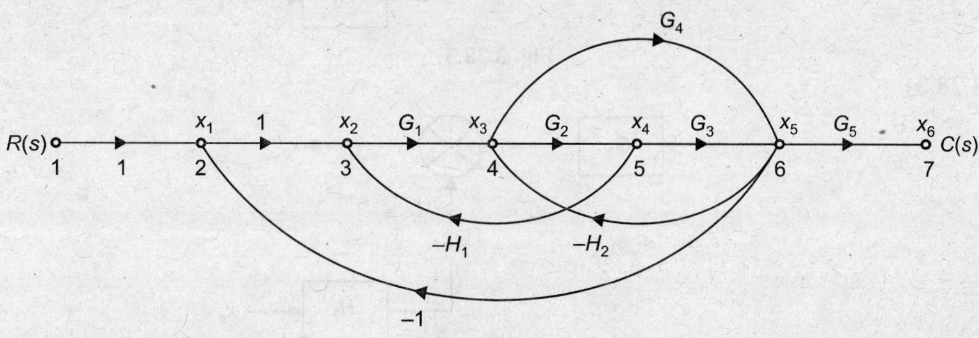

Fig. 3.28.1

Solution:

Step 1: System equation for each node.

The system equation at node 2, the incoming branches are from $R(s)$ and x_5

$$x_1 = 1R(s) - 1x_5 = R(s) - x_5 \quad (3.28.1)$$

at node 3, there are two incoming branches

$$x_2 = 1x_1 - H_1 x_4 = x_1 - H_1 x_4 \quad (3.28.2)$$

at node 4, there are two incoming branches

$$x_3 = G_1 x_2 - H_2 x_5 \quad (3.28.3)$$

Similarly at node 5 and 6 the system equation are

$$x_4 = G_2 x_3 \quad (3.28.4)$$

$$x_5 = G_3 x_4 + G_4 x_3 \quad (3.28.5)$$

at node 7,

$$x_6 = G_5 x_5 \quad (3.28.6)$$

Step 2: Draw the block diagram for each system equation

for Eq. (3.28.1)

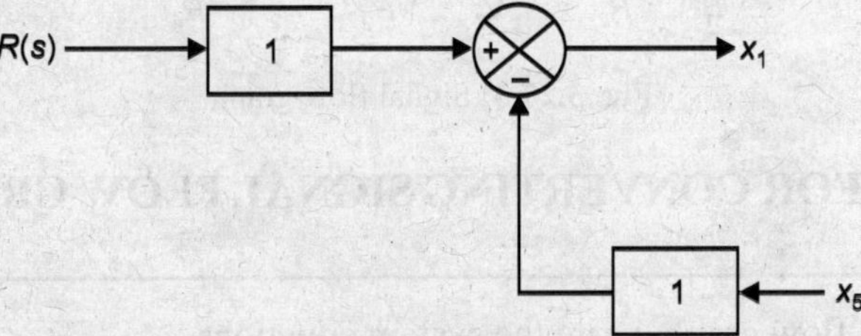

Fig. 3.28.2

for Eq. (3.28.2)

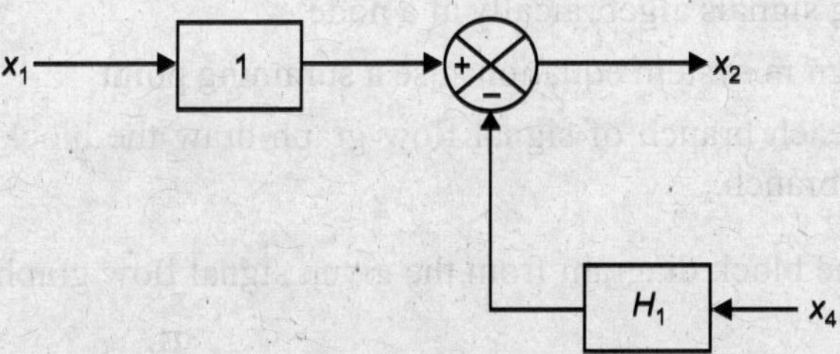

Fig. 3.28.3

for Eq. (3.28.3)

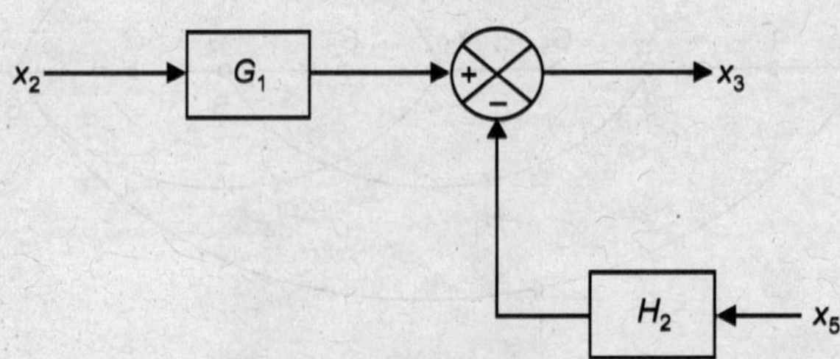

Fig. 3.28.4

for Eq. (3.28.4)

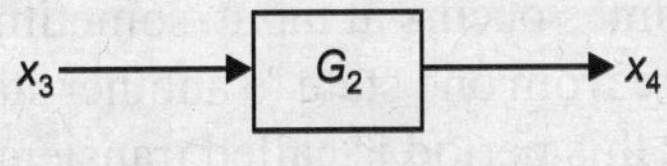

Fig. 3.28.5

for Eq. (3.28.5)

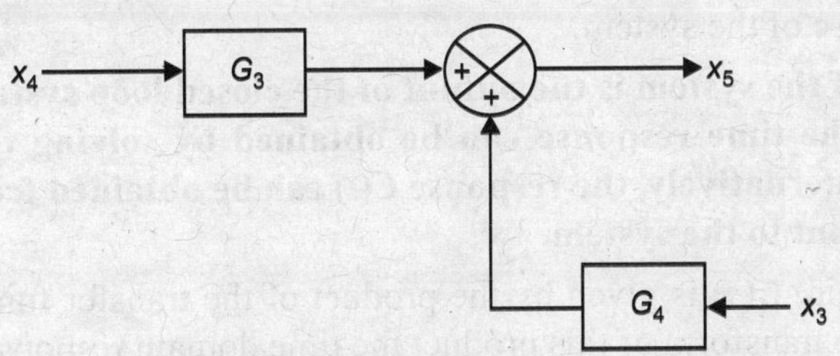

Fig. 3.28.6

for Eq. (3.28.6)

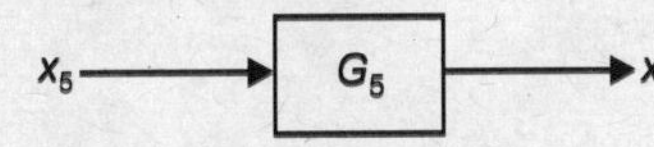

Fig. 3.28.7

Step 3: **Combine all the block diagrams i.e. from (3.28.2) to (3.28.7)**

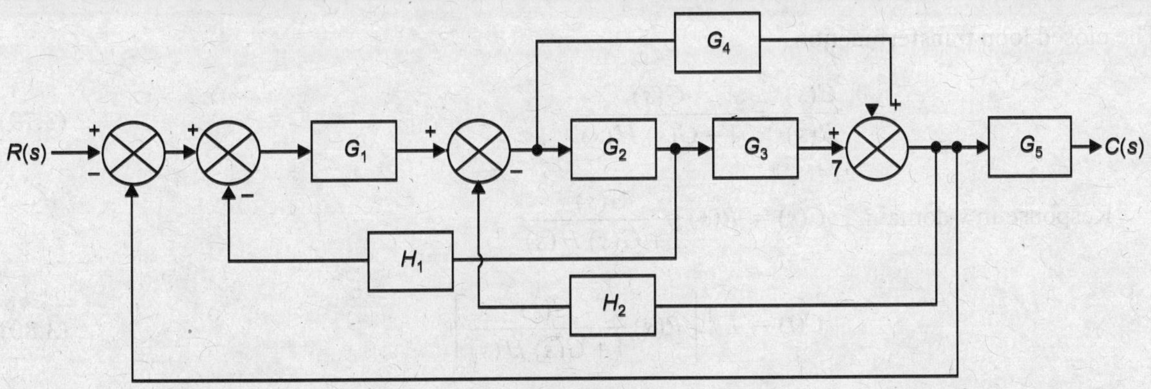

Fig. 3.28.8

Result: The Fig. (3.28.7) is the complete block diagram of given signal flow graph.

III SIGNALS

3.10 INTRODUCTION TO TYPES OF SIGNALS

Any system containing energy storing element like inductor, capacitor, mass and inertia etc. Posses certain energy. These energy storing elements are the part of the control system and cannot be avoided.

If the energy state of the system is distributed then it takes a certain time to change from one state to another state. This disturbance sometimes occurs at input, sometimes occurs at output and sometimes at both ends. The time required to change from one state to another state is known as transient time and the values of current and voltages during this period is called transient response. These transient may have oscillations which may be either sustained or decaying in nature. This will depend upon the parameters of the system. For any system we obtain a linear differential equation. The solution of linear differential equation gives the response of the system.

The time response of the system is the output of the closed loop system as a function of time. It is denoted by *C(t)*. The time response can be obtained by solving the differential equation governing the system. Alternatively, the response *C(t)* can be obtained from the transfer function of the system and the input to the system.

The output in s-domain, $C(s)$ is given by the product of the transfer function and the input, $R(s)$. On taking inverse Laplace transform of this product the time domain response, $C(t)$ can be obtained.

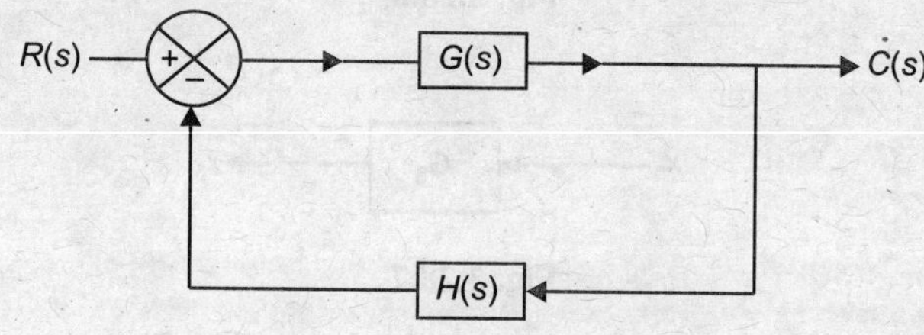

Fig. 3.19: Closed loop.

The closed loop transferfunction,

$$\frac{C(s)}{R(s)} = \frac{G(s)}{1+G(s)\,H(s)} \tag{3.79}$$

Response in s-domain, $C(s) = R(s)\dfrac{G(s)}{1+G(s)\,H(s)}$

$$C(t) = L^{-1}\left[R(s)\frac{C(s)}{1+G(s)\,H(s)}\right] \tag{3.80}$$

The time response of a control system consists of two parts: the transient and the steady state response. The transient response of the system shows the response of the system when the input changes from one state to another state.

3.11 TEST SIGNALS

The knowledge of input signal is required to predict the response of a system. In most of the system the input signals are not known ahead of time and also it is difficult to express the input signal mathematically by simple equation. The characteristics of actual input signals are a sudden shock, a sudden changes a constant velocity and a constant acceleration. Hence test signals which resembles these characteristics are used as input signals to predict the performance of the system. The commonly used test input signals are impulse, step, ramp, and parabolic signals.

The standard test signals are

1. (a) Step Signal
   (b) Unit Step Signal
2. (a) Ramp Signal
   (b) Unit Ramp Signal
3. (a) Parabolic Signal
   (b) Unit Parabolic Signal
4. (a) Impulse Signal
   (b) Unit impulse Signal

Since the test signals are simple functions of time, they can be easily generated in laboratories. The mathematical and experimental analysis of control system using these signals can be carried out easily. The use of the test signals can be justified because of a correlation existing between the response characteristics of a system to a test input signal and capability of the system to cope with actual input signals.

Step signal: The step signal is a signal whose value changes from zero to A at $t = 0$ and remains constant at A for $t > 0$. The step signal resembles an actual steady input to a system. A special case of step signal is unit step in which A is unity. The mathematical representation of the step signal is

$$r(t) = Au(t)$$

where
$$u(t) = 1; \quad t \geq 0$$
$$u(t) = 0; \quad t < 0$$

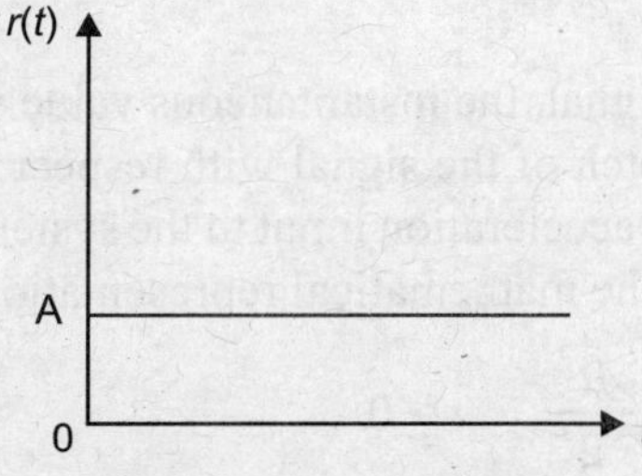

Fig. 3.20: Ramp signal.

Let $u(t)$ be defined in the interval $0 \leq t \leq \infty$, Laplace transform is obtained by multiplying $u(t)$ by e^{-st} and integrate between the limit 0 to ∞.

$$L[u(t)] = \int_0^\infty u(t)\, e^{-st}\, dt$$

$$= \int_0^\infty 1 \cdot e^{-st}\, dt = \left[\frac{e^{-st}}{-s}\right]_0^\infty = \frac{1}{s}$$

Step function is also called displacement function. Step function can be described as sudden application of input signal to a system.

If input is $R(s)$; then

$$R(s) = \frac{1}{s}$$

Ramp signal: The ramp signal is a signal whose value increase linearly with time from an initial value of zero at $t = 0$. The ramp signal reasembles a constant velocity input to the signal in which the value of A is unity. The mathematical representation of the ramp signal is.

$$r(t) = At; \quad \text{when } t \geq 0$$

$$r(t) = 0, \quad \text{when } t < 0$$

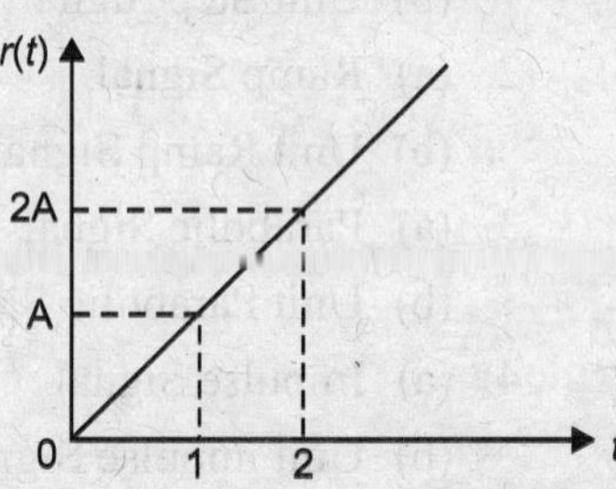

Fig. 3.21: Ramp signal.

Let $r(t)$ be defined in the interval $0 \leq t \leq \infty$ Laplace transform is obtained by multiplying $r(t)$ by e^{-st} and integrate between the limit 0 to ∞.

$$Lr(t) = \int_0^\infty r(t)\, e^{-st} dt \qquad \text{where } r(t) = At$$

$$Lr(t) = \int_0^\infty At\, e^{-st} dt = \frac{A}{s^2}$$

$$R(s) = \frac{A}{s^2}$$

If A is unity then [i.e. $A = 1$]

$$R(s) = \frac{1}{s^2}$$

Parabolic signal: In parabolic signal, the instantaneous value varies as square of the time from an initial value of zero at $t = 0$. The sketch of the signal with respect to time resembles a parabolas. The parabolic signal resembles a constant acceleration input to the system. A special case of parabolic is unit parabolic signal in which A is unity. the mathematical representation of the parabolic signal is

$$r(t) = \frac{At^2}{2}; \quad t \geq 0$$

$$r(t) = 0; \quad t < 0$$

$$Lr(t) = \int_0^\infty r(t)\, e^{-st}$$

where $r(t) = \frac{At^2}{2}$

$$Lr(t) = \int_0^\infty \frac{At^2}{2} e^{-st}\, dt$$

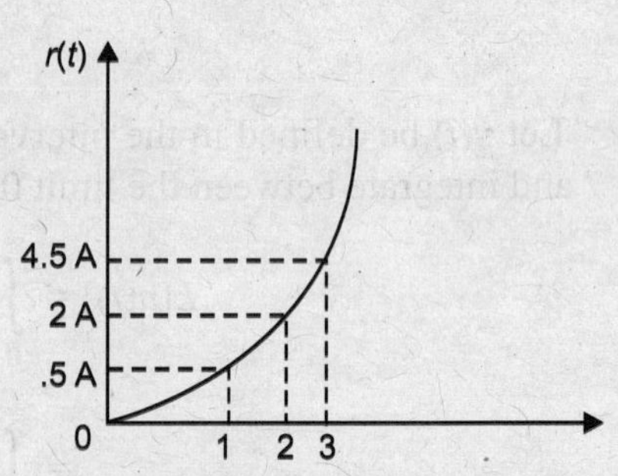

Fig. 3.22: Parabolic signal.

$$Lr(t) = \frac{A}{s^3}$$

$$R(s) = \frac{A}{s^3}$$

For unit parabolic function $A = 1$

So $$R(s) = \frac{1}{s^3}$$

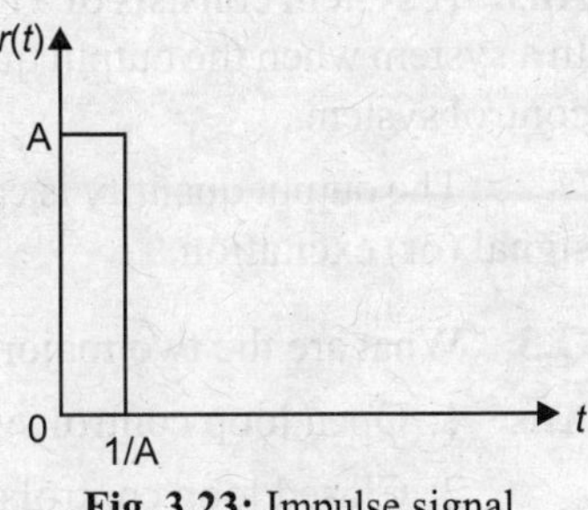

Fig. 3.23: Impulse signal.

Impulse signal : A signal which is variable for very short duration is called impulse signal Ideal impulse signal is a unit impulse signal which is defined as a signal having zero values at all times expect at $t = 0$. At $t = 0$, the magnitude is infinite. It is denoted by $\delta(t)$ and mathematically it is expressed as

$$\delta(t) = 0 \quad \text{for } t \neq 0$$

and $$\underset{t_1 \to 0}{Lt} \int_{-t_1}^{+t_1} \delta(t)\, dt = 1$$

or $$\delta(t) = \begin{bmatrix} 0;\ t \neq 0 \\ \infty;\ t = 0 \end{bmatrix} \int_{-\infty}^{\infty} \delta(t)\, dt = 1$$

Since a perfect impulse cannot be achieved in practice it is usually approximated by a pulse of small width but mathematically an impulse function is the derivative of a step function. Laplace transform of the impulse function is unity.

The response of the system, with input impulse signal is called weighting function of the system. It is given by the inverse Laplace transform of the system transfer function.

$$C(t) = L^{-1}\left[R(s)\frac{G(s)}{1+G(s)\,H(s)}\right] = L^{-1}\left[\frac{G(s)}{1+G(s)\,H(s)}\right]$$

[$\because R(s) = 1$ for impulse]

Table 3.4.1: Standard test signals

Name of the signal	Time domain equation of signal, $r(t)$	Laplace transform of the signal, $R(s)$
Step *i/p* Unit step *i/p*	A 1	A/s $1/s$
Ramp *i/p* Unit Ramp *i/p*	At t	A/s^2 $1/s^2$
Parabolic Unit parabolic	$At^2/2$ t^2/t	A/s^3 $1/s^3$
Impulse	$\delta(t)$	1

SHORT QUESTION AND ANSWERS

Q.1 What is system?

Ans. When a number of elements (or) components are connected in a sequence to perform a specified function, the group thus formed is called a system.

Q.2 What is control system?

Ans. A system consists of a number of components connected together to perform a specified function. In a system when the output quantity is controlled by varying the input quantity, then the system is called control system.

The output quantity is called controlled variable or response and input quantity is called command signal (or) excitation.

Q.3 What are the two major type of control system?

Ans. 1. Open loop control system

2. Closed loop control system.

Q.4 Define open loop control system.

Ans. The control system in which the output quantity has no effect upon the input quantity are called open loop control system. This means that the output is not feedback to the input for correction.

Q.5 Define closed loop control system.

Ans. The control system in which the output has an effect upon the input quantity so as to maintain the desired output value are called closed loop control system.

Q.6 What is feedback? What type of feedback is employed in control system?

Ans. The feedback is a control action in which the output is sampled and a proportional signal is given to input for automatic correction of any changes in desired output.

Negative feedback is employed in control system (usually).

Q.7 What are the components of feedback control system?

Ans. The components of feedback control system are plants, feedback path elements, error detector and controller.

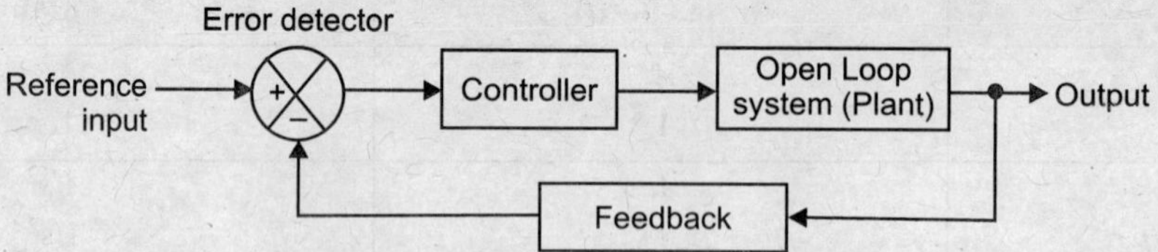

Q.8 Why negative feedback is preferred in a closed loop system?

Ans. The negative feedback results in better stability in steady state and rejects any disturbance signals. It also has low sensitivity to parameter variations. Hence negative feedback is preferred in closed loop systems.

Q.9 What are the characteristics of negative feedback?

Ans. The characteristics of negative feedback are,

(a) accuracy in tracking steady state value

(b) Rejection of disturbance signals.

(c) Low sensitivity to parameter variations.

(d) Reduction in gain at the expense of better stability.

Q.10 What is the effect of positive feedback on stability?

Ans. The positive feedback increases the error signal and drives the output to instability. But sometimes the positive feedback is used in minor loops in control system to amplify certain internal signals or parameters.

Q.11 Distinguish between open loop and closed loop system?

Ans.

Open loop	Closed loop
(a) Inaccurate and unreliable	(a) Accurate and reliable
(b) Simple and economical	(b) Complex and costlier
(c) The changes in output due to external disturbances are not corrected automatically	(c) The changes in output due to external disturbances are corrected automatically.
(d) They are generally stable	(d) Great efforts are neede to design a stable system

Q.12 What is servomechanism?

Ans. The servomechanism is a feedback control system component which corrects the position of mechanical load to desired output.

Q.13 State the principle of superposition and homogeneity (or) state the principle of superposition?

Ans. The principle of superposition and homogeneity states that if the system has responses $y_1(t)$ and $y_2(t)$ for the inputs $X_1(t)$ and $X_2(t)$ respectively then the system responses to the linear combination of these input $a_1X_1(t) + a_2X_2(t)$ is given by linear combination of the individual outputs $a_1y_1(t) + a_2y_2(t)$, where a_1 ande a_2 are constants.

Q.14 Define linear system?

Ans. A system is said to be linear if it obeys the principle of superposition and homogeneity principle of superposition states that the response of a system to a weighed sum of signals is equal to the corresponding weighed sum of the responses of the system to each of the individual input signals.

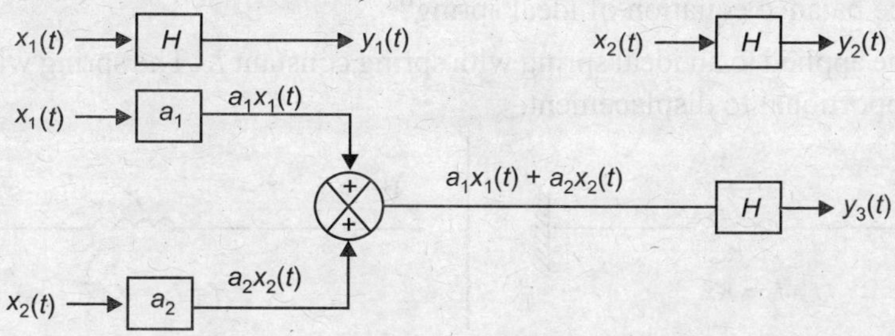

CHAPTER 3

Q.15 What is time invariant system?

Ans. A system is said to be time invariant if its input-output characteristics do not change with time. A linear time invariant system can be represented by constant coefficient differential equations. (In linear time varying system the coefficients of the differential equations governing the system are function of time).

Q.16 Define transfer function?

Ans. The transfer function of a system is defined as the ratio of Laplace transform of output to Laplace transform of input with zero initial conditions. It is also defined of the Laplace transform of the impulse response of system with zero initial conditions.

Q.17 State whether transfer function technique is applicable to non-linear system and whether the transfer function is independent of the input of a system?

Ans. (i) The transfer function technique is not applicable to non-linear system.

(ii) The transfer function of a system is independent of input and depends only on system parameters but the output of a system depends on input.

Q.18 What are the basic elements used for modelling mechanical translational system?

Ans. The model of mechanical translational system can be obtained by using three basic elements mass, spring and dashpot.

Q.19 Write the force balance equation of ideal mass element?

Ans. Let a force be applied to an ideal mass M. The mass will offer an opposing force, f_m which is proportional to acceleration.

$$\text{Therefore, } f = f_m = M\frac{d^2x}{dt^2}$$

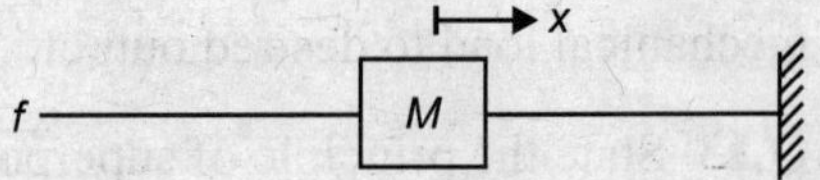

Q.20 Write the force balance equation of ideal dashpot.

Ans. Let a force f be applied to an ideal dashpot, with viscous frictional coefficient B. The dashpot will offer an opposing force, f_b which is proportional to velocity.

x

f

B

$$f = f_b = \frac{Bdx}{dt}$$

x_1 x_2

f

$$f = f_b = B\frac{d}{dt}(x_1 - x_2)$$

Q.21 Write the force balance equation of ideal spring?

Ans. Let a force f be applied to an ideal spring with spring constant K. The spring will offer as opposing force f_K which is proportional to displacement.

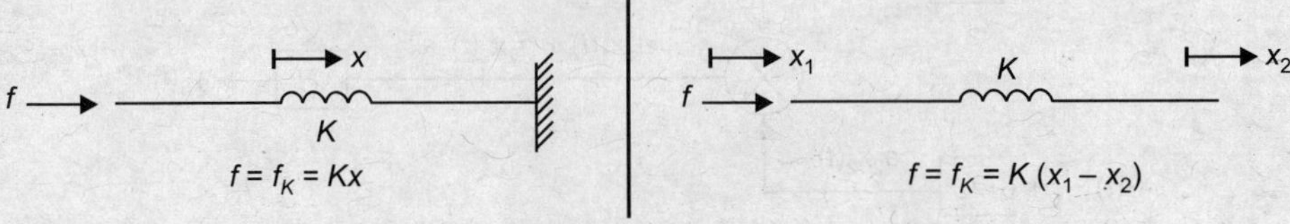

Q.22 What are the basic elements used for modelling mechanical rotational system?

Ans. The model of mechanical rotational system can be obtained by using three basic elements mass with moment of inertia, J dashpot with rotational frictional coefficient, B and spring with stiffness, K.

Q.23 Name the two types of electrical analogous for mechanical system.

Ans. The two types of analogies for the mechanical system are force-voltage and force current analogy.

Q.24 Write the analogous electrical elements in force-voltage analogy for the elements of mechanical translational system

Ans.

Force, $f \rightarrow$ voltage, e | Mass, $M \rightarrow$ Inductance, L

Velocity, $v \rightarrow$ Current, i | Stiffness, $K \rightarrow$ Inverse of capacitance, $1/C$

Displacement, $x \rightarrow$ Charge, q | Newton's second $\rightarrow$ KVL

Frictional coefficient $B \rightarrow$ Resistance, R | $\sum f = 0 \quad \sum V = 0$

Q.25 Write the analogous electrical elements in force current analogy for the elements of mechanical translational system.

Ans.

Force, $f \rightarrow$ current, i | Mass, $M \rightarrow$ Capacitance, C

Velocity, $v \rightarrow$ Voltage, v | Stiffness, $K \rightarrow$ Inverse of inductance, $1/C$

Displacement, $x \rightarrow$ Flux, ϕ | Newton's second $\rightarrow$ KCL

frictional coefficient, $B \rightarrow$ Conductance

$G = 1/R$ | $\sum f = 0 \quad \sum i = 0$

Q.26 What is block diagram? What are the basic components of block diagram?

Ans. A block diagram of system is pictorial representation of the functions performed by each component of the system and shows the flow of signals. The basic elements of block diagram are block, branch point and summing point.

Q.27 What is the basic for framing the rules of block diagram reduction technique?

Ans. The ruless for block diagram reduction techniques are framed such that any modification made on the diagram does not alter the input output relation.

Q.28 Write the rule for moving the summing point ahead of a block.

Ans.

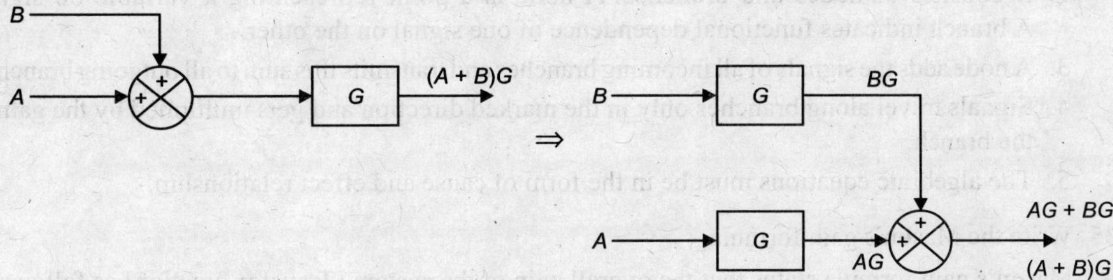

CHAPTER 3

Q.29 Write the rule for eliminating negative feedback loop?

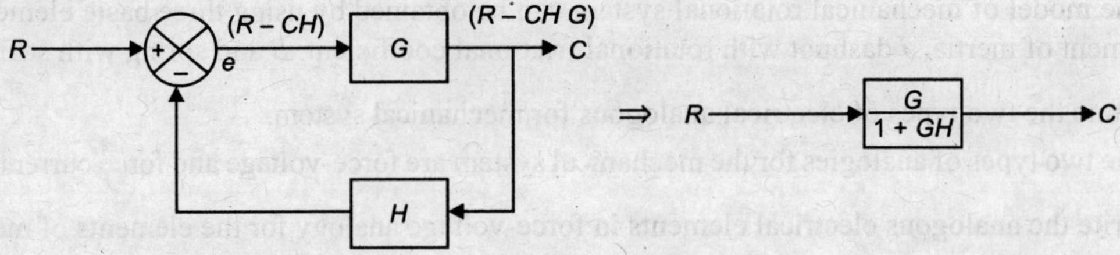

Ans. ***Proof:***

$$e = (R - CH)G \Rightarrow C + CHG = RG$$

$$C = RG - CHG \Rightarrow C\,(1 + GH) = RC$$

$$\frac{C}{R} = \frac{G}{1 + GH}$$

Q.30 What is a signal flow graph?

Ans. A signal flow graph is a diagram that represents a set of simultaneous linear algebraic equations. By taking Laplace transform, the time domain differential equations governing a control system can be transfereed to a set of algebraic equations in S-domain. The signal flow graph of the system can be constructed using these equations.

Q.31 What is transmittance?

Ans. The transmittance is the gain acquired by the signal when it travels from one node to another node in signal flow graph.

Q.32 What is Sink and Source?

Ans. Source is the input node in the signal flow graph and it has only outgoing branches. Sink is a output node in the signal flow graph and it has only incoming branches.

Q.33 Define non-touching loop.

Ans. The loops are said to be non-touching if they do not have common nodes.

Q.34 What are the basic proporties of signal flow graph?

Ans.
1. Signal flow graph is applicable to linear system.
2. It consists of nodes and branches. A node is a point representing a variable on signal. A branch indicates functional dependence of one signal on the other.
3. A node adds the signals of all incoming branches and transmits this sum to all outgoing branches.
4. Signals travel along branches only in the marked direction and gets multiplied by the gain of the branch.
5. The algebraic equations must be in the form of cause and effect relationship.

Q.35 Write the Mason's gain formula.

Ans. Mason's gain formula states that the overall gain of the system [Transfer function] as follows, overall gain, $T = \frac{1}{\Delta} \sum_{K} P_k \Delta_k$

$T = T(s)$ = Transfer function of the system

K = Number of forward paths in the signal flow graph.

P_R = Forward path gain of Kth forward path.

Δ = 1– [Sum of individual loop gains] + [Sum of gain products of all possible combinations of two non-touching loops] – [Sum of gain products of all possible combinations of three non-touching loops] + …

ΔK = [Δ for that part of the graph which is not touching K^{th} forward path].

Q.36 For the given signal flow graph, identify the number of forward path and number of individual loop.

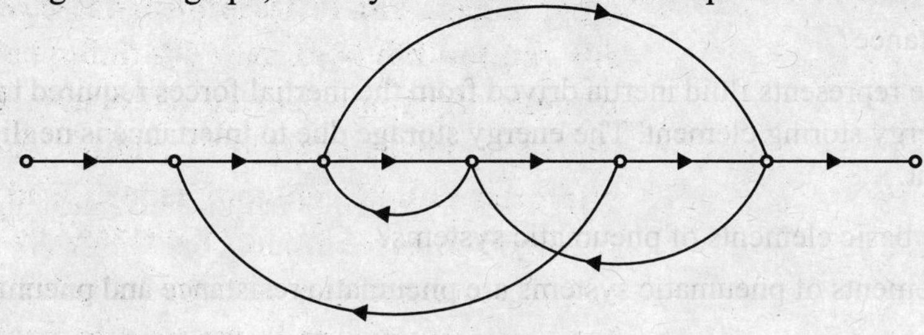

Ans. No. of forward paths = 2

No. of individual loop = 3.

Q.37 What are the basic elements of thermal system?

Ans. The basic elements of thermal system are thermal resistance and thermal capacitance.

Q.38 Define thermal resistance?

Ans. The thermal resistance for heat transfer between two substances is defined as the ratio of change in temperature and change in heat flow rate.

$$\text{Thermal resistance, } R = \frac{\text{Change in temperature, °C}}{\text{Change in heat flow rate, Kcal/sec.}}$$

Q.39 Define thermal capacitance?

Ans. Thermal capacitance is defined as the ratio of change in heat stored and change in temperature.

$$\text{Thermal capacitance, } C = \frac{\text{Change in heat stored, Kcal}}{\text{Change in temperature, °C}}$$

Q.40 Mention the electrical analogous of simple thermal system?

Ans. The electrical analogous of simple Ist order thermal system is R–C parallel circuit.

Q.41 What are the basic elements of hydraulic system?

Ans. The basic elements of hydraulic system are resistance, inductance and capacitance.

Q.42 Define hydraulic resistance?

Ans. The resistance for liquid flow is defined as the change in the level difference necessary to cause the unit change in flow rate.

$$R = \frac{\text{Change in level difference, } m}{\text{Change in flow node, m}^2/\text{sec}}$$

Q.43 Define hydraulic capacitance?

Ans. The capacitance C of tank is defined as the change in quantity of stored liquid necessary to cause a unit change in head (height).

$$C = \frac{\text{Change in liquid stored, m}^3}{\text{Change in head, m.}}$$

Q.44 What is inertance?

Ans. The inertance represents fluid inertia drived from the inertial forces required to accelerate a fluid in a pipe. It is a energy storing element. The energy storage due to intertance is negligible compared to capacitance element.

Q.45 What are the basic elements of pneumatic systems?

Ans. The basic elements of pneumatic systems are pneumatic resistance and pneumatic capacitance.

Q.46 Define pneumatic resistance?

Ans. The gas flow resistance, R may be defined as follows:

$$R = \frac{\text{Change in gas pressure difference N/m}^2}{\text{Change in gas flow rate, m}^2/\text{sec}}$$

EXERCISE PROBLEMS

1. Write the circuit differential equations governing the mechanical system. Also draw the force-voltage and force-current analogous circuit for Fig. 1.

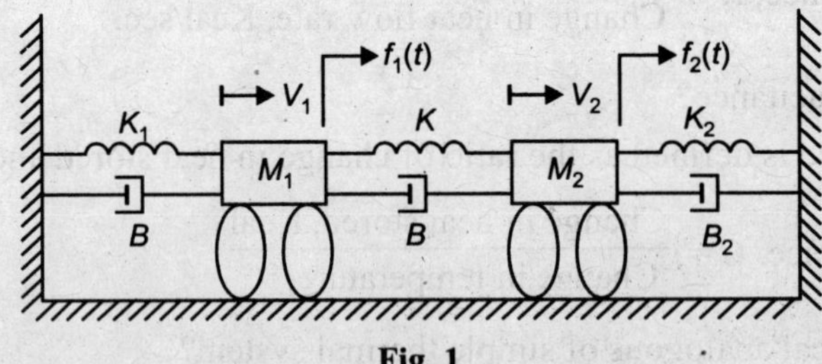

Fig. 1

Ans. $$f_1(t) = M_1 \frac{dV_1}{dt} + B_1 V_1 + B(V_1 - V_2) + K_1 \int V_1\, dt + K \int (V_1 - V_2)\, dt$$

$$f_2(t) = M_2 \frac{dV_2}{dt} + B_2 V_2 + B(V_2 - V_1) + K_2 \int V_2\, dt + K \int (V_2 - V_1)\, dt$$

Note: $V = \dfrac{dx}{dt}$.

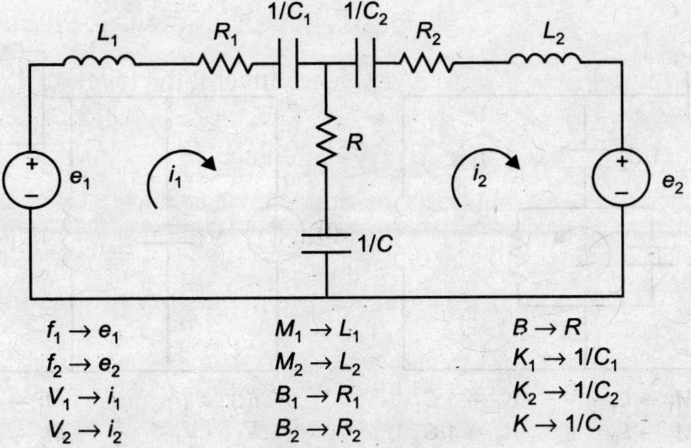

Fig. 2: Voltage analogous circuit.

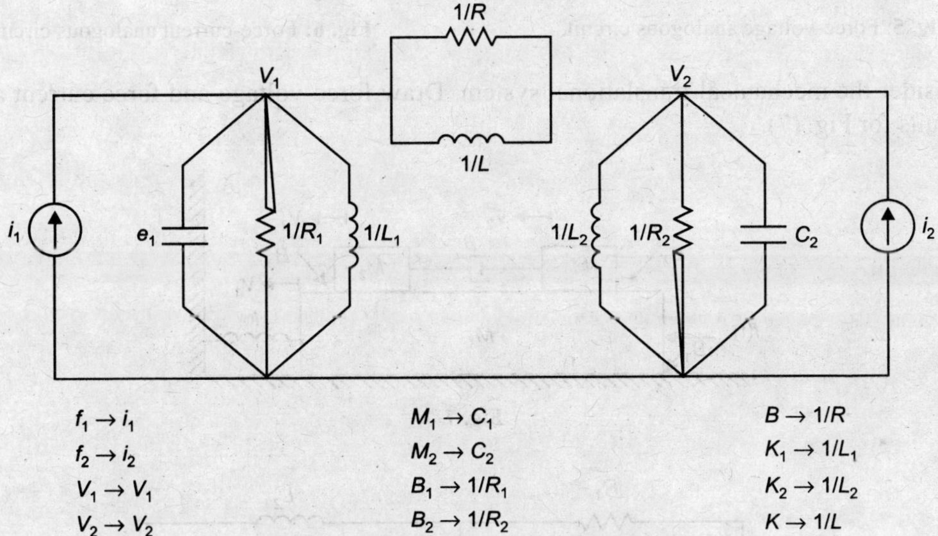

Fig. 3: Force-current analogous circuit.

2. Write the differential equations governing the mechanical system. Draw the force-voltage and force-current analogous circuits for Fig. (4).

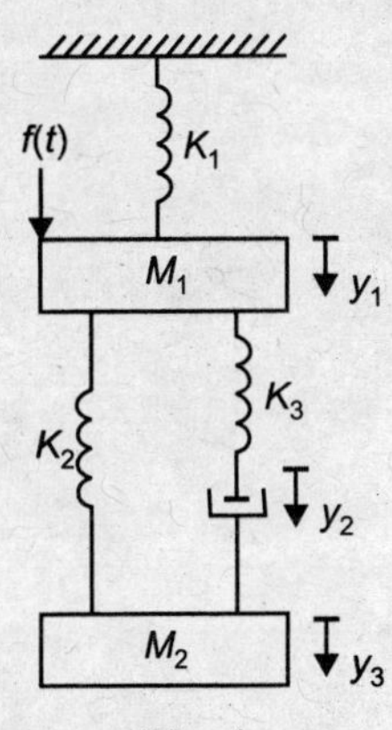

Fig. 4

Ans.

$$M_1 \frac{dV_1}{dt} + K_1 \int V_1\, dt + K_2 \int (V_1 - V_3)\, dt + K_3 \int (V_1 - V_2)\, dt = f(t)$$

$$K_3 \int (V_2 - V_1)\, dt + B\,(V_2 - V_3) = 0$$

$$M_2 \frac{dV_3}{dt} + B\,(V_3 - V_2) + K_2\,(V_3 - V_1)\, dt = 0$$

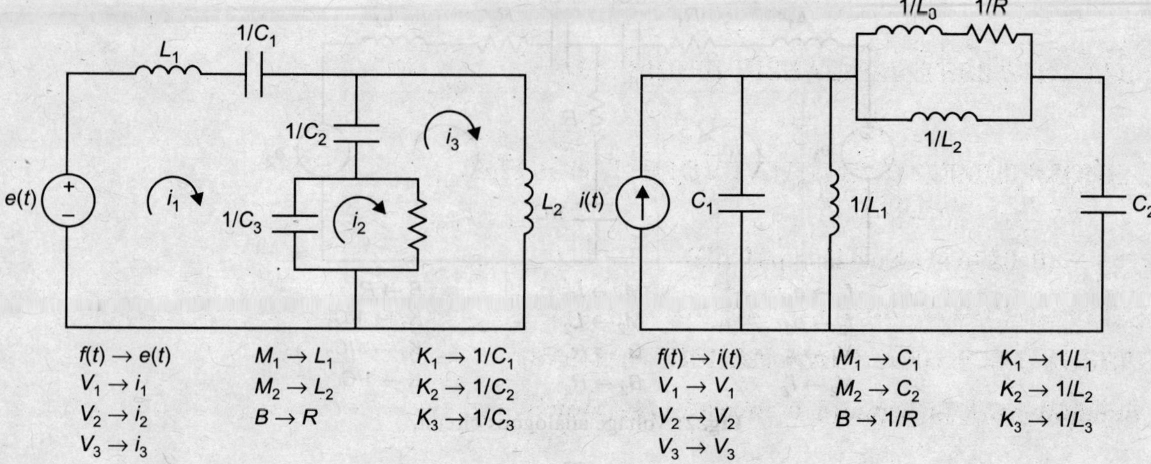

Fig. 5: Force-voltage analogous circuit.

Fig. 6: Force-current analogous circuit.

3. Consider the mechanical translational system. Draw force-voltage and force current analogous circuits for Fig. (7).

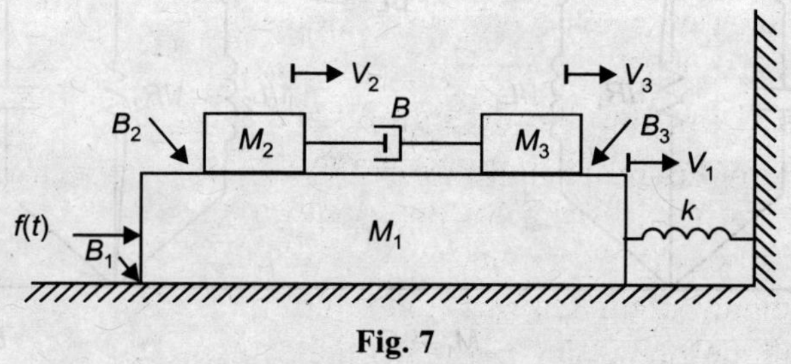

Fig. 7

Ans.

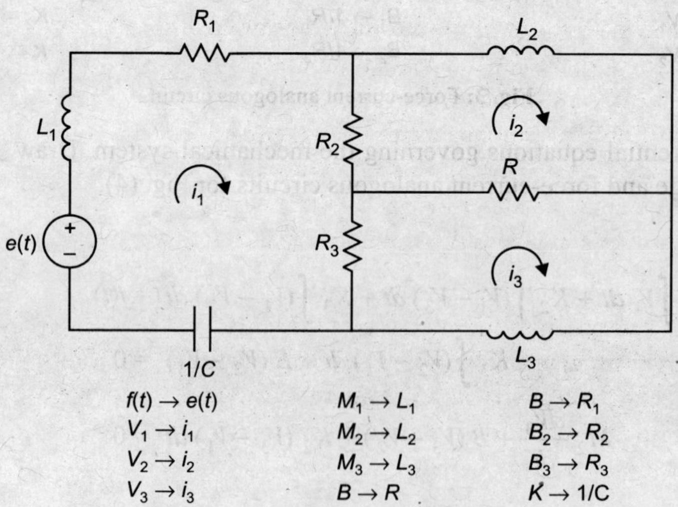

Fig. 8: Force-voltage analogous circuit.

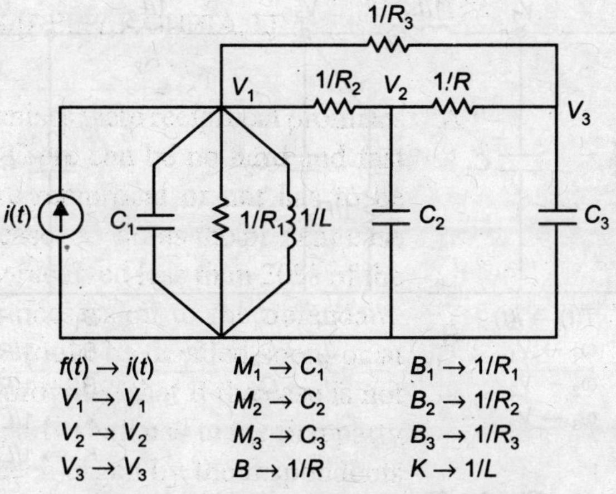

Fig. 9: Force-current analogous circuit.

4. Write the differential equations governing the rotational mechanical system. Also draw the torque-voltage and torque-current analogous circuits for Fig. (10).

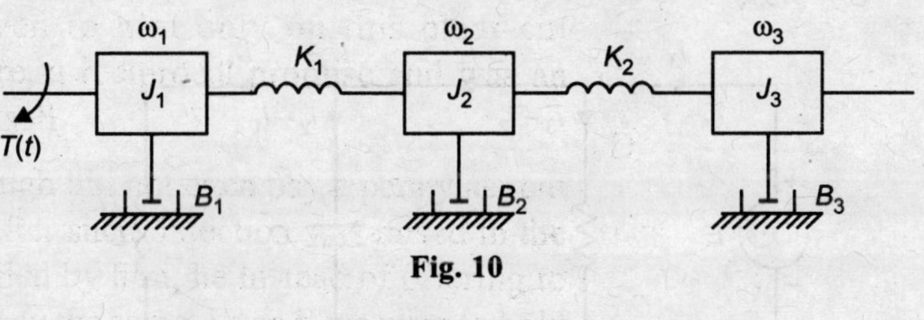

Fig. 10

Ans.

$$J_1 \frac{d\omega_1}{dt} + B_1\omega_1 + K_1 \int(\omega_1 - \omega_2)\, dt = T(t)$$

$$J_2 \frac{d\omega_2}{dt} + B_2\, \omega_2 + K_1 \int(\omega_2 - \omega_1)\, dt + K_2 \int(\omega_2 - \omega_3)\, dt = 0$$

$$J_3 \frac{d\omega_3}{dt} + B_3\, \omega_3 + K_2 \int(\omega_3 - \omega_2)\, dt = 0$$

L_1, R_1, L_2, R_2, L_3, $e(t)$, i_1, C_1, i_2, C_2, i_3, R_3

$T(t) \to e(t)$	$J_1 \to L_1$	$B_1 \to R_1$
$\omega_1 \to i_1$	$J_2 \to L_2$	$B_2 \to R_2$
$\omega_2 \to i_2$	$J_3 \to L_3$	$B_2 \to R_3$
$\omega_3 \to i_3$		$K_1 \to 1/C_1$
		$K_2 \to 1/C_2$

Fig. 11: Torque-voltage analogous.

CHAPTER 3

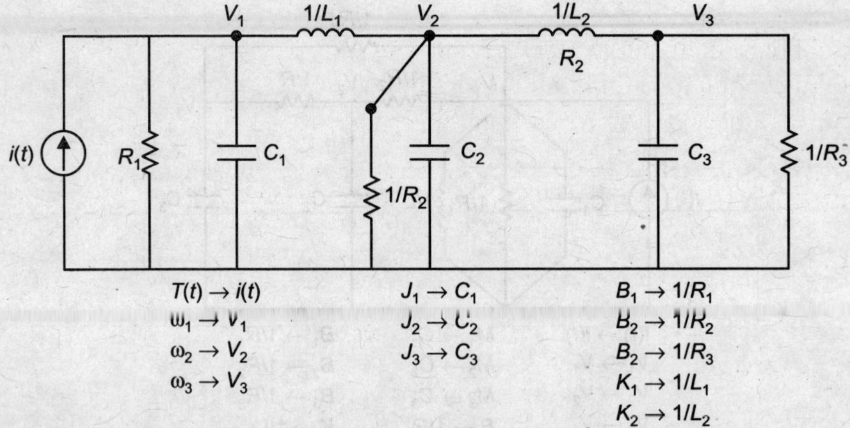

Fig. 12: Torque-current analogous circuit.

5. In an electrical circuit the elements resistance, capacitance, and inductance and connected in parallel across the voltage source E. Draw translation mechanical analogous system and rotational mechanical analogous system for Fig. (13).

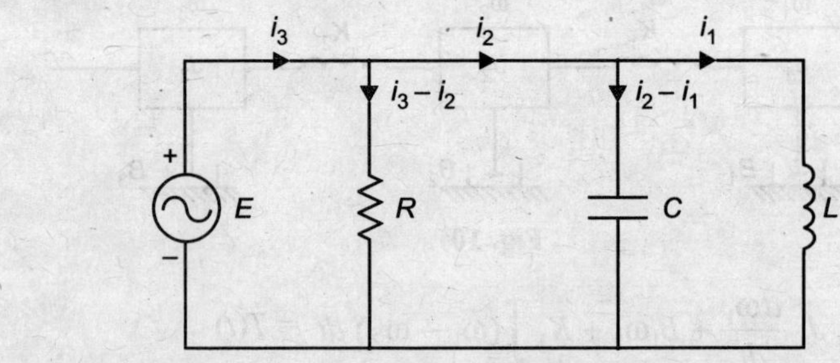

Fig. 13

Ans.

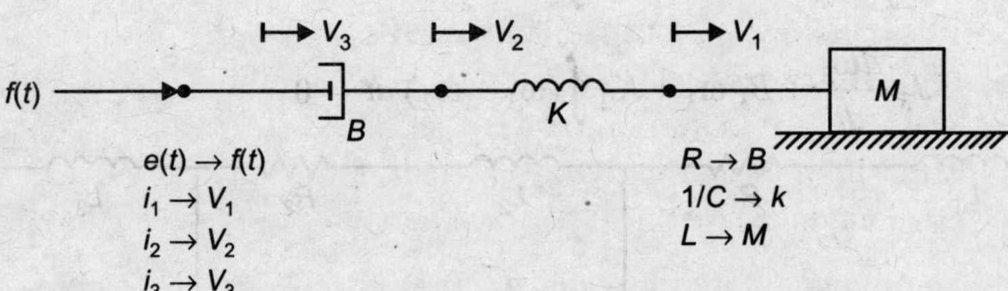

Fig. 14: Analogous mechanical traslational system.

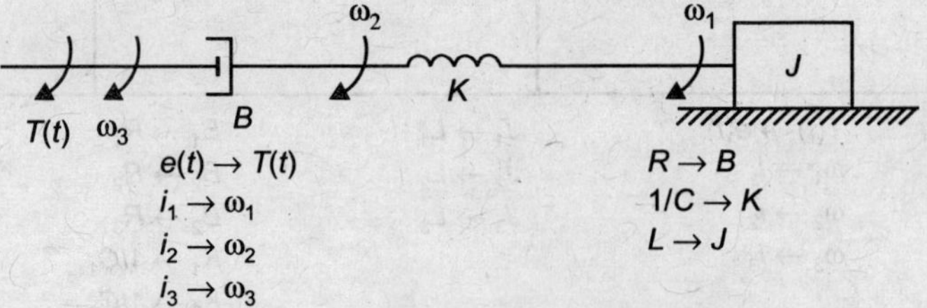

Fig. 15: Analogous mechanical rotational system.

6. Use reduction method to find out *C*/*R* by for the following diagram for Fig. (16).

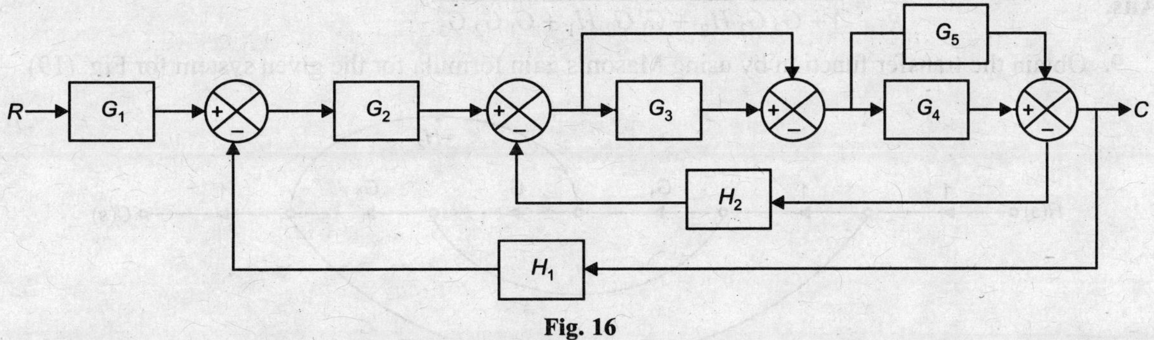

Fig. 16

Ans. $$\frac{C}{R} = \frac{G_1 G_2 (1+G_3)(G_4+G_5)}{[1+(1+G_3)(G_4+G_5)H_2+(1+G_3)(G_4+G_5)G_2 H_1]}$$

7. By using the following block diagram obtain *C*/*R* for Fig. (17).

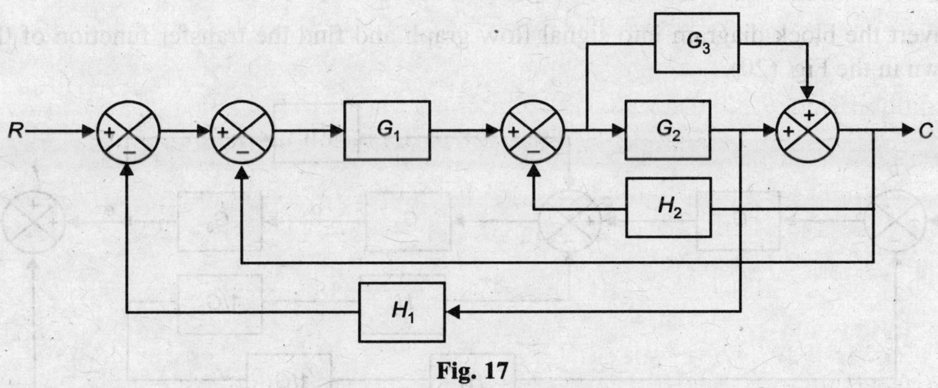

Fig. 17

Ans. $$\frac{C}{R} = \frac{G_1 G_2 + G_1 G_3}{[1+G_1 G_2 H_1 + G_1 G_2 + G_1 G_2 + G_1 G_3 + G_2 H_2 + G_3 H_2]}$$

8. Obtain *C*/*R* for Fig. (18).

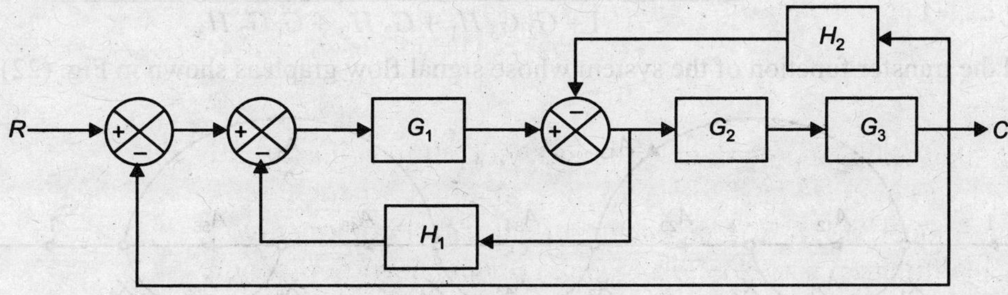

Fig. 18

Ans. $\dfrac{C}{R} = \dfrac{G_1 G_2 G_3}{1 + G_2 G_3 H_2 + G_1 G_2 H_1 + G_1 G_2 G_3}$

9. Obtain the transfer function by using Mason's gain formula for the given system for Fig. (19).

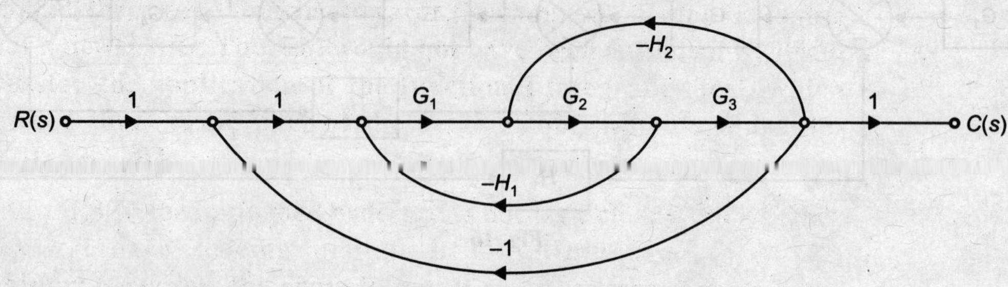

Fig. 19

Ans. $\dfrac{C(s)}{R(s)} = \dfrac{G_1 G_2 G_3}{1 + G_1 G_2 H_1 + G_2 G_3 H_2 + G_1 G_2 G_3}$

10. Convert the block diagram into signal flow graph and find the transfer function of the system shown in the Fig. (20).

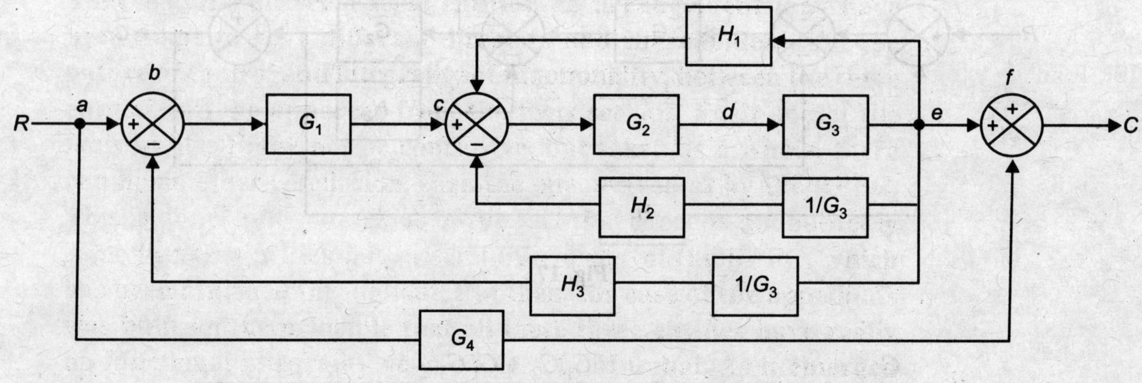

Fig. 20

Ans. $\dfrac{G_1 G_2 G_3 + G_4 + G_2 G_3 G_4 H_1 + G_2 G_4 H_2 + G_1 G_2 G_4 H_3}{1 + G_2 G_3 H_1 + G_2 H_2 + G_1 G_2 H_3}$

11. Find the transfer function of the system whose signal flow graph as shown in Fig. (22).

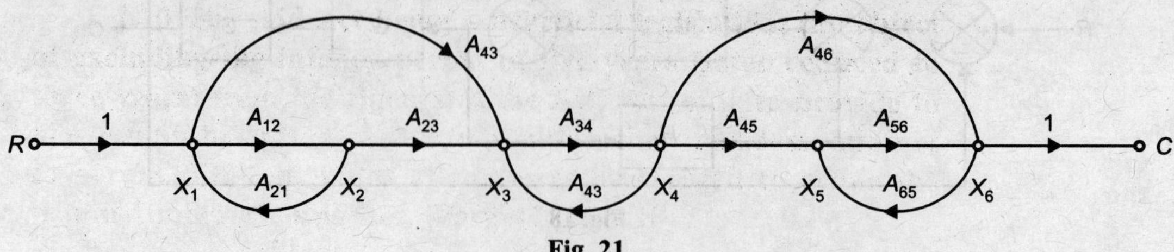

Fig. 21

Ans. $\dfrac{C}{R} = \dfrac{A_{12}\,A_{23}\,A_{34}\,A_{45}\,A_{56} + A_{13}\,A_{34}\,A_{45}\,A_{56} + A_{12}\,A_{23}\,A_{34}\,A_{46} + A_{13}\,A_{34}\,A_{46}}{1-(A_{12}\,A_{21} + A_{34}\,A_{43} + A_{56}\,A_{65}) + (A_{12}\,A_{21}\,A_{34}\,A_{43} + A_{12}\,A_{21}\,A_{56}\,A_{65} + A_{34}\,A_{43}\,A_{56}\,A_{56}) - (A_{12}\,A_{21}\,A_{34}\,A_{43}\,A_{56}\,A_{65})}$

12. Obtain the transfer function of the system shown in Fig. (22).

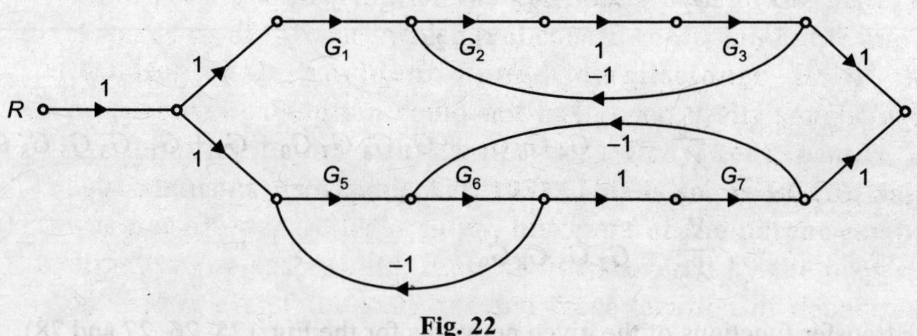

Fig. 22

Ans.

$$\frac{C}{R} = \frac{G_1\,G_2\,G_3\,G_4\,(1 + G_5\,G_6 + G_6\,G_7) + G_4\,G_5\,G_6\,G_7\,(1 + G_1\,G_2 + G_2\,G_3)}{1 + G_1\,G_2 + G_2\,G_3 + G_5\,G_6 + G_6\,G_7 + G_1\,G_2\,G_3\,G_6 + G_5\,G_6\,G_2\,G_3 + G_1\,G_2\,G_6\,G_7 + G_2\,G_3\,G_6\,G_7}$$

13. Find X_8/X_1 for Fig. (23).

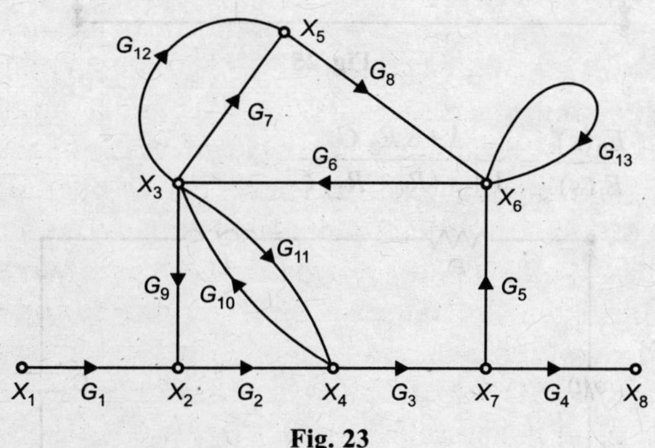

Fig. 23

Ans.
$$\frac{X_8}{X_1} = \frac{[G_1\,G_2\,G_3\,G_4][1-(G_7\,G_{12} + G_6\,G_7\,G_8\,G_{13}) + G_7\,G_{12}\,G_{13}]}{1-\{G_2\,G_9\,G_{10} + G_{10}\,G_{11} + G_2\,G_3\,G_5\,G_6\,G_9 + G_3\,G_5\,G_6\,G_{11} + G_7\,G_{12} + G_6\,G_7\,G_8 + G_{13}\} + G_2\,G_9\,G_{10}\,G_{13} + G_{10}\,G_{11}\,G_{13} + G_7\,G_{12}\,G_{13}}$$

14. Obtain $\dfrac{X_8}{X_1}$ and $\dfrac{X_8}{X_2}$ from the signal flow graph shown in Fig. 24.

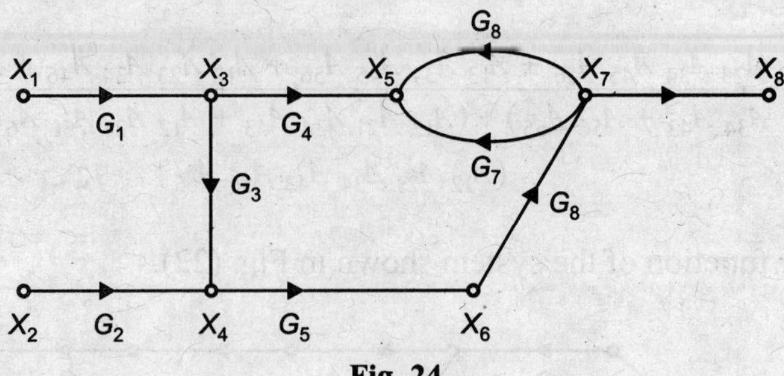

Fig. 24

Ans.

$$\frac{X_8}{X_1} = G_1\,G_4\,G_6\,G_9 \;+ G_1\,G_4\,G_7\,G_9 + G_9 + G_1\;\;G_3\,G_5\,G_8\,G_9$$

$$\frac{X_8}{X_2} = G_2\,G_5\,G_8\,G_9$$

15. Find the transfer functions of the given networks for the Fig. (25, 26, 27 and 28).

(a)

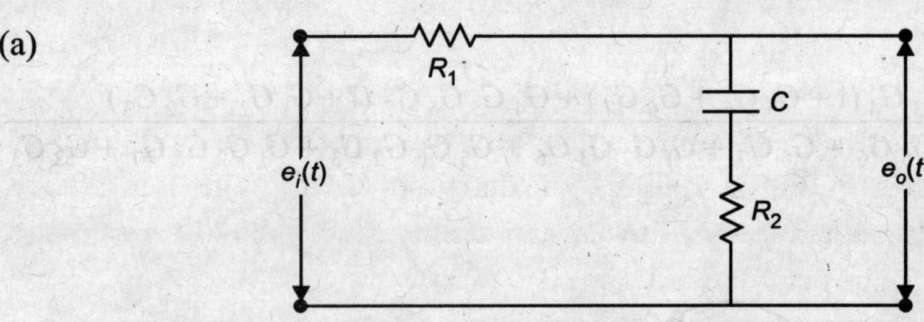

Fig. 25

Ans.

$$\frac{E_0(s)}{E_i(s)} = \frac{1 + s\,R_2\,C}{1 + s\,(R_1 + R_2)\,C}$$

(b)

Fig. 26

Ans.

$$\frac{E_0(s)}{E_i(s)} = \frac{1}{s^2\;LC + sRC + 1}$$

(c)

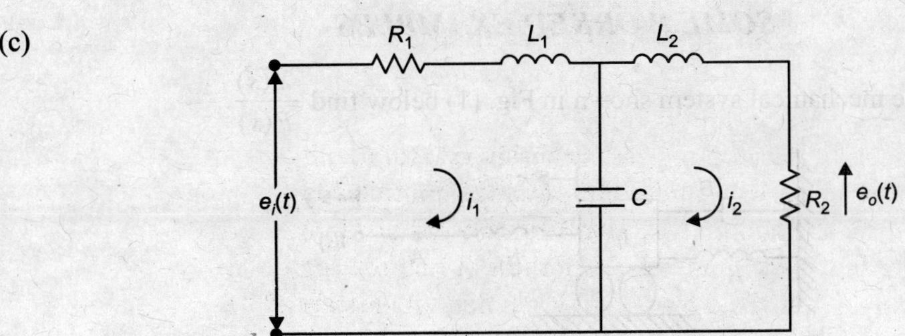

Fig. 27

Ans. $$\frac{E_0(s)}{E_i(s)} = \frac{sR_2C}{(s^2L_1C + sR_1C + 1)(s^2L_2C + sR_2C + 1) - 1}$$

(d)

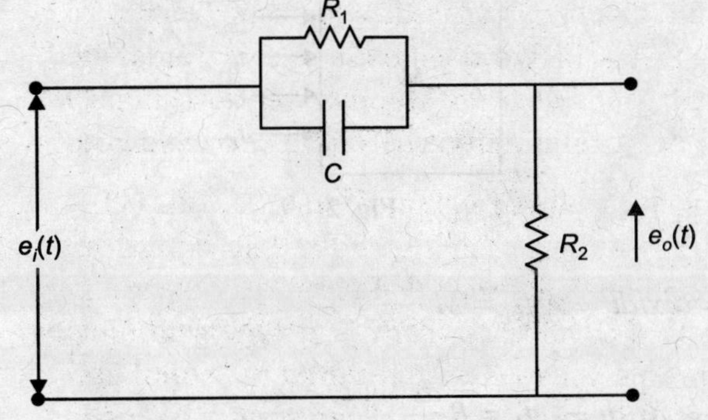

Fig. 28

Ans. $$\frac{E_0(s)}{E_i(s)} = \frac{sR_1R_2C + R_2}{sR_1R_2C + (R_1 + R_2)}$$

16. Find the transfer function of the circuit given below?

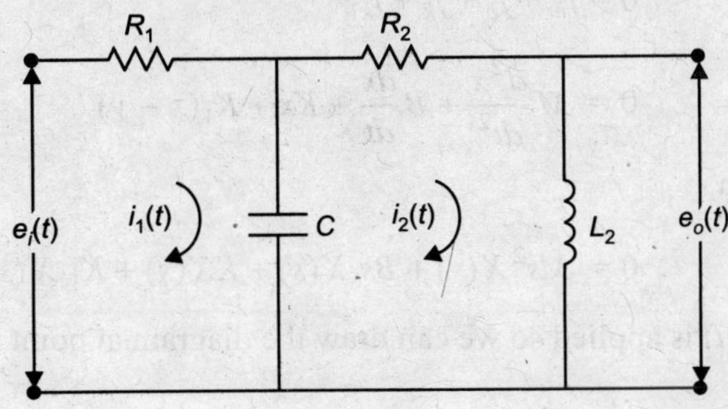

Fig. 29

Ans. $$\frac{E_o(s)}{E_i(s)} = \frac{s^2L_2C}{[sR_1C + 1][s^2(L_1 + L_2)C\, sR_2C + 1] - 1}$$

SOME WORKED EXAMPLES

EXAMPLE 1: For the mechanical system shown in Fig. (1) below find $\frac{X(s)}{F(s)}$.

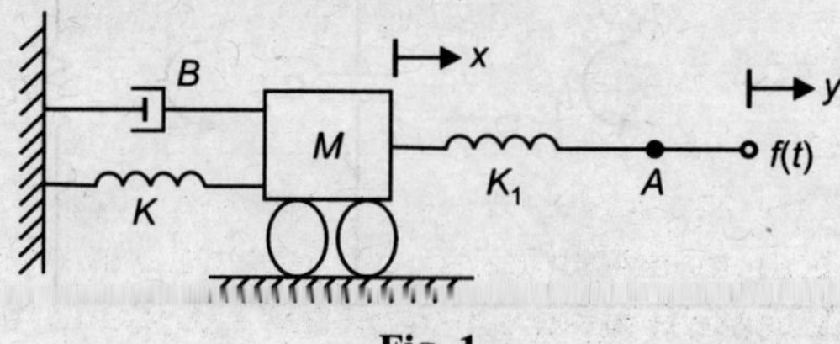

Fig. 1

Solution: The forces which are acting on mass M is f_m, f_b, f_k and f_{k1}.

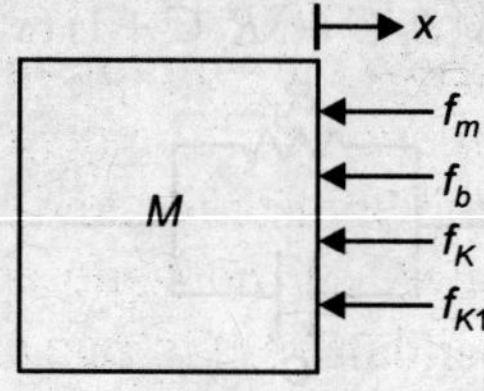

Fig. 2

$$f_m \propto d^2x/dt^2 \Rightarrow f_m = M\frac{d^2x}{dt^2}$$

$$f_b \propto dx/dt \Rightarrow f_b = B\frac{dx}{dt}$$

$$f_k \propto x \Rightarrow f_k = K_x$$

$$f_{k1} \propto (x-y) \Rightarrow f_{k1} = K_1(x-y)$$

Applied force on M is zero, so the force balance equation becomes.

$$0 = f_m + f_b + f_k + f_{k1}$$

$$0 = M\frac{d^2x}{dt^2} + B\frac{dx}{dt} + Kx + K_1(x-y)$$

on taking laplace transform,

$$0 = Ms^2X(s) + Bs.X(s) + KX(s) + K_1\,X(s) - K_1\,Y(s) \qquad (1)$$

At point A, the force $f(t)$ is applied so we can draw the diagram at point A,

Fig. 3

For the above Fig. (3), the force balance equation is,

$$f_{k1} \propto (y - x)$$

$$f_{k1} = K_1(y - x)$$

We know applied force at point A is $f(t)$

$$\therefore \quad f(t) = f_{k1}$$

$$f(t) = K_1(y - x)$$

on taking laplace transform,

$$F(s) = K_1\, Y(s) - K_1\, X(s) \tag{2}$$

Substituting equation (2) in (1) we obtain,

$$\boxed{\frac{X(s)}{F(s)} = \frac{1}{(Ms^2 + Bs + K)}} \tag{3}$$

Result: The equation (3) is the required transfer function.

EXAMPLE 2: Determine the transfer function of the circuit given below.

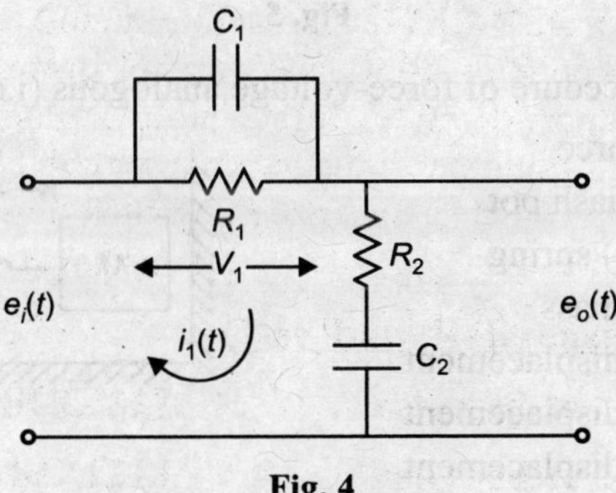

Fig. 4

Solution: The required transfer function is $\dfrac{E_o(s)}{E_i(s)}$

From above Fig. (4), we can write,

$$e_o(t) = R_2 i_1(t) + \frac{1}{C_2}\int i_2(t)$$

on taking laplace transform

$$E_o(s) = R_2\, I_1(s) + \frac{I_1(s)}{s\, C_2}$$

$$E_o(s) = I_1(s)\left[R_2 + \frac{1}{sC_2}\right]$$

$$E_o(s) = I_1(s)\left[\frac{sR_2\, C_2 + 1}{s\, C_2}\right] \tag{1}$$

Similarly for $e_i(t)$, $$E_i(s) = I_1(s)\left[\left(\frac{R_1}{R_1\,C_1\,s+1}\right)+\left(\frac{R_2\,C_2 s+1}{s\,C_2}\right)\right] \quad (2)$$

Divide equation (1) by (2)

$$\boxed{\frac{E_o(s)}{E_i(s)} = \frac{(1+R_1C_1s)\,(1+R_2C_2s)}{(1+R_1C_1s)\,(1+R_2C_2s)+R_1C_2s}} \quad (3)$$

Result: The Eq. (3) is the required transfer function.

EXAMPLE 3: In an electrical circuit, the elements resistance, capacitance and inductance are connected in parallel across the voltage source E. Draw the translational mechanical analogous circuit.

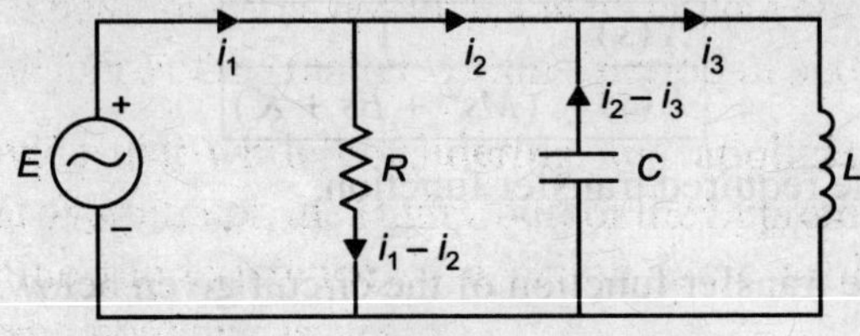

Fig. 5

Solution: This is the reverse procedure of force-voltage analogous (i.e.,) voltage-force analogous.

E is analogous to $f \rightarrow$ (i.e) force
R is analogous to $B \rightarrow$ (i.e) dash pot
C is analogous to $1/K \rightarrow$ (i.e) spring
L is analogous to $M \rightarrow$ (i.e) mass
i_1 is analogous to $x_1 \rightarrow$ (i.e) displacement
i_2 is analogous to $x_2 \rightarrow$ (i.e) displacement
i_3 is analogous to $x_3 \rightarrow$ (i.e) displacement

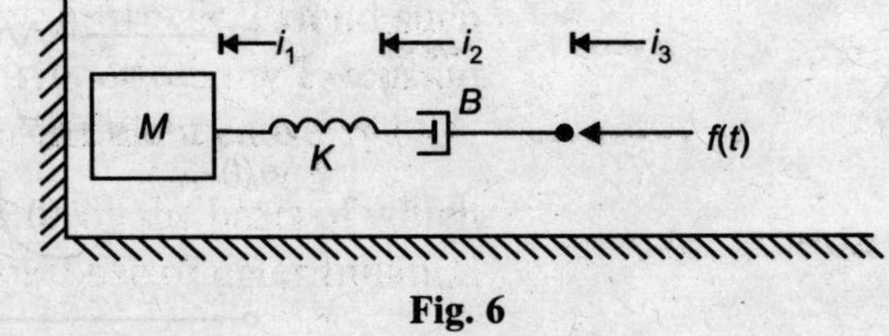

Fig. 6

by replacing the equivalent electrical to mechanical analogous, the following circuit to obtained.

EXAMPLE 4: Convert the below signal flow graph into block diagrams.

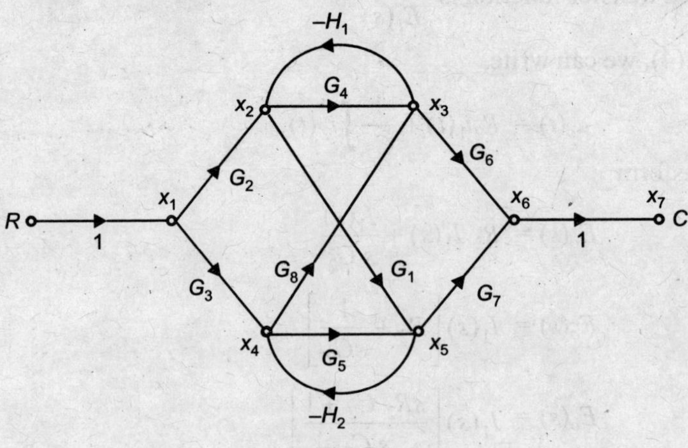

Fig. 7

Solution:

Step 1: Write the system equation for every node.

at node x_1	$x_1 = R(1)$	(1)
at node x_2	$x_2 = -x_3H_1 + x_1G_2$	(2)
at node x_3	$x_3 = x_2G_4 + x_4G_8$	(3)
at node x_4	$x_4 = x_1G_3 - x_5H_2$	(4)
at node x_5	$x_5 = x_2G_1 + x_4G_5$	(5)
at node x_6	$x_6 = x_3G_6 + x_5G_7$	(6)
at node x_7	$x_7 = x_6(1)$	(7)

Step 2: Combine all the equations for drawing block diagram by combining the Eq. (1) to (7), let draw the block diagram.

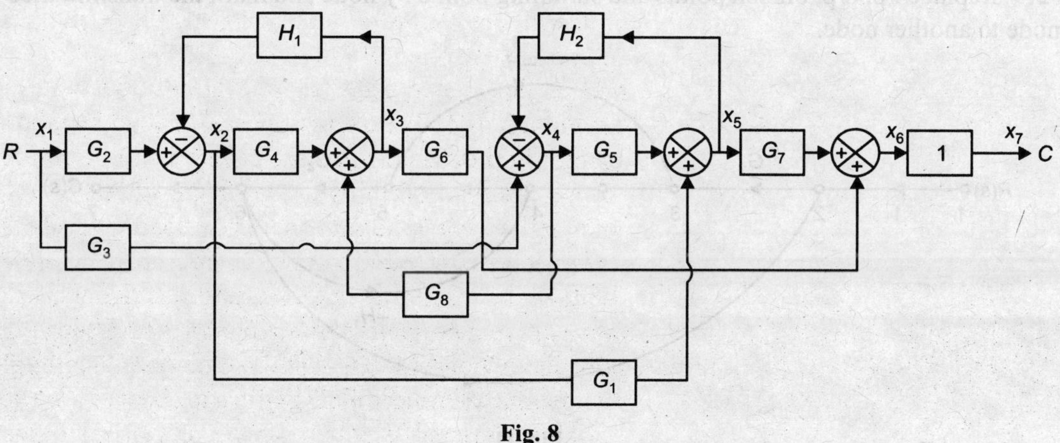

Fig. 8

The above Fig. (8) is the complete block diagram for the given signal flow graph.

EXAMPLE 5: Convert the given block diagram into signal flow graph:

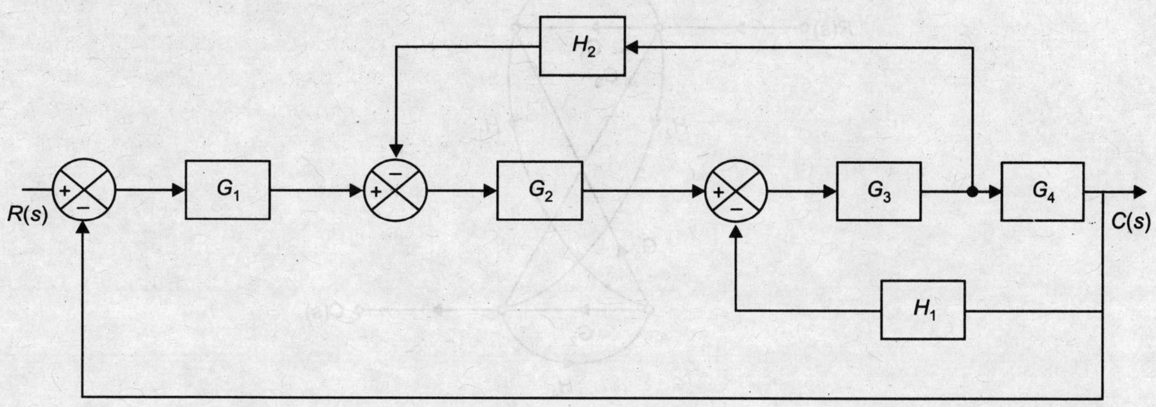

Fig. 9

Solution:

Step 1: Assign node number of input, output, summing points and branch points.

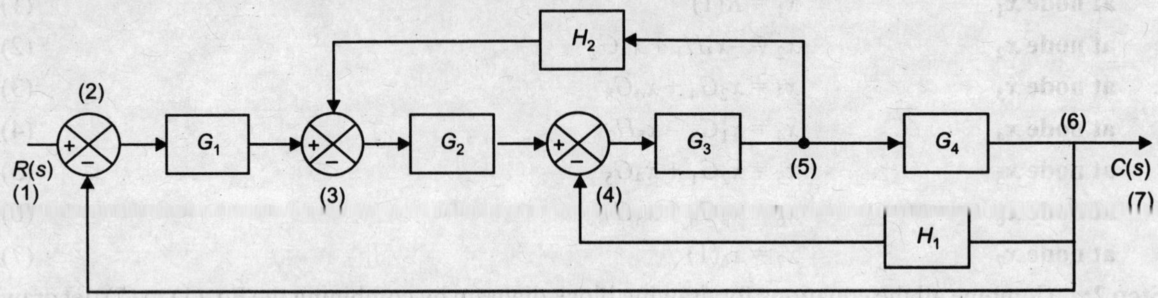

Fig. 10

Step 2: Replace i/p, o/p, branch points and summing points by node and mark the transmittance from one node to another node.

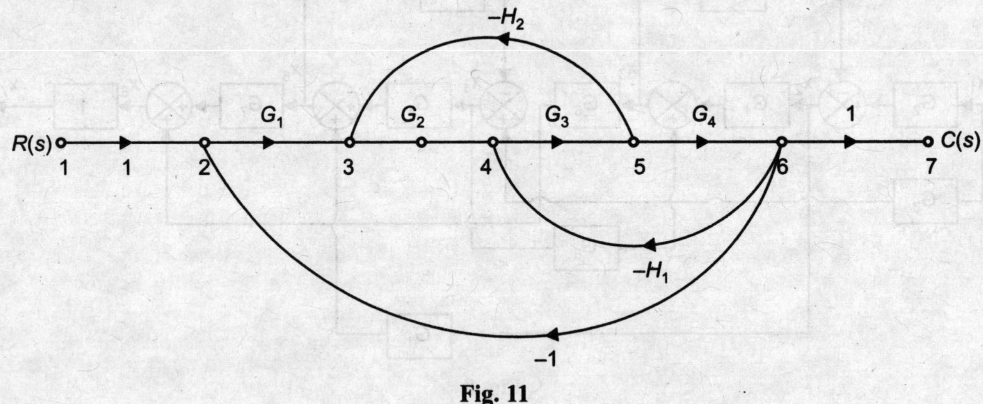

Fig. 11

EXAMPLE 6: Derive the transfer function of the given signal flow graph.

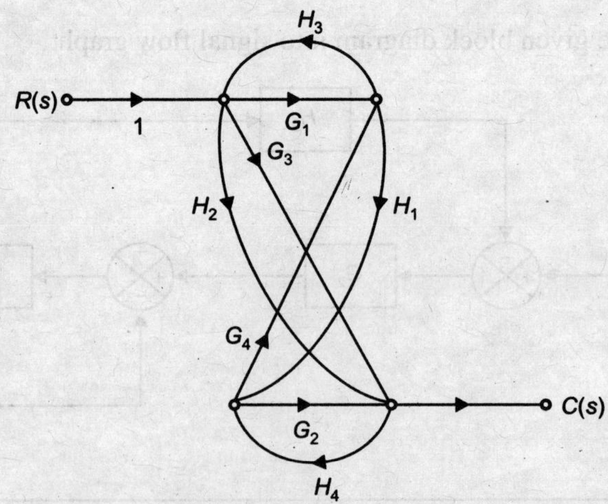

Fig. 12

Solution:

Step 1: The number of forward paths, $K = 2$

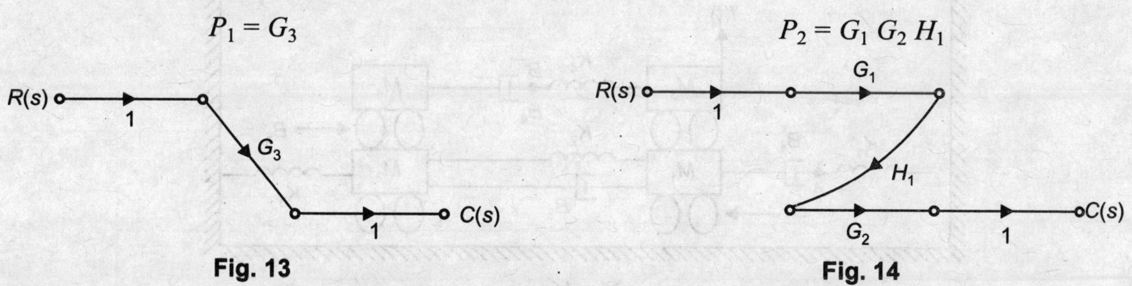

Fig. 13 **Fig. 14**

Step 2: The total number of individual loops are 4.

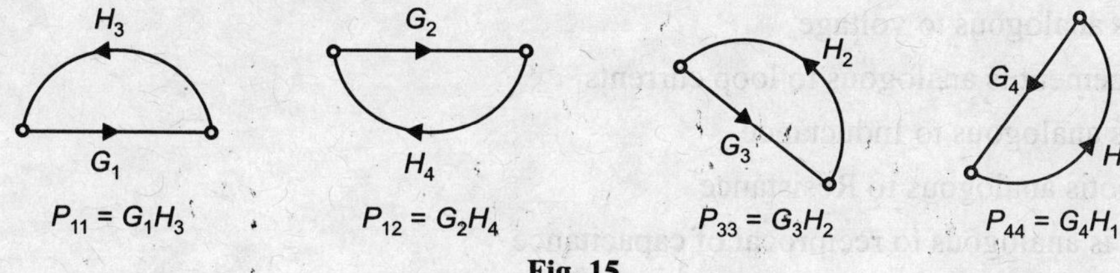

Fig. 15

Step 3: The total number of non touching loops are

$$P_{12} = G_1 G_2\, H_3\, H_4 \text{[combine 1st and 2nd loops]}$$

$$P_{34} = G_3\ G_4\ H_1\ H_2 \text{ [combine 3}^{\text{rd}}\text{ and 4}^{\text{th}}\text{ loop]}$$

Step 4: Find Δ

$$\Delta = 1 - \text{[sum of individual loop gain] + [sum of two non touching loop pairs] ...}$$

$$\Delta = 1 - [P_{11} + P_{22} + P_{33} + P_{44}] + [P_{12} + P_{34}]$$

$$\therefore \quad \Delta = 1 - [G_1H_3 + G_2H_2 + G_3H_2 + G_4H_1] + [G_1G_2H_3H_4 + G_3G_4H_1H_2]$$

Step 5: Find Δ_K

$$\Delta_2 = 1 - 0 = 1 \text{ [No non touching loop for path 2]}$$

$$\Delta_1 = 1 - G_4H_1 \text{ [The loop } G_1H_1 \text{ is not touching with path 1]}$$

The mason's gain formula,

$$T(s) = \frac{1}{\Delta}\Sigma P_k \Delta_k$$

where $K = 2$

$$T(s) = \frac{1}{\Delta}[P_1\Delta_1 + P_2\Delta_2]$$

$$\boxed{T(s) = \frac{1[G_3(1 - G_4H_1) + G_1G_2H_1(1)}{[1 - (G_1H_3 + G_2H_4 + G_3H_2 + G_4H_1) + (G_1G_2H_3H_4 + G_3G_4H_1H_2)]}}$$

CHAPTER 3

EXAMPLE 7: For the mechanical system shown in below Fig. (16), draw the force-voltage analogous circuit.

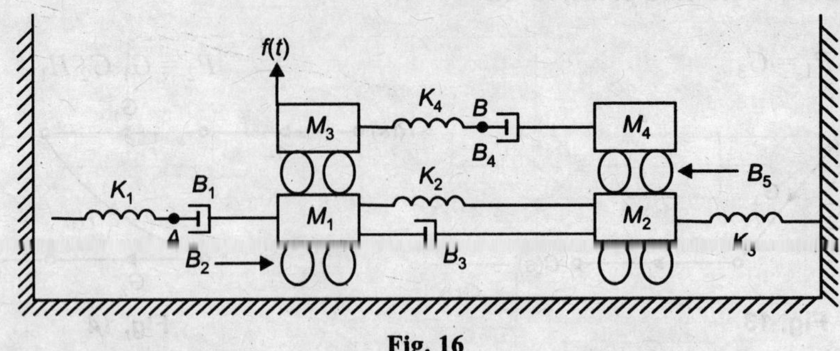

Fig. 16

Solution: Force-voltage analogous: In force voltage analogous we know

→ force is analogous to voltage

→ Displacement is analogous to loop currents

→ Mass is analogous to Inductance

→ Dash potis analogous to Resistance

→ Spring is analogous to reciprocal of capacitance

by use of above points,

$M_1 \rightarrow L_1$	$B_1 \rightarrow R_1$	$K_1 \rightarrow 1/C_1$	$x_1 \rightarrow i_1$	
$M_2 \rightarrow L_2$	$B_2 \rightarrow R_2$	$K_3 \rightarrow 1/C_2$	$x_2 \rightarrow i_2$	
$M_3 \rightarrow L_3$	$B_3 \rightarrow R_3$	$K_3 \rightarrow 1/C_3$	$x_3 \rightarrow i_3$	$f(t) = 0(t)$
$M_4 \rightarrow L_4$	$B_4 \rightarrow R_4$	$K_4 \rightarrow 1/C_4$	$x_4 \rightarrow i_4$	
	$B_5 \rightarrow R_5$		$x_5 \rightarrow i_5$	
			$x_6 \rightarrow i_6$	

Let draw the circuit,

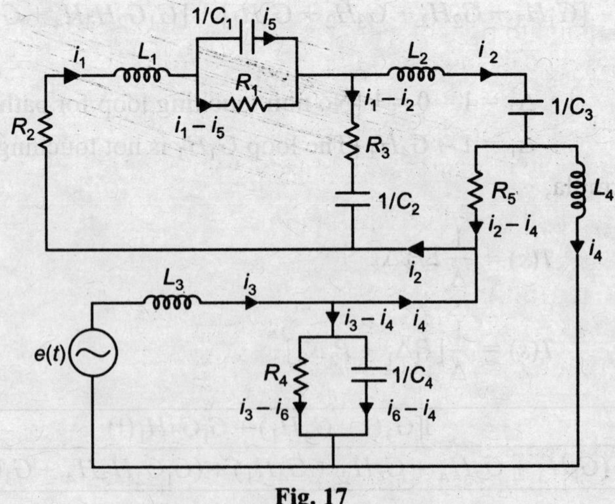

Fig. 17

EXAMPLE 8: Find the transfer function of the system using block diagram reduction for the Fig. shown below.

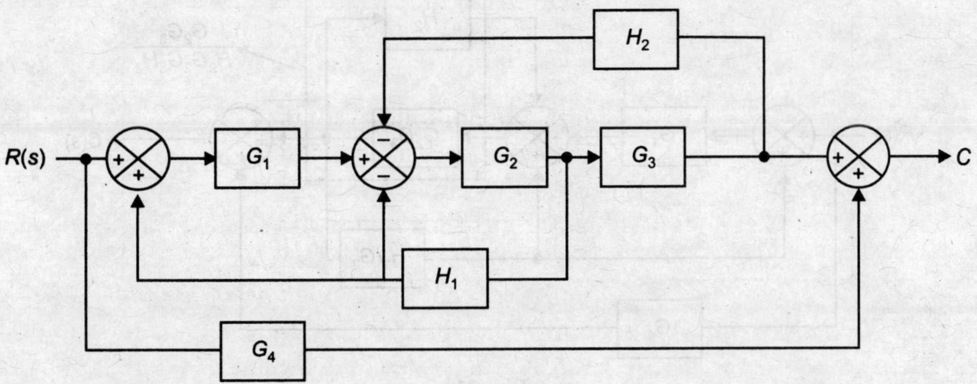

Fig. 18

Solution:

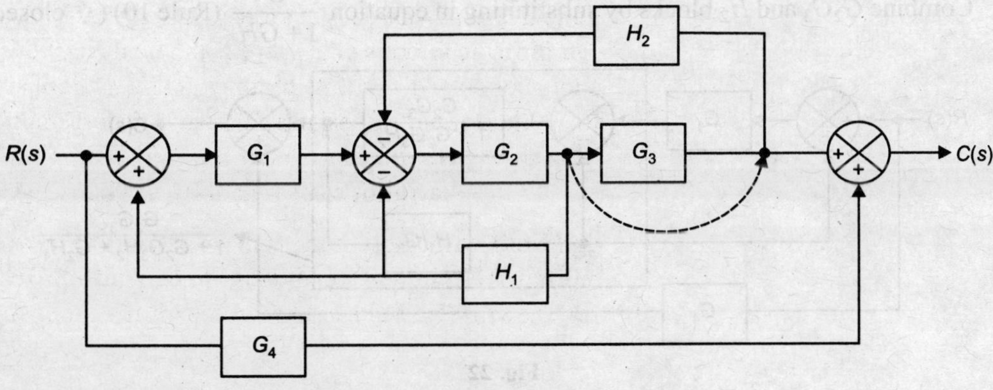

Fig. 19

Step 1: Shift the branch point to right of gain G_3. (Rule 3) ($BR\ 1/G$)

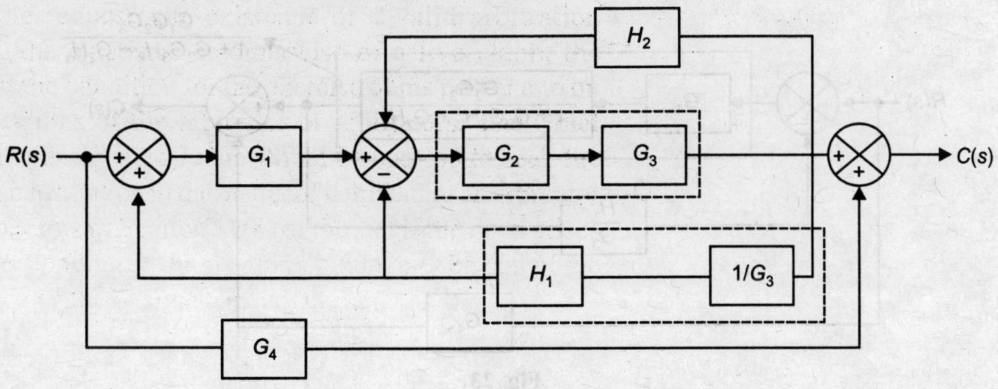

Fig. 20

Step 2: Multiply G_2 and G_3 block then multiply H_1 and $1/G_3$ blocks (Rule 1).

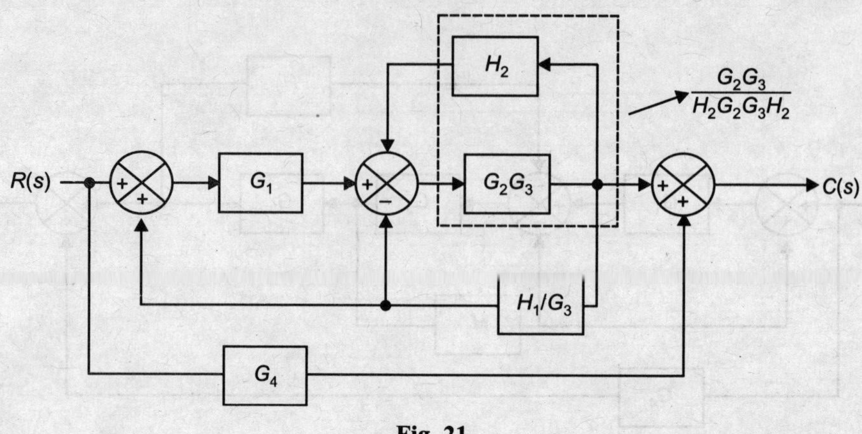

Fig. 21

Step 3: Combine G_2G_3 and H_2 blocks by substituting in equation $\frac{G}{1+GH}$ (Rule 10) ($\because$ closed loop).

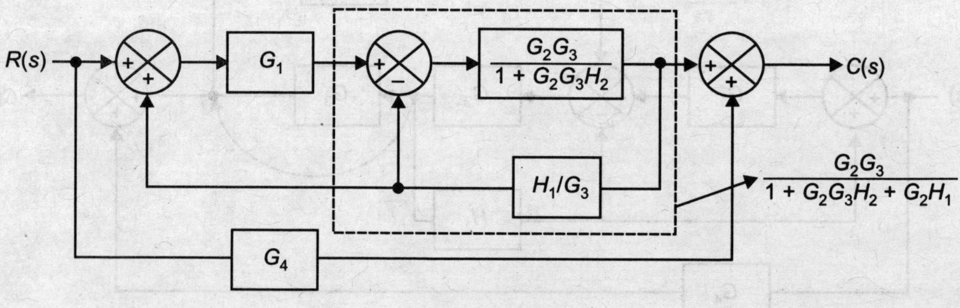

Fig. 22

Step 4: Combine $\frac{G_2G_3}{1+G_2G_3H_2+G_2H_1}$ and G_1 block (Rule 1)

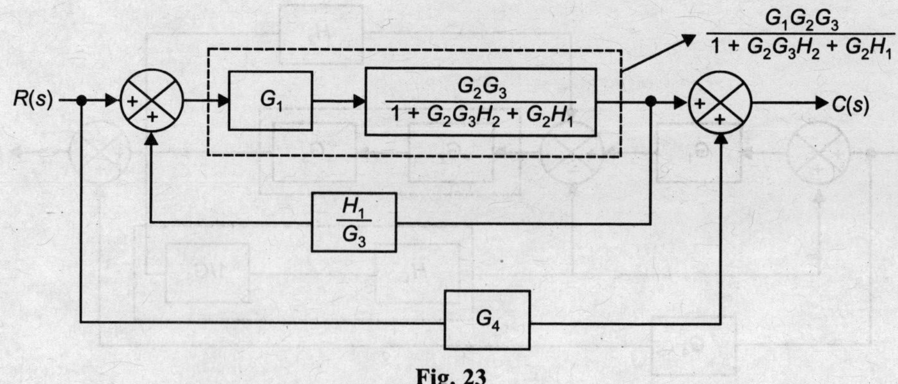

Fig. 23

Step 5: Combine the blocks H_1/G_3 and $\dfrac{G_2G_3}{1+G_2G_3H_2+G_2H_1}$ by rule 10.

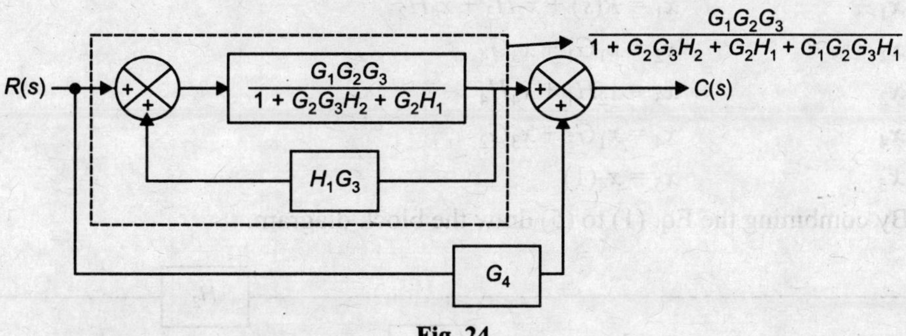

Fig. 24

Step 6: Combine the blocks using rule 2.

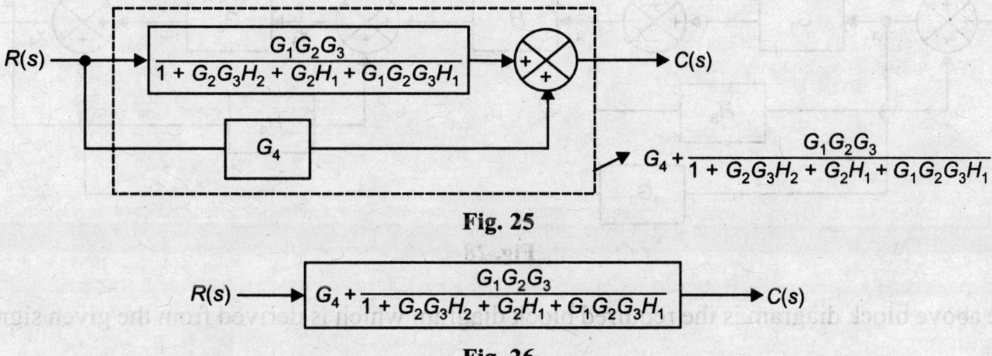

Fig. 25

R(s) → $G_4 + \dfrac{G_1G_2G_3}{1+G_2G_3H_2+G_2H_1+G_1G_2G_3H_1}$ → C(s)

Fig. 26

the required transfer function is,

$$\frac{C(s)}{R(s)} = G_4 + \left[\frac{G_1G_2G_3}{1+G_2G_3H_2+G_2H_1+G_1G_2G_3H_1}\right]$$

EXAMPLE 9: Convert the given signal flow graph into block diagram.

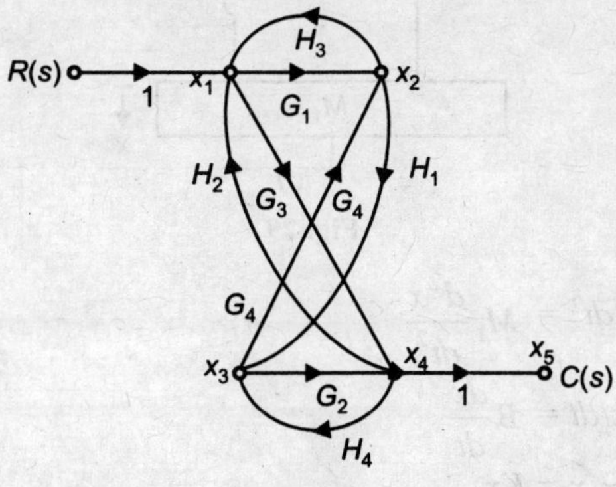

Fig. 27

Solution:

Step 1: Write the system equation for every node.

at node x_1 $\quad x_1 = R(s) + x_2H_3 + x_4H_2 \quad$ (1)

at node x_2 $\quad x_2 = x_1G_1 + x_3G_4 \quad$ (2)

at node x_3 $\quad x_3 = x_2H_1 + x_4H_4 \quad$ (3)

at node x_4 $\quad x_4 = x_1G_3 + x_3G_2 \quad$ (4)

at node x_5 $\quad x_5 = x_4(1) \quad$ (5)

Step 2: By combining the Eq. (1) to (5) draw the block diagram.

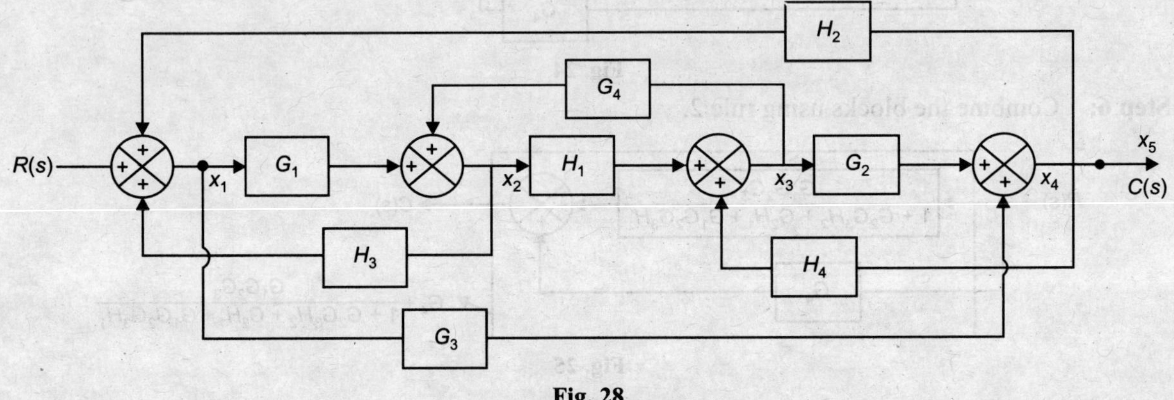

Fig. 28

The above block diagram is the required block diagram which is derived from the given signal flow graph.

EXAMPLE 10: For the mechanical system shown in Fig. (29), derive the transfer function. Also draw the force-voltage and force-current analogous circuits.

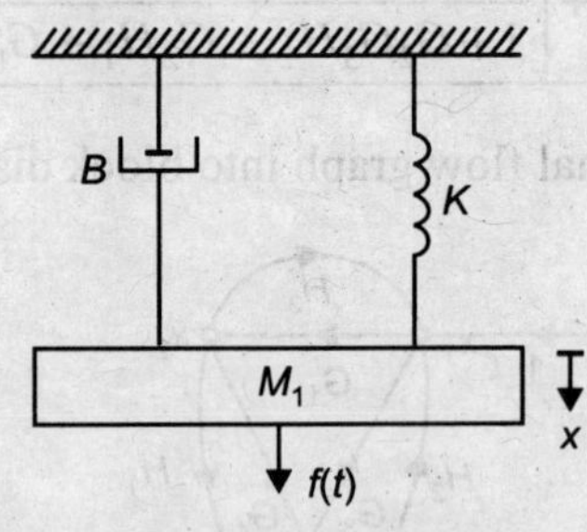

Fig. 29

Solution:

$$f_{m1} \propto d^2x/dt^2 = M_1\frac{d^2x}{dt^2}$$

$$f_b \propto dx/dt = B\frac{dx}{dt}$$

$$f_k \propto x = Kx$$

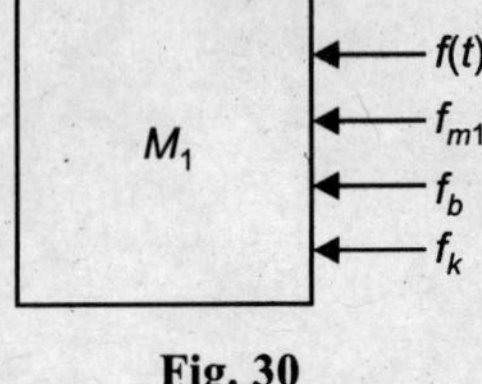

Fig. 30

By Newtons' second law,

the force balance equation is,

$$f(t) = f_{m1} + f_b + f_k$$

$$f(t) = M_1 \frac{d^2x}{dt^2} + B\frac{dx}{dt} + K_1 s$$

on taking laplace transform,

$$F(s) = M_1 s^2 X(s) + Bs\, X(s) + K\, X(s)$$

$$F(s) = X(s)\,[M_1 s^2 + Bs + K]$$

$$\therefore \quad \frac{X(s)}{F(s)} = \frac{1}{[M_1 s^2 + Bs + K]} \qquad (1)$$

The force voltage analogous circuit (refer Fig. 31)

In force-voltage analogy;

$$M \to L$$

$$B \to R$$

$$K \to 1/C$$

$$f(t) \to V$$

$$x \text{ (or) } y \to i$$

The force current analogous circuit (refer Fig. 32)

In force current analogy.

$$M \to C$$

$$B \to 1/R$$

$$K \to 1/L$$

$$f(t) \to i$$

$$x \text{ (or) } y \to V$$

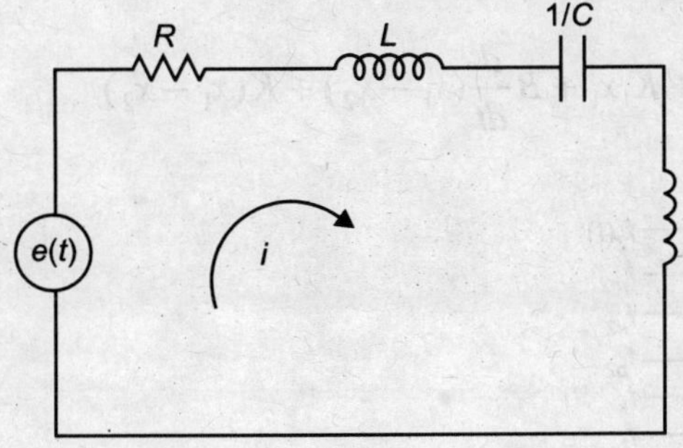

Fig. 31

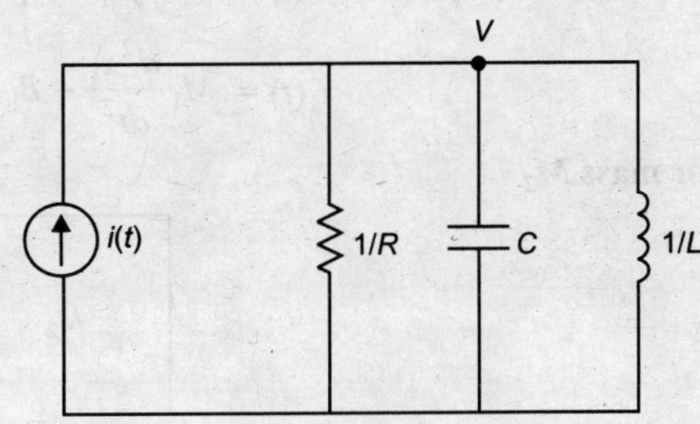

Fig. 32

CHAPTER 3

EXAMPLE 11: Write the circuit differential equations governing the mechanical system. Also draw the force voltage and force current analogous circuit.

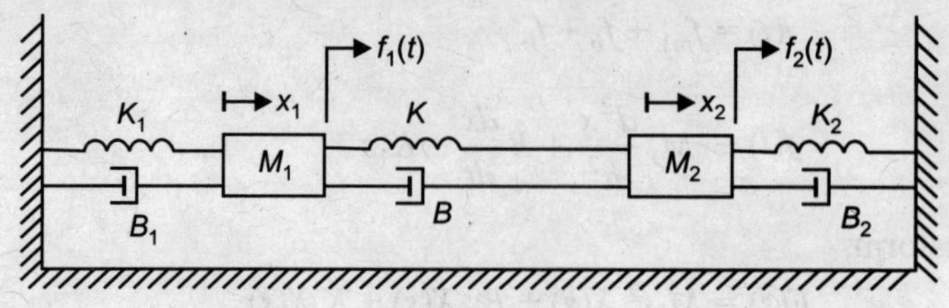

Fig. 33

Solution: (i) Differential equation

For mass M,

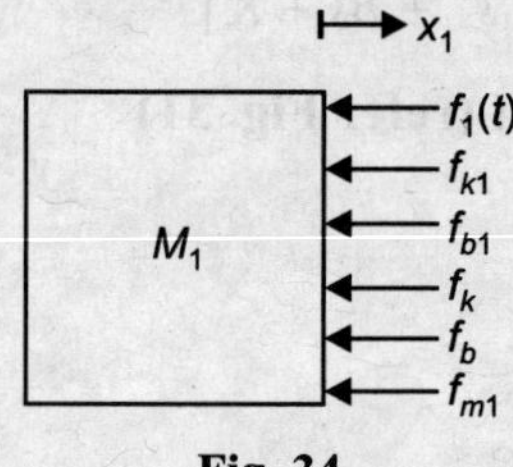

Fig. 34

where,
$$f_{m1} \propto \frac{d^2x_1}{dt^2} \Rightarrow M_1 \frac{d^2x_1}{dt^2}$$
$$f_{k1} \propto x_1 \Rightarrow K_1x_1$$
$$f_{b1} \propto \frac{dx_1}{dt} \Rightarrow B_1 \frac{dx_1}{dt}$$
$$f_k \propto (x_1 - x_2) \Rightarrow K(x_1 - x_2)$$
$$f_b \propto \frac{d}{dt}(x_1 - x_2) \Rightarrow B\frac{d}{dt}(x_1 - x_2)$$

The force balance equation is,
$$f_1(t) = f_{m1} + f_{b1} + f_{k1} + f_b + f_k$$

$$\therefore \quad f_1(t) = M_1 \frac{d^2x_1}{dt^2} + B_1 \frac{dx_1}{dt^2} + K_1x_1 + B\frac{d}{dt}(x_1 - x_2) + K(x_1 - x_2) \quad (1)$$

For mass M_2

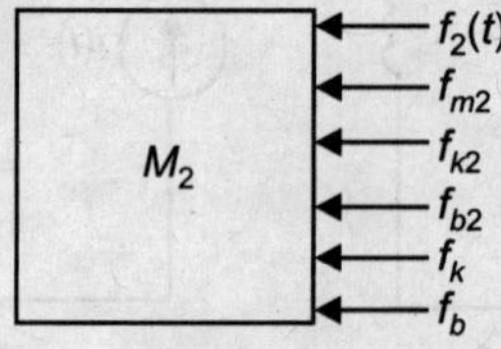

Fig. 35

where, $$f_{m2} \propto \frac{d^2x_2}{dt^2} \Rightarrow M_2\frac{d^2x_2}{dt^2}$$

$$f_{b2} \propto dx_2/dt \Rightarrow B_2\frac{dx_2}{dt}$$

$$f_{k2} \propto x_2 \Rightarrow K_2x_2$$

$$f_b \propto \frac{d}{dt}(x_2 - x_1) \Rightarrow B\frac{d}{dt}(x_2 - x_1)$$

$$f_K \propto (x_2 - x_1) \Rightarrow K(x_2 - x_1)$$

The force balance equation is,

$$f_2(t) = f_{m2} + f_{b2} + f_{K2} + f_b + f_K$$

$$\therefore \quad f_2(t) = M_2\frac{d^2x_2}{dt^2} + B_2\frac{dx_2}{dt^2} + K_2x_2 + B\frac{d}{dt}(x_2 - x_1) + K(x_2 - x_1) \quad (2)$$

(ii) Force-voltage analogous circuit:

$$M_1 \rightarrow L_1$$
$$M_2 \rightarrow L_2$$
$$K_1 \rightarrow 1/C_1$$
$$B_1 \rightarrow R_1$$
$$K_2 \rightarrow 1/C_2$$
$$B_2 \rightarrow R_2$$
$$K \rightarrow 1/C$$
$$B \rightarrow R$$
$$f_1(t) \rightarrow V_1(t)$$
$$f_2(t) \rightarrow V_2(t)$$
$$x_1 \rightarrow i_1(t)$$
$$x_2 \rightarrow i_2(t)$$

so, let draw the force-voltage analogous circuit for the Fig. (36).

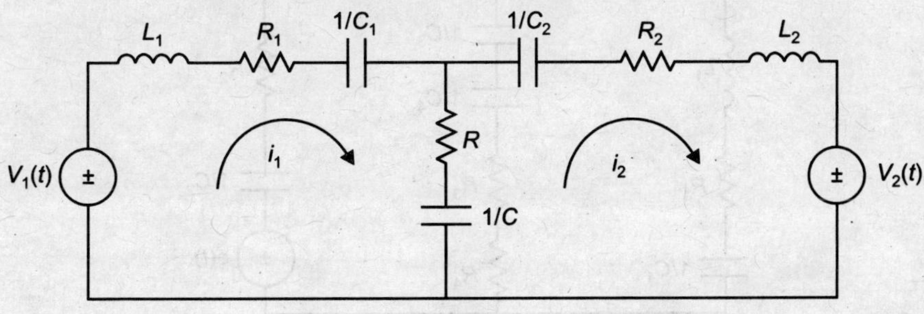

Fig. 36

CHAPTER 3

(iii) Force-current analogous circuit

$M_1 \rightarrow C_1$ $\quad K_2 \rightarrow 1/L_2$ $\quad f_1(t) \rightarrow i_1(t)$

$M_2 \rightarrow C_2$ $\quad B_2 \rightarrow 1/R_2$ $\quad f_2(t) \rightarrow i_2(t)$

$K_1 \rightarrow 1/L_1$ $\quad K \rightarrow 1/L$ $\quad x_1 \rightarrow V_1(t)$

$B_1 \rightarrow 1/R_1$ $\quad B \rightarrow 1/R$ $\quad x_2 \rightarrow V_2(t)$

so, let draw the force-current analogous

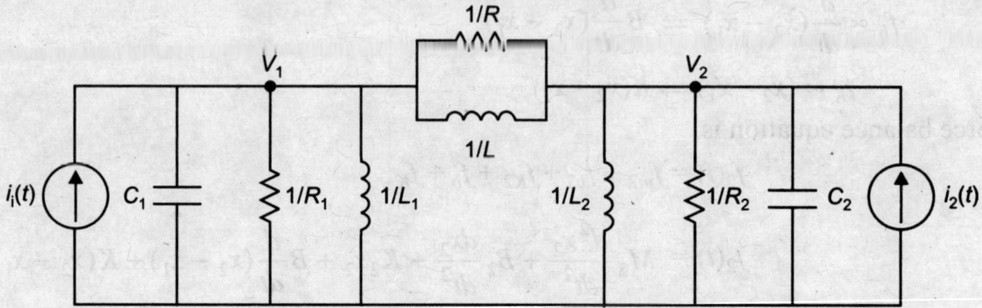

Fig. 37

EXERCISE PROBLEMS

1. For the mechanical system derive the transfer function. Also draw the force-voltage and force current analogous circuits.

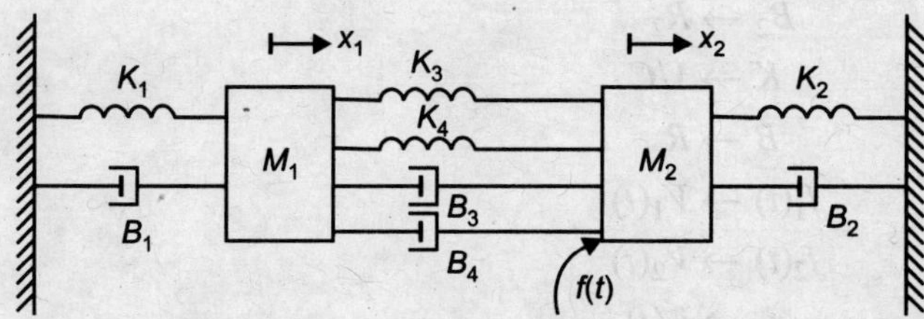

Fig. 1

Ans.

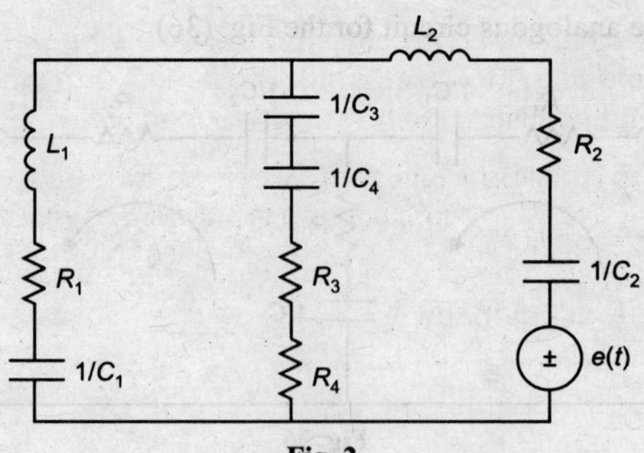

Fig. 2

$$\frac{X_2(s)}{F(s)} = \frac{M_1s^2 + s(B_1 + B_3 + B_4) + (K_1 + K_3)}{(M_1s^2 + s(B_1 + B_3 + B_4) + (K_1 + K_3)(M_2s^2 + B_2s + K_2 + K_3 + K_4)}$$

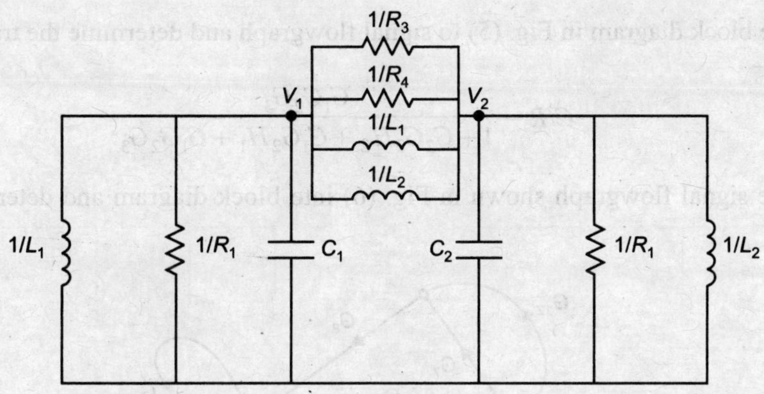

Fig. 3

2. Use block diagram reduction technique to find out C/R.

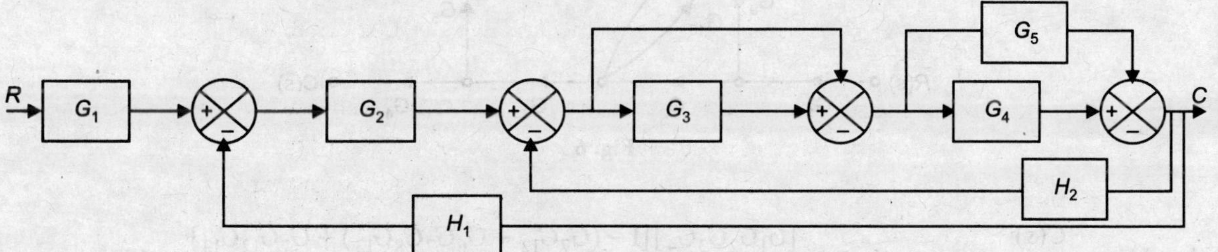

Fig. 4

Ans.
$$\frac{G_1G_2(1+G_3)(G_4+G_5)}{[1+(1+G_3)(G_4+G_5)H_2+(1+G_3)(G_4+G_5)G_2H_2]}$$

3. Convert the block diagram shown in Fig. (4) to signal flow graph and determine the transfer function.

Ans.
$$\frac{G_1G_2(1+G_3)(G_4+G_5)}{[1+(1+G_3)(G_4+G_5)H_2+(1+G_3)(G_4+G_5)G_2H_1]}$$

4. Convert the block diagram shown in Fig. 5 to signal flow graph and determine the transfer function.

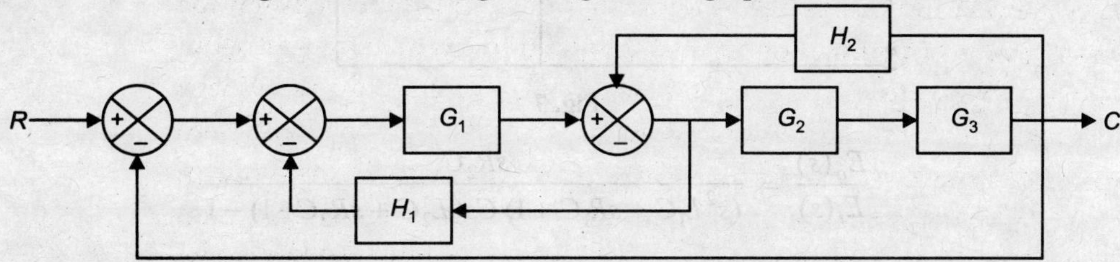

Fig. 5

Ans. $$C/R = \frac{G_1G_2G_3}{1 + G_2G_3H_2 + G_1G_2H_1 + G_1G_2G_3}$$

5. Convert the block diagram in Fig. (5) to signal flowgraph and determine the transfer function.

Ans. $$C/R = \frac{G_1G_2G_3}{1 + G_2G_3H_2 + G_1G_2H_1 + G_1G_2G_3}$$

6. Convert the signal flowgraph shown in Fig. (6) into block diagram and determine the transfer function.

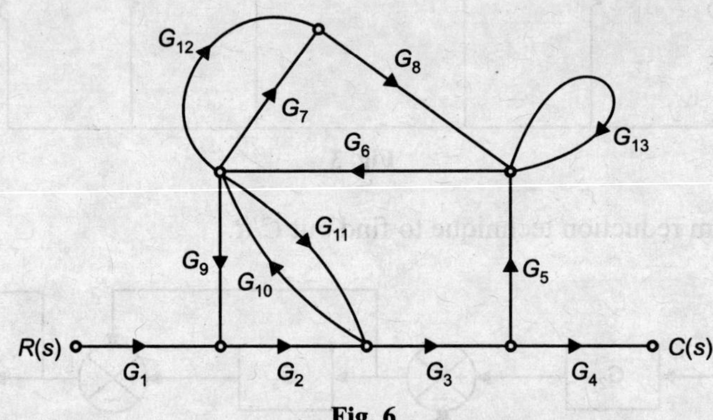

Fig. 6

Ans. $$\frac{C(s)}{R(s)} = \frac{[G_1G_2G_3G_4][1-(G_7G_{12} + G_6G_7G_8G_{13}) + G_7G_{12}G_{13}]}{1-\{G_2G_7G_{10} + G_{10}G_{11} + G_2G_3G_5G_6G_9 + G_3G_5G_6G_{11} + G_7G_{12} + G_6G_7G_8 + G_{13}\} + G_2G_9G_{10}G_{13} + G_{10}G_{13}G_{11}G_{13} + G_7G_{12}G_{13}}$$

7. Find transfer function of network shown in Fig. 7.

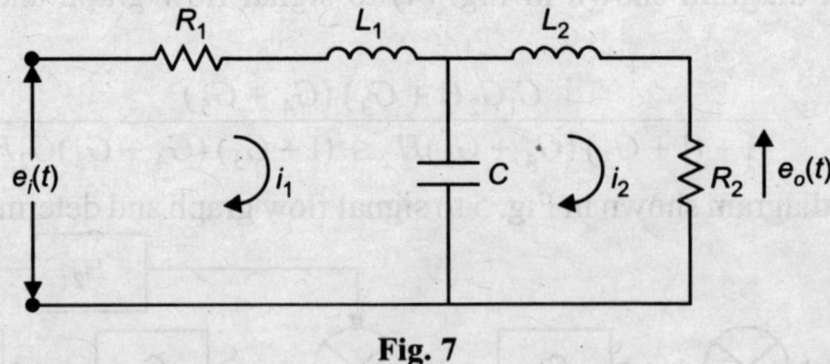

Fig. 7

Ans. $$\frac{E_o(s)}{E_i(s)} = \frac{sR_2C}{(s^2L_1C + sR_1C + 1)\ Cs^2L_2C + sR_2C + 1) - 1}$$

8. Obtain $E_o(s)/E_i(s)$

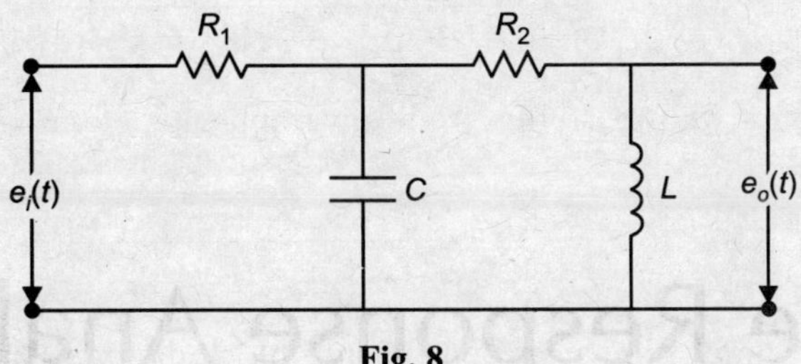

Fig. 8

Ans. $$\frac{E_o(s)}{E_i(s)} = \frac{sL_2C}{(sR_1C+1)\,[s^2(L_1+L_2)]}$$

9. Draw the block diagram for the given signal flowgraph.

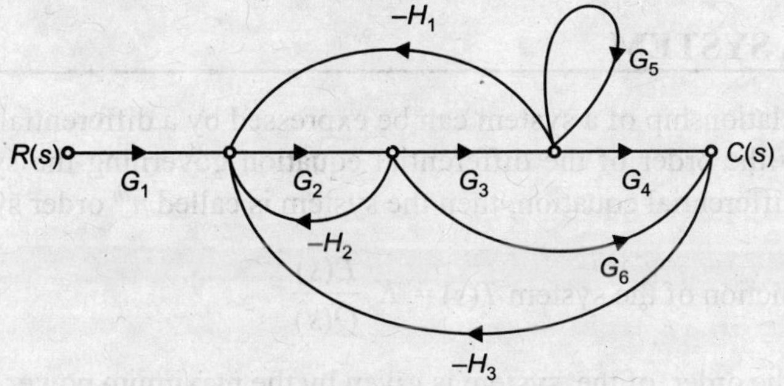

Fig. 9

Ans.

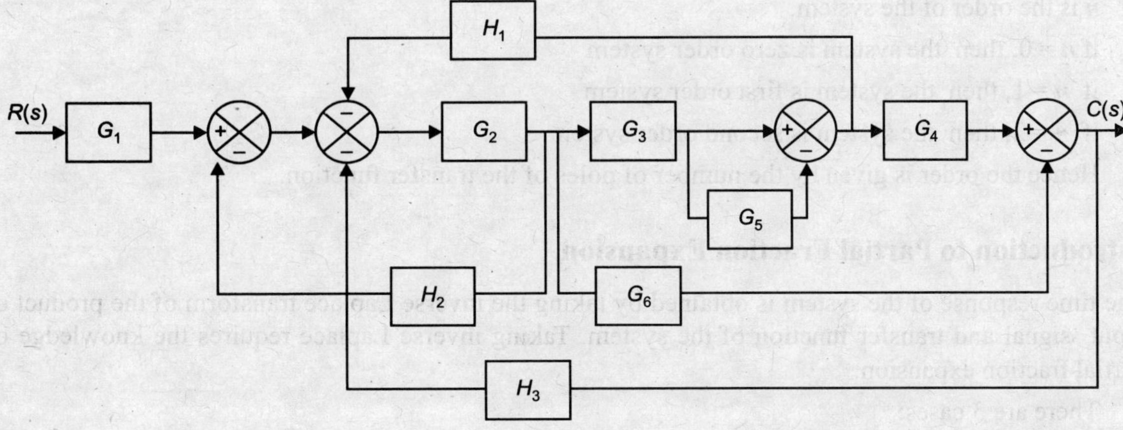

Fig. 10

CHAPTER

4

Time Response Analysis

I. TIME RESPONSE ANALYSIS

4.1 ORDER OF A SYSTEM

The input and output relationship of a system can be expressed by a differential equation. The order of the System is given by the order of the differential equation governing the system. If the system is governed by n^{th} order differential equation, then the system is called n^{th} order system.

Let the Transfer function of the system $T(s) = K.\dfrac{P(s)}{Q(s)}$

In above equation the order of the system is given by the maximum power of s in the denominator polynomial, $Q(s)$

If $Q(s) = a_0 s^n + a_1 s^{n-1} + a_2 s^{n-2} + ...a_{n-1} s + a_n$ then,

n is the order of the system.

if $n = 0$, then the system is zero order system

if $n = 1$, then the system is first order system

if $n = 2$, then the system is second order system

Hence the order is given by the number of poles of the transfer function.

Introduction to Partial Fraction Expansion

The time response of the system is obtained by taking the inverse Laplace transform of the product of input signal and transfer function of the system. Taking inverse Laplace requires the knowledge of partial fraction expansion.

There are 3 cases:

Case 1: Functions with separate poles

Case 2: Functions with multiple poles

Case 3: Functions with complex conjugate poles

Case 1: When the transfer functions has the distinct poles (Functions with separate poles)

$$T(s) = K/s(s + P_1)(s + P_2)$$

By partial fraction expansion, $T(s)$ can be expressed as

$$T(s) = \frac{K}{s(s+P_1)(s+P_2)} = \frac{A}{s} + \frac{B}{s+P_1} + \frac{C}{s+P_2}$$

The residues A, B and C are given by,

$$A = T(s) \cdot s\big|_{s=0}$$

$$B = T(s)(s+P_1)\big|_{s=-P_1}$$

$$C = T(s)(s+P_2)\big|_{s=-P_2}$$

By using partial fraction find out the values of A, B and C then substitute those values in the equation of $T(s)$.

Case 2: When the transfer function has multiples poles (functions with multiple poles)

$$T(s) = \frac{K}{s(s+P_1)(s+P_2)^2}$$

By partial fraction expansion, $T(s)$ can be expressed as

$$T(s) = \frac{K}{s(s+P_1)(s+P_2)^2} = \frac{A}{s} + \frac{B}{s+P_1} + \frac{C}{(s+P_2)^2} + \frac{D}{s+P_2}$$

The residues A, B, C and D are given by

$$A = T(s) \cdot s\big|_{s=0}$$

$$B = T(s)(s+P_1)\big|_{s=-P_1}$$

$$C = T(s)(s+P_2)^2\big|_{s=-P_2}$$

$$D = \frac{d}{ds}[T(s)(s+P_2)^2]\Big|_{s=-P_2}$$

Find out A, B, C and D by using partial fraction then substitute in the $T(s)$ equation

Case 3: When the transfer function has complex conjugate poles (functions with complex conjugate poles)

$$T(s) = \frac{K}{(s+P_1)(s^2+bs+c)}$$

By partial fraction expansion, $T(s)$ can be expressed as

$$T(s) = \frac{K}{(s+P_1)(s^2+bs+s)} = \frac{A}{s+P_1} = \frac{Bs+C}{s^2+bs+c}$$

CHAPTER 4

The residue A given by, $A = T(s) \cdot (s + P_1)$

The residues B and C are solved by cross multiplying the equation A. and then equating the coefficient of s and power of s. Finally express $T(s)$ as shown below,

$$T(s) = \frac{A}{s + P_1} + \frac{Bs}{s^2 + bs + c} + \frac{C}{s^2 + bs + c}$$

$$= \frac{A}{s + P_1} + \frac{BS}{\left(s + \frac{b}{2}\right)^2 + \left(c - \frac{b^2}{4}\right)} + \frac{C}{\left(s + \frac{b}{2}\right)^2 + \left(c - \frac{b}{4}\right)^2}$$

Find out A, B and C by using partial fractioin then substitute in the $T(s)$ equation.

Note: After substituting A, B, C ... (residues) in $T(s)$ equation, the inverse lapalace transform will give the time response of a system for all the cases [case (1), case (2) and case (3)].

4.2 TIME RESPONSE OF FIRST ORDER SYSTEM FOR UNIT STEP INPUT

The closed loop transfer function of first order system,

$$\frac{C(s)}{R(s)} = \frac{1}{1 + T_s} \tag{4.1}$$

If the input is unit step then, $r(t) = 1$ and $R(s) = 1/s$. [Refer table (1.2.1)]

The response in S-domain,

$$C(s) = R(s)\frac{1}{(1 + Ts)}$$

$$= \frac{1}{s} \cdot \frac{1}{(1 + Ts)}$$

$$= \frac{1}{sT\left(\frac{1}{T} + s\right)} = \frac{1/T}{s\,(s + 1/T)} \tag{4.2}$$

By partial fraction response expansion,

$$C(s) = \frac{1/T}{s\left(s + \frac{1}{T}\right)} = \frac{A}{s} + \frac{B}{\left(s + \frac{1}{T}\right)} \tag{4.3}$$

A is obtained by multiplying $C(s)$ by s and letting $s = 0$.

$$A = C(s) \cdot s = \frac{1/T}{s\left(s + \frac{1}{T}\right)} \cdot s = \left.\frac{1/T}{s + \frac{1}{T}}\right|_{s=0}$$

$$A = \frac{1/T}{1/T} = 1 \text{ ie. } A = 1$$

B is obtained by multiplying $C(s)$ by $(s + 1/T)$ and letting $s = -1/T$

$$B = C(s) \cdot \left(S + \frac{1}{T}\right) = \frac{1/T}{s\,(s + 1/T)}\left(s + \frac{1}{T}\right) = \left.\frac{1/T}{s}\right|_{s=-1/T}$$

$$= \frac{1/T}{-1/T} = -1 \quad \text{ie., } B = -1$$

Substitute A and B values in Eq. (2.5.3)

$$C(s) = \frac{1}{s} - \frac{1}{s + 1/T} \tag{4.4}$$

The response in time domain is obtained by taking inverse laplace transform of Eq. (4.4),

$$C(t) = L^{-1}[C(s)] = L^{-1}\left[\frac{1}{s} - \frac{1}{s + \frac{1}{T}}\right] = 1 - e^{-t/T} \tag{4.5}$$

Substitute various values for time (t) in Eq. (2.5.5)

when	$t = 0$,	$C(t) = 1 - e^{0}$	$= 0$
when	$t = T$,	$C(t) = 1 - e^{-1}$	$= 0.632$
when	$t = 2T$,	$C(t) = 1 - e^{-2}$	$= 0.865$
When	$t = 3T$,	$C(t) = 1 - e^{-3}$	$= 0.95$
When	$t = 4T$,	$C(t) = 1 - e^{-4}$	$= 0.9817$
when	$t = 5T$,	$C(t) = 1 - e^{-5}$	$= 0.993$
When	$t = \infty$,	$C(t) = 1 - e^{-\infty}$	$= 1$

Here T is called time constant of the system. After a time of $5T$, the system is assumed to have attained steady state. the input and output of the system is given below.

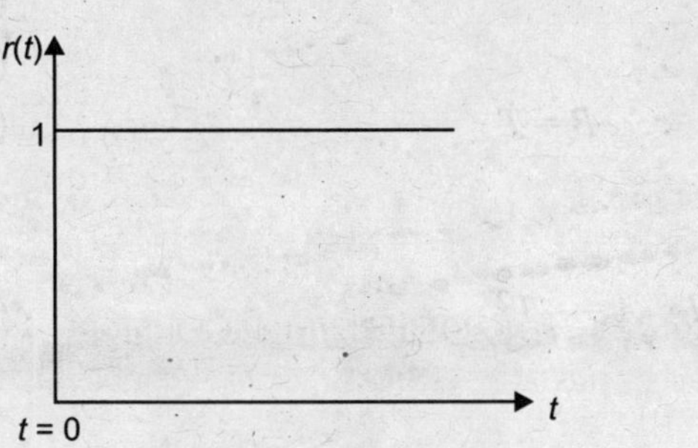

Fig. 4.1: Unit step input.

⇒

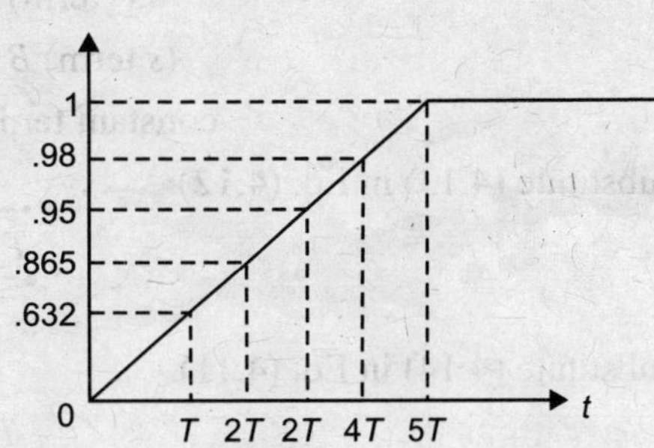

Fig. 4.2: Response of 1st order system to unit step input.

CHAPTER 4

4.3 TIME RESPONSE OF THE FIRST ORDER SYSTEM WITH UNIT RAMP FUNCTION

The closed loop transfer function of first order system,

$$\frac{C(s)}{R(s)} = \frac{1}{1+Ts} \tag{4.6}$$

In the input is unit ramp then,

$$r(t) = t, \quad R(s) = 1/s^2 \text{ [refer Table (1.4.1)]}$$

∴ The response in s-domain,

$$C(s) = R(s)\frac{1}{(1+Ts)}$$

$$C(s) = \frac{1}{s^2}\cdot\frac{1}{(1+Ts)}$$

$$C(s) = \frac{1}{s^2\,(1+Ts)} = \frac{1}{s^2 T\,[s+1/T]}$$

$$C(s) = \frac{1/T}{s^2\,[s+1/T]} \tag{4.7}$$

By Partial fraction

$$C(s) = \frac{1/T}{s^2\,[s+1/T]} = \frac{A}{s} + \frac{B}{s^2} + \frac{C}{(s+1/T)}$$

$$\therefore \quad A(s)\,(s + 1/T) + B(s + 1/T) + Cs^2 = 1 \tag{4.8}$$

$$As^2 + \frac{As}{T} + Bs + \frac{B}{T} + s^2C = 1 \tag{4.9}$$

$$s^2[A + C] + s\,[B + A/T] + B/T = 1 \tag{4.10}$$

Equate s^2, s and constant terms from both L.H.S and R.H.S

$$(s^2 \text{ term}) \; A + C = 0 \tag{4.11}$$

$$(s \text{ term}) \; B + A/T = 0 \tag{4.12}$$

$$\text{constant term)} \; B/T = 1 \quad \therefore \quad B = T \tag{4.13}$$

Substitute (4.13) in Eq. (4.12)

$$T + A/T = 0$$

$$A/T = -T \quad A = -T^2 \tag{4.14}$$

Substitute (4.14) in Eq. (4.11)

$$-T^2 + C = 0$$

and

$$C = T^2 \tag{4.15}$$

Substitute the values of A, B and C in Eq. $C(s)$

$$C(s) = \frac{1}{T}\left[\frac{-T^2}{s} + \frac{T}{s^2} + \frac{T^2}{s+1/T}\right]$$

$$\therefore \quad C(s) = \left[\frac{-T}{s} + \frac{1}{s^2} + \frac{T}{s+\frac{1}{T}}\right]$$

The response in time domain is obtained by taking inverse laplace transform of $C(s)$,

$$C(t) = L^{-1}[C(s)] = L^{-1}\left[\frac{-T}{s} + \frac{1}{s^2} + \frac{T}{s+1/T}\right]$$

$$C(t) = -T + t + T \times e^{-t/T} \tag{4.16}$$

The Eq. (4.16) in the time response of a first order system when unit ramp function is given as input.

4.4 TIME RESPONSE OF FIRST ORDER SYSTEM WITH UNIT PARABOLIC FUNCTION

The closed loop transfer function of first order system,

$$\frac{C(s)}{R(s)} = \frac{1}{1+Ts} \tag{4.17}$$

If the input is unit parabolic then,

$$r(t) = \frac{t^2}{2}, \quad R(s) = 1/s^3 \text{ [refer Table 1.2.1)]}$$

∴ The response in s-domain,

$$C(s) = R(s) \cdot \frac{1}{1+Ts}$$

$$C(s) = \frac{1}{s^3} \cdot \frac{1}{1+Ts}$$

$$C(s) = \frac{1}{s^3\,(1+Ts)}$$

$$C(s) = \frac{1}{s^3\,T(s+1/T)}$$

$$C(s) = \frac{1/T}{s^3\,(s+1/T)} \tag{4.18}$$

By taking partial fraction,

$$\frac{A}{s} + \frac{B}{s^2} + \frac{C}{s^3} + \frac{D}{s+1/T} = 1 \tag{4.19}$$

$$As^2\,(s + 1/T) + Bs(s + 1/T) + C(s + 1/T) + Ds^3 = 1$$

CHAPTER 4

$$As^3 + As^2/T + Bs^2 + Bs/T + Cs + C/T + Ds^3 = 1 \qquad (4.20)$$

Equate the coefficients from both L.H.S and R.H.S of s^3, s^2, s and constant terms

$$A + D = 0 \qquad (4.21) \text{ [equate } s^3]$$

$$A/T + B = 0 \qquad (4.22) \text{ [equate } s^2]$$

$$B/T + C = 0 \qquad (4.23) \text{ [equate } s]$$

$$C/T = 1 \Rightarrow C = T \qquad (4.24) \text{ [equate constant]}$$

Substitue C in Eq. (4.23)

$$B/T + T = 0, \quad B = -T^2$$

Substitute B values in Eq. (4.22)

$$A/T + (-T)^2 = 0$$

$$A/T = T^2$$

$$A = T^3$$

Substiturte A in Eq. (4.21)

$$T^3 + D = 0 \quad \Rightarrow D = -T^3$$

Substitute A, B, C and D values in Eq. (4.19),

$$\left[\frac{T^3}{s} + \frac{-T^2}{s^2} + \frac{T}{s^3} + -\frac{T^3}{s + 1/T}\right] \times \frac{1}{T} = C(s)$$

$$\frac{T^2}{s} + \frac{-T}{s^2} + \frac{1}{s^3} + \frac{-T^2}{s + 1/T} = C(s)$$

By takine laplace transform,

$$C(t) = T^2 - Tt + t^2 - T^2 e^{-t/T} \qquad (4.25)$$

The Eq. (4.25) is the time response of first order system when unit parabolic signal is given as input.

4.5 TIME RESPONSE OF FIRST ORDER SYSTEM WITH UNIT IMPULSE FUNCTION

The closed loop transfer function of first order system,

$$\frac{C(s)}{R(s)} = \frac{1}{1 + Ts} \qquad (4.26)$$

If the input is unit impulse function then,

$$R(s) = 1 \quad \text{(Refer Table 1.4.1)}$$

∴ The response in s-domain

$$C(s) = \frac{R(s) \times 1}{1 + T_s}$$

$$C(s) = 1 \times \frac{1}{1+Ts} = \frac{1}{1+Ts} \qquad [\because R(s) = 1]$$

$$C(s) = \frac{1}{T[s+1/T]} = \frac{1/T}{(s+1/T)} \qquad (4.27)$$

By applying partial fraction in Eq. (4.27)

$$C(s) = \frac{A}{(s+1/T)} \qquad (4.28)$$

Substitute $s = -1/T$ in Eq. (2.8.3)

$$1 = \frac{A}{s+1/T} \Rightarrow 1 = \frac{A}{-1/T+1/T}$$

$$\therefore \qquad A = 1$$

$$C(s) = \frac{1}{s+1/T} \times \frac{1}{T}$$

By taking laplace transform,

$$L^{-1}[C(s)] = L^{-1}\left[\frac{1}{T} \times \frac{1}{s+1/T}\right]$$

$$\therefore \qquad C(t) = \frac{1}{T} e^{-t/T} \qquad (4.29)$$

The Eq. (4.29) is the time response of first order system when unit impulse is given as input.

4.6 COMPARISON OF FIRST ORDER RESPONSE WITH UNIT STEP, RAMP, PARABOLIC AND IMPULSE FUNCTIONS

The time response of first order system with unit step input,

$$C(t) = 1 - e^{-t/T} \qquad (4.30)$$

The time response of first order system with unit ramp input,

$$C(t) = t - T + te^{-t/T} \qquad (4.31)$$

The time response of first order system with unit parabolic input,

$$C(t) = T^2 - Tt + t^2 - Te^{-t/T} \qquad (4.32)$$

The time response of first order system with unit impulse input,

$$C(t) = \frac{1}{T} e^{-t/T} \qquad (4.33)$$

From Eq. (4.30) to (4.33) it is clear that the unit step input is the derivative of unit ramp input and unit impulse input is the derivative of unit step input and vice versa. Thus, the response to the derivative of an input signal can be obtained by differentiating the response of the system to the original signal. This is the property of the linear time-invariant systems.

4.7 SECOND ORDER SYSTEM

The standard form of closed loop transfer function of second order system is given by

$$\frac{C(s)}{R(s)} = \frac{\omega_n^2}{s^2 + 2\zeta\omega_n s + \omega_n^2} \tag{4.34}$$

where,

$\omega_n \rightarrow$ Undamped natural frequency, rad/sec

$\zeta \rightarrow$ Damping ratio

The damping ratio is defined as the ratio of actual damping to the critical damping. The response $C(t)$ of second order system depends on the value of damping ratio. Depending on the value of ζ, the system can be classified in to the following four cases,

Case 1: Undamped system, $\zeta = 0$

Case 2: Under damped system, $0 < \zeta < 1$

Case 3: Critically damped system, $\zeta = 1$

Case 4: Over damped system, $\zeta > 1$

The characteristics Eq. of the 2nd order system is

$$s^2 + 2\zeta\omega_n s + \omega_n^2 = 0 \tag{4.35}$$

It is a quadratic Eq. and the roots of this Eq. is given by

$$s_1, s_2 = \frac{-2\omega\zeta_n \pm \sqrt{4\zeta^2\omega_n^2 - 4\omega_n^2}}{2} = \frac{-2\zeta\omega_n \pm \sqrt{4\omega_n^2(\zeta^2 - 1)}}{2}$$

$$s_1, s_2 = -\zeta\omega_n \pm \omega_n\sqrt{\zeta^2 - 1} \tag{4.36}$$

when $\zeta = 0$, $s_1\ s_2 = \pm j\omega_n$; {Roots are purely imaginary and the system is undamped}

when $\zeta = 1$, $s_1, s_2 = -\omega_n$; {Roots are real and equal and the system is critically damped}

when $\zeta > 1$, $s_1, s_2 = -\zeta\omega_n \pm \omega_n\sqrt{\zeta^2 - 1}$; {Roots are real and unequal and the system is over damped}

when $0 < \zeta < 1$, $s_1, s_2 = -\zeta\omega_n \pm \omega_n\sqrt{\zeta^2 - 1}$

$$= -\zeta\omega_n \pm \omega_n\sqrt{(-1)(1-\zeta^2)} = -\zeta\omega_n \pm j\omega_n\sqrt{1-\zeta^2}$$

$$= -\zeta\omega_n \pm j\omega_d$$ {Roots are complex conjugate the system is under damped}

where, $\omega_d = \omega_n\sqrt{1-\zeta^2}$

Here ω_d: called damped frequency of oscillation of the system and its unit is radian/second.

4.7.1 Response of Undamped Second Order System For Unit Step Input

The standard form of closed loop transfer function of second order system is

$$\frac{C(s)}{R(s)} = \frac{\omega_n^2}{s^2 + 2\zeta\omega_n s + \omega_n^2} \tag{4.37}$$

For undamped system, $\zeta = 0$

$$\therefore \quad \frac{C(s)}{R(s)} = \frac{\omega_n^2}{s^2 + \omega_n^2} \tag{4.38}$$

when the input is unit step, $r(t) = 1$

and $R(s) = 1/s$

$\therefore$ The response in s-domain,

$$C(s) = R(s) \cdot \frac{\omega_n^2}{s^2 + \omega_n^2}$$

$$C(s) = \frac{1}{s} \cdot \frac{\omega_n^2}{s^2 + \omega_n^2} \tag{4.39}$$

By partial fraction expansion,

$$C(s) = \frac{\omega_n^2}{s(s^2 + \omega_n^2)} = \frac{A}{s} + \frac{B}{s^2 + \omega_n^2} \tag{4.40}$$

A is obtained by multiplying $C(s)$ by s and letting $s = 0$

$$A = C(s) \times s \Rightarrow \frac{\omega_n^2}{\not{s}(s^2 + \omega_n^2)} \times \not{s} \quad \left. \frac{\omega_n^2}{s^2 + \omega_n^2} \right|_{s=0} = \frac{\omega_n^2}{\omega_n^2} = 1$$

$$\boxed{A = 1}$$

B is obtained by multiplying $C(s)$ by $(s^2 + \omega_n^2)$ and letting $s^2 = -\omega_n^2$ (or) $s = j\omega_n$

$$B = C(s)\,(s^2 + \omega_n^2) = \frac{\omega_n^2}{s(s^2 + \omega_n^2)} \times (s^2 + \omega_n^2) \Rightarrow \left. \frac{\omega_n^2}{s} \right|_{s = -\omega_n^2}$$

$$\therefore \quad B = \frac{\omega_n^2}{j\omega_n} = -j\omega_n = -s$$

$$\boxed{B = -s}$$

Substitute A and B values in Eq. (4.40)

$$C(s) = \frac{A}{s} + \frac{s}{(s^2 + \omega_n^2)} = \frac{1}{s} - \frac{s}{(s^2 + \omega_n^2)} \tag{4.41}$$

CHAPTER 4

The time domain response of Eq. (4.42) is obtained by taking inverse laplace transform,

$$L^{-1}[C(s)] = L^{-1}\left[\frac{1}{s} - \frac{s}{s^2 + \omega_n^2}\right] = 1 - \cos \omega_n t \tag{4.42}$$

The Eq. (4.42) is the time response of second order system when unit step input is given under ζ = 0.

Substitute various time for time in Eq. (4.42)

when, $t = 0 \Rightarrow C(t) = 1 - \cos \omega_n t \Rightarrow 1 - \cos 0 \Rightarrow 1 - 1 = 0$

$t = \pi/2 \Rightarrow C(t) = 1 - \cos \omega_n t \Rightarrow 1 - \cos \pi/2 \Rightarrow 1 - 0 = 1$

$t = \pi \Rightarrow C(t) = 1 - \cos \omega_n t \Rightarrow 1 - \cos \pi \Rightarrow 1 - (-1) = 2$

$t = 3\pi/2 \Rightarrow C(t) = 1 - \cos \omega_n t \Rightarrow 1 - \cos (3\pi/2) \Rightarrow 1 - 0 = 1$

$t = 2\pi \Rightarrow C(t) = 1 - \cos \omega_n t \Rightarrow 1 - \cos 2\pi \Rightarrow 1 - 1 = 0$

The input and output of the system shown in Fig. (4.3) and (4.4)

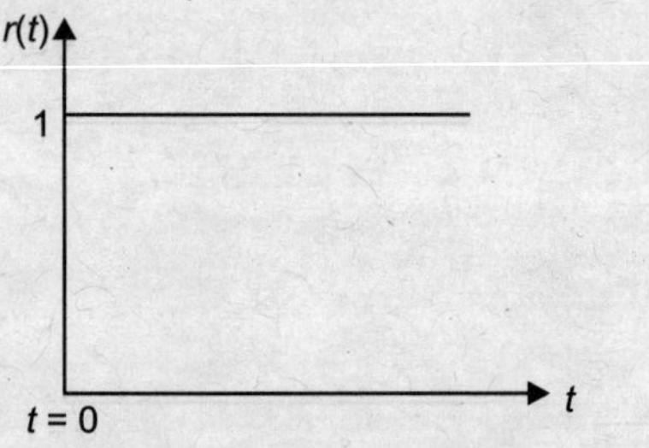

Fig. 4.3: Unit step i/p

$\Rightarrow$

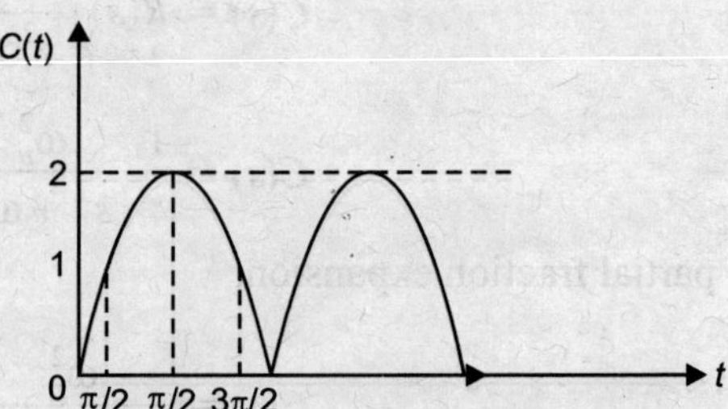

Fig. 4.4: Response of undamped second order system for unit step input.

Conclusion: The response of undamped second order system for unit step input is completely oscillatory.

Note: Every practical system has some amount of damping. Hence undamped system does not exist in practice.

4.7.2 Time Response of Second Order System For Under damped Case and When the input is Unit Step

The standard form of closed loop transfer function of second order system is

$$\frac{C(s)}{R(s)} = \frac{\omega_n^2}{s^2 + 2\zeta\omega_n s + \omega_n^2} \tag{4.43}$$

For underdamped system, $0 < \zeta < 1$. and roots of the denominator [characteristic equation] are complex conjugate.

The roots of the denominator are,

$$s = -\zeta\omega_n \pm \omega_n\sqrt{\zeta^2 - 1}$$

since $\zeta < 1, \zeta^2 < 1$

$$B = -\zeta\omega_n \pm \omega_n \sqrt{(-1)(1-\zeta^2)} = -\zeta\omega_n \pm j\omega_n \sqrt{1-\zeta^2}$$

The damped frequency of oscillation,

$$\omega_d = \omega_n \sqrt{1-\zeta^2}$$

$$\therefore \quad s = -\zeta\omega_n \pm j\omega_d$$

The response in s-domain,

$$C(s) = R(s)\frac{\omega_n^2}{s^2 + 2\zeta\omega_n s + \omega_n^2}$$

For unit step input, $r(t) = 1$ and $R(s) = 1/s$

$$\therefore \quad C(s) = \frac{\omega_n^2}{s(s^2 + 2\zeta\omega_n s + \omega_n^2)} \qquad (4.44)$$

By partial fraction expansion,

$$C(s) = \frac{A}{s} + \frac{Bs + C}{s^2 + 2\zeta\omega_n s + \omega_n^2} \qquad (4.45)$$

$$A = s\,C(s)\big|_{s=0} = \frac{\omega_n^2}{\omega_n^2} = 1$$

To solve B and C

$$\frac{\omega_n^2}{s(s^2 + 2\zeta\omega_n s + \omega_n^2)} = \frac{1}{s} + \frac{Bs + C}{s^2 + 2\zeta\omega_n s + \omega_n^2}$$

on cross multiplication and equating the coefficients of like power of s we get,

$$\omega_n^2 = s^2 + 2\zeta\omega_n s + \omega_n + (Bs + C)s$$

$$\omega_n^2 = s^2 + 2\zeta\omega_n s + \omega_n^2 + Bs^2 + Cs$$

Equating coefficients of s^2 we get,

$$0 = 1 + B$$

$$\therefore \quad B = -1$$

Equating coefficients of s we get,

$$0 = 2\zeta\omega_n + C$$

$$\therefore \quad C = 2\zeta\omega_n$$

Substitute A, B and C values in Eq. (4.45)

$$C(s) = \frac{1}{s} - \frac{(s + 2\zeta\omega_n)}{s^2 + 2\zeta\omega_n s + \omega_n^2} \qquad (4.46)$$

CHAPTER 4

Adding and subtracting $s^2\ \omega_n^2$ to the denominator of second term in Eq. (4.46)

$$= \frac{1}{s} - \frac{(s + 2\zeta\omega_n)}{s^2 + 2\zeta\omega_n s + \omega_n^2 + \zeta^2\omega_n^2 - \zeta^2\omega_n^2}$$

$$= \frac{1}{s} - \frac{(s + 2\zeta\omega_n)}{(s^2 + 2\zeta\omega_n s + \zeta^2\omega_n^2) + (\omega_n^2 - \zeta^2\omega_n^2)}$$

$$= \frac{1}{s} - \frac{(s + 2\zeta\omega_n)}{(s + \zeta\omega_n)^2 + \omega_n^2(1 - \zeta^2)}$$

as we know $\omega_d = \omega_n\sqrt{1 - \zeta^2}$

$$C(s) = \frac{1}{s} - \frac{(s + 2\zeta\omega_n)}{(s + \zeta\omega_n)^2 + \omega_d^2}$$

$$= \frac{1}{s} - \frac{(s + \zeta\omega_n)}{(s + \zeta\omega_n)^2 + \omega_d^2} - \frac{(\zeta\omega_n)}{(s + \zeta\omega_n)^2 + \omega_d^2} \tag{4.47}$$

Multiplying and dividing by ω_d in the third term of the Eq. (4.47)

$$C(s) = \frac{1}{s} - \frac{(s + \zeta\omega_n)}{(s + \zeta\omega_n)^2 + \omega_d^2} - \left(\frac{\zeta\omega_n}{\omega_d}\right)\left(\frac{\omega_d}{(s + \zeta\omega_n)^2 + \omega_d^2}\right) \tag{4.48}$$

on taking inverse Laplace transform

$$C(t) = 1 - e^{-\zeta\omega_n t}\cos\omega_d t - \frac{\zeta\omega_n}{\omega_d}e^{-\zeta\omega_n t}\sin\omega_d t$$

$$C(t) = 1 - e^{-\zeta\omega_n t}\left[+\cos\omega_d t + \frac{\zeta\omega_n}{\omega_n\sqrt{1 - \zeta^2}}\sin\omega_d t\right]$$

$$= 1 - e^{-\zeta\omega_n t}\left[\cos\omega_d t + \frac{\zeta}{\sqrt{1 - \zeta^2}}\sin\omega_d t\right]$$

$$= 1 - \frac{e^{-\zeta\omega_n t}}{\sqrt{1 - \zeta^2}}\left[\sqrt{1 - \zeta^2}\cos\omega_d t + \zeta\sin\omega_d t\right]$$

$$C(t) = \frac{e^{-\zeta\omega_n t}}{\sqrt{1 - \zeta^2}}\left[\sin\omega_d t \cdot \zeta + \cos\omega_d t \cdot \sqrt{1 - \zeta^2}\right]$$

$$C(t) = 1 - \frac{e^{-\zeta\omega_n t}}{\sqrt{1 - \zeta^2}}[\sin\omega_d t\cos\theta + \cos\omega_d t\sin\theta]$$

(Refer note)

$$C(t) = 1 - \frac{e^{-\zeta\omega_n t}}{\sqrt{1-\zeta^2}} \sin(\omega_d t + \theta) \qquad (4.49)$$

The Eq. (4.49) is the time response of second order system when unit step input is given [for under dampled case]

where,

$$\theta = \tan^{-1}\left[\frac{\sqrt{1-\zeta^2}}{\zeta}\right]$$

The response of underdamped second order system oscillates before setting to a final value. The oscillations depends on the value of damping ratio.

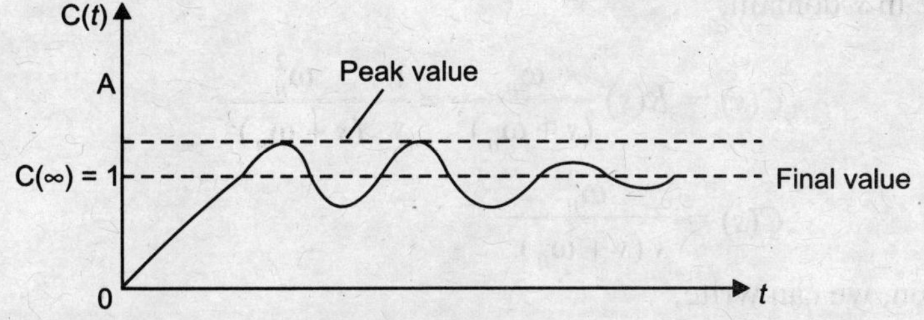

Fig. 4.5: Response of under damped second order system for unit step input.

Note: On constructing right angle triangle with ζ and $\sqrt{1-\zeta^2}$, we get,

$$\sin\theta = \sqrt{1-\zeta^2}$$

$$\cos\theta = \zeta$$

1

$\sqrt{1-\zeta^2}$

ζ

Fig. 4.6

$$\tan\theta = \frac{\sqrt{1-\zeta^2}}{\zeta}$$

$$\theta = \tan^{-1}\left[\frac{\sqrt{1-\zeta^2}}{\zeta}\right]$$

4.7.3 Time Response of Second Order System for Critically Damped Case When Input is Unit Step Input

The standard of closed loop transfer function of second order system is

$$\frac{C(s)}{R(s)} = \frac{\omega_n^2}{s^2 + 2\, s\omega_n s + \omega_n^2}$$

For critical damping $\zeta = 1$

$$\therefore \quad \frac{C(s)}{R(s)} = \frac{\omega_n^2}{s^2 + 2\zeta\omega_n s + \omega_n^2} = \frac{\omega_n^2}{(s+\omega_n)^2} \tag{4.50}$$

when input is unit step, $r(t) = 1$ and $R(s) = 1/s$

$\therefore$ The response in s-domain,

$$C(s) = R(s)\frac{\omega_n^2}{(s+\omega_n)^2} = \frac{1}{s}\cdot\frac{\omega_n^2}{(s+\omega_n)^2}$$

$$C(s) = \frac{\omega_n^2}{s\,(s+\omega_n)^2} \tag{4.51}$$

By partial fraction, we can write,

$$C(s) = \frac{\omega_n^2}{s(s+\omega_n)^2} = \frac{A}{s} + \frac{B}{(s+\omega_n)^2} + \frac{C}{(s+\omega_n)} \tag{4.52}$$

$$A = s\,C(s)\big|_{s=0} = \left.\frac{\omega_n^2}{(s+\omega_n)^2}\right|_{s=0} = \frac{\omega_n^2}{\omega_n^2} = 1$$

$$A = 1$$

$$B = (s+\omega_n)^2\, C(s)\big|_{s=-\omega_n} = \left.\frac{-\omega_n^2}{s}\right|_{s=-\omega_n} = -\omega_n$$

$$B = -\,\omega_n$$

$$C = \frac{d}{ds}\left[(s+\omega_n)^2\, C(s)\right]\Big|_{s=-\omega_n} = \frac{d}{ds}\left[\frac{\omega_n^2}{s}\right]\Big|_{s=-\omega_n}$$

$$= \left.\frac{-\omega_n^2}{s^2}\right|_{s=-\omega} \quad \frac{-\omega_n^2}{(-\omega_n)^2} = -1$$

$$C = -\,1$$

Substitute A, B and C values in Eq. (4.52)

$$C(s) = \frac{A}{s} + \frac{B}{(s+\omega_n)^2} + \frac{C}{(s+\omega_n)} = \frac{1}{s} - \frac{\omega_n}{(s+\omega_n)^2} - \frac{1}{(s+\omega_n)} \tag{4.53}$$

The response in time domain, on taking inverse laplace transform,

$$C(t) = L^{-1}[C(s)] = L^{-1}\left[\frac{1}{s} \frac{-\omega_n}{(s+\omega_n)^2} - \frac{1}{(s+\omega_n)}\right]$$

$$C(t) = 1 - \omega_n\, t\, e^{-\omega_n t} - e^{-\omega_n t}$$

$$C(t) = 1 - e^{-\omega_n t}(1 + \omega_n t)$$

Conclusion: The response of critically damped second order system have no oscillations.

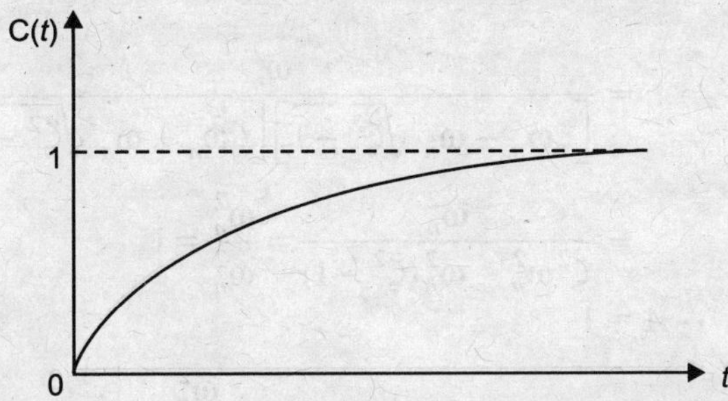

Fig. 4.7: Response of critically damped second order system for unit step input.

4.7.4 Response of Second Order System For Over Damped Case and When Input is Unit Step Input

The standard form of closed loop transfer function of second order system is

$$\frac{C(s)}{R(s)} = \frac{\omega_n^2}{s^2 + 2\zeta\omega_n s + \omega_n^2} \tag{4.54}$$

For overdamped system $\zeta > 1$. The roots of the denominator of transfer function are real and distinct. Let the roots of the denominator be s_a, s_b

$$s_a, s_b = -\zeta\omega_n \pm \omega_n\sqrt{\zeta^2 - 1} = -[\zeta\omega_n \pm \omega_n\sqrt{\zeta^2 - 1}] \tag{4.55}$$

Let, $s_1 = -s_a$ and $s_2 = -s_b$

$\therefore$

$$s_1 = \zeta\omega_n - \omega_n\sqrt{\zeta^2 - 1} \tag{4.56}$$

$$s_2 = \zeta\omega_n + \omega_n\sqrt{\zeta^2 - 1} \tag{4.57}$$

The closed loop transfer function can be written in terms of s_1 and s_2 as shown below,

$$\frac{C(s)}{R(s)} = \frac{\omega_n^2}{s^2 + 2\zeta\omega_n s + \omega_n^2} = \frac{\omega_n^2}{(s+s_1)(s+s_2)} \tag{4.58}$$

For unit step input $r(t) = 1$ and $R(s) = 1/s$

CHAPTER 4

$$\therefore \qquad C(s) = R(s) \cdot \frac{\omega_n^2}{(s+s_1)(s+s_2)} = \frac{\omega_n^2}{s(s+s_1)(s+s_2)} \tag{4.59}$$

By partial fraction expansion we can write,

$$C(s) = \frac{\omega_n^2}{s(s+s_1)(s+s_2)} = \frac{A}{s} + \frac{B}{(s+s_1)} + \frac{C}{(s+s_2)} \tag{4.60}$$

$$A = sC(s)\big|_{s=0} = s\,\frac{\omega_n^2}{s(s+s_1)(s+s_2)}\bigg|_{s=0} = \frac{\omega_n^2}{s_1\,s_2}$$

$$= \frac{\omega_n^2}{\left[\zeta\omega_n - \omega_n\sqrt{\zeta^2-1}\right]\left[\zeta\omega_n + \omega_n\sqrt{\zeta^2-1}\right]} = 1$$

$$= \frac{\omega_n^2}{\zeta^2\omega_n^2 - \omega_n^2(\zeta^2-1)} = \frac{\omega_n^2}{\omega_n^2} = 1$$

$$A = 1$$

$$B = (s+s_1)\,C(s)\big|_{s=-s_1} = \frac{\omega_n^2}{s\,(s+s_2)}\bigg|_{s=-s_1}$$

$$= \frac{\omega_n^2}{-s_1\,(-s_1+s_2)}$$

$$= \frac{-\omega_n^2}{s_1\left[-\zeta\omega_n + \omega_n\sqrt{s^2-1+\zeta\omega_n+\omega_n\sqrt{\zeta^2-1}}\right]}$$

$$= \frac{-\omega_n^2}{\left[2\,\omega_n\sqrt{\zeta^2-1}\right]s_1} = \frac{\omega_n}{2\sqrt{\zeta^2-1}} \times \frac{1}{s_1}$$

$$B = \frac{-\omega_n}{2\sqrt{\zeta^2-1}} \cdot \frac{1}{s_1}$$

$$C = C(s)\,(s+s_2)\big|_{s=-s_2} = \frac{\omega_n^2}{s\,(s+s_1)}\bigg|_{s=s_2}$$

$$C = \frac{\omega_n^2}{-s_2\,(-s_2+s_1)}$$

$$C = \frac{\omega_n^2}{-s_2\left[-\zeta\omega_n - \omega_n\sqrt{\zeta^2-1} + \zeta\omega_n - \omega_n\sqrt{\zeta^2-1}\right]}$$

$$C = \frac{\omega_n}{2\sqrt{\zeta^2-1}} \cdot \frac{1}{s_2}$$

Substitute A, B and C values in Eq. (4.60)

$$\therefore \quad = \frac{1}{s} - \frac{\omega_n}{2\sqrt{\zeta^2-1}} \cdot \frac{1}{s_1} \frac{1}{(s+s_1)} + \frac{\omega_n}{2\sqrt{\zeta^2-1}} \cdot \frac{1}{s_2} \frac{1}{(s+s_2)}$$

on taking inverse Laplace transform we get,

The response in time domain,

$$c(t) = 1 - \frac{\omega_n}{2\sqrt{\zeta^2-1}} \cdot \frac{1}{s_1} e^{-s_1 t} + \frac{\omega_n}{2\sqrt{\zeta^2-1}} \cdot \frac{1}{s_2} e^{-s_2 t}$$

$$c(t) = 1 - \frac{\omega_n}{2\sqrt{\zeta^2-1}} \left[\frac{e^{-s_1 t}}{s_1} - \frac{e^{-s_2 t}}{s_2} \right] \tag{4.61}$$

where $\quad s_1 = \zeta\omega_n - \omega_n\sqrt{\zeta^2-1}$

and $\quad s_2 = \zeta\omega_n + \omega_n\sqrt{\zeta^2-1}$

The Eq. (4.61) is the time response of second order system when input is unit step input under overdamped.

The response of overdamped second order system has no oscillations but it takes longer time for the response to reach the final steady state value.

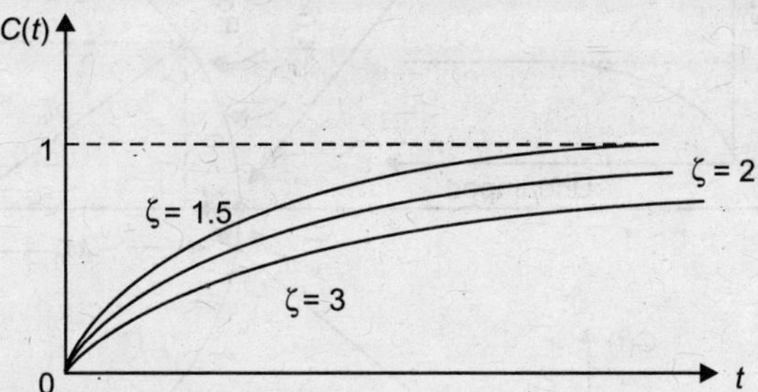

Fig. 4.8: Response of overdamped second order system for unit step input.

4.7.5 Location of Roots of Characteristic Equation and Time Response of Second Order System

$$\frac{C(s)}{R(s)} = \frac{\omega_n^2}{s^2 + 2\zeta\omega_n s + \omega_n^2}$$

$$s^2 + 2\zeta\,\omega_n\,s + \omega_n^2 = 0 \tag{4.62}$$

The roots of Eq. of (4.62)

$$s_1 = -\zeta\omega_n + j\omega_n\sqrt{1-\zeta^2}$$

$$s_2 = -\zeta\omega_n + j\omega_n\sqrt{1-\zeta^2}$$

The real part of the roots ($-s\omega_n$) represents the damping and imaginary part ($\omega_n \sqrt{1-\zeta^2}$) represents the damped frequency.

From Fig. (4.9)

1. ω_n is the distance of root from origin
2. $\cos\theta = \dfrac{\zeta\omega_n}{\omega_n} = \zeta$ i.e. $\theta = \cos^{-1}\zeta$

   (when the roots are in Let-half of *s*-plane)
3. ω_d is the imaginary part of the root and is known as damped frequency (or) conditional frequency.

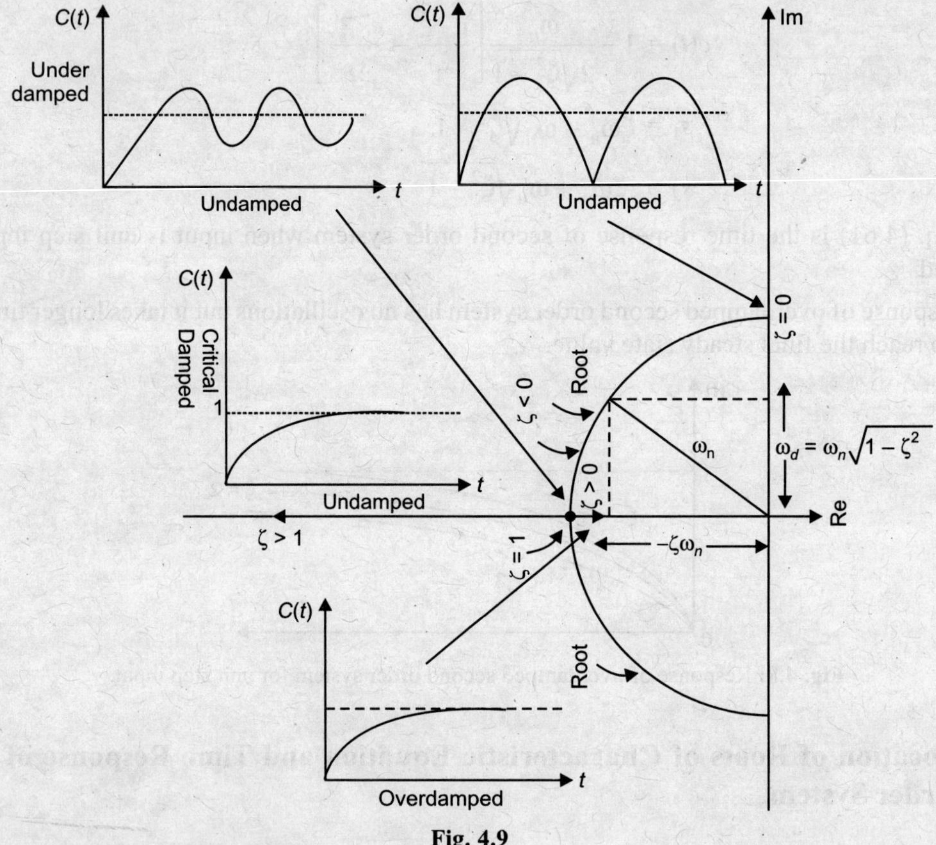

Fig. 4.9

4.7.6 Time Domain Specification

The desired performance characteristics of control systems are specified in terms of time domain specifications. Systems with energy storage elements cannot respond instantaneously and will exhibit transient responses, whenever they are subjected to inputs (or) disturbance.

The desired performance characteristics of a system of any order may be specified in terms of the transient response to a unit step input signal. The response of a second order system for unit-step input with various values of damping ratio is shown below

The transient response of a system to a unit step input depends on the initial conditions. Therefore to compare the time response of various systems it is necessary to start with standard initial conditions. The most practical standard is to start with the system at rest and so output and all time derivatives before $t = 0$ will be zero. The transient response of a practical control system often exhibits damped oscillation before reaching steady state. A typical damped oscillatory response of a system is shown below Fig. (4.11).

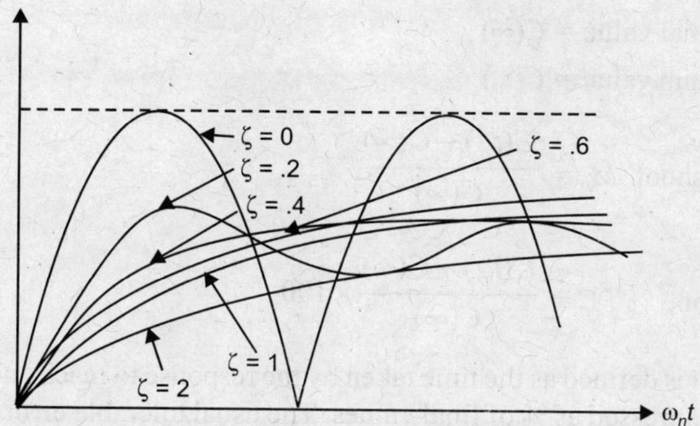

Fig. 4.10: Unit step response of second order system.

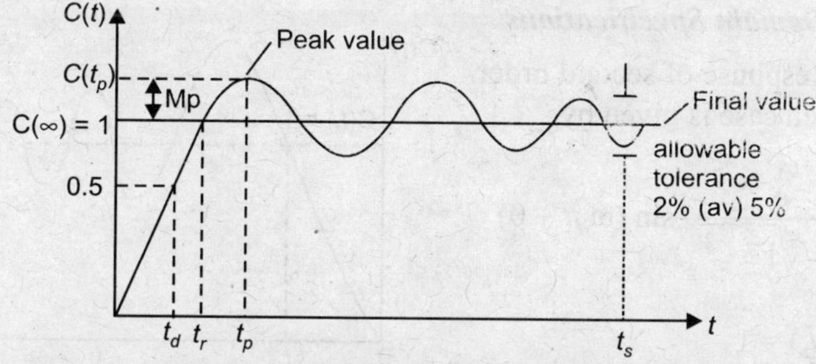

Fig. 4.11: Damped oscillatory response of second order system.

The transient response characteristics of a control system to a unit step input is specified in terms of the following time domain specifications.

1. Delay time, t_d
2. Rise time, t_r
3. Peak time, t_p
4. Maximum overshoot, M_p
5. Settling time, $t(s)$

The time domain specifications are defined as follows:

1. **Delay time (t_d):** It is the time taken for the response to reach 50% of the final value, for the very first time.

CHAPTER 4

2. **Rise time (t_r):** It is the time taken for the response to raise from 0 to 100% for the very first time. For underdamped system, the rise time is calculated from 0 to 100%. But for overdamped system it is the time taken by the response to raise from 10% to 90%. For critically damped system, it is the time taken for response to raise from 5% to 95%.
3. **Peak time (t_p):** It is the time taken for the response to reach peak value for the very first time. (or) it is the time taken for the response to reach the peak overshoot, M_p.
4. **Peak overshoot (M_p):** It is defined as the ratio of the maximum peak value measured from peak value to the final value

$$\text{Let final value} = C(\infty)$$

$$\text{Maximum value} = C(t_p)$$

$$\text{Peak overshoot, } M_p = \frac{C(t_p) - C(\infty)}{C(\infty)} \tag{4.63}$$

$$\text{\% Peak overshoot, \% } M_p = \frac{C(t_p) - C(\infty)}{C(\infty)} \times 100 \tag{4.64}$$

5. **Settling time (t_s):** It is defined as the time taken by the response to reach and stay within a specified error. It is usually expressed as % of final values. The usual tolerable error is 2% or 5% of the final value.

Expression for Time Domain Specifications

Rise time (t_r): Response of second order system for underdamped case is given by,

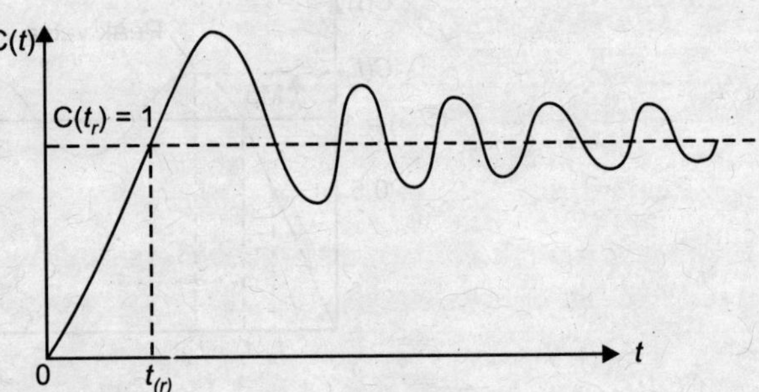

Fig. 4.12

$$C(t) = 1 - \frac{e^{-\zeta\omega_n t}}{\sqrt{1-\zeta^2}} \cdot \sin(\omega_d t + \theta)$$

at $t = t_r$,

$$C(t) = C(t_r) = 1$$

$$\therefore \quad C(t_r) = 1 - \frac{e^{-\zeta\omega_n \cdot t_r}}{\sqrt{1-\zeta^2}} \cdot \sin(\omega_d t_r + \theta) = 1$$

$$\therefore \quad = \frac{-e^{-\zeta\omega_n t_r}}{\sqrt{1-\zeta^2}} \cdot \sin(\omega_d t_r + \theta) = 0$$

Since $-e^{-\zeta\omega_n t_r} \neq 0$, the term, $\sin(\omega_d t_r + \theta) = 0$

$\sin \phi = 0$, when $\phi = 0, \pi, 2\pi, 3\pi$

$$\therefore \quad \omega_d t_r + \theta = \pi$$

$$\omega_d t_r = \pi - \theta$$

$$\therefore \qquad \text{Rise time, } t_r = \frac{\pi - \theta}{\omega_d}$$

Here, $\theta = \tan^{-1} \dfrac{\sqrt{1-\zeta^2}}{\zeta}$ (Refer note 1)

and damped frequency of oscillation, $\omega_d = \omega_n \sqrt{1-\zeta^2}$

$$\text{Rise time } t_r = \frac{\pi - \tan^{-1} \dfrac{\sqrt{1-\zeta^2}}{\zeta}}{\omega_n \sqrt{1-\zeta^2}} \qquad (4.65)$$

Note 1: On constructing right angle triangle with ζ and $\sqrt{1-\zeta^2}$, we get

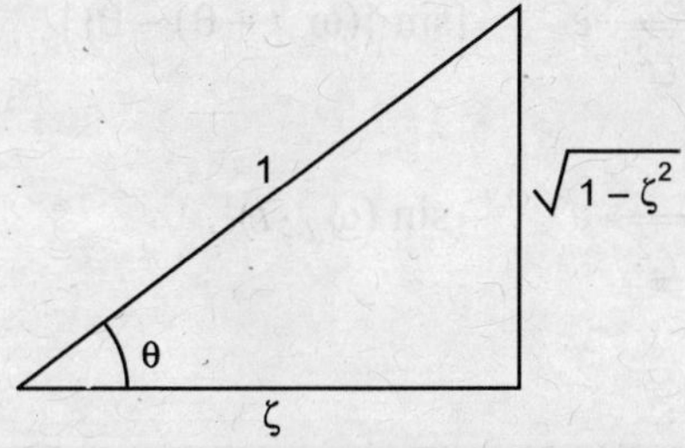

Fig. 4.13

Note 2: $\tan^{-1} \dfrac{\sqrt{1-\zeta^2}}{\zeta}$ should be measured in radians.

Peak time (t_p): To find the expression for peak time, t_p, differentiate $C(t)$ with respect to t and equate to 0.

i.e. $$\left.\frac{d}{dt} C(t)\right|_{t=t_p} = 0$$

The unit step response of second order system is given by

$$C(t) = 1 - \frac{e^{-\zeta\omega_n t}}{\sqrt{1-\zeta^2}} \sin(\omega_d t + \theta)$$

Differentiating $C(t)$ with respect to t.

$$\frac{d}{dt} C(t) = \frac{-e^{\zeta\omega_n t}}{\sqrt{1-\zeta^2}} (-\zeta\omega_n) \sin(\omega_d t + \theta) + \left(\frac{-e^{-\zeta\omega_n t}}{\sqrt{1-\zeta^2}}\right) \cos(\omega_d t + \theta)\, \omega_d$$

Put, $\omega_d = \omega_n \sqrt{1-\zeta^2}$

$$\therefore \quad \frac{d}{dt}c(t) = \frac{e^{-\zeta\omega_n t}}{\sqrt{1-\zeta^2}}(\zeta\omega_n)\sin(\omega_d t+\theta) - \frac{\omega_n\sqrt{1-\zeta^2}}{\sqrt{1-\zeta^2}}e^{-\zeta\omega_n t}\cos(\omega_d t+\theta)$$

$$= \frac{\omega_n \cdot e^{-\zeta\omega_n t}}{\sqrt{1-\zeta^2}}\left[\zeta\sin(\omega_d t+\theta) - \sqrt{1-\zeta^2}\cos(\omega_d t+\theta)\right]$$

$$= \frac{\omega_n}{\sqrt{1-\zeta^2}} \cdot e^{-\zeta\omega_n t}\left[\cos\theta\sin(\omega_d t+\theta) - \sin\theta\cos(\omega_d t+\theta)\right]$$

$$= \frac{\omega_n}{\sqrt{1-\zeta^2}} \cdot e^{-\zeta\omega_n t}\left[\sin(\omega_d t+\theta)\cos\theta - \cos(\omega_d t+\theta)\sin\theta\right]$$

$$= \frac{\omega_n}{\sqrt{1-\zeta^2}} \cdot e^{-\zeta\omega_n t}\left[\sin\{(\omega_d t+\theta)-\theta\}\right]$$

$$= \frac{\omega_n}{\sqrt{1-\zeta^2}} \cdot e^{-\zeta\omega_n t} \cdot \sin(\omega_d \cdot t)$$

at $t = t_p$, $\frac{d}{dt}C(t) = 0$

$$\therefore \quad \left.\frac{d}{dt}C(t)\right|_{t=t_p} = \frac{\omega_n}{\sqrt{1-\zeta^2}} \cdot e^{-\zeta\omega_n t_p}\sin(\omega_d t_p) = 0$$

$$\therefore \quad \frac{\omega_n}{\sqrt{1-\zeta^2}} \cdot e^{-\zeta\omega_n t_p} \cdot \sin(\omega_d t_p) = 0$$

since, $0^{-\zeta\omega_n t_p} \neq 0$, the term, $\sin(\omega_d t_p) = 0$ when $\omega_d t_p = 0, \pi, 2\pi, 3\pi$

$$\therefore \quad \omega_d t_p = \pi$$

$\therefore$ Peak time, $$t_p = \pi/\omega_d \tag{4.66}$$

The damped frequency of oscillation, $\omega_d = \omega_n\sqrt{1-\zeta^2}$

Peak time, $$t_p = \frac{\pi}{\omega_n\sqrt{1-\zeta^2}} \tag{4.67}$$

Note 3: On constructing right angle triangle with ζ and $\sqrt{1-\zeta^2}$, we get

$$\sin\theta = \sqrt{1-\zeta^2}$$

$$\cos\theta = \zeta$$

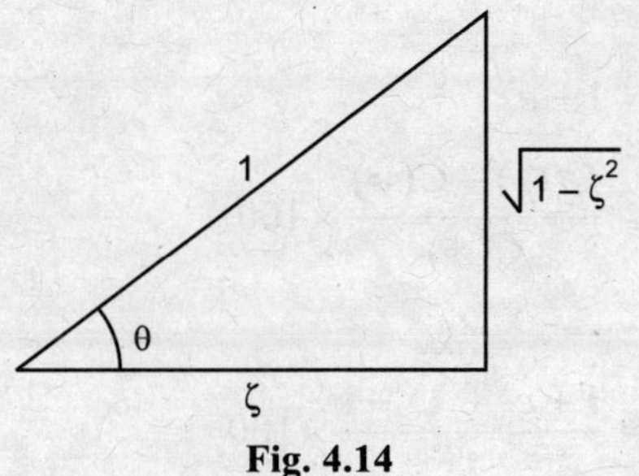

Fig. 4.14

Peak Overshoot (M_p)

% Peak overshoot, % $M_p = \dfrac{C(t_p) - C(\infty)}{C(\infty)} \times 100$

where $C(t_p)$ = Peak response at $t = t_p$

The out step response of second order system is given by

$$C(t) = 1 - \frac{e^{-\zeta\omega_n t}}{\sqrt{1-\zeta^2}} \sin(\omega_d t + \theta)$$

at $t = \infty$, $C(t) = C(\infty) = 1 - \dfrac{e^{-\infty}}{\sqrt{1-\zeta^2}} \sin(\omega_d t + \theta) = 1 - 0 = 1$ (4.68)

at $t = t_p = C(t) = C(t_p) = 1 - \dfrac{e^{-\zeta\omega_n t_p}}{\sqrt{1-\zeta^2}} \sin(\omega_d t_p + \theta)$

put $t_p = \pi/\omega_d$

$$\therefore \qquad C(t_p) = 1 - \frac{e^{-\zeta\omega_n \cdot \frac{\pi}{\omega_d}}}{\sqrt{1-\zeta^2}} \sin\left(\omega_d \frac{\pi}{\omega_d} + \theta\right)$$

The damped frequency of oscillation,

$$\omega_d = \omega_n\sqrt{1-\zeta^2}$$

$$C(t_p) = 1 - \frac{e^{-\zeta\omega_n \cdot \frac{\pi}{\omega_n\sqrt{1-\zeta^2}}}}{\sqrt{1-\zeta^2}} \sin(\pi + \theta)$$

$$C(t_p) = 1 + \frac{e^{\frac{-\zeta\pi}{\sqrt{1-\zeta^2}}}}{\sqrt{1-\zeta^2}} \sin\theta \qquad [\text{since } (\pi + \theta) = -\sin\theta]$$

$$C(t_p) = 1 + \frac{e^{\frac{-\zeta\pi}{\sqrt{1-\zeta^2}}}}{\sqrt{1-\zeta^2}} \cdot \sqrt{1-\zeta^2}$$

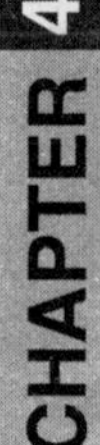

$$C(t_p) = 1 + e^{\frac{-\zeta\pi}{\sqrt{1-\zeta^2}}} \quad (4.69)$$

$$\% M_p = \frac{C(t_p) - C(\infty)}{C(\infty)} \times 100$$

$$\% M_p = \frac{1 + e^{\frac{-\zeta\pi}{\sqrt{1-\zeta^2}}} - 1}{1} \times 100$$

Percentage peak overshoot $\% M_p = e^{\frac{-\zeta\pi}{\sqrt{1-\zeta^2}}} \times 100$ (4.70)

Note 4: On constructing right angle triangle with ζ and $\sqrt{1-\zeta^2}$, are get

$$\sin\theta = \sqrt{1-\zeta^2}$$

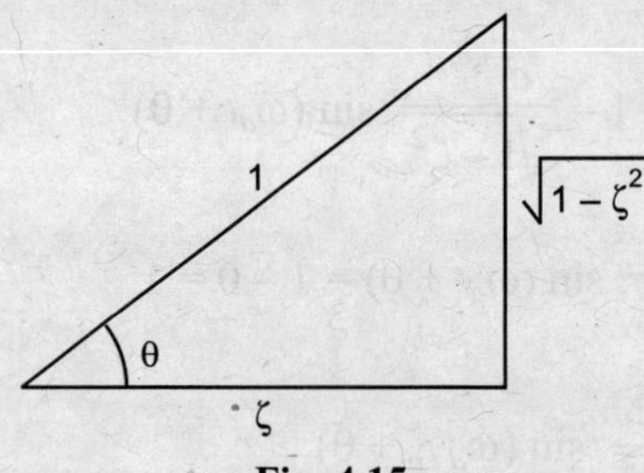

Fig. 4.15

Settling Time (t_s):

The response of second order system has two components they are,

1. Decaying exponential component, $\frac{e^{-\zeta\omega_n t}}{\sqrt{1-\zeta^2}}$
2. Sinusoidal component, sin $(\omega_d t + \theta)$.

In this decaying exponential term reduces the oscillations produced by sinusoidal component. Hence the settling time is decided by the exponential component. The settling time can be found out by equating exponential component to percentage toterance errors.

For 2% toterance error band, at $t = t_s$,

$$\frac{e^{-\zeta\omega_n t_s}}{\sqrt{1-\zeta^2}} = 0.02$$

For least values of ζ, $e^{-\zeta\omega_n t_s} = 0.02$. On taking natural logarithm we get,

$$-\zeta\omega_n t_s = l_n\,(0.02)$$

$$-\zeta\omega_n t_s = -4$$

$\therefore$ Settling time,
$$t_s = \frac{4}{\zeta\omega_n} = 4T; \text{ for 2\% error.} \quad (4.71)$$

where,
$$T = \frac{1}{\zeta\omega_n} = \text{Time constant of the system} \quad (4.72)$$

For 5% error, $e^{-\zeta\omega_n t_s} = 0.05$

on taking natural logarithm we get,

$$-\zeta\,\omega_n t_s = l_n\,(0.05)$$
$$-\zeta\,\omega_n t_s = -3$$

$\therefore$ Settling time,
$$t_s = \frac{3}{\zeta\omega_n} = 3T; \text{ for 5\% errors} \quad (4.73)$$

In general for a specified percentage error settling time,

$$t(s) = \frac{l_n(\%\text{ error})}{\zeta\omega_n} \quad (4.74)$$

4.7.7 Time Response of Second Order System with Unit Ramp Input

The closed loop transfer function of first order system,

$$\frac{C(s)}{R(s)} = \frac{\omega_n^2.}{s^2 + 2\,\zeta\omega_n s + \omega_n^2}$$

It is input is unit step, then,

$$R(s) = 1/s^2$$

$\therefore$ The response in s-domain,

$$C(s) = \frac{R(s)\cdot\omega_n^2}{s^2 + 2\,\zeta\omega_n s + \omega_n^2} = \frac{\omega_n^2}{s^2(s^2 + 2\zeta\omega_n s + \omega_n^2)}$$

By partial fraction,

$$\frac{\omega_n^2}{s^2\,(s^2 + 2\zeta\omega_n s + \omega_n^2)} = \frac{A}{s} + \frac{B}{s^2} + \frac{Cs + D}{s^2 + 2\zeta\omega_n s + \omega_n^2} \quad (4.75)$$

$$\omega_n^2 = As(s^2 + 2\zeta\omega_n\,s + \omega_n^2) + B\,(s^2 + 2\zeta\omega_{n\,s} + \omega_n^2) + (Cs + D)\,s^2$$

Equating the coefficient of s^3, s^2, s and constant terms.

$$A + C = 0 \quad (4.76)$$
$$2\zeta\omega_n\,A + B + D = 0 \quad (4.77)$$
$$A\omega_n^2 + 2\,\zeta\omega_n\,B = 0 \quad (4.78)$$
$$B\,\omega_n^2 = \omega_n^2 \quad (4.79)$$

CHAPTER 4

From equations (4.76) to (4.79)

$$A = -\frac{2\zeta}{\omega_n}; \quad C = \frac{2\zeta}{\omega_n}$$

$$B = 1; \quad D = -1 + 4\zeta^2$$

Substitute all these values in Eq. (4.75)

$$C(s) \; \frac{-2\zeta}{\omega_n s} + \frac{1}{s^2} + \frac{\frac{2\zeta}{\omega_n}\cdot s + (4\zeta^2 - 1)}{s^2 + 2\zeta\omega_n s + \omega_n^2}$$

$$= \frac{1}{s^2} - \frac{2\zeta}{\omega_n s} + \frac{2\zeta}{\omega_n}\cdot\frac{(s + \zeta\omega_n - \zeta\omega_n)}{(s+\zeta\omega_n)^2 + \omega_n^2(1-\zeta^2)} + \frac{4\zeta^2 - 1}{(s+\zeta\omega_n)^2 + \omega_n(1-\zeta^2)}$$

$$= \frac{1}{s^2} - \frac{2\zeta}{\omega_n s} + \frac{2\zeta}{\omega_n}\cdot\frac{(s+\zeta\omega_n)}{(s+\zeta\omega_n)^2 + \omega_n^2(1-\zeta^2)} - \frac{2\zeta}{\omega_n}\cdot\frac{\zeta\omega_n}{(s+\zeta\omega_n)^2 + \omega_n^2(1-\zeta^2)} + \frac{4\zeta^2 - 1}{(s+\zeta\omega_n)^2 + \omega_n(1-\zeta^2)}$$

$$C(s) = \frac{1}{s^2} - \frac{2\zeta}{\omega_n s} + \frac{2\zeta}{\omega_n}\cdot\frac{(s+\zeta\omega_n)}{(s+\zeta\omega_n)^2 + \omega_n^2(1-\zeta^2)} + \frac{2\zeta^2 - 1}{(2+\zeta\omega_n)^2 + \omega_n^2(1-\zeta^2)}$$

We know,

$$\omega_d^2 = \omega_n^2(1-\zeta^2)$$

$$C(s) = \frac{1}{s^2} - \frac{2\zeta}{\omega_n s} + \frac{2\zeta}{\omega_n}\cdot\frac{(s+\zeta\omega_n)}{(s+\zeta\omega_n)^2 + \omega_d^2} + \frac{2\zeta^2 - 1}{(s+\zeta\omega_n)^2 + \omega_d^2}\cdot\frac{\omega_d}{\omega_d}$$

$$C(s) = \frac{1}{s^2} - \frac{2\zeta}{\omega_n s} + \frac{2\zeta}{\omega_n}\cdot\frac{(s+\zeta\omega_n)}{(s+\zeta\omega_n)^2 + \omega_d^2} + \frac{2\zeta^2 - 1}{\omega_d}\cdot\frac{\omega_d}{(s+\zeta\omega_n)^2 + \omega_d^2} \qquad (4.80)$$

On taking inverse Laplace transform of Eq. (4.80)

$$C(t) = t - \frac{2\zeta}{\omega_n} + \frac{2\zeta}{\omega_n}\cdot e^{-\zeta\omega_n t}\cos\omega_d t + \frac{2\zeta^2 - 1}{\omega_d} e^{-\zeta\omega_n t}\sin\omega_d t$$

$$= t - \frac{2\zeta}{\omega_n} + \frac{2\zeta\sqrt{1-\zeta^2}}{\omega_n\sqrt{1-\zeta^2}} e^{-\zeta\omega_n t}\cos\omega_d t + \frac{2\zeta^2 - 1}{\omega_n\sqrt{1-\zeta^2}} e^{-\zeta\omega_n t}\sin\omega_d t$$

$$= t - \frac{2\zeta}{\omega_n} + \frac{e^{-\zeta\omega_n t}}{\omega_n\sqrt{1-\zeta^2}}\left[2\zeta\sqrt{1-\zeta^2}\cos\omega_d t + (2\zeta^2 - 1)\sin\omega_d t\right]$$

Put,

$$2\zeta^2 - 1 = \cos\phi$$

$$2\zeta\sqrt{1-\zeta^2} = \sin\theta$$

$$C(t) = t - \frac{2\zeta}{\omega_n} + \frac{e^{-\zeta\omega_n t}}{\omega_n\sqrt{1-\zeta^2}}\,[\sin\phi\cos\omega_d t + \cos\phi\sin\omega_d t\,]$$

$$= t - \frac{2\zeta}{\omega_n} + \frac{e^{-\zeta\omega_n t}}{\omega_n\sqrt{1-\zeta^2}}\sin(\omega_d t + \phi)$$

$$C(t) = t - \frac{2\zeta}{\omega_n} + \frac{e^{-\zeta\omega_n t}}{\omega_n\sqrt{1-\zeta}}\sin\left[\left(\omega_n\sqrt{1-\zeta^2}\right)t + \phi\right] \quad (4.81)$$

where,

$$\phi = \cos^{-1}(2\zeta^2 - 1)$$

or

$$\phi = \tan^{-1}\frac{2\zeta\sqrt{1-\zeta^2}}{2\zeta^2 - 1}$$

The Eq. (4.81) is the time response of second order system under unit ramp input.

4.7.8 Time Response of Second Order System with Unit Impulse Input

Since, input is unit impulse $R(s) = 1$

$$C(s) = \frac{\omega_n^2}{s^2 + 2\zeta\omega_n s + \omega_n^2} \quad (4.82)$$

Break the Eq. (4.82) by partial fraction.

$$\frac{\omega_n^2}{s^2 + 2\zeta\omega_n s + \omega_n^2} = \frac{A}{s - s_1} + \frac{B}{s - s_2} \quad (4.83)$$

The roots of the characteristics Eq. is,

$$s_1 = -\zeta\omega_n + j\omega_n\sqrt{1-\zeta^2}$$

$$s_2 = -\zeta\omega_n - j\omega_n\sqrt{1-\zeta^2}$$

$$\omega_d = \omega_n\sqrt{1-\zeta^2}$$

$$s_1 = -\zeta\omega_n + j\omega_d$$

$$s_2 = -\zeta\omega_n - j\omega_d$$

From Equ. (4.83)

$$A = \left.\frac{\omega_n^2}{s - s_2}\right|_{s=s_1}, \quad = \frac{\omega_n^2}{-\zeta\omega_n + j\omega_d + \zeta\omega_n + j\omega_d} = \frac{\omega_n^2}{2\,j\omega_d}$$

CHAPTER 4

$$B = \frac{\omega_n^2}{-j\,2\omega_d}$$

Substitute the values of A and B in (4.83)

$$C(s) = \frac{\omega_n^2}{s^2 + 2\zeta\omega_n s + \omega_n^2} = \frac{\omega_n^2}{j\,2\omega_d}\cdot\frac{1}{s-s_1} - \frac{\omega_n^2}{j\,2\omega_d}\cdot\frac{1}{s-s_2} \tag{4.84}$$

On taking inverse Laplace transform for Eq. (4.84)

$$L^{-1}[C(s)] = C(t) = \frac{\omega_n^2}{2\,j\omega_d}[e^{s_1 t} - e^{s_2 t}]$$

$$= \frac{\omega_n^2}{j\,2\omega_d}[e^{(-\zeta\omega_n + j\omega_d)t} - e^{(-\zeta\omega_n - j\omega_d)t}]$$

$$= \frac{\omega_n^2}{j\,2\omega_d}[e^{-\zeta\omega_{nt}}\cdot e^{j\omega_d t} - e^{-\zeta\omega_n t}\cdot e^{-j\omega_d t}]$$

$$= \frac{\omega_n^2\cdot e^{-\zeta\omega_n t}}{j\,2\omega_d}[e^{j\omega_d t} - e^{-j\omega_d t}]$$

$$= \frac{\omega_n^2\cdot e^{-\zeta\omega_n t}}{j\,2\omega_d}[\cos\omega_d t + j\sin\omega_d t - \cos\omega_d t + j\sin\omega_d t]$$

$$= \frac{\omega_n^2\cdot e^{-\zeta\omega_n t}}{j\,2\omega_d}[j\,2\sin\omega_d t] = \frac{\omega_n^2\cdot e^{-\zeta\omega_n t}}{\omega_d}\cdot\sin\omega_d t \tag{4.85}$$

Put ω_d in Eq. (4.85)

$$C(t) = \frac{\omega_n^2\cdot e^{-\zeta\omega_n t}}{\omega_n\sqrt{1-\zeta^2}}\sin\left(\omega_n\sqrt{1-\zeta^2}\right)t$$

$$C(t) = \frac{\omega_n}{\sqrt{1-\zeta^2}}\cdot e^{-\zeta\omega_n t}\cdot\sin\left(\omega_n\sqrt{1-\zeta^2}\right)t \tag{4.86}$$

(i) $\zeta < 1$:

$$C(t) = \frac{\omega_n}{\sqrt{1-\zeta^2}}\cdot e^{-\zeta\omega_n t}\sin\left(\omega_n\sqrt{1-\zeta^2}\right)t$$

(ii) $\zeta = 1$:

$$C(t) = \underset{\zeta\to 1}{Lt}\frac{\omega_n}{\sqrt{1-\zeta^2}}e^{-\zeta\omega_n t}\sin\left(\omega_n\sqrt{1-\zeta^2}\right)t$$

$$C(t) = t\cdot e^{-\omega_n t}\cdot\omega_n^2$$

(iii) $\zeta > 1$

Equation (4.81) can be written as

$$C(s) = \frac{\omega_n^2}{(s+\zeta\omega_n)^2 - \omega_n^2(\zeta^2-1)} \tag{4.87}$$

Apply partial fraction and put $\omega_d^2 = \omega_n^2(\zeta^2-1)$

$$\frac{\omega_n^2}{(s+s\omega_n)^2 - \omega_d^2} = \frac{\omega_n^2}{(s+\zeta\omega_n+\omega_d)(s+\zeta\omega_n-\omega_d)}$$

$$= \frac{A}{(s+\zeta\omega_n+\omega_d)} + \frac{B}{(s+\zeta\omega_n-\omega_d)} \tag{4.88}$$

From Eq. (4.87)

$$A = \left.\frac{\omega_n^2}{s+\zeta\omega_n-\omega_d}\right|_{s=-\zeta\omega_n-\omega_d}$$

$$A = \frac{-\omega_n}{2\sqrt{\zeta^2-1}}$$

Similarly,
$$B = \frac{-\omega_n}{2\sqrt{\zeta^2-1}}$$

Substitute the values in Eq. (4.87)

$$C(s) = \frac{\omega_n^2}{(s+\zeta\omega_n)^2 - \omega_d^2}$$

$$= \frac{-\omega_n}{2\sqrt{\zeta^2-1}} \cdot \frac{1}{s+\zeta\omega_n+\omega_d} + \frac{\omega_n}{2\sqrt{\zeta^2-1}} \cdot \frac{1}{s+\zeta\omega_n-\omega_d}$$

Take inverse Laplace,

$$L^{-1}[C(s)] = C(t) = \frac{\omega_n}{2\sqrt{\zeta^2-1}}\left[-e^{-[\zeta\omega_n+\omega_d]t} + e^{-(\zeta\omega_n-\omega_d)t}\right]$$

$$C(t) = \frac{\omega_n}{2\sqrt{\zeta^2-1}}\left[e^{-\left(\zeta-\sqrt{\zeta^2-1}\right)\omega_n t} - e^{-\left(\zeta+\sqrt{\zeta^2-}\right)\omega_n t}\right] \tag{4.88}$$

The Eq. (4.88) in the time response of second order system under unit impulse input.

4.7.9 Time Response of Higher Order Control System

The input, output and transfer function are given by

$$C(s) = R(s) \cdot T(s)$$

where,

$C(s) \rightarrow$ Output of the system

$R(s) \rightarrow$ Input of the system

$T(s) \rightarrow$ Transfer function of the system

$$T(s) = \frac{K\,(s^m + a_1 s^{m-1} + \ldots a_m)}{(s^n + b_1\, s^{n-1} + \ldots b_n)}$$

$$C(s) = R(s) \cdot \frac{K\,(s^m + a_1\, s^{m-1} + \ldots + a_m)}{(s^n + b_1\, s^{n-1} + \ldots b_n)}$$

Apply partial fraction,

$$C(s) = \frac{K_1}{s + P_1} + \frac{K_2}{(s + P_2)} + \ldots \frac{K_n}{(s + P_n)}$$

For time response take inverse Laplace of each part of above equation

$$L^{-1}\, C(s) = L^{-1}\left(\frac{K_1}{s + P_1}\right) + L^{-1}\left(\frac{K_2}{s + P_2}\right) + \ldots L^{-1}\left(\frac{K_n}{s + P_n}\right)$$

$$C(t) = K_1\, e^{-P_1 t} + K_2 e^{-P_2 t} + \ldots + K_n e^{-P_n t} \tag{4.89}$$

The Eq. (4.89) is the time response of higher order system.

SOME WORKED EXAMPLES

EXAMPLE 4.1: Obtain the response of unity feedback system whose open loop transfer function is $G(s) = \dfrac{4}{s\,(s + 5)}$ and the input is unit step.

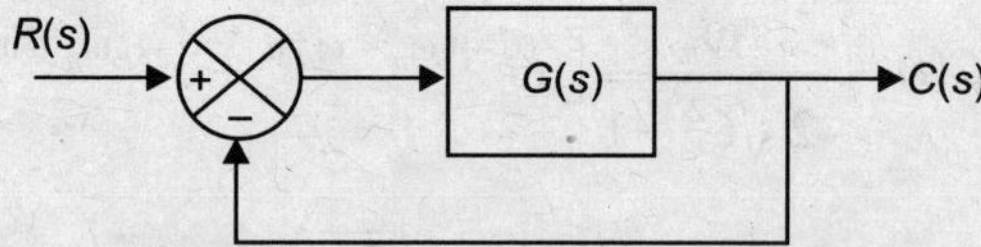

Fig. 4.1.1: Closed loop system.

Solution: The closed loop system is shown in Fig. (4.1.1)

The closed loop transfer function,

$$\frac{C(s)}{R(s)} = \frac{G(s)}{1 + G(s)}$$

$$\frac{C(s)}{R(s)} = \frac{4/s(s+5)}{1+\dfrac{4}{s(s+5)}} = \frac{4/s(s+5)}{\dfrac{s(s+5)+4}{s(s+5)}} = \frac{4}{s(s+5)+4}$$

$$\frac{C(s)}{R(s)} = \frac{4}{s^2+5s+4} = \frac{4}{(s+4)(s+1)}$$

The response in s-domain

$$C(s) = R(s)\cdot\frac{4}{(s+1)(s+4)}$$

Since the input is unit step, $R(s) = 1/s$

$$\therefore \qquad C(s) = \frac{4}{s(s+1)(s+4)}$$

By Partial fraction expansion, we can write,

$$C(s) = \frac{4}{s(s+1)(s+4)} = \frac{A}{s} + \frac{B}{(s+1)} + \frac{C}{(s+4)}$$

$$A = C(s)\big|_{s=0} = \left.\frac{4}{(s+1)(s+4)}\right|_{s=0} = \frac{4}{1\times 4} = 1 \Rightarrow A = 1$$

$$B = C(s)(s+1)\big|_{s=-1} = \frac{4}{s(s+4)} = \frac{4}{-1(-1+4)} = \frac{-4}{3} \Rightarrow B = -4/3$$

$$C = C(s)(s+4)\big|_{s=-4} = \left.\frac{4}{s(s+1)}\right|_{s=-4}$$

$$= \frac{4}{-4(-4+1)} = \frac{1}{3} \Rightarrow C = 1/3$$

Substitute A, B and C values in $C(s)$,

$$C(s) = \frac{1}{s} - \frac{4}{3(s+1)} + \frac{1}{3(s+4)}$$

The time domain response $C(t)$ is obtained by taking inverse Laplace transform of $C(s)$.

The response in time domain,

$$C(t) = L^{-1}[C(s)] = L^{-1}\left[\frac{1}{s} - \frac{4}{3}\frac{1}{s+1} + \frac{1}{3}\frac{1}{s+4}\right]$$

$$= 1 - \frac{4}{3}e^{-t} + \frac{1}{3}.e^{-4t} = 1 - \frac{1}{3}[4e^{-t} - e^{-4t}]$$

CHAPTER 4

Results: The response of unity feedback system is:

$$C(t) = 1 - \frac{1}{3}[4e^{-t} - e^{-4t}]$$

EXAMPLE 4.2: A positional control system with velocity feedback is shown in Fig. (4.2.1) what is the the response of the system for unit step input.

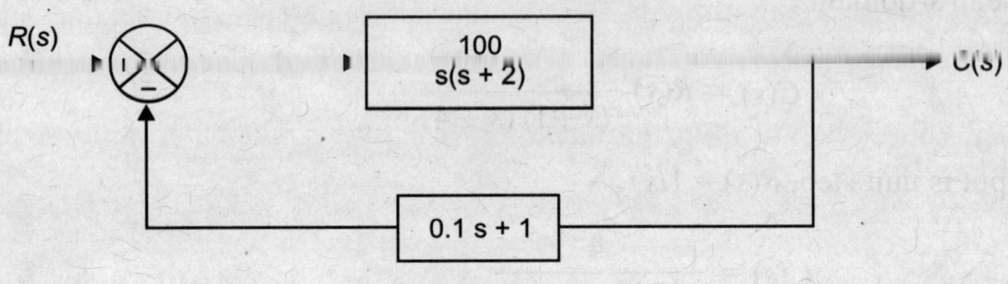

Fig. 4.2.1

Solution: The closed loop transfer function,

$$\frac{C(s)}{R(s)} = \frac{G(s)}{1 + G(s)\,H(s)}$$

Given that $G(s) = \dfrac{100}{s(s+2)}$ and $H(s) = .1s + 1$

$$= \frac{\dfrac{100}{s\,(s+2)}}{1 + \left[\dfrac{100}{s(s+2)}\right][0.1s+1]} = \frac{100/s(s+2)}{\dfrac{s(s+2) + 100\,(.1\,s+1)}{s(s+2)}}$$

$$= \frac{100}{s^2 + 2s + 10s + 100} = \frac{100}{s^2 + 12s + 100}$$

Here, $(s^2 + 12s + 100)$ is the characteristics polynomial. The roots of the characteristic polynomial are

$$s_1, s_2 = \frac{-12 \pm \sqrt{144 - 400}}{2} = \frac{-12 \pm j\,16}{2} = -6 \pm j\,8$$

The roots are complex conjugate. The system is underdamped and so the response of the system will have damped oscillations.

The response in s-domain

$$C(s) = R(s)\frac{100}{s^2 + 12s + 100}$$

since inut is unit step, $R(s) = 1/s$

$$\therefore \quad C(s) = \frac{1}{s} \cdot \frac{100}{s^2 + 12s + 100} = \frac{100}{s(s^2 + 12s + 100)}$$

By Partial fraction expansion we can write,

$$C(s) = \frac{100}{s(s^2 + 12s + 100)} = \frac{A}{s} + \frac{Bs + C}{s^2 + (12s + 100)}$$

The residue A is obtained by multiplying $C(s)$ by s and letting $s = 0$.

$$A = C(s)\big|_{s=0} = \frac{100}{s^2 + 12s + 100}\bigg|_{s=0} = \frac{100}{100} = 1 \Rightarrow A = 1$$

The residue B and C are evaluated by cross multiplying the following equation and equating the coefficients of like power of s.

$$\frac{100}{s(s^2 + 12s + 100)} = \frac{A}{s} + \frac{Bs + C}{s^2 + 12s + 100}$$

$$100 = A(s^2 + 12s + 100) + (Bs + C)\, s$$

$$100 = As^2 + 12As + 100\,A + Bs^2 + Cs$$

on equating the coefficients of s^2 we get,

$$A + B = 0$$

$$\therefore \quad B = -A = -1$$

$$A = 1$$

on equating the coefficients of s we get,

$$0 = 12A + C \quad \therefore C = -12A = -12$$

$$C = -12$$

Substitute A, B and C values is Eq. (1)

$$\therefore \quad C(s) = \frac{1}{s} + \frac{-s - 12}{s^2 + 12s + 100}$$

$$= \frac{1}{s} - \frac{s + 12}{s^2 + 12s + 36 + 64} = \frac{1}{s} - \frac{s + 6 + 6}{(s + 6)^2 + 64}$$

$$= \frac{1}{s} - \frac{(s + 6)}{(s + 6)^2 + 64} - \frac{6}{(s + 6)^2 + 64}$$

$$C(s) = \frac{1}{s} - \frac{(s + 6)}{(s + 6)^2 + 8^2} - \frac{6}{8} \cdot \frac{8}{(s + 6)^2 + 8^2}$$

CHAPTER 4

The time domain response is obtained by taking inverse Laplace transform of $C(s)$

The response in time domain

$$C(t) = L^{-1}[C(s)] = L^{-1}\left[\frac{1}{s} - \frac{s+6}{(s+6)^2+8^2} - \frac{6}{8}\cdot\frac{8}{(s+6)^2+8^2}\right]$$

$$C(t) = 1 - e^{-6t}\cos t - \frac{6}{8}e^{-6t}\sin 8t$$

$$= 1 - e^{-6t}\left[\frac{6}{8}\sin 8t + \cos 8t\right]$$

The result can be converted to another standard form by constructing right angle triangle with ζ and $\sqrt{1-\zeta^2}$. The damping ratio ζ is evaluated by comparing the closed loop transfer function of the system with standard form of second order transfer function.

$$\frac{C(s)}{R(s)} = \frac{\omega_n^2}{s^2+2\zeta\omega_n s+\omega_n^2} = \frac{100}{s^2+12s+100}$$

on companing we get,

$$\omega_n^2 = 100$$

$$\boxed{\omega_n = 100}$$

$$2\zeta\omega_n = 12$$

$$\zeta = \frac{12}{\omega_n} = \frac{12}{2\times 10} = .6 \Rightarrow \boxed{\zeta = .6}$$

constructing right angle triangle with ζ and $\sqrt{1-\zeta^2}$ we get,

$$\sin\theta = 0.8;\ \cos\theta = 0.6$$

$$\tan\theta = \frac{0.8}{0.6}$$

$$\theta = \tan^{-1}\left[\frac{0.8}{0.6}\right] = 53° = .925 \text{ rad}$$

$$\theta = .925 \text{ rad}$$

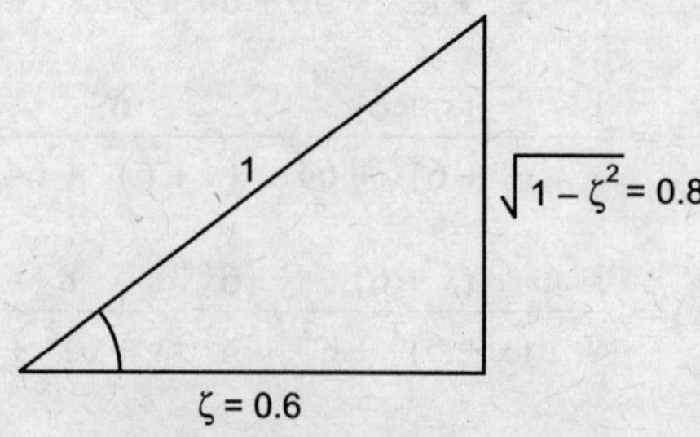

Fig. 4.2.2: Closed loop system.

The response in time domain

$$C(t) = 1 - e^{-6t}\left[\frac{6}{8}\sin 8t + \cos 8t\right]$$

$$= 1 - e^{-6t}\frac{10}{8}e^{-6t}\left[\frac{6}{8}\sin 8t + \frac{8}{10}\cos 8t\right]$$

$$= 1 - \frac{10}{8}e^{-6t}\,[\sin 8t \times 0.6 + \cos 8t \times 0.8]$$

$$= 1 - 1.25\,e^{-6t}\,[\sin 8t + \cos\theta + \cos 8t \sin\theta]$$

$$= 1 - 1.25\,e^{-6t}\,[\sin 8t + \theta]$$

$$C(t) = 1 - 1.25e^{-6t}\sin\,[8t + .925]$$

Result: The response in time domain,

$$C(t) = 1 - e^{-6t}\left[\frac{6}{8}\sin 8t + \cos 8t\right]$$

$$C(t) = 1 - 1.25e^{-6t}\sin\,(8t + .925)$$

EXAMPLE 4.3: Measurements conducted on a servomechanism shows the system response to be $C(t) = 1 + 0.2.\,e^{-60t} - 1.2\,e^{-10t}$ when subject to a unit step input. obtain an expression for closed loop transfer function. determine the undamped natural frequancy and damping ration.

Solution: Given that $C(t) = 1 + 0.2\,e^{-60t} - 1.2\,e^{-10t}$

on taking Laplace transform of $c(t)$ we get,

$$L[C(t)] = C(s) = \frac{1}{s} + 0.2\frac{1}{(s+60)} - 1.2\frac{1}{(s+10)}$$

$$= \frac{(s+60)\,(s+10) + 0.2\,s\,(s+10) - 1.2\,s\,(s+60)}{s(s+60)\,(s+10)}$$

$$= \frac{s^2 + 70s + 600 + 0.2\,s^2 + 2s - 1.2s^2 - 72s}{s\,(s+60)\,(s+10)} = \frac{600}{s(s+60)\,(s+10)}$$

$$C(s) = \frac{600}{s(s+60)\,(s+10)} = \frac{1}{s}\,\frac{600}{(s+60)\,(s+10)}$$

Since input is unit step, $R(s) = 1/s$

$$C(s) = R(s)\frac{600}{s^2 + 70s + 600}$$

The closed loop transfer function of the system,

$$\frac{C(s)}{R(s)} = \frac{600}{s^2 + 70s + 600}$$

The damping ratio and natural frequency of oscillation can be estimated by comparing the system transfer function with standard form of second order transfer function.

$$\frac{C(s)}{R(s)} = \frac{\omega_n^2}{s^2 + 2\zeta\omega_n s + \omega_n^2} = \frac{600}{s^2 + 70s + 600}$$

Compare L.H.S and R.H.S

$$\omega_n^2 = 600$$

$$\omega_n = \sqrt{600} = 24.49 \text{ rad/sec}$$

$$2\zeta\,\omega_n = 70$$

$$\zeta = \frac{70}{2\omega_n} = \frac{70}{2 \times 24.49} = 1.43.$$

Result:

- The closed loop transfer function of the system

$$\frac{C(s)}{R(s)} = \frac{600}{s^2 + 70s + 600}$$

- Natural frequency of oscillation, $\omega_n = 24.49$ rad/sec.
- Damping ratio, $\zeta = 1.43$ (over damped).

EXAMPLE 4.4: The unity feedback system is characterized by an open loop transfer function $G(s) = \dfrac{K}{s(s+10)}$.

Determine the gain K, so that the system will have a damping ratio of 0.5 for this value of K. Determine settling time, peak overshoot and time to peak overshoot for a unit step input.

Solution: This is the unity feedback system.

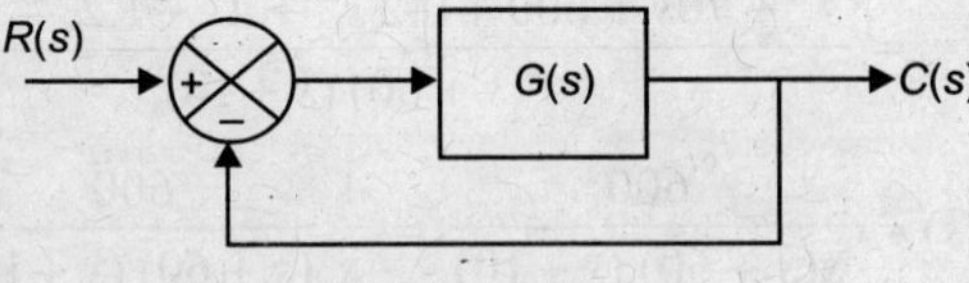

Fig. 4.4.1

The closed loop transfer function,

$$\frac{C(s)}{R(s)} = \frac{G(s)}{1 + G(s)} \qquad [\because H(s) = 1]$$

Given that, $G(s) = \frac{K}{s(s+10)}$.

So,

$$\frac{C(s)}{R(s)} = \frac{K/s(s+10)}{1+K/s(s+10)} = \frac{K}{s(s+10)+K} = \frac{K}{s^2+10s+K}$$

The value of K can be evaluated by comparing the system transfer function with standard form of second order transfer function.

$$\frac{C(s)}{R(s)} = \frac{\omega_n^2}{s^2+2\zeta\omega_n s+\omega_n^2} = \frac{K}{s^2+10s+K}$$

Compare the co-efficient of ω_n in both RHS and LHS

$\omega_n^2 = K$ $\quad\quad$ $2\zeta\omega_n = 10$

$\omega_n = \sqrt{K}$ $\quad\quad$ $\omega_n = 10/\text{rad/sec}$

Now $\zeta = .5$ and $\omega_n = \sqrt{K}$

$$\text{so, } 2 \times .5 \times \sqrt{K} = 10$$

$$\sqrt{K} = 10, \quad \therefore \text{ gain, } K = 100$$

Percentage peak overshoot, $\% M_p = e^{-\zeta\pi/\sqrt{1-\zeta^2}} \times 100$

$$= e^{-0.5\pi/\sqrt{1-(0.5)^2}} \times 100$$

$$= 0.163 \times 100$$

$$\% M_p = 16.3\%$$

peak time, $t_p = \pi/\omega_d = \frac{\pi}{\omega_n\sqrt{1-\zeta^2}}$

$$= \frac{\pi}{10\sqrt{1-(0.5)^2}} = 0.363 \text{ sec}$$

Result: The value of gain, $K = 100$

Percentage peak overshoot $\% M_p = 16.3\%$

Peak time $t_p = 0.363$ sec.

EXAMPLE 4.5: The open loop transfer function of a unity feedback system is given by $G(s) = K/s(sT+1)$ where K and T are positive constant. By what factor should the amplifier gain K be reduced, so that the peak overshoot of unit step response of the system is reduced step response of the system is reduced from 75% to 25%.

Solution: Unit feedback system is,

CHAPTER 4

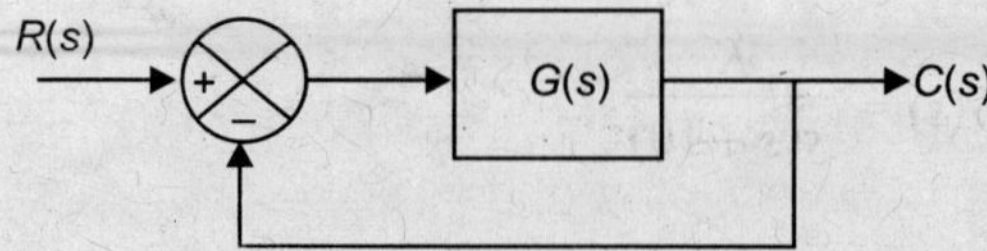

Fig. 4.5.1: Closed loop system.

The closed loop transfer function,

$$\frac{C(s)}{R(s)} = \frac{G(s)}{1+G(s)} \qquad [\because H(s)=1]$$

Given that, $G(s) = K/s(sT+1)$

$$\frac{C(s)}{R(s)} = \frac{K/s(sT+1)}{1+K/s(sT+1)} = \frac{K}{s(sT+1)+K} = \frac{K}{s^2T+s+K} = \frac{K/T}{s^2+\frac{1}{T}s+\frac{K}{T}}$$

Expression for ζ and ω_n can be obtained by comparing the transfer function with the standard form of second order transfer function.

$$\frac{C(s)}{R(s)} = \frac{\omega_n^2}{s^2+2\zeta\omega_n s+\omega_n^2} = \frac{K/T}{s^2+\frac{1}{T}\cdot s+\frac{K}{T}}$$

Compare the co-efficient of ω_n in RHS and LHS

$$\omega_n^2 = K/T \qquad\qquad 2\zeta\omega_n = 1/T$$

$$\omega_n = \sqrt{K/T} \qquad\qquad \zeta = \frac{1}{2\omega_n T}$$

$$\zeta = \frac{1}{2\sqrt{\frac{K}{T}}\cdot T} = \frac{1}{2\sqrt{KT}}$$

The peak overshoot, M_p is reduced by increasing the damping ratio ζ. The damping ratio is increased by reducing the gain K.

when $M_p = 0.75$, Let $\zeta = \zeta_1$ and $K = K_1$

when $M_p = 0.25$, Let $\zeta = \zeta_2$ and $K = K_2$

Peak overshoot, $M_p = e^{-\zeta\pi/\sqrt{1-\zeta^2}}$ taking logarithm on both sides,

$$\ln M_p = \frac{-\zeta\pi}{\sqrt{1-\zeta^2}}$$

on squarring, we get,

$$(ln\ M_p)^2 = \frac{\zeta^2\pi^2}{1-\zeta^2}$$

on cross multiplication we get,

$$(1 - \zeta^2)\,(ln\ M_p)^2 = \zeta^2\,\pi^2$$

$$(ln\ M_p)^2 - \zeta^2\,(ln\ M_p)^2 = \zeta^2\,\pi^2$$

$$(ln\ M_p)^2 = \zeta\,\pi^2 + \zeta^2\,(ln\ M_p)^2$$

$$(ln\ M_p)^2 = \zeta^2[\pi^2 + (\ln M_P)^2] \qquad (1)$$

$$\therefore \qquad \zeta^2 = \frac{(ln\ M_p)^2}{\pi^2 + (ln\ M_p)^2}$$

But $$\zeta = \frac{1}{2\sqrt{KT}}$$

$$\zeta^2 = \frac{1}{4KT} \qquad (2)$$

On equating Eq. (1) and (2)

$$\frac{1}{4KT} = \frac{(ln\ M_p)^2}{\pi^2 + (ln\ M_p)^2}$$

$$\frac{1}{K} = \frac{4T\ (ln\ M_p)^2}{\pi^2 + (ln\ M_p)^2}$$

$$K = \frac{\pi^2 + (ln\ M_p)^2}{4T\ (ln\ M_p)^2}$$

when, $K = K_1$, $M_p = 0.75$

$$K_1 = \frac{\pi^2 + (ln\,0.75)^2}{4T\ (ln\,0.75)^2} = \frac{9.952}{0.331T} = \frac{30.06}{T}$$

when, $K = K_2$, $M_p = 0.25$

$$K_2 = \frac{\pi^2 + (ln\,0.25)^2}{4T\ (ln\,0.25)^2} = \frac{11.79}{7.68T} = \frac{1.53}{T}$$

$$\frac{K_1}{K_2} = \frac{(1/T)\,30.06}{(1/T)\,1.53} = 19.6$$

$$K_1 = 19.6\,K_2 \text{ (or) } K_2 = \frac{1}{19.6}K_1$$

To reduce peak overshoot from 0.75 to 0.25, K should be reduced by 19.6 times. (approximately 20 times).

Results: The value of gain, K should be reduced approximately 20 times to reduce peak overshoot from 0.75 to 0.25.

EXAMPLE 4.6: A positional control system with velocity feedback is shown. What is the response $C(t)$ to the unit step input. Giventhat $\zeta = 0.5$. Also calculate, rise time, peak time, maximum overshoot and settling time.

CHAPTER 4

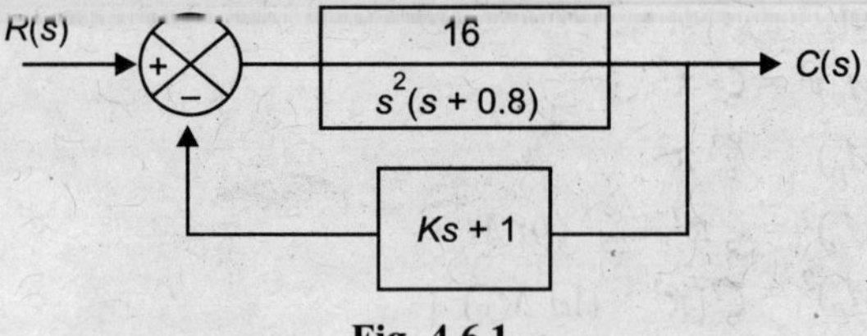

Fig. 4.6.1

Solution: The closed loop transfer function,

$$\frac{C(s)}{R(s)} = \frac{G(s)}{1+G(s)\,H(s)}$$

Given that, $G(s) = 16/s\ (s + 0.8)$ and

$H(s) = Ks + 1$

$$\frac{C(s)}{R(s)} = \frac{16/s(s+0.8)}{1+\dfrac{16}{s(s+0.8)}(Ks+1)} = \frac{16}{s^2+(0.8+16K)\,s+16}$$

The values of K and ω_n are obtained by comparing the system transfer function with standard form of second order transfer function.

$$\frac{C(s)}{R(s)} = \frac{\omega_n^2}{s^2+2\,\zeta\omega_n s+\omega_n^2} = \frac{16}{s^2+(0.8+16K)\,s+16}$$

Compare LHS and RHS

$\omega_n^2 = 16$ $\qquad 0.8 + 16\,K = 2\,\zeta\,\omega_n$

$\omega_n = 4$ rad/sec $\qquad K = \dfrac{2\,\zeta\omega_n - 0.8}{16} = \dfrac{2\times 0.5\times 4 - 0.8}{16} = 0.2$

$$\frac{C(s)}{R(s)} = \frac{16}{s^2+(0.8+16\times 0.2)\,s+16} = \frac{16}{s^2+4s+16} \quad 3$$

Given that the damping ratio, $\zeta = 0.5$. Hence the system is underdamped and so the response of the system will have damped oscillations. The roots of characteristics polynomial will be complex conjugate.

The response in s-domain, $C(s) = R(s)\dfrac{16}{s^2+4s+16}$

For unit step input, $R(s) = 1/s$

$$C(s) = \frac{1}{s}\cdot\frac{16}{s^2+4s+16} = \frac{16}{s(s^2+4s+16)}$$

apply partial fraction,

$$C(s) = \frac{16}{s(s^2+4s+16)} = \frac{A}{s}+\frac{Bs+C}{s^2+4s+16}$$

The residue A is obtained by multiplying $C(s)$ by s and letting $s = 0$

$$A = C(s) \cdot s\big|_{s=0} = \left.\frac{16}{s^2 + 4s + 16}\right|_{s=0} = \frac{16}{16} = 1 \Rightarrow A = 1$$

The residues B and C are evaluated by cross multiplying the following Eq. and equating the co-efficients of like power of s.

$$\frac{16}{s(s^2 + 4s + 16)} = \frac{A}{s} + \frac{Bs + C}{s^2 + 4s + 16}.$$

on cross multiplication we get,

$$16 = A(s^2 + 4s + 16) + (Bs + C)s$$

$$16 = As^2 + 4As + 16A + Bs^2 + Cs.$$

Equate the co-efficients of s^2 terms, s terms and constant terms, we get

$O = A + B$ $\qquad O = 4A + C$

$\therefore$ $B = -A = -1$ $\qquad \therefore C = -4A = -4$

$\boxed{A = 1}$

$$C(s) = \frac{1}{s} + \frac{-s-4}{s^2 + 4s + 16} = \frac{1}{s} - \frac{s+4}{s^2 + 4s + 4 + 12}$$

$$= \frac{1}{s} - \frac{s+2+2}{(s+2)^2 + 12} = \frac{1}{s} - \frac{s+2}{(s+2)^2 + 12} - \frac{2}{\sqrt{12}}\frac{\sqrt{12}}{(s+2)^2 + 12}$$

Apply inverse Laplace transform of $C(s)$ to get time domain response,

$$C(t) = L^{-1}[C(s)] = L^{-1}\left[\frac{1}{s} - \frac{s+2}{(s+2)^2 + 12} - \frac{2}{\sqrt{12}}\frac{\sqrt{12}}{(s+2)^2 + 12}\right]$$

$$= 1 - e^{-2t}\cos\sqrt{12}t - \frac{2}{2\sqrt{3}} \cdot e^{-2t} \cdot \sin\sqrt{12}t$$

$$= 1 - e^{-2t}\left[\frac{1}{\sqrt{3}}\sin\left(\sqrt{12}t\right) + \cos\left(\sqrt{12}t\right)\right]$$

to get the result, use right angle triangle with ζ and $\sqrt{1-\zeta^2}$,

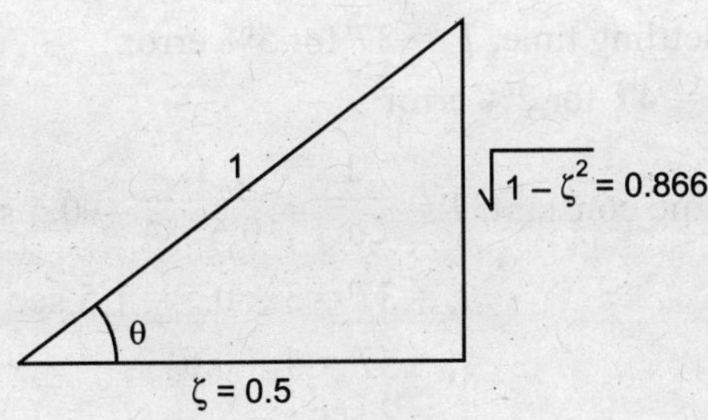

Fig. 4.6.2

$$\sin\theta = 0.866 = \frac{\sqrt{3}}{2}$$

$$\cos\theta = 0.5 = \frac{1}{2}$$

$$\tan\theta = 1.732$$

$$\theta = \tan^{-1} - 60° = 1.047 \text{ rad}$$

The response in time domain is,

$$C(t) = 1 - e^{-2t}\left[\frac{1}{\sqrt{3}} \times 2 \times \sin\sqrt{12}t \times \frac{1}{2} + \frac{2}{\sqrt{3}} \times \cos\sqrt{12}t \times \frac{\sqrt{3}}{2}\right]$$

$$= \left(1 - e^{-2t}\right)\left(\frac{2}{\sqrt{3}}\right)\left[\sin\sqrt{12}t\cos\theta + \cos\theta\sqrt{12}t\sin\theta\right]$$

$$= 1 - \frac{2}{\sqrt{3}}e^{-2t}\left[\sin\left(\sqrt{12}t + \theta\right)\right]$$

$$C(t) = 1 - \frac{2}{\sqrt{3}}e^{-2t}\left[\sin\left(\sqrt{12}t + 1.047\right)\right]$$

Damped frequency of oscillation

$$\omega_d = \omega_n\sqrt{1-\zeta^2} = 4\sqrt{1-(0.5)^2} = 3.464 \text{ rad/sec}$$

$$\therefore \quad \text{Rise time, } t_r = \frac{\pi - \theta}{\omega_d} = \frac{\pi - 1.047}{3.464} = 0.6046 \text{ sec}$$

$$\text{Peak time, } t_p = \frac{\pi}{\omega_d} = \frac{\pi}{3.464} = 0.907 \text{ sec}$$

$$\text{\% Maximum overshoot, } \%M_p = e^{\frac{-\zeta\pi}{\sqrt{1-\zeta^2}}} \times 100$$

$$= e^{\frac{-0.5\times\pi}{\sqrt{1-(0.5)^2}}} \times 100$$

$$= 0.163 \times 100 = 16.3\%$$

Settling time, $T_s = 3T$ for 5% error

$= 4T$ for 2% error

$$\text{Time constant, } T = \frac{1}{\zeta\omega_n} = \frac{1}{0.5 \times 4} = 0.5 \text{ sec}$$

For 5% error, settling time, $t_s = 3T = 3 \times 0.5 = 1.5$ sec

For 2% error, settling time, $t_s = 4T = 4 \times 0.5$

$= 2$ sec

Result: The time domain response,

$$C(t) = 1 - e^{-2t}\left[\frac{1}{\sqrt{3}}\sin\left(\sqrt{12}t\right) + \cos\left(\sqrt{12}t\right)\right]$$

or
$$C(t) = 1 - \frac{2}{\sqrt{3}}e^{-2t}\left[\sin\left(\sqrt{12}t + 1.047\right)\right]$$

Rise time, $t_r = 0.6046$ sec

Peak time $t_p = 0.907$ sec

Percentage maximum overshoot, % $M_p = 16.3\%$

Settling time, $t_s = 1.5$ sec for 5% error

= 2 sec for 2% error

EXAMPLE 4.7: A unity feedback control system is characterized by the following open loop transfer function $G(s) = (0.4s + 1)/s(s + 0.6)$. Determine its transient response for unit step input and sketch response. Evaluate the maximum overshoot and the corresponding peak time.

Solution: The closed loop transfer function, $\dfrac{C(s)}{R(s)} = \dfrac{G(s)}{1 + G(s)\,H(s)}$

Given that, $G(s) = (0.4s + 1)/s(s + 0.6)$

For unity feedback system, $H(s) = 1$

$$\frac{C(s)}{R(s)} = \frac{G(s)}{1+G(s)} = \frac{\dfrac{0.4s+1}{s\,(s+0.6)}}{1+\dfrac{0.4s+1}{s\,(s+0.6)}} = \frac{0.4s+1}{s\,(s+0.6)\;0.4s+1}$$

$$= \frac{0.4s+1}{s^2+0.6s+0.4s+1} = \frac{0.4s+1}{s^2+s+1}$$

The s-domain response, $C(s) = R(s)\cdot\dfrac{0.4s+1}{s^2+s+1}$.

For step input, $R(s) = \dfrac{1}{s}$

$$C(s) = \frac{1}{s}\cdot\frac{0.4s+1}{s^2+s+1} = \frac{0.4s+1}{s(s^2+s+1)}$$

Apply partial fraction, $C(s) = \dfrac{0.4s+1}{s(s^2+s+1)}\;\dfrac{A}{s} + \dfrac{Bs+C}{s^2+s+1}$

multiply $C(s)$ by s and letting $s = 0$,

$$A = C(s)\cdot\Big|_{s=0} = \left.\frac{0.4s+1}{s^2+s+1}\right|_{s=0} = 1$$

The residues B and C are solved by cross multiplying the following equation and equating the coefficients of like power of s.

$$\frac{0.4s+1}{s(s^2+s+1)} = \frac{A}{s} + \frac{Bs+C}{s^2+s+1}$$

on cross multiplication we get

$$0.4s + 1 = A(s^2 + s + 1) + (Bs + C)s$$

$$0.4s + 1 = As^2 + As + A + Bs^2 + Cs.$$

on equating coefficients of s^2,

$$0 = A + B$$

$$B = -A = -1$$

on equating coefficients of s,

$$0.4 = A + C$$

$$C = 0.4 - A = -0.6$$

$$C(s) = \frac{1}{s} + \frac{-s-0.6}{s^2+s+1} = \frac{1}{s} - \frac{s+0.6}{s^2+s+0.25+0.75}$$

$$= \frac{1}{s} - \frac{s+0.5+0.1}{(s+0.5)^2+0.75} = \frac{1}{s} - \frac{s+0.5}{(s+0.5)^2+0.75} - \frac{0.1}{\sqrt{0.75}}\frac{\sqrt{0.75}}{(s+0.5)^2+0.75}$$

take inverse Laplace transform of $C(s)$ to get time domain response,

$$C(t) = L^{-1}[C(s)] = L^{-1}\left[\frac{1}{s} - \frac{s+0.5}{(s+0.5)^2+0.75} - \frac{0.1}{\sqrt{0.75}} \cdot \frac{\sqrt{0.75}}{(s+0.5)^2+0.75}\right]$$

$$= 1 - e^{-0.5t} \cdot \cos\sqrt{0.75}t - \frac{0.1}{\sqrt{0.75}} e^{-0.5t} \cdot \sin\sqrt{.75}t$$

$$= 1 - e^{-0.5t}\left[.1155\sin\left(\sqrt{.75}t\right) + \cos\left(\sqrt{.75}t\right)\right]$$

The transient response is the part of the output which vanishes t tends to infinity. Here as t tends to infinity the exponential component $e^{0.5t}$ tends to zero. Hence the transient response is given by the damped sinusoidal component.

The transient response of $C(t) = e^{-0.5t}\left[0.1155\sin\left(\sqrt{0.75}t + \cos\sqrt{0.75}t\right)\right]$

The value of ζ and ω_n can be estimated by comparing the characteristics equation of the system transfer function with standard form of second order characteristics equation.

$$s^2 + 2\,\zeta\,\omega_n\, s + \omega_n^2 = s^2 + s + 1$$

on comparing we get,

$$\omega_n^2 = 1 \qquad 2\,\zeta\omega_n = 1$$

$$\omega_n = 1 \text{ rad/sec} \qquad \zeta = \frac{1}{2\omega_n} = \frac{1}{2} = 0.5$$

Maximum overshoot, $M_p = e^{\frac{-\zeta\pi}{\sqrt{1-\zeta^2}}}$

$$= e^{\frac{-0.5\,\pi}{\sqrt{1-(0.5)^2}}} = 0.163$$

% Maximum overshoot, % $M_p = M_p \times 100 = 0.163 \times 100 = 16.3\%$

Peak time $t_p = \frac{\pi}{\omega_d} = \frac{\pi}{\omega_n\sqrt{1-\zeta^2}} = \frac{\pi}{1\times\sqrt{1-(0.5)^2}} = 3.628$ sec

The response of the system underdamped, it is shown below,

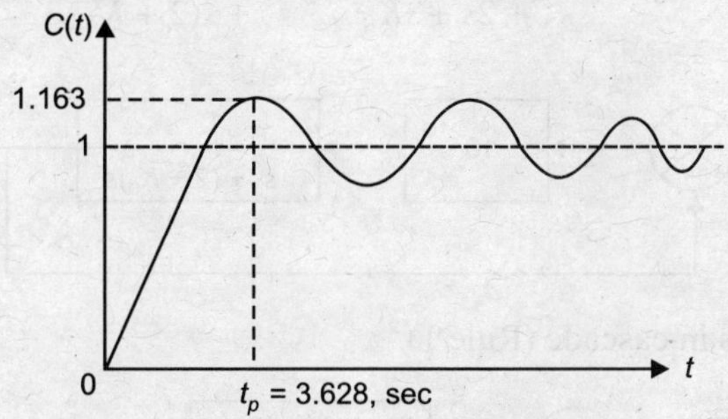

Fig. 4.7.1: Response of underdamped system.

Result: Transient response of the system = $e^{-0.5t}\left[0.1155\sin\left(\sqrt{0.75t}\right)+\cos\left(\sqrt{0.75t}\right)\right]$

% Maximum peak overshoot = 16.3%

Peak time = 3.628 sec.

EXAMPLE 4.8: A unit feedback control system has an amplifier with gain $K_A = 10$ and gain ratio, $G(s) = 1/s\,(s+2)$ in the feed forward path. A derivative feedback, $H(s) = sK_0$ is introduced as a loop $G(s)$. Determine derivative feedback constant, K_0 so that the system damping factor is 0.6.

Solution: The given system can be represented by the block diagram shown in Fig. (4.8.1).

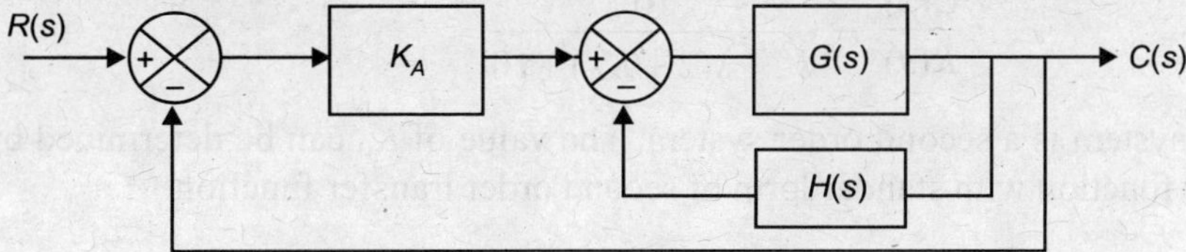

Fig. 4.8.1

Here, $K_A = 10$; $G(s) = \frac{1}{s(s+2)}$ and $H(s) = s \cdot K_0$

The closed loop transfer function of the system can be obtained by block diagram reduction techniques.

CHAPTER 4

Step 1: Reducing the inner feedback loop.

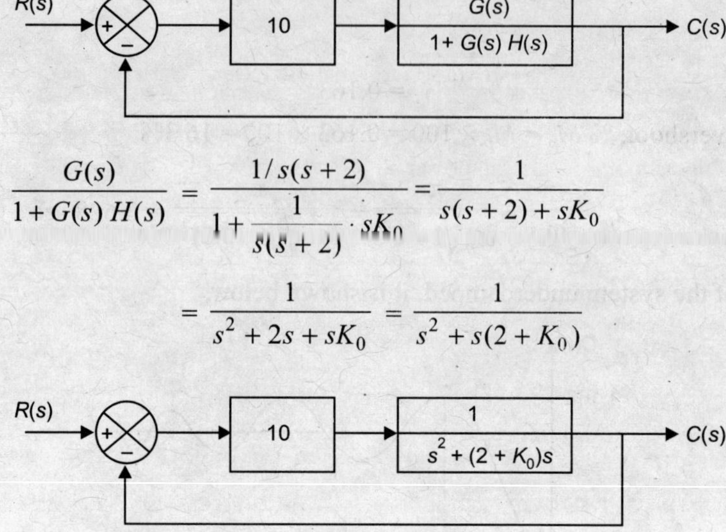

$$\frac{G(s)}{1+G(s)\,H(s)} = \frac{1/s(s+2)}{1+\dfrac{1}{s(s+2)}\cdot sK_0} = \frac{1}{s(s+2)+sK_0}$$

$$= \frac{1}{s^2+2s+sK_0} = \frac{1}{s^2+s(2+K_0)}$$

Step 2: Combining blocks in cascade (Rule 1).

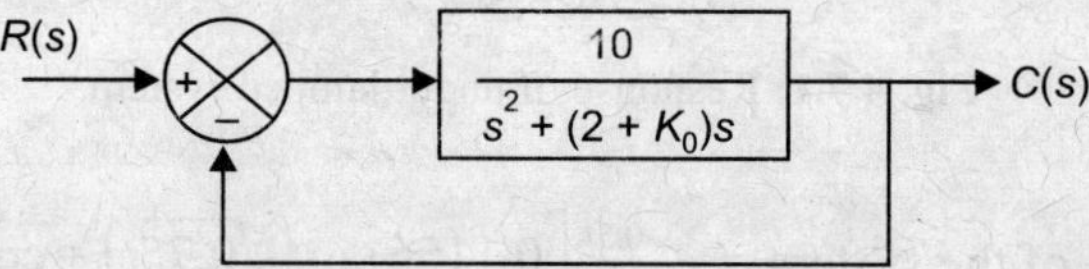

Step 3: Reducing the unity feedback path (Rule 10),

R(s) → $\dfrac{10}{s^2+s(2+K_0)+10}$ → C(s)

The closed loop transfer function is,

$$\frac{C(s)}{R(s)} = \frac{10}{s^2+s(2+K_0)+10}$$

The given system is a second order system. The value of K_0 can be determined by comparing the system transfer function with standardorm of second order transfer function.

$$\frac{C(s)}{R(s)} = \frac{\omega_n^2}{s^2+2\zeta\omega_n s+\omega_n^2} = \frac{10}{s^2+s(2+K_0)+10}$$

on comparing we get,

$\omega_n^2 = 10$ $\qquad\qquad 2+K_0 = 2\,\zeta\omega_n$

$\omega_n = \sqrt{10}$ $\qquad K_0 = 2\,\zeta\omega_n - 2$

$= 3.162$ rad/sec $\qquad = (2 \times 0.6 \times 3.162) - 2 = 1.7944 \Rightarrow K_0 = 1.7944$

Result: The value of constant, $K_0 = 1.7944$

EXAMPLE 4.9: A unity feedback control system has an open loop transfer function, $G(s) = 10/s(s+2)$. Find the rise time, percentage overshoot, peak time and settling time for a step input of 12 units.

Solution: The unity feedback system is shown below.

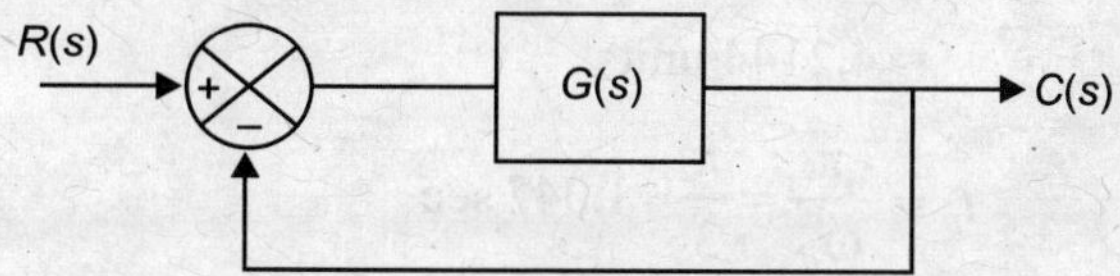

Fig. 4.9.1: Unity Feedback system.

The closed loop transfer function,

$$\frac{C(s)}{R(s)} = \frac{G(s)}{1+G(s)}$$

Given that, $G(s) = 10/s(s+2)$

$$\frac{C(s)}{R(s)} = \frac{10/s(s+2)}{1+\dfrac{10}{s(s+2)}} = \frac{10}{s(s+2)+10} = \frac{10}{s^2+2s+10}$$

The values of damping ratio ζ and natural frequency of oscillation ω_n are obtained by comparing the system function with standard form of second order transfer function.

$$\frac{C(s)}{R(s)} = \frac{\omega_n^2}{s^2+2\,\zeta\omega_n\,s+\omega_n^2} = \frac{10}{s^2+2s+10}$$

On comparing we get,

$\omega_n^2 = 10$ $\qquad 2\,\zeta\omega_n = 2$

$\omega_n = \sqrt{10}$ $\qquad \zeta = \dfrac{2}{2\,\omega_n} = \dfrac{1}{3.162} = 0.316$

$= 3.162$ rad/sec

$$\theta = \tan^{-1}\frac{\sqrt{1-\zeta^2}}{\zeta} = \tan^{-1}\frac{\sqrt{1-(0.316)^2}}{0.316} = 1.249 \text{ rad}$$

$$\omega_d = \omega_n\sqrt{1-\zeta^2} = 3.162\sqrt{1-(0.316)^2} = 3 \text{ rad/sec}$$

Rise time, $\qquad t_r = \dfrac{\pi-\theta}{\omega_d} = \dfrac{\pi-1.249}{3} = 0.63$ sec

Percentage peak over short $= e^{-\zeta\pi/\sqrt{1-\zeta^2}} \times 100$

$$= e^{\frac{-0.316\pi}{\sqrt{1-(0.316)^2}}} \times 100$$

$$= 0.3512 \times 100 = 35.12\%$$

Peak overshoot, M_p $= \frac{35.12}{100} \times 12$ units

$= 4.2144$ units

Peak time $t_p = \frac{\pi}{\omega_d} = \frac{\pi}{3} = 1.047$ sec

Time constant, $T = \frac{1}{\zeta\omega_n} = \frac{1}{0.316 \times 3.162} = 1$ sec

For 5% error, settling time, $t_s = 3T = 3$ sec

For 2% error, settling time, $t_s = 4T = 4$ sec

Result: Rise time, $t_r = 0.63$ sec

Percentage overshoot, $\% M_p = 35.12\%$

Peak overshoot = 4.2144 units (for a input of 12 units)

Peak time, $t_p = 1.047$ sec

Settling time, $t_s = 3$ sec for 5% error

$= 4$ sec for 2% error

EXAMPLE 4.10: A closed loop servo is represented by the differential equation $\frac{d^2c}{dt^2} + 8\frac{dc}{dt} = 64\,e$. Where c is the displacement of the output shaft, r is the displacement of the input and $e = r - c$. Determine undamped natural frequency, damping ratio and perdentage maximum overshoot for unit step unit.

Solution:

$$\frac{d^2c}{dt^2} + 8\frac{dc}{dt} = 64\,e \tag{4.10.1}$$

and

$$e = r - c \tag{4.10.2}$$

Put $e = r - c$ in Eq. (2.10.1)

$$\frac{d^2c}{dt^2} + 8\frac{dc}{dt} = 64\,(r - c) \tag{4.10.3}$$

Let $L[c] = C(s)$

and $L[r] = R(s)$

take Laplace transform for Eq. (2.10.3)

$s^2C(s) + 8s\,C(s) + 64\,[R(s) - C(s)]$

$$s^2 C(s) + 8s C(s) + 64 C(s) = 64 R(s)$$

$$(s^2 + 8s + 64) C(s) = 64 R(s)$$

$$\frac{C(s)}{R(s)} = \frac{64}{s^2 + 8s + 64}$$

The ratio $C(s)/R(s)$ is the closed loop transfer function of the system. On comparing the system transfer function with standard form of second order transfer function, we can estimate the values of ζ and ω_n.

$$\frac{C(s)}{R(s)} = \frac{\omega_n^2}{s^2 + 2\zeta\omega_n s + \omega_n^2} = \frac{64}{s^2 + 8s + 64}$$

on comparing, we get

$\omega_n^2 = 64$ $\qquad$ $2\zeta\omega_n = 8$

$\omega_n = 8$ rad/sec $\qquad$ $\zeta = \frac{8}{2\omega_n} = \frac{8}{2 \times 8} = 0.5$

Percentage peak overshoot, $\% M_p = e^{\frac{-\zeta\pi}{\sqrt{1-(0.5)^2}}} \times 100 = 16.3\%$

Result: Undamped natural frequency of oscillation

$\omega_n = 8$ rad/sec

Damping ratio, $\zeta = 0.5$

Percentage peak overshoot, $\% M_p = 16.3\%$

4.7.10 Type Number of Control System

The type number is specified for loop transfer function $G(s) \cdot H(s)$. The number of poles of the loop transfer function using at the origin decides the type number of the system. In general if N is the number of poles at the origin then the type number is N.

The loop transfer function can be expressed as a ratio of two polynomials in s,

$$G(s) \cdot H(s) = K\frac{P(s)}{Q(s)} = \frac{K(s+z_1)(s+z_2)(s+z_3)}{s^n (s+P_1)(s+P_2)(s+P_3)} \qquad (4.90)$$

where $Z_1, Z_2, Z_3 \rightarrow$ are zeros of transfer function.

$P_1, P_2, P_3 \rightarrow$ are poles of transfer function

$K \rightarrow$ is constant

$n \rightarrow$ is number of poles at the origin

The values of N in the denominator polynomial decides the type number of the system.

If $n = 0$, then the system is type – 0 system

If $n = 1$, then the system is type –1 system

CHAPTER 4

If $n = 2$, then the system is type –2 system

If $n = 3$, then the system is type –3 system etc.,

EXAMPLE 4.11: A second order system has *s* transfer function given by $G(s) = 25/s^2 + 8s + 25$ Find the value of damping ratio.

Solution: $R(s) = 1/s$ (for unit step input)

$$G(s) = \frac{C(s)}{R(s)} \Rightarrow C(s) = G(s)\,R(s)$$

$$C(s) = \frac{25}{s(s^2 + 8s + 25)} \qquad (4.11.1)$$

Compare the Eq. (4.11.1) with,

$$s^2 + 2\,\zeta\,\omega_n\,s + \omega_n^2 = 0$$

$$2\,\zeta\omega_n = 8$$

$$\omega_n = \sqrt{25}$$

$$= 5 \text{ rad/sec}$$

$$\zeta = .8$$

Result: The value of damping ratio is .8.

EXAMPLE 4.12: The block diagram of a feedback system is shown, below, Fig. (4.12.1)

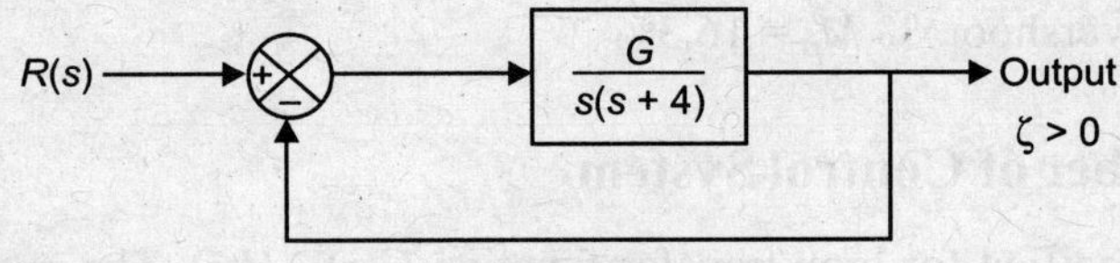

Fig. 4.12.1

(a) Find closed loop transfer function

(b) Find the minimum value of *G* for which the step response of the system would exhibit an overshoot as shown below (Fig. 4.12.2).

(c) For *G* equal to twice this minimum value, find the time period '*T*' indicated in Fig. (4.12.2).

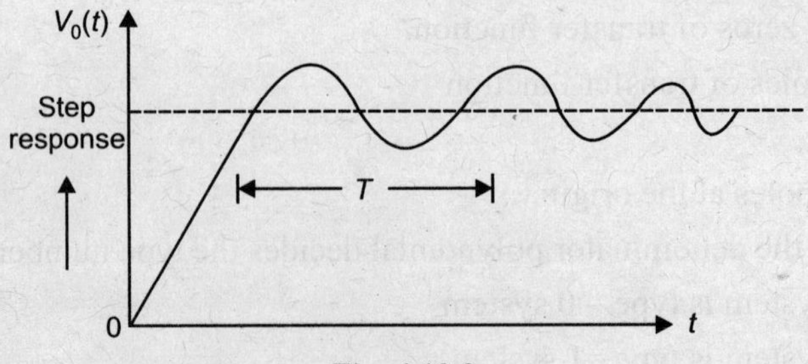

Fig. 4.12.2

Solution:

(a) Closed loop transfer function.

$$\frac{C(s)}{R(s)} = \frac{G/s(s+4)}{1+\dfrac{G}{s(s+4)}}$$

$$\frac{C(s)}{R(s)} = \frac{G/s(s+4)}{\dfrac{s(s+4)+G}{s(s+4)}}$$

$$= \frac{G}{s(s+4)+G}$$

$$\frac{C(s)}{R(s)} = \frac{G}{s^2+4s+G}$$

(b) Characteristic equation, $s^2 + 4s + G = 0$. Compare the characteristics equation $s^2 + 4s + G = 0$ with the equation of $s^2 + 2\zeta\omega_n s + \omega_n^2$

so we get,

$$s^2 + 4s + G = s^2\, 2\,\zeta\omega_n s + \omega_n^2$$

compare the coefficient of s^2, s and constant in both RHS and LHS

$$2\zeta\omega_n = 4$$

$$\omega_n^2 = G \rightarrow \omega_n = \sqrt{G}$$

Let we assume $\zeta = .5$

$$2\,\zeta\omega_n = 4$$

$$2\,\zeta\sqrt{G} = 4$$

$$2 \times .5 \times \sqrt{G} = 4$$

$$\sqrt{G} = \frac{3}{2 \times .5} \Rightarrow \sqrt{G} = 3$$

$$G = 9$$

(c)

$$G^1 = 2 \times 9 = 18$$

G = Twice the minimum value

$$\omega_n = \sqrt{9} = 3 \text{ rad/sec}$$

$$\zeta = 3/\,2\omega_n$$

$$= 3/2 \times 3.53 = 0.424$$

CHAPTER 4

$$\omega_d = \omega_n\sqrt{1-\zeta^2}$$

$$= \omega_n\sqrt{1-\zeta^2}$$

$$= 3.53\sqrt{1-(0.424)^2} = 3.197$$

$$\frac{2\pi}{T} = \omega_d$$

$$\frac{2\pi}{T} = 3.197$$

$$T = 1.96 \text{ sec}$$

Result:

- The closed loop transfer function

$$\frac{C(s)}{R(s)} = \frac{G}{s^2 + 3s + G}$$

- Minimum value of G when the step response of the system would exhibit an overshoot $G = 9$
- $T = 1.96$ sec when $G' = 2G$.

EXAMPLE 4.13: The open loop transfer function of a servo system with feedback is a given by $G(s) = 10/(s+2)(s+5)$ Determine the damping ratio, undamped natural frequency of oscillation. What is the percentage overshoot of the response to a unit step input.

Solution: Given that, $G(s) = 10/(s+2)(s+5)$

$$H(s) = 1$$

$$T(s) = \frac{C(s)}{R(s)} = \frac{G(s)}{1+G(s)}$$

$$= \frac{\dfrac{10}{(s+2)(s+5)}}{1+\dfrac{10}{(s+2)(s+5)}}$$

$$T(s) = \frac{10}{(s+2)(s+5)+10}$$

The characteristic Eq. is,

$$(s+2)(s+5) + 10 = 0$$

$$s^2 + 7s + 20 = 0 \qquad (4.13.1)$$

Compare the Eq. (2.14.1) with,

$$s^2 + 2\,\zeta\omega_n s + \omega_n^2 = 0$$

$$s^2 + 2\,\zeta\omega_n\, s + \omega_n^2 = s^2 + 7s + 20$$

compare the coefficients of s^2, s and constant from both the RHS and LHS

$$\omega_n^2 = 20$$

$$\omega_n = \sqrt{20}$$

$$= 4.472$$

$$\omega_n = 4.472 \text{ rad/sec}$$

and

$$2\zeta\omega_n = 7$$

$$2 \times \zeta \times 4.472 = 7$$

$$\zeta = 7/2 \times 4.472$$

$$\zeta = .7826$$

Percentage maximum overshoot,

$$\%\, M_p = e^{-\frac{\pi\zeta}{\sqrt{1-\zeta^2}}}$$

$$\%\, M_p = e^{-\frac{\pi \times 0.7826}{\sqrt{1-(0.7826)}}} \times 100$$

$$\%\, M_p = 1.92\%$$

Result: Damping ratio, $\zeta = 0.7826$. Natural frequency $\omega_n = 4.472$ rad/sec

Percentage maximum overshoot, % M_p = 1.92 %

EXAMPLE 4.14: For a RLC series circuit, the value of $R = 1\text{k}\Omega$, L = 10mH and C = 0.01 μF and the supply voltage V = 6V. DC supply. For this circuit, findout the undamped natural frequency and damping ratio of the circuit assume initial conditions are zero.

Solution:

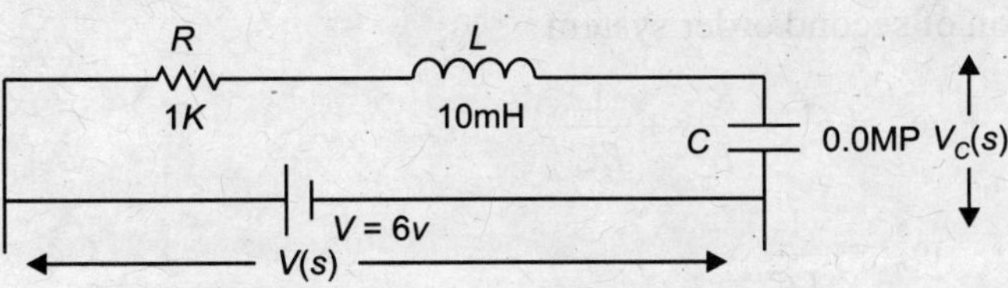

Fig. 4.14.1

CHAPTER 4

For the above circuit first determine transfer function.

$$\text{Transfer function } C(s) = \frac{O/P}{I/P}$$

$$= \frac{O/P}{I/P} = \frac{V_C(s)}{V(s)}$$

$$V_c = \frac{1}{C}\int i\,dt$$

$$V_c(s) = \frac{I(s)}{Cs} \tag{4.14.1}$$

$$V = R_i + \frac{Ldi}{dt} + \frac{1}{C}\int i\,dt$$

$$V(s) = RI(s) + Ls\,I(s) + \frac{I(s)}{Cs}$$

$$V(s) = I(s)\left[R + Ls + \frac{1}{Cs}\right]$$

$$V(s) = \left[\frac{RCs + LCs^2 + 1}{Cs}\right] I(s) \tag{4.14.2}$$

Now divide Eq. (4.14.1) by (4.14.2)

$$\frac{V_c(s)}{V(s)} = \frac{I(s)/Cs}{I(s)\left[\frac{RCs + LCs^2 + 1}{Cs}\right]}$$

$$\frac{V_c(s)}{V(s)} = \frac{1}{RCs + LCs^2 + 1}$$

$$T(s) = \frac{\frac{1}{LC}}{\frac{RCs}{LC} + s^2 + \frac{1}{LC}}$$

Compare the Eq. (4.14.3) with

Characteristic equation of second order system

$$s^2 + 2\,\zeta\omega_n s + \omega_n^2 = s^2 + \frac{R}{L}s + \frac{1}{LC}$$

$$\omega_n^2 = \frac{1}{LC}$$

$$\omega_n = \frac{1}{\sqrt{LC}} \tag{2.14.4}$$

$$2\zeta\omega_n = \frac{R}{L}$$

$$\zeta = \frac{R}{L2\omega_n}$$

where, $$\omega_n = \frac{1}{\sqrt{LC}}$$

$$\zeta = \frac{R}{L \times 2 \times \frac{1}{\sqrt{LC}}}$$

$$= \frac{R\sqrt{LC}}{2L} = \frac{R\sqrt{LC}}{2\sqrt{L}\sqrt{L}}$$

$$\zeta = \frac{R}{2}\sqrt{\frac{C}{L}} \quad (2.14.5)$$

(i) $$\omega_n = \frac{1}{\sqrt{LC}} = 10^5 \text{ rad/sec} \quad \left[\frac{1}{\sqrt{10 \times 10^{-3} \times -01 \times 10^{-1}}}\right]$$

$$\zeta = \frac{1000}{2}\sqrt{\frac{0.01 \times 10^{-6}}{10 \times 10^{-3}}} \quad [\because R = 1000\ \Omega]$$

$$\zeta = 0.51$$

Result:

1. Damping ratio $\zeta = .5$
2. Natural frequency $\omega_n = 10^5$ rad/sec

EXAMPLE 4.15: Determine the value of K and H of the closed loop system Fig. (4.15.1). So that the maximum overshoot in unit step response is 25% and the peaim is 2 sec. Assume $J = 1$ kg-m^2.

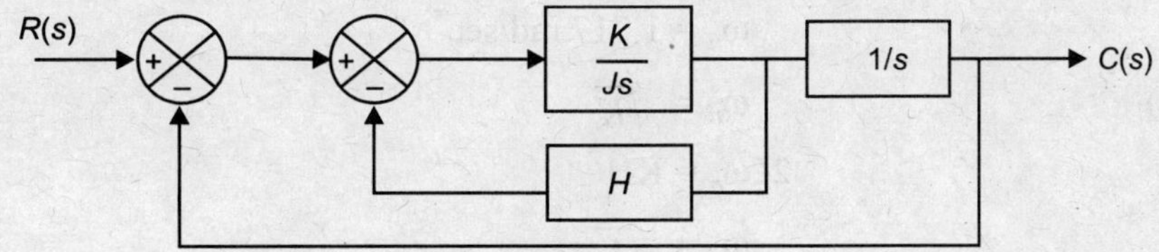

Fig. 4.15.1

Solution: $$\frac{C(s)}{R(s)} = \frac{K}{Js^2 + KHs + K}$$

Put $J = 1$ (It is given in the problem)

$$\frac{C(s)}{R(s)} = \frac{K}{s^2 + KHs + K}$$

CHAPTER 4

Characteristics Eq. is $s^2 + KHs + K$.

Compare characteristics Eq. with $s^2 + 2\,\zeta\omega_n s + \omega_n^2 = 0$

$$s^2 + 2\,\zeta\omega_n s + \omega_n^2 = s^2 + KHs + K$$

Compare the s term coefficient in both the sides in RHS and LHS.

$$\omega_n = \sqrt{K} \text{ and } 2\,\zeta\omega_n = KH \tag{4.15.1}$$

Given that $M_p = 0.25$; $[\therefore \%\, M_p = 25\%]$

$$M_p = e^{\frac{-\pi\zeta}{\sqrt{1-\zeta^2}}} = 0.25 \tag{4.15.2}$$

Solve the Eq. (4.15.2) to get the value of ζ. By taking log in both sides, we get,

$$\zeta = 0.404$$

Peak time
$$t_p = \frac{\pi}{\omega_n\sqrt{1-\zeta^2}}$$

$$\omega_n\sqrt{1-\zeta^2} = \frac{\pi}{t_p}$$

We know, $t_p = 2$[from problem it is given]

$$\omega_n\sqrt{1-\zeta^2} = \frac{\pi}{2}$$

Now substitute ζ value,

$$\omega_n\sqrt{1-(0.404)^2} = \frac{\pi}{2}$$

$$\omega_n = 1.717 \text{ rad/sec}$$

From (4.15.1)
$$\omega_n = \sqrt{K}$$

and
$$2\zeta\omega_n = \text{KH}$$

$$\omega_n = \sqrt{K}$$

$$K = \omega_n^2$$

$$= (1.73)^2 = 2.95 \text{ Nm.}$$

and
$$2\,\zeta\omega_n = KH$$

$$H = \frac{2\,\zeta\omega_n}{K} = \frac{2 \times 0.404 \times 1.717}{2.95} = .47$$

$$H = .47$$

Result: $K = 2.95$ Nm

$H = 0.47$

EXAMPLE 4.16: A second order control system is represented by a transfer function given below.

$$\frac{\theta_0(s)}{T(s)} = \frac{1}{Js^2 + fs + k}$$

where, $\theta(s)$ is proportional output and T is applied to the system and test results are given below.

(a) $\% M_p = 10\%$

(b) $t_p = 1$ sec

(c) The steady state error of the output is 0.5 radian. Determine the value of J, K and f.

Solution: Given that, $T(s) = 10/s$

$$\frac{\theta_0(s)}{T(s)} = \frac{1}{Js^2 + fs + K} \tag{4.16.1}$$

Put $T(s) = 10/s$ in Eq. (2.18.1)

$$\theta_0(s) = \frac{10}{s\,(Js^2 + fs + K)}$$

$$\text{steady state value} = \lim_{s\to 0} s \cdot \theta_0(s)$$

$$0.5 = \lim_{s\to 0} s \cdot \frac{10}{s\,(Js^2 + fs + K)}$$

$$0.5 = \frac{10}{K}$$

$$K = 20$$

$$\% M_p = 10\%$$

$$M_p = 0.1$$

$$M_p = e^{-\pi\zeta/\sqrt{1-\zeta^2}} \times 100$$

$$0.1 = e^{-\pi\zeta/\sqrt{1-\zeta^2}}$$

$$\zeta = .59$$

$$t_p = 1 \text{ sec} \qquad \text{(Given)}$$

we know that, $$t_p = \frac{\pi}{\omega_n\sqrt{1-\zeta^2}}$$

Put $t_p = 1$, then we get $$1 = \frac{\pi}{\omega_n\sqrt{1-\zeta^2}}$$

$$1 = \frac{\pi}{\omega_n \sqrt{1-(.59)^2}}$$

$$\omega_n = 3.89 \text{ rad/sec}$$

From (4.16.1) Eq. characteristic Eq. is

$$Js^2 + fs + K = 0$$

$$s^2 + \frac{f}{J}s + \frac{K}{J} = 0 \tag{4.16.2}$$

Compare Eq. (4.16.1) with $s^2 + 2\,\zeta\omega_n s + \omega_n^2 = 0$

$$\omega_n^2 = K/J \qquad\qquad 2\,\zeta\omega_n = \frac{f}{T}$$

$$J = \frac{K}{\omega_n^2} = 20/(3.89)^2 \qquad\qquad f = 2\,\zeta\omega_n\, J = 2 \times .59 \times 3.89 \times 1.32$$

$$J = 1.32 \qquad\qquad f = 6.06$$

Result:

$$J = 1.32$$

$$K = 20$$

$$f = 6.06$$

EXAMPLE 4.17: A feedback system is described by the following transfer function $G(s) = \frac{12}{s^2 + 4s + 16}$; $H(s) = Ks$.

The damping factor of the system is 0.6. Determine the overshoot of the system and the value of K.

Solution:

$$\frac{C(s)}{R(s)} = \frac{G(s)}{1 + G(s)\,H(s)}$$

$$\frac{C(s)}{R(s)} = \frac{12/s^2 + 4s + 16}{1 + \frac{12}{s^2 + 4s + 16} \cdot Ks} = \frac{\frac{12}{s^2 + 4s + 16}}{\frac{(s^2 + 4s + 16) + 12\,Ks}{(s^2 + 4s + 16)}}$$

$$= \frac{12}{s^2 + (4 + 16K)\,s + 16} \tag{4.17.1}$$

Characteristics Eq. is $s^2 + (4 + 16K)s + 16 = 0$

Compare the characteristics Eq. with $s^2 + 2\,\zeta\,\omega_n\,s + \omega_n^2 = 0$

$$s^2 + (4 + 16K)s + 16 = s^2 + 2\,\zeta\omega_n\,s + \omega_n^2$$

Compare the s^2, s and constant terms in both the RHS and LHS

$$\omega_n^2 = 16 \qquad 2\zeta\omega_n = 4 + 16K$$

$$\omega_n = 4 \text{ rad/sec} \qquad 2 \times 0.8 \times 4 = 4 + 16K$$

$$K = .15$$

$$\frac{C(s)}{R(s)} = \frac{16}{s^2 + (4 + 16 \times .15)s + 16}$$

$$\% M_p = e^{-\frac{\pi\zeta}{\sqrt{1-\zeta^2}}} \times 100$$

$$= e^{\frac{\pi\, 0.8}{\sqrt{1-(0.8)^2}}} \times 100$$

$$\% M_p = 1.5\%$$

Result:

$$K = 0.15$$

$$M_p = 1.5\%$$

EXAMPLE 4.18: A system having a forward path transfer function $G(s) = 20/s(s + 2)$ and unity feedback. Determine the value of undamped natural frequency, damping ratio. If tachometer is introduced, the feedback path transfer function becomes $(1 + Ks)$. What should be the value of K to obtain damping ratio of 0.6. Also calculate the percentage peak overshoot, first undershoot, t_p and settling time within the 2% of final value.

Solution: The characteristic equation with unity feedback is,

$$1 + G(s) = 0 \quad \text{and} \quad H(s) = 1$$

$$G(s) = 20/s(s + 2)$$

So,

$$1 + \frac{20}{s\,(s+2)} = 0$$

$$1 + \frac{20}{(s^2 + 2s)} = 0$$

$$s^2 + 2s + 20 = 0 \tag{4.18.1}$$

Compare the Eq. (4.18.1) with $s^2 + 2\,\zeta\omega_{n\,s} + \omega_n^2 = 0$

$$\omega_n^2 = 20 \qquad 2\,\zeta\omega_n = 2$$

$$\omega_n = 4.47 \text{ rad/sec} \qquad 2 \times \zeta \times 4.47 = 2 \Rightarrow 8.94\ \zeta = 2$$

$$\zeta = 2/8.94$$

$$\zeta = .22$$

Consider

$$H(s) = 1 + Ks$$

CHAPTER 4

Here,

$$1 + G(s)\,H(s) = 0 \quad (2.18.2)$$

Put $G(s) = 20/s(s + 2)$ and $H(s) = 1 + Ks$ in Eq. (4.18.2)

$$1 + \frac{20}{s(s+2)}(1 + Ks) = 0$$

$$s^2 + s(2 + 20K) + 20 = 0 \quad (4.18.3)$$

Compare (4.18.3) with $s^2 + 2\zeta\omega_n\, s + \omega_n^2 = 0$

$$s^2 + (2 + 20K)\,s + 20 = s^2 + 2\,\zeta\omega_n s + \omega_n^2$$

Compare LHS and RHS

$\omega_n^2 = 20$	$2\,\zeta\omega_n = 2 + 20K$
$\omega_n = 4.47$ rad/sec	$2 \times 0.6\ \times 4.47 = 2 + 20K \Rightarrow 5.36 = 2 + 20K$
	$\Rightarrow 20K = 3.364 \Rightarrow K = .168$

Percentage maximum overshoot $\%\,M_p = e^{\frac{-\pi\zeta}{\sqrt{1-\zeta^2}}} \times 100$

when $\zeta = .22$,

$$\%\,M_p = e^{\frac{-\pi(0.22)}{\sqrt{1-(0.22)^2}}} \times 100$$

$$\%\,M_p = 49.3\%$$

when $\zeta = 0.6$,

$$\%\,M_p = e^{\frac{-\pi(0.6)}{\sqrt{1-(0.6)^2}}} \times 100$$

$$\%\,M_p = 9.48\%$$

$$\text{First undershoot} = e^{\frac{-2\pi}{\sqrt{1-\zeta^2}}}$$

$$= e^{\frac{-2\pi\,(0.6)}{\sqrt{1-(0.6)^2}}} = 0.009$$

Finding Peak Time:

when, $\zeta = .22$

$$t_p = \frac{\pi}{\omega_n\sqrt{1-\zeta^2}} = \frac{\pi}{4.47\sqrt{1-(.22)^2}}$$

$$t_p = .720 \text{ sec}$$

when $\zeta = 0.6$

$$t_p = \frac{\pi}{4\sqrt{1-(0.6)^2}} = 0.98 \text{ sec}$$

$$t_p = 0.98 \text{ sec}$$

Finding Settling Time t_s:

$$t_s = \frac{4}{\zeta\omega_n}$$

when $\zeta = .22$

$$t_s = \frac{4}{.22 \times 4.47} = \frac{4}{.983} = 4\,\text{sec}$$

when $\zeta = 0.6$

$$t_s = \frac{4}{4.47 \times 0.6} = \frac{4}{2.682} = 1.49$$

$$t_s = 1.49 \text{ sec}$$

Results:

1. The value of K when $\zeta = 0.6$ $K = .168$
2. Undamped natural frequency $\omega_n = 4.47$ rad/sec
3. Damping ratio $\zeta = .22$
4. % Peak overshoot % $M_p = 9.48\%$
5. First undershoot = .009
6. Peak time = .72 sec
7. Settling time within 2% of final value.
   (i) when $\zeta = .22$ $t_s = 4$ sec
   (ii) when $\zeta = 0.6$ $t_s = 1.49$ sec

4.8 EFFECT OF FEEDBACK ON TIME CONSTANT OF A CONTROL SYSTEM

Consider the open loop transfer function as shown in below Fig. (4.16).

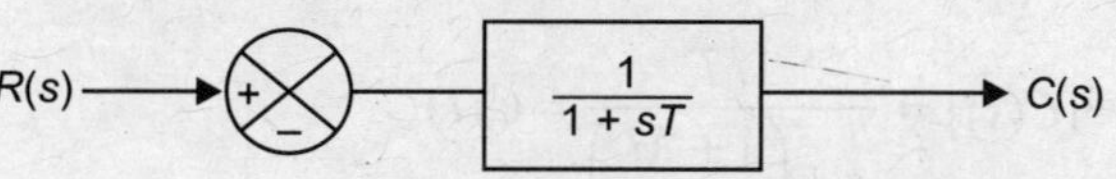

Fig. 4.16

$$\frac{C(s)}{R(s)} = \frac{1}{1+sT} \tag{4.91}$$

Let $R(s) = \frac{1}{s}$.

Put $R(s) = \frac{1}{s}$ in Eq. (4.91)

$$C(s) = \frac{R(s)}{1+sT}$$

$$C(s) = \frac{1}{1+sT} \cdot \frac{1}{s} \tag{4.92}$$

apply inverse Laplace for Eq. (4.92)

CHAPTER 4

$$L^{-1}[C(s)] = L^{-1}\left[\frac{1}{s\,(1+sT)}\right]$$

$$C(t) = 1 - e^{-t/T} \tag{4.93}$$

In Eq. (4.93) T is the constant.

Now consider the closed loop system as shown in below Fig. (4.92)

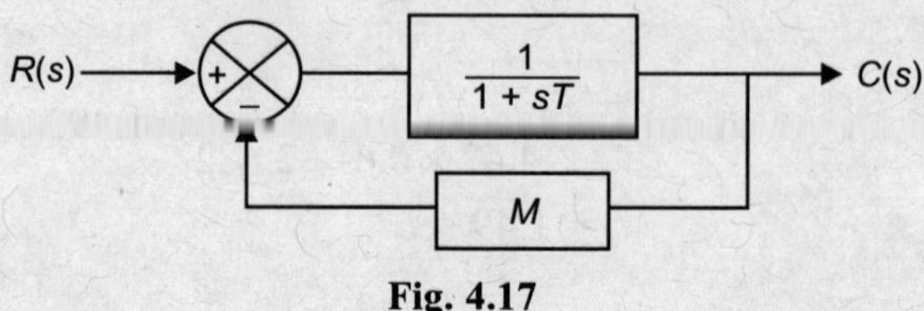

Fig. 4.17

If $H(s) = M$, then the overall transfer function is,

$$\frac{C(s)}{R(s)} = \frac{1/1+sT}{1+\dfrac{1}{1+sT}\cdot M}$$

$$\frac{C(s)}{R(s)} = \frac{1/1+sT}{\dfrac{1+sT+M}{1+sT}} = \frac{1}{1+sT+M}$$

$$\frac{C(s)}{R(s)} = \frac{1/T}{s+\left(\dfrac{1+M}{T}\right)} \tag{4.94}$$

apply inverse Laplace for Eq. (4.94)

$$L^{-1}[C(s)] = \frac{1/T}{s+\left(\dfrac{1+M}{T}\right)} = C(t)$$

$$C(t) = \frac{1}{1+M}\left[1-e^{\frac{-t}{T/1+M}}\right] \tag{4.95}$$

From the Eq. (4.95) time constant is,

Time constant = $T/1 + M$

Conclusion: The time constant of closed loop system is less than the time constant of open loop system.

Less time constant means response is faster. Therefore, feedback improves the time response of the system.

4.9 STEADY STATE ERROR

The steady state error is the value of error signal $e(t)$, when t tends to infinity the steady state error is a measure of system accuracy. These errors arise from the nature of inputs, type of system and from non-linearity of system components. The steady state performance of a stable control system is generally judged by its steady state error to step, ramp and parabolic inputs.

Consider a closed loop system shown in Fig. (2.11.1).

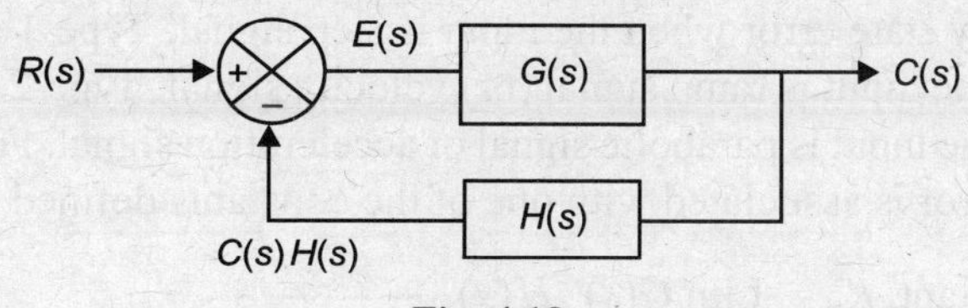

Fig. 4.18

From Fig. (4.18),

The error signal, $E(s) = R(s) - C(s)\,H(s)$ (4.96)

The output signal, $C(s) = E(s)\,G(s)$ (4.97)

Substitute the Eq. (4.97) in Eq. (4.96)

$$E(s) = R(s) - [E(s)\,G(s)] \cdot H(s)$$

$$E(s) + E(s)\,G(s)\,H(s) = R(s)$$

$$E(s)\,[1 + G(s)\,H(s)] = R(s)$$

$$E(s) = \frac{R(s)}{1 + G(s)\,H(s)} \tag{4.98}$$

Let $e(t)$ = error signal in time domain

$$e(t) = L^{-1}\,[E(s)] = L^{-1}\left[\frac{R(s)}{1 + G(s)\,H(s)}\right] \tag{4.99}$$

Let e_{ss} = steady state error

The steady state error is defined as the value of $C(t)$ when t tends to infinity.

$$e_{ss} = \lim_{t \to \infty} e(t) \tag{4.100}$$

The final Value theorem states that,

$$F(s) = L[f(t)] \text{ then,}$$

$$\lim_{t \to \infty} f(t) = \lim_{s \to \infty} s \cdot F(s) \tag{4.101}$$

Using final value theorem,

The steady state error,

$$e_{ss} = \lim_{t \to \infty} e(t) = \lim_{s \to 0} s \cdot E(s)$$

$$= \lim_{s \to 0} \frac{s \cdot R(s)}{1 + G(s)\,H(s)} \tag{4.102}$$

4.9.1 Static Error Constants

When a control system is excited with standard input signal, the steady state error may be zero, constant or infinity. The value of steady state error depends on the type number and the input signal. Type-0 system will have a steady state error when the input is step signal. Type-1 system will have a constant steady state error when the input is ramp signal (or) velocity signal. Type-2 system will have a constant steady state error when the input is parabolic signal or acceleration signal. For the three cases mentioned above the steady state error is associated with one of the constants defined as follows.

$$\text{Positional error constant, } K_p = \lim_{s\to 0} G(s)\cdot H(s) \tag{4.103}$$

$$\text{Velocity error constant, } K_V = \lim_{s\to 0} s\cdot G(s)\cdot H(s) \tag{4.104}$$

$$\text{Acceleration error constant, } K_a = \lim_{s\to 0} s^2\, G(s)\cdot H(s) \tag{4.105}$$

The K_p, K_v and K_a are in general called static error constants.

4.9.2 Steady State Error when the Input is Step Signal

Steady state error, $e_{ss} = \lim_{s\to 0} \dfrac{sR(s)}{1+G(s)+H(s)}$ when the input is unit step, $R(s) = 1/s$

$$e_{ss} = \lim_{s\to 0} \frac{s\cdot 1/s}{1+G(s)\,H(s)}$$

$$= \lim_{s\to 0} \frac{1}{1+G(s)\,H(s)} = \frac{1}{1+\lim_{s\to 0} G(s)\,H(s)} = \frac{1}{1+K_P} \tag{4.106}$$

where, $K_P = \lim_{s\to 0} G(s)\cdot H(s)$. The constant K_P is called positional error constant.

Type-0 System:

$$K_P = \lim_{s\to 0} G(s)\,H(s) = \lim_{s\to 0} K\frac{(s+Z_1)(s+Z_2)(s+Z_3)\ldots}{(s+P_1)(s+P_2)(s+P_3)\ldots}$$

$$= K\frac{Z_1\cdot Z_2\cdot Z_3}{P_1\cdot P_2\cdot P_3} = \text{constant}$$

$$e_{ss} = \frac{1}{1+K_P} = \text{constant}$$

In type-0 system when the input is unit step there will be constant steady state error.

Type -1 System

$$K_P = \lim_{s\to 0} G(s)\cdot H(s) = \lim_{s\to 0} K\frac{(s+Z_1)(s+Z_2)(s+Z_3)\ldots}{s(s+P_1)(s+P_2)(s+P_3)\ldots} = \infty$$

$$e_{ss} = \frac{1}{1+K_P} = \frac{1}{1+\infty} = 0$$

In system with type number 1 and above, for unit step input the value of K_P is infinity and so the steady state error is zero.

4.9.3 Steady State Error when the Input is Unit Ramp Signal

Steady state error, $e_{ss} = \lim\limits_{s\to 0} \dfrac{sR(s)}{1+G(s)\,H(s)}$

when the input is unit ramp, $R(s) = 1/s^2$

$$e_{ss} = \lim_{s\to 0} \frac{s\cdot\frac{1}{s^2}}{1+G(s)\,H(s)} = \lim_{s\to 0} \frac{1}{s+sG(s)\,H(s)}$$

$$= \lim_{s\to 0} \frac{1}{sG(S)\,H(s)} = \frac{1}{K_V}$$

where, $K_V = \lim\limits_{s\to 0} s\cdot G(s)\cdot H(s)$ The constant K_V is called velocity error constant.

Type - 0 system:

$$K_V = \lim_{s\to 0} sG(s)\,H(S)\ \lim_{s\to 0} sK\frac{(s+Z_1)\,(s+Z_2)\,(s+Z_3)\ldots}{(s+P_1)\,(s+P_2)\,(s+P_3)\ldots} = 0 \Rightarrow K_V = 0$$

$$e_{ss} = \frac{1}{K_V} = \frac{1}{0} = \infty$$

In type-0 systems when the input is unit ramp, the steady state error is infinity.

Type - 1 system:

$$K_V = \lim_{s\to 0} s\cdot G(s)\cdot H(s) = \lim_{s\to 0} sK\frac{(s+Z_1)\,(s+Z_2)\,(s+Z_3)\ldots}{s\,(s+P_1)\,(s+P_2)\,(s+P_3)\ldots}$$

$$= K\frac{Z_1\cdot Z_2\cdot Z_3\ldots}{P_1\cdot P_2\cdot P_3\ldots} = \text{constant}$$

$$e_{ss} = 1/K_V = \text{constant}$$

In type 1 system when the input is unit ramp there will be a constant steady state error.

Type - 2 system:

$$K_V = \lim_{s\to 0} s\cdot G(s)\,H(s) = \lim_{s\to 0} sK\frac{(s+Z_1)\,(s+Z_2)\,(s+Z_3)\ldots}{s^2(s+P_1)\,(s+P_2)\,(s+P_3)\ldots} = \infty$$

$$e_{ss} = \frac{1}{K_V} = \frac{1}{\infty} = 0$$

CHAPTER 4

In system with type number 2 and above, for unit ramp input, the value of K_V is infinity so the steady state error is zero.

4.9.4 Steady State Error when the Input is Unit Parabolic Signal

Steady state error, $e_{ss} = \lim_{s\to 0} \frac{sR(s)}{1+G(s)\,H(s)}$

when the input is unit parabola, $R(s) = 1/s^3$

$$e_{ss} = \lim_{s\to 0} \frac{s \cdot \frac{1}{s^3}}{1+G(s)\,H(s)} = \lim_{s\to 0} \frac{1}{s^2 + s^2\,G(s)\,H(s)}$$

$$= \frac{1}{\lim_{s\to 0} s^2\,G(s)\,H(s)} = \frac{1}{K_a}$$

where, $K_a = \lim_{s\to 0} s^2\,G(s)\cdot H(s)$. The constant K_a is called acceleration error constant.

Type - 0 system:

$$K_a = \lim_{s\to 0} s^2 \cdot G(s)\cdot H(s)$$

$$= \lim_{s\to 0} s^2 K \frac{(s+Z_1)(s+Z_2)(s+Z_3)\ldots}{(s+P_1)(s+P_2)(s+P_3)\ldots} = 0 \Rightarrow K_a = 0$$

$$e_{ss} = \frac{1}{K_a} = \frac{1}{0} = \infty$$

In type - 0 systems for unit parabolic input, the steady state error is infinity.

Type - 1 system:

$$K_a = \lim_{s\to 0} s^2\,G(s)\,H(s) = \lim_{s\to 0} s^2 K \frac{(s+Z_1)(s+Z_2)(s+Z_3)\ldots}{s\,(s+P_1)(s+P_2)(s+P_3)\ldots} = 0 \Rightarrow K_a = 0$$

$$e_{ss} = \frac{1}{K_a} = \frac{1}{0} = \infty$$

Hence in type - 1 systems for unit parabolic input the steady state error is infinity.

Type - 2 systems:

$$K_a = \lim_{s\to 0} s^2\,G(s)\,H(s)$$

$$= \lim_{s\to 0} s^2 K \frac{(s+Z_1)(s+Z_2)(s+Z_3)\ldots}{s^2\,(s+P_1)(s+P_2)(s+P_3)\ldots}$$

$$K_a = K \cdot \frac{Z_1 \cdot Z_2 \cdot Z_3 \ldots}{P_1 \cdot P_2 \cdot P_3 \ldots} = \text{constant}$$

$$e_{ss} = \frac{1}{K_a} = \text{constant}$$

Hence in type - 2 system when the input is unit parabolic signal there will be a constant steady state error.

Type - 3 system:

$$K_a = \lim_{s \to 0} s^2\, G(s)\, H(s)$$

$$= \lim_{s \to 0} s^2 K \frac{(s+Z_1)(s+Z_2)(s+Z_3)\ldots}{s^3\,(s+P_1)(s+P_2)(s+P_3)\ldots}$$

$$e_{ss} = \frac{1}{K_a} = \frac{1}{\infty} = 0$$

In systems with type number 3 and above for unit parabolic input the value of K_a is infinity and so the steady state error is zero.

Table 4.2: The steady state error for various types of inputs

Type number	Steady state error when the input signal is (e_{ss})		
	Unit step	**Unit ramp**	**Unit parabolic**
0	$1/1+K_P$	∞	∞
1	0	$1/K_V$	∞
2	0	0	$1/K_a$
3	0	0	0

4.9.5 Generalized Error Coefficient

The drawback in static error coefficient is that it does not show the variation of error with time and input should be a standard input. The generalized error coefficients gives the steady state error as a function of time. Also using the generalized error coefficients, the steady state error can be found for any type of input.

The error signal in s-domain, $E(s)$ can be expressed a product of two s-domain function.

$$E(s) = \frac{R(s)}{1+G(s)\,H(s)} = \frac{1}{1+G(s)\,H(s)} \cdot R(s) = F(s) \cdot R(s) \tag{4.107}$$

where, $F(s) = \dfrac{1}{1+G(s)\,H(s)}$

Let

$$e(t) = L^{-1}[E(s)] \quad \text{(error signal in time domain)}$$
$$f(t) = L^{-1}[F(s)]$$
$$r(t) = L^{-1}[R(s)] \quad \text{(input signal in time domain)}$$

The convolution theorem states that the inverse Laplace transform of the product of two s-domain functions is equal to convolution of their time domain function.

i.e.,

$$L^{-}[F(s)\,R(s)] = \int_{-\infty}^{\infty} f(T)\cdot r(t-T)\cdot dT$$

Hence by convolution theorem the error function in time domain can be written as,

$$e(t) = \int_{-\infty}^{\infty} f(T)\cdot r\,(t-T)\cdot dT \tag{4.108}$$

where $T \rightarrow$ is a dummy variable. It is assumed that the input signal starts only at $t = 0$ and does not exist before $t = 0$. Hence the limit of integral can be changed as 0 to ∞.

$$e(t) = \int_{0}^{\infty} f(T)\,r(t-T)\,dT \tag{4.109}$$

Using Taylor's series expansion the signal $r(t-T)$ can be expressed as,

$$r(t-T) = r(t) - T\dot{r}\,(t) + \frac{T^2}{2!}\ddot{r}(t) - \frac{T^3}{3!}\dddot{r}\,(t) + \ldots + (-)^n\,\frac{T^n}{n!}\,r^n\,(t) \tag{4.110}$$

where, $r'(t)$ = 1st derivative of $r(t)$.

$r''(t)$ = 2nd derivative of $r(t)$.

and $r^n(t)$ = nth derivative of $r(t)$

On substituting the Taylor's series expansion of $r(t-T)$, the error $e(t)$ can be written as,

$$e(t) = \int_{0}^{\infty} f(T)\left[r(T) - Tr'(t) + \frac{T^2}{2!}r''\,(t) - \frac{T^3}{3!}r'''(t) + \ldots + (-1)^n\,\frac{T^n}{n!}\,r^{n'}\,r^n t\ldots\right] \times dT$$

$$e(t) = \int_{0}^{\infty} f(T)\,r(r)\;dT - \int_{0}^{\infty} f(T)\,Tr'(t)\,dT + \int_{0}^{\infty} f(T)\frac{T^2}{2!}r''(t)\cdot dT -$$

$$\int_{0}^{\infty} f(T)\frac{T^3}{3!}r'''\,(t)\cdot dT + \ldots + \int_{0}^{\infty} f(T)\,(-)^n\,\frac{T^n}{n!}\,r^{n'}\,r^n\,(t)\,dT \mathrel{\therefore}\ldots\infty$$

where $r(t), r'(t), r''(t), \dots\ r^{n'}(t)$ are constants. After integration with respect to T, the error signal can be written as,

$$e(t) = r(t)\int_0^\infty f(T)\,dT - r'(t)\int_0^\infty Tf(T)\,dT + \frac{r''(t)}{2!}\int_0^\infty T^2\,f(T)\cdot dT -$$

$$= \frac{r'''(t)}{3!}\int_0^\infty T^3\,f(T)\,dT + \dots (-1)^n\,\frac{r^{n'}(t)}{n!}\int T^n\,f(T)\cdot dT \dots$$

Let $$C_0 = +\int_0^\infty f(T)\cdot dT$$

$$C_1 = \int_0^\infty T\cdot f(T)\cdot dT$$

$$C_2 = +\int_0^\infty T^2\,f(T)\cdot dT$$

$$C_3 = -\int_0^\infty T^3\,f(T)\cdot dT$$

$$\vdots$$

$$C_n = (-1)^n\int_0^\infty T^n\,f(T)\cdot dT \tag{4.111}$$

$$e(t) = r(t)\,C_0 + r'(t)\,C_1 + r''(t)\,\frac{C_2}{2!} + r'''\,(t)\,\frac{C_3}{3!} + \dots + r^{n'}\,C(t)\,\frac{C_n}{n} + \dots$$

$$= C_0\,r(t) + C_1 r'(t) + \frac{C_2}{2!}\,r''(t) + \frac{C_3}{3!}\,r'''(t) + \dots + \frac{C_n}{n!}\,r^{n'}\,r(t)\dots \tag{4.112}$$

The coefficients C_0, C_1, C_2 … C_n are called the generalized error coefficients (or) dynamic error coefficients.

The steady state error e_{ss} is obtained by taking limit $r \to \infty$ on $e(t)$,

$$\therefore \qquad e_{ss} = \lim_{t\to\infty} e(t)$$

4.9.6 Evaluation of Generalized Error Coefficients (or) Dynamic Error Coefficients

The generalized error coefficients is given by,

$$C_n = (-)^n\int_0^\infty T^n f(T)\,dT$$

where, $F(s) = \dfrac{1}{1 + G(s)\,H(s)}$

we know that $L[f(t)] = F(s)$,

By the definition of Laplace transform

$$F(s) = \int_0^\infty f(T) \cdot e^{-sT} \cdot dT \tag{4.113}$$

take $\lim_{s \to 0}$ on both sides for Eq. (4.113)

$$\lim_{s \to 0} F(s) = \lim_{s \to 0} \int_0^\infty f(T) \cdot e^{-sT} \cdot dT$$

$$= \int_0^\infty f(T) \lim_{s \to 0} e^{-sT} \cdot dT = \int_0^\infty f(T) \cdot dT = C_0$$

$$C_0 = \lim_{s \to 0} F(s) \tag{4.114}$$

on differentiating Eq. (4.113) with 1espect to s,

$$\frac{d}{ds} F(s) = \frac{d}{ds} \int_0^\infty f(T) \cdot e^{-sT} \cdot dT$$

$$= \int_0^\infty f(T) \cdot \frac{d}{ds}(e^{-sT}) \cdot dT = \int_0^\infty f(T)(-T)\, e^{-sT} \cdot dT$$

$$\frac{d}{ds} F(s) = -\int_0^\infty T\, f(T) \cdot e^{-sT} \cdot dT \tag{4.115}$$

take $\lim_{s \to 0}$ on both sides,

$$\lim_{s \to 0} \frac{d}{ds} F(s) = \lim_{s \to 0} -\int_0^\infty T\, f(T) \cdot e^{-sT} \cdot dT$$

$$= -\int_0^\infty T\, f(T) \lim_{s \to 0} e^{-sT}\, dT = -\int_0^\infty T\, f(t)\, dT = C_1$$

$$C_1 = \lim_{s \to 0} \frac{d}{ds} F(s) \tag{4.116}$$

differentiate Eq. (4.115), w.r. to s

$$\frac{d}{ds}\left[\frac{d}{ds} F(s)\right] = \frac{d}{ds}\left[-\int_0^\infty T\, f(T) \cdot e^{-sT}\, dT\right]$$

$$\frac{d^2}{ds^2}F(s) = \left[-\int_0^{\infty} T\, f(T)\frac{d}{ds}(e^{-sT})\,dT\right] = -\int_0^{\infty} T\, f(T)\,(-T)\,e^{-sT}\cdot dT$$

$$\frac{d^2}{ds^2}F(s) = \int_0^{\infty} T^2 f(T)\, e^{-sT}\cdot dT \qquad (4.117)$$

Apply the limit $s \to 0$

$$\underset{s\to 0}{\text{Lim}}\frac{d^2}{ds^2}F(s) = \underset{s\to 0}{\text{Lim}}\int_0^{\infty} T^2 f(T)\cdot e^{-sT}\cdot dT$$

$$\int_0^{\infty} T^2 f(T)\,\underset{s\to 0}{\text{Lim}}\, e^{-sT}\cdot dT = \int_0^{\infty} T^2 f(T)\cdot dT = C_2$$

$$C_2 = \underset{s\to 0}{\text{Lim}}\frac{d^2}{ds^2}F(s) \qquad (4.118)$$

Similarly

$$C_n = \underset{s\to 0}{\text{Lim}}\frac{d^n}{ds^n}F(s) \qquad (4.119)$$

4.9.7 Correlation between Static and Dynamic Error Coefficients

The values of dynamic error coefficients can be used to calculate static error coefficients. The following expressions shows the relationship between them.

$$C_0 = 1/1 + K_p \qquad (4.120)$$

$$C_1 = 1/K_V \qquad (4.121)$$

$$C_2 = 1/K_a \qquad (4.122)$$

Proof:

$$C_0 = \underset{s\to 0}{\text{Lim}}\, F(s) = \underset{s\to 0}{\text{Lim}}\frac{1}{1+G(s)\,H(s)}$$

$$= \frac{1}{1+\underset{s\to 0}{\text{Lim}}\, G(s)\,H(s)} = \frac{1}{1+K_P}$$

4.9.8 Alternate Method for Generalized Error Coefficients

The error signal is s-domain, $E(s) = \dfrac{R(s)}{1+G(s)\,H(s)}$

$$\frac{E(s)}{R(s)} = \frac{1}{1+G(s)\,H(s)} \qquad (4.123)$$

The (4.123) equation can be expressed as a power series of s as shown in Eq. (4.124).

$$\frac{E(s)}{R(s)} = \frac{1}{1+G(s)\,H(s)} = C_0 + C_1 s + \frac{C_2}{2!}s^2 + \frac{C_3}{3!}s^3 + \ldots \tag{4.124}$$

$$E(s) = C_0\,R(s) + C_1 s\,R(s) + \frac{C_2}{2!}s^2\,R(s) + \frac{C_3}{3!}s^3\,R(s) + \ldots \tag{4.125}$$

Take inverse Laplace transform of Eq. (4.125)

$$e(t) = C_0\,r(t) + C_1\,r'(t) + \frac{C_2}{2!}r''(t) + \frac{C_3}{3!}r'''(t) + \ldots \tag{4.126}$$

The Eq. (4.126) is same as that of Eq. (4.119). The coefficients C_0, C_1, $C_2 \ldots$ are called generalized error coefficients (or) dynamic error coefficients. The steady State error e_{ss} is obtained by taking limit $t \to \infty$ on $e(t)$.

$$e_{ss} = \lim_{t\to\infty} e(t)$$

EXAMPLE 4.19: For a unity feedback control system the open loop transfer function $G(s) = 10\,(s+2)/s^2\,(s+1)$. Find (a) the position, velocity and acceleration error constants, (b) the steady state error when the input is $R(s) = \frac{3}{s} - \frac{2}{s^2} + \frac{1}{3s^3}$.

Solution:

(a) *Static error constants*

For a unity feedback system, $H(s) = 1$

Position error constant,

$$K_P = \lim_{s\to0} G(s)\,H(s) = \lim_{s\to0} G(s)$$

$$= \lim_{s\to0} \frac{10\,(s+2)}{s^2\,(s+1)} = \infty$$

Velocity error constant,

$$K_V = \lim_{s\to0} s\cdot G(s)\,H(s)$$

$$= \lim_{s\to0} s\cdot G(s)$$

$$= \lim_{s\to0} s\cdot\frac{10\,(s+2)}{s^2\,(s+2)} = \infty$$

Acceleration error constant,

$$K_a = \lim_{s\to0} s^2\cdot G(s)\,H(s)$$

$$= \lim_{s\to0} s^2\cdot G(s)$$

$$= \lim_{s\to0} s^2\,\frac{10\,(s+2)}{s^2\,(s+1)} = \frac{10\times2}{1} = 20.$$

(b) *Steady state error*

Ist Method: The steady error for non-standard input is obtained using generalized error series,

The error signal, $e(t) = r(t)\,C_0 + r'(t)\,C_1 + r''(t)\,\frac{C_2}{2!} + \ldots + r^{n'}(t)\,\frac{C_n}{n!} + \ldots$

Given that, $$R(s) = \frac{3}{s} - \frac{2}{s^2} + \frac{1}{3s^3}$$

Input signal in time domain, $r(t) = L^{-1}[R(s)]$

$$= L^{-1}\left[\frac{3}{s} - \frac{2}{s^2} + \frac{1}{3s^2}\right] = \frac{1}{3} - 2t + \frac{1t^2}{3 \times 2!}$$

$$= \frac{1}{3} - 2t + \frac{t^2}{6}$$

$$r'(t) = \frac{d}{dt}r(t) = -2 + \frac{1}{6}2t = -2 + \frac{t}{3}$$

$$r''(t) = \frac{d^2}{dt^2}r(t) = \frac{d}{dt}r'(t) = \frac{1}{3}$$

$$r'''(t) = \frac{d^3}{dt^3}r(t) = \frac{d}{dt}r''(t) = 0$$

The derivatives of $r(t)$ is zero after second derivative. Hence we have to evaluate only three constants C_0, C_1 and C_2.

The generalized error constants are given by $C_0 = \underset{S\to 0}{\text{Lim}}\,F(s)$; $C_1 = \underset{S\to 0}{\text{Lim}}\,\frac{d}{ds}F(s)$; $C_2 = \underset{S\to 0}{\text{Lim}}\,\frac{d^2}{ds^2}F(s)$.

$$F(s) = \frac{1}{1+G(s)\,H(s)} = \frac{1}{1+G(s)} = \frac{1}{1+\dfrac{10(s+2)}{s^2\,(s+1)}}$$

$$= \frac{s^2\,(s+1)}{s^2\,(s+1)+10\,(s+2)} = \frac{s^3+s^2}{s^3+s^2\,10s+20}$$

$$C_0 = \underset{s\to 0}{\text{Lim}}\,F(s) = \underset{s\to 0}{\text{Lim}}\left[\frac{s^3+s^2}{s^3+s^2+10s+20}\right] = 0$$

$$C_1 = \underset{s\to 0}{\text{Lim}}\,\frac{d}{ds}F(s) = \underset{s\to 0}{\text{Lim}}\,\frac{d}{ds}\left[\frac{s^3+s^2}{s^3+s^2+10s+20}\right]$$

CHAPTER 4

$$= \lim_{s\to 0}\left[\frac{(s^3+s^2+10s+20)(3s^2+2s)-(s^3+s^2)(3s^2+2s+10)}{(s^3+s^2+10s+20)^2}\right]$$

$$= \lim_{s\to 0}\left[\frac{20s^3+70s^2+40s}{(s^3+s^2+10s+20)^2}\right]=0$$

$$C_2 = \lim_{s\to 0}\frac{d^2}{ds^2}F(s) = \lim_{s\to 0}\frac{d}{ds}\left[\frac{d}{ds}F(s)\right]$$

$$= \lim_{s\to 0}\frac{d}{ds}\left[\frac{(20s^3+70s^2+40s)}{(s^3+s^2+10s+20)^2}\right]$$

$$= \lim_{s\to 0}\left[\frac{\begin{array}{c}(s^3+s^2+10s+20)^2\,(60s^2+140s+40)-\\(20s^3+70s^2+40s)\,2\times(s^3+s^2+10s+20)\,(3s^2+2s+10)\end{array}}{(s^3+s^2+10s+20)^4}\right]$$

$$= \frac{20^2\times 40}{20^4}=\frac{1}{10}$$

Error signal, $e(t) = r(t)\,C_0 + r'(t)\,C_1 + r''(t)\,\frac{C_2}{2!}$

$$= \left(\frac{1}{3}-2t+\frac{t^2}{6}\right)\times 0+\left(-2+\frac{t}{3}\right)\times 0+\frac{1}{3}\times\frac{1}{10}\times\frac{1}{2!}=\frac{1}{60}$$

Steady state error,

$$e_{ss} = \lim_{t\to\infty} e(t) = \lim_{t\to\infty}\frac{1}{60}=\frac{1}{60}$$

IInd Method: The error signal in a s-domain, $E(s) = \dfrac{R(s)}{1+G(s)\,H(s)}$

Given that $R(s) = \dfrac{3}{s}-\dfrac{2}{s^2}+\dfrac{1}{3s^3}$

$$G(s) = \frac{10\,(s+2)}{s^2\,(s+1)} \text{ and } H(s)=1$$

$$E(s) = \frac{\dfrac{3}{s}-\dfrac{2}{s^2}+\dfrac{1}{3s^3}}{1+\dfrac{10\,(s+2)}{s^2\,(s+1)}} = \frac{\dfrac{3}{s}-\dfrac{2}{s^2}+\dfrac{1}{3s^3}}{\dfrac{s^2\,(s+1)+10\,(s+2)}{s^2\,(s+1)}}$$

$$= \frac{3}{s}\left[\frac{s^2(s+1)}{s^2(s+1)+10(s+2)}\right] - \frac{2}{s^2}\left[\frac{s^2(s+1)}{s^2(s+1)+10(s+2)}\right] + \frac{1}{3s^3}\left[\frac{s^2(s+1)}{s^2(s+1)+10(s+2)}\right]$$

The steady state error e_{ss} can be obtained from final value theorem.

Steady state error, $e_{ss} = \lim\limits_{t\to\infty} e(t) = \lim\limits_{s\to\infty} s \cdot E(s)$

$$e_{ss} = \lim_{t\to\infty} s\left\{\frac{3}{5}\left[\frac{s^2(s+1)}{s^2(s+1)+10(s+2)}\right] - \frac{2}{s^2}\left[\frac{s^2(s+1)}{s^2(s+1)10(s+2)}\right] + \frac{1}{3s^3}\left[\frac{s^2(s+1)}{s^2(s+1)10(s+2)}\right]\right\}$$

$$= \lim_{s\to 0}\left\{\frac{3s^2(s+1)}{s^2(s+1)+10(s+2)} - \frac{2s(s+1)}{s^2(s+1)+10(s+2)} + \frac{(s+1)}{3s^2(s+1)+30(s+2)}\right\}$$

$$= 0 - 0 + \frac{1}{60} = \frac{1}{60}$$

$$e_{ss} = \frac{1}{60}$$

IIIrd Method:

Error signal in s-domain, $E(s) = \dfrac{R(s)}{1+G(s)H(s)}$

$$\frac{E(s)}{R(s)} = \frac{1}{1+G(s)H(s)},$$

Given that $G(s) = \dfrac{10(s+2)}{s^2(s+1)}$ and $H(s) = 1$

$$\frac{E(s)}{R(s)} = \frac{1}{1+\dfrac{10(s+2)}{s^2(s+1)}} = \frac{s^2(s+1)}{s^2(s+1)+10(s+2)} = \frac{s^3+s^2}{s^3+s^2+10s+20}$$

$$= \frac{s^2+s^3}{20+10s+s^2+s^3} = \frac{s^2}{20}+\frac{s^3}{40}+\ldots$$

$$E(s) = R(s)\left[\frac{s^2}{20}+\frac{s^3}{40}+\ldots\right]$$

$$= \frac{1}{20}s^2 R(s) + \frac{1}{40}s^3 R(s) + \ldots$$

CHAPTER 4

On taking inverse Laplace transform,

$$e(t) = \frac{1}{20} r''(t) + \frac{1}{40} r'''(t) + \ldots$$

Given that $$R(s) = \frac{3}{s} - \frac{2}{s^2} + \frac{1}{3s^3}$$

$$r(t) = L^{-1}[R(s)] = L^{-1}\left[\frac{3}{s} - \frac{2}{s^2} + \frac{1}{3s^3}\right]$$

$$= \frac{1}{3} - 2t + \frac{1t^2}{3 \times 2!} = \frac{1}{3} - 2t + \frac{t^2}{6}$$

$$r'(t) = \frac{d}{dt^2} r(t) = -2 + \frac{1}{6} 2t = -2 + \frac{t}{3}$$

$$r''(t) = \frac{d^2}{dt^3} r(t) = \frac{d}{dt} r''(t) = \frac{1}{3}$$

$$r'''(t) = \frac{d^3}{dt^3} r(t) = \frac{d}{dt} r^2(t) = 0$$

Error signal in time domain, $e(t) = \frac{1}{20} \cdot \frac{d}{dt} r''(t)$

$$= \frac{1}{20}\left(\frac{1}{3}\right) = \frac{1}{60}$$

Steady state error, $$e_{ss} = \lim_{t \to \infty} e(t) = \lim_{t \to \infty} \frac{1}{60} = \frac{1}{60}$$

Result:

(a) Position error constant, $K_P = \infty$

Velocity error constant, $K_V = \infty$

Acceleration error constant, $K_a = 20$

(b) Steady state error when

$$R(s) = \frac{3}{s} - \frac{2}{s^2} + \frac{1}{3s^3} \text{ is } 1/60$$

i.e., $e_{ss} = \frac{1}{60}$

EXAMPLE 4.20: For servomechanisms with open loop transfer function given below explain what type of input signal give rise to a constant steady state error and calculate their values.

(i) $$G(s) = \frac{20\,(s+2)}{s\,(s+1)\,(s+3)}$$

(ii) $G(s) = \frac{10}{(s+2)(s+3)}$

(iii) $G(s) = \frac{10}{s^2(s+1)(s+2)}$

Solution:

(i) $G(s) = \frac{20(s+2)}{s(s+1)(s+3)}$

For unity feedback system, $\therefore H(s) = 1$.

The open loop system has a pole at origin. Hence it is a Type-1 system. In system with type number-1, the velocity (ramp) input will give a constant steady state error.

The steady state error with unit velocity input, $e_{ss} = \frac{1}{K_V}$.

Velocity error constant, $K_V = \lim_{s\to 0} s \cdot G(s) \cdot H(s)$

$$= \lim_{s\to 0} s \cdot G(s)$$

$$= \lim_{s\to 0} s \frac{20(s+2)}{s(s+1)(s+3)} = \frac{20\times 2}{1\times 3} = \frac{40}{3}.$$

Steady state error, $e_{ss} = 1/K_V = \frac{3}{40} = 0.075$.

(ii) $G(s) = 10/(s+2)(s+3)$.

For unity feedback system, $H(s) = 1$

The open loop system has no pole at origin. Hence it is a type-0 system. In systems with type number-0, the step input will give a constant steady state error.

The steady state error with unit step input, $e_{ss} = 1/1 + K_P$ position error constant,

$$K_P = \lim_{s\to 0} G(s)\, H(s) = \lim_{s\to 0} G(s)$$

$$= \lim_{s\to 0} \frac{10}{(s+2)(s+3)} = \frac{10}{2\times 3} = \frac{5}{3}$$

Steady state error, $e_{ss} = 1/1 + K_P = \frac{1}{1+5/3} = \frac{3}{3+5} = 0.375$

(iii) $G(s) = \frac{10}{s^2(s+1)(s+2)}$

For unity feedback system, $H(s) = 1$. The open loop system has two poles at origin. Hence it is a Type-2 system. In systems with type number-2, the acceleration (Parabolic) input will give a constant steady state error.

The steady state error with unit acceleration in put,

Acceleration error constant, $K_a = \lim_{s\to 0} s^2 \cdot G(s)\cdot H(s)$

$$= \lim_{s\to 0} s^2\; G(s) = \lim_{s\to 0} s^2 \cdot \frac{10}{s^2\,(s+1)\,(s+2)} = \frac{10}{2\times 1} = 5$$

Steady state error, $e_{ss} = 1/K_a = 1/5 = 0.2$

Results: Steady state error in system (i) with unit velocity input = 0.075 (ii) with unit step input = 0.375 (iii) with unit acceleration input = 0.2.

EXAMPLE 4.21: The open loop transfer function of a servo system with unity feedback is $G(s) = 10/s(0.1\,s + 1)$. Evaluate the static error constants of the system. Obtain the steady state error of the system. When subjected to an input given by the polynomial $r(t) = a_0 + a_1 t + \frac{a_2}{2}\cdot t^2$.

Solution: Finding static error constant

For unity feedback system, $H(s) = 1$.

loop transfer function, $G(s)\,H(s) = G(s)$

The static error constants are K_P, K_V and K_a

Position error constant, $K_P = \lim_{s\to 0} G(s) = \lim_{s\to 0} \frac{10}{s(0.1s+1)} = \infty$

Velocity error constant, $K_V = \lim_{s\to 0} s\cdot G(s) = s\cdot \frac{10}{s\,(0.1s+1)} = 10$

Acceleration error constant, $K_a = \lim_{s\to 0} s^2 \cdot G(s) = \lim_{s\to 0} s^2 \cdot \frac{10}{s\,(0\cdot 1s+1)} = 0$

Finding Steady State Error

Ist Method: The steady state error for non-standard input is obtained using generalized error series, given below.

The error signal, $e(t) = r(t)\,C_0 + \dot{r}(t)\,C_1 + \ddot{r}(t)\,\frac{C_2}{2!} + \ldots + r^n(t)\,\frac{C_n}{n!} + \ldots$

Given that, $r(t) = a_0 + a_1 t + \frac{a_2}{2}\,t^2$

$\therefore$ $\dot{r}(t) = a_1 + a_2 t$

$\ddot{r}(t) = a_2$

$\dddot{r}(t) = 0$

The derivatives of $r(t)$ is zero after second derivatives. Hence we have to evaluate only three constants C_0, C_1 and C_2.

The generalized error constants are given by,

$$C_0 = \lim_{s\to 0} F(s);\ C_1 = \lim_{s\to 0} \frac{d}{ds} F(s);\ C_2 = \lim_{s\to 0} \frac{d^2}{ds^2} F(s).$$

$$F(s) = \frac{1}{1+G(s)\,H(s)} = \frac{1}{1+G(s)} = \frac{1}{1+\dfrac{10}{s\,(0.1s+1)}} = \frac{s\,(0.1s+1)}{s\,(0.1s+1)+10}$$

$$F(s) = \frac{0.1s^2+s}{0.1s^2+s+10}$$

$$C_0 = \lim_{s\to 0} F(s) = \lim_{s\to 0} \frac{0.1s^2+s}{0.1s^2+s+10} = 0$$

$$C_1 = \lim_{s\to 0} \frac{d}{ds} F(s) = \lim_{s\to 0} \frac{d}{ds}\left[\frac{0.1s^2+s}{0.1s^2+s+10}\right]$$

$$= \lim_{s\to 0}\left[\frac{(0.1s^2+s+10)\,(0.2s+1)-(0.1s^2+s)\,(0.25+1)}{(0.1s^2+s+10)^2}\right]$$

$$= \lim_{s\to 0} \frac{2s+10}{(0.1s^2+s+10)} = \frac{10}{10^2} = 0.1$$

$$C_2 = \lim_{s\to 0} \frac{d^2}{ds^2} F(s) = \lim_{s\to 0} \frac{d}{ds}\left[\frac{d}{ds} F(s)\right] = \lim_{s\to 0} \frac{d}{ds}\left[\frac{2s+10}{(0.1s^2+s+10^2)}\right]$$

$$= \lim_{s\to 0}\left[\frac{(0.1s^2+s+10)^2\,(2)-(2s+10)\,(2)\,(0.1s^2+s+10)\,(0.2s+1)}{(0.1s^2+s+10)^4}\right]$$

$$= \frac{10^2\times 2 - 10\times 2\times 10\times 1}{10^4} = 0.$$

Error signal, $e(t) = r(t)\,C_0 + \dot{r}(t)\,C_1 + \ddot{r}(t)\,\dfrac{C_2}{2!}$

$= \dot{r}(t)\,C_1$ ($\therefore$ C_0 and C_2 are zero)

$= (a_1 + a_2\,t)\,0.1$

Steady state error is obtained by letting t to ∞

$\therefore$ Steady state error, $e_{ss} = \lim_{t\to\infty} e(t) = \lim_{t\to\infty}\left[(a_1+a_2 t)\,0.1\right] = \infty$

IInd Method:

The error signal in s-domain, $E(s) = \dfrac{R(s)}{1+G(s)\,H(s)}$

Given that $$r(t) = a_0 + a_1 t + \frac{a_2}{2} t^2$$
$$G(s) = 10/s(0.1S + 1)$$
$$H(s) = 1$$

On taking Laplace transform of $r(t)$ we get $R(s)$.

$$R(s) = \frac{a_0}{s} + \frac{a_1}{s^2} + \frac{a_2}{2} \cdot \frac{2!}{s^3} = \frac{a_0}{s} + \frac{a_1}{s^2} + \frac{a_2}{s^3}$$

$$E(s) = \frac{R(s)}{1 + G(s)\,H(s)} = \frac{\frac{a_0}{s} + \frac{a_1}{s^2} + \frac{a_2}{s^3}}{1 + \frac{10}{s(0.1s+1)}} = \frac{\frac{a_0}{s} + \frac{a_1}{s^2} + \frac{a_2}{s^3}}{\frac{s\,(0.1s+1)+10}{s(0.1s+1)}}$$

$$= \frac{a_0}{s}\left[\frac{s(0.1s+1)}{s(0.1s+1)+10}\right] + \frac{a_1}{s^2}\left[\frac{s(0.1s+1)}{s(0.1s+1)+10}\right] + \frac{a_2}{s^3}\left[\frac{s(0.1s+1)}{s(0.1s+1)+10}\right]$$

The steady state error e_{ss} can be obtained from find value theorem,

Steady state error, $e_{ss} = \lim_{t\to\infty} e(t) \lim_{s\to 0} s \cdot E(s)$

$$e_{ss} = \lim_{s\to 0} s\left\{\frac{a_0}{s}\left[\frac{s\,(0.1s+1)}{s\,(0.1s+1)+10}\right] + \frac{a_1}{s^2}\left[\frac{s\,(0.1s+1)}{s(0.1s+1)+10}\right] + \left[\frac{s(0.1s+1)}{s(0.1s+1)+10}\right]\frac{a_2}{s^3}\right.$$

$$= 0 + \frac{a_1}{10} + \infty = \infty$$

$$e_{ss} = \infty$$

IIIrd Method:

Error signal in s-domain,

$$E(s) = \frac{R(s)}{1 + G(s)\,H(s)}$$

$$\therefore \quad \frac{E(s)}{R(s)} = \frac{1}{1 + G(s)\,H(s)}$$

Given that $$G(s) = \frac{10}{s\,(0.1s+1)} \text{ and } H(s) = 1$$

$$\therefore \quad \frac{E(s)}{R(s)} = \frac{1}{1 + \frac{10}{s\,(0.1s+1)}} = \frac{s\,(0.1s+1)}{s\,(0.1s+1)+10}$$

$$\frac{E(s)}{R(s)} = \frac{.1s^2 + s}{0.1s^2 + s + 10} = \frac{s + 0.1s^2}{10 + s + 0.1s^2}$$

$$= \frac{s}{10} - \frac{s^3}{1000} + \ldots$$

$$\therefore \qquad E(s) = \frac{s}{10}R(s) - \frac{s^3}{1000}R(s) + \ldots$$

$$\begin{array}{r|l}
 & \frac{s}{10} - \frac{s^3}{1000} \\
\hline
10s + s + 0.1s^2 & s + 0.1s^2 \\
 & \underline{s + \frac{s^2}{10} + \frac{s^3}{100}} \\
 & \underline{\frac{-s^3}{100}} \\
 & \underline{-\frac{s^3}{100} - \frac{s^4}{1000} - \frac{s^5}{1000}} \\
 & \underline{\frac{s^4}{1000} + \frac{s^5}{1000}}
\end{array}$$

On taking inverse Laplace transform,

$$e(t) = \frac{1}{10}\dot{r} - \frac{1}{1000}\dddot{r}(t) + \ldots$$

Given that $r(t) = a_0 + a_1 t + \frac{a_2}{2}t^2$

$$\therefore \qquad \dot{r} = \frac{d}{dt}r(t) = a_1 + a_2 t$$

$$\ddot{r}(t) = \frac{d}{dt}\dot{r}(t) = a_2$$

$$\dddot{r}(t) = \frac{d}{dt}\ddot{r}(t) = 0$$

$\therefore$ Error signal in time domain,

$$e(t) = \frac{1}{10}\dot{r}(t) = \frac{1}{10}(a_1 + a_2 t)$$

Steady state error, $$e_{ss} = \underset{t\to\infty}{\text{Lt}}\ e(t) = \underset{t\to\infty}{\text{Lt}}\ \frac{1}{10}(a_1 + a_2 t) = \infty$$

Results:

1. Position error constant, $K_P = 0$
2. Velocity error constant, $K_V = 10$
3. Acceleration error constant, $K_a = 0$
4. Steady state error is ∞ i.e., $e_{ss} = \infty$

$$\left(\text{when input } r(t) = a_0 + a_1 + \frac{a_2 t^2}{2}\right),$$

$$e_{ss} = \infty$$

EXAMPLE 4.22: A unity feedback system with a closed loop transfer function $\frac{C(s)}{R(s)} = \frac{K_s + b}{s^2 + as + b}$. Determine the open loop transfer function $G(s)$. Show that the steady state error with unit ramp input is given by $\left(\frac{a-K}{b}\right)$.

Solution: For unity feedback system, $H(s) = 1$

The closed loop transfer function,

$$\frac{C(s)}{R(s)} = \frac{G(s)}{1+G(s)\,H(s)} = \frac{G(s)}{1+G(s)}$$

Let,

$$M(s) = \frac{C(s)}{R(s)} = \frac{G(s)}{1+G(s)}$$

on cross multiplication we get,

$$G(s) = M(s)\,[1 + G(s)] = M(s) + M(s)\,G(s)$$

$$G(s) - M(s)\,G(s) = M(s)$$

$$G(s)\,[1 - M(s)] = M(s)$$

$$\therefore \quad G(s) = \frac{M(s)}{1-M(s)}$$

Given that

$$M(s) = \frac{K_s + b}{s^2 + as + b}$$

$\therefore$ Open loop transfer function,

$$G(s) = \frac{\frac{Ks+b}{s^2+as+b}}{1 - \frac{Ks+b}{s^2+as+b}} = \frac{Ks+b}{(s^2+as+b)-(Ks+b)}$$

$$= \frac{Ks+b}{s^2+as+b-Ks-b} = \frac{Ks+b}{s^2+(a-K)\,s} = \frac{Ks+b}{s[s+(a-K)]}$$

Velocity error constant,

$$K_v = \underset{s\to 0}{\text{Lt}}\ s\cdot G(s)\,H(s) = \underset{s\to 0}{\text{Lt}}\ s\cdot G(s)$$

$$= \underset{s\to 0}{\text{Lt}}\ s\cdot\frac{Ks+b}{s\,[s+(a-K)]} = \frac{b}{a-K}$$

Steady state error with velocity input,

$$e_{ss} = \frac{1}{K_V} = \frac{a-K}{b}$$

Results: Open loop transfer function,

$$G(s) = \frac{Ks+b}{s[s+(a-K)]}$$

Steady state error with velocity input,

$$e_{ss} = \left[\frac{a-K}{b}\right]$$

EXAMPLE 4.23: A unity feedback system has the forward transfer function $G(s) = \dfrac{K_1\,(2s+1)}{s(5s+1)\,(1+s)^2}$.

The input $r(t) = 1 + 6t$ is applied to the systems. Determine the minimum value of K_1 if the steady state error is to be less than 0.1.

Solution: Given that input $r(t) = 1 + 6t$.

On taking Laplace transform of $r(t)$ we get $R(s)$

$$\therefore \qquad R(s) = L[r\,(t)] = l[1+6t] = \frac{1}{s} + \frac{6}{s^2}$$

The error signal in s-domain

$$E(s) = \frac{R(s)}{1+G(s)\,H(s)}$$

$$\therefore \qquad E(s) = \frac{\dfrac{1}{s}+\dfrac{6}{s^2}}{1+\dfrac{K_1\,(2s+1)}{S\,(5s+1)\,(1+s)^2}} \qquad \text{(Here } H(s) = 1\text{)}$$

$$= \frac{\dfrac{1}{s}+\dfrac{6}{s^2}}{\dfrac{s(5s+1)\,(1+s)^2 + K_1(2s+1)}{s(5s+1)\,(1+s)^2}}$$

$$= \frac{1}{s}\left[\frac{3\,(5s+1)\,(1+s)^2}{s(5s+1)\,(1+s)^2 + K_1(2s+1)}\right] + \frac{6}{s^2}\left[\frac{s\,(5s+1)\,(1+s)^2}{s(5s+1)\,(1+s^2)+K_1(2s+1)}\right]$$

The steady state error e_{ss} can be obtained from final value theorem.

$$e_{ss} = \underset{t\to\infty}{\text{Lt}}\ e(t) = \underset{s\to 0}{\text{Lt}}\ s\cdot E(s)$$

$$= \underset{s\to 0}{\text{Lt}}\ s\left\{\frac{1}{s}\left[\frac{s\,(5s+1)\,(1+s)^2}{s\,(5s+1)\,(1+s)^2 + K_1\,(2s+1)}\right] + \frac{6}{s^2}\left[\frac{s\,(5s+1)\,(1+s)^2}{s\,(5s+1)\,(1+s)^2 + K_1\,(2s+1)}\right]\right\}$$

$$= \underset{s\to 0}{\text{Lt}} \left\{ \frac{s\,(5s+1)\,(1+s)^2}{s\,(5s+1)\,(1+s)^2 + K_1\,(2s+1)} + \frac{6\,(5s+1)\,(1+s)^2}{s\,(5s+1)\,(1+s)^2 + K_1\,(2s+1)} \right\}$$

$$= 0 + \frac{6}{K_1} = \frac{6}{K_1}$$

Given that $e_{ss} < 0.1$

$$\therefore \qquad 0.1 = \frac{6}{K_1} \Rightarrow K_1 = \frac{6}{0.1} \Rightarrow K_1 = 60$$

Results: For steady state error

$e_{ss} < 0.1$, the value of K_1 should be greater than 60.

EXAMPLE 4.24: The open loop transfer function of unity feedback system is given by

$$G(s) = \frac{50}{(1+0.1s)\,(s+10)}$$

Determine the static error coefficients K_P, K_v, and K_a.

Solution:

$$K_P = \underset{s\to 0}{\text{Lt}}\; G(s)\,H(s) = \underset{s\to 0}{\text{Lt}} \frac{50}{(1+0.1s)\,(s+10)} = 5$$

$$K_P = 5$$

$$K_v = \underset{s\to 0}{\text{Lt}}\; s \cdot \frac{50}{(1+0.1s)\,(s+10)} = 0$$

$$K_v = 0$$

$$K_a = \underset{s\to 0}{\text{Lt}}\; s^2 \cdot \frac{50}{(1+0.1s)\,(s+10)} = 0$$

$$K_a = 0$$

Result: 1. $K_P = 5$,

2. $K_v = 0$.

SHORT QUESTION AND ANSWERS

Q.1 What is time Response?

Ans. The time response is the output of the closed loop system as a function of time. It is denoted by $C(t)$. It is given by inverse Laplace of the product of input and transfer function of the system.

The closed loop transfer function,

$$\frac{C(s)}{R(s)} = \frac{G(s)}{1+G(s)\,H(s)}$$

Response is s-domain,

$$C(s) = \frac{R(s)\,G(s)}{1+G(s)\,H(s)}$$

Response in time domain

$$C(t) = L^{-1}[C(s)] = L^{-1}\left[\frac{R(s)\,G(s)}{1+G(s)\,H(s)}\right]$$

Q.2 What is transient and steady state response?

Ans. The transient response is the response of the system when the input changes from one state to another. The response of the system $t \to 0$ is called steady state response.

Q.3 What is the importance of test signals?

Ans. The test signals resembles, the characteristics of actual input signals. The test signals are used to predetermine the performance of the system. If the response of a system is satisfactory for a test signal, then the system will be suitable for practical applications.

Q.4 Name the test signals used in control system.

Ans. The commonly used test input signals in control system are impulse, step, ramp, acceleration and sinusoidal signals.

Q.5 Define step signal.

Ans. The step signal is a signal whose values changes from 0 to A and remains constant at A for $t > 0$. The mathematical representation of step signal is

$$r(t) = Au(t)$$

where, $u(t) = 1,\ t \geq 0$

$$u(t) = 0,\ t < 0$$

CHAPTER 4

Q.6 Define ramp signal.

Ans.

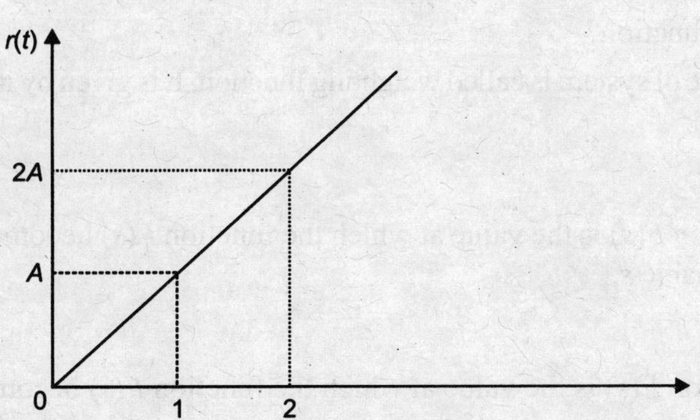

A ramp signal is a signal whose value increases linearly with time from an initial value of zero at $t = 0$. It is mathematically represented as

$$r(t) = At, \text{ when } t \geq 0$$

$$r(t) = 0; \text{ when } t < 0$$

Q.7 Define parabolic signal

Ans.

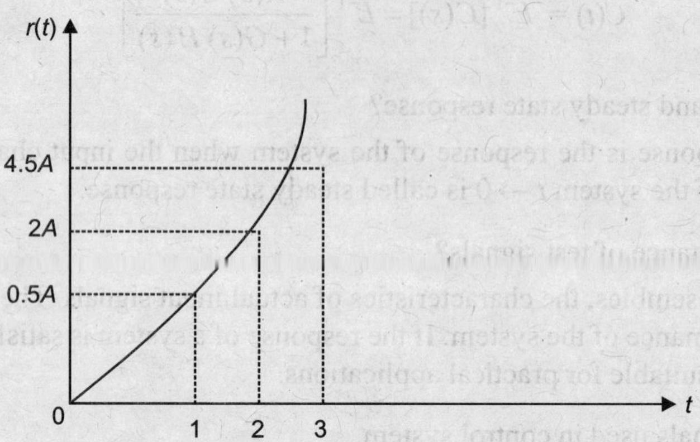

It is a signal in which the instantaneous values varies as square of the time from an initial value of zero at $t = 0$. It is mathematically represented as

$$r(t) = \frac{At^2}{2}, t \geq 0$$

$$r(t) = 0, t < 0$$

Q.8 What is an impulse signal?

Ans. A signal which is available for very short duration is called impulse signal. Ideal impulse signal is a unit impulse signal which is defined as a signal having zero values at all time except at $t = 0$; At $t = 0$ the magnitude becomes infinite. It is denoted by $\delta(t)$ and mathematically expressed as

$$\delta(t) = 0 \text{ for } t \neq 0 \text{ and } \underset{t_1 \to 0}{\text{Lt}} \int_{-t_1}^{+t_1} \delta(t)\, dt = 1$$

Q.9 What is weighting function?

Ans. The impulse response of system is called weighting function. It is given by inverse Laplace transform of system transfer function.

Q.10 Define Pole?

Ans. The pole of a function $F(s)$ is the value at which the function $F(s)$ becomes infinite, where $F(s)$ is a function of complex variable s.

Q.11 Define zero?

Ans. The zero of a function $F(s)$ is the value at which the function $F(s)$ becomes zero, where $F(s)$ is a function of complex variable s.

Q.12 What is the order of a system?

Ans. The order of the system is given by the order of the differential equation governing the system. It is also given by the maximum power of s in the denominator polynomial of transfer function. The maximum power of s also gives the number of poles of the system and so the order of the system also given by number of poles of the transfer function.

Q.13 Define damping ratio?

Ans. The damping ratio is defined as the ratio of the actual damping to critical damping.

Q.14 Give the expression for damping ratio of mechanical and electrical system?

Ans. The damping ratio of second order mechanical translational system, $\zeta = \dfrac{B}{2\sqrt{MK}}$.

The damping ratio of second order mechanical rotational system, $\zeta = \dfrac{B}{2\sqrt{JK}}$.

The damping ratio of second order electrical system, $\zeta = \dfrac{R}{2\sqrt{\dfrac{L}{C}}}$

Q.15 How the second order system is classified?

Ans. Depending on the value of damping? Depending on the value of damping, the system can be classified inot the following four Cases,

Case 1: Undamped system, $\zeta = 0$

Case 2: Underdamped system, $0 < \zeta < 1$

Case 3: Critically damped system, $\zeta = 1$

Case 4: Over damped system, $\zeta > 1$.

Q.16 What will be the nature of response of a second order system with different types of damping?

Ans. For undamped system the response is oscillatory.

For underdamped system the response is damped oscillatory.

For critically damped system the response is exponentially rising.

For overdamped system the response is exponentially rising but the rise time will be very large.

Q.17 Sketch the response of a second order underdamped system.

Ans.

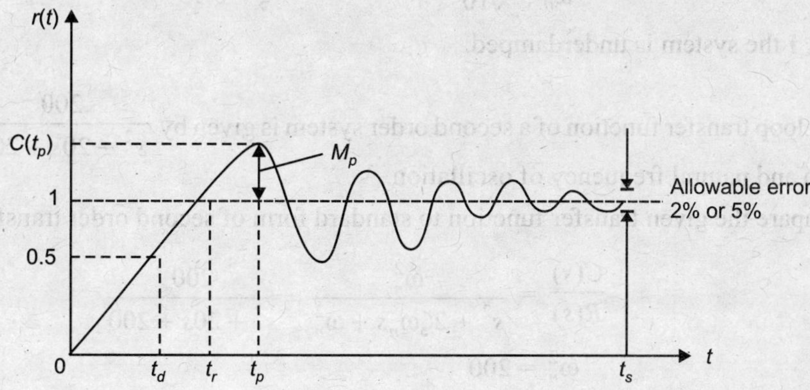

CHAPTER 4

Q.18 What is damped frequency of oscillation?

Ans. In underdamped system the response is damped oscillatory. The frequency of damped oscillation is given by $\omega_d = \omega_n \sqrt{1-\zeta^2}$

Q.19 Give the expression for natural frequency of oscillations of electrical and mechanical system.

Ans. The natural frequency of oscillation of second order mechanical translational system

$$\omega_n = \sqrt{\frac{K}{M}}$$

The natural frequency of oscillation of second order mechanical rotational system

$$\omega_n = \sqrt{\frac{K}{J}}$$

The natural frequency of oscillation of second order electrical system

$$\omega_n = \frac{1}{\sqrt{LC}}$$

Q.20 The closed loop transfer function of second order system is

$$\frac{C(s)}{R(s)} = \frac{10}{s^2 + 6s + 10}$$

What is the type of damping in the system?

Ans. Let us compare the given transfer function with the standard form of second order transfer function

$$\therefore \quad \frac{C(s)}{R(s)} = \frac{\omega_n^2}{s^2 + 2\zeta\omega_n s + \omega_n^2} = \frac{10}{s^2 + 6s + 10}$$

$$\therefore \quad \omega_n^2 = 10 \qquad 2\zeta\omega_n = 6$$

$$\omega_n = \sqrt{10} \qquad \zeta = 0.95$$

Since $\zeta < 1$ the system is underdamped.

Q.21 The closed loop transfer function of a second order system is given by $\frac{200}{s^2 + 20s + 200}$. Determine the damping ratio and natural frequency of oscillation.

Ans. Let us compare the given transfer function to standard form of second order transfer function.

$$\frac{C(s)}{R(s)} = \frac{\omega_n^2}{s^2 + 2\zeta\omega_n s + \omega_n^2} = \frac{200}{s^2 + 20s + 200}$$

$$\therefore \quad \omega_n^2 = 200$$

$$\omega_n = \sqrt{200} = 14.14 \text{ rad/sec}$$

$$2\zeta\omega_n = 200$$
$$\zeta = .707$$

Result: Damping ratio = .707

Natural frequency ω_n = 14.14 rad/sec

Q.22 List the time domain specifications

Ans. The time domain specifications are

(i) Delay time, (ii) Rise time, (iii) Peak time, (iv) Maximum peak overshoot, (v) Settling time.

Q.23 Define delay time.

Ans. It is the time taken for response to reach 50% of the final value for the very first time.

Q.24 Define rise time.

Ans. It is the time taken for response to meet 100% of the final value, for the very first time.

Q.25 Define peak time.

Ans. It is the time taken for the response to reach the peak value for the very first time (or) It is the time taken for the response to reach peak overshoot, M_P.

Q.26 Define peak overshoot.

Ans. It is defined as the ratio of the maximum peak value measured from final value to final value. Let final value = $C(\infty)$, maximum value = $C(t_p)$

$$\text{Peak overshoot, } M_p = \frac{C(t_p) - C(\infty)}{C(\infty)}$$

Q.27 Define settling time.

Ans. It is defined as the time taken by the response to reach and stay within a specified error and the error is usually specified as % of final value. The usual tolerable error in 2% or 5% of the final value.

Q.28 What is type of a system what is its significance?

Ans. The type number is given by number of poles of loop transfer function at the origin. The type number of the system decides the steady state error.

Q.29 Distinguish between type and order of a system.

Ans. 1. Type number is specified for loop transfer function but order can be specified for any transfer function. (open loop or closed loop transfer function).

2. The type number is given by number of poles of loop transfer function lying at origin of s-plane but the order is given by the number of poles of transfer function.

Q.30 For the system with following transfer function, determine type and orderof the system.

$$G(s)\,H(s) = \frac{K}{s(s+1)\,(s^2+6s+8)}$$

Ans. Type –1, order: 4

CHAPTER 4

Q.31 What is steady state error?

Ans. The steady state error is the value of error signal $e(t)$, when t tends to infinity. The steady state error is a measure of system accuracy. The errors arise from the nature of inputs, type of system and from non-linearity of system components.

Q.32 What are static error constants?

Ans. The K_p, K_v and K_a are called static error constants. These constants are associated with steady state error in a particular type of system and for a standard input.

Q.33 Define positional error constant?

Ans. The positional error constant

$$K_p = \lim_{s \to 0} G(s)\, H(s)$$

The steady state error in type – 0 system when the input is unit step is given by $1/1 + K_p$

Q.34 Define velocity error constant?

Ans. The velocity error constant

$$K_v = \lim_{s \to 0} s \cdot G(s)\, H(s)$$

The steady state erin type – 1 system for unit ramp input is given by $1/K_v$.

Q.35 Define acceleration error constant?

Ans. The acceleration error constant

$$K_a = \lim_{s \to 0} s^2\, G(s) \cdot H(s)$$

The steady state error in type–2 system for unit parabolic input is given by $1/K_a$.

Q.36 A unity feedback system has a open loop transfer function of $G(s) = \dfrac{10}{(s+1)\,(s+2)}$. Determine the steady state error for unit step input?

Ans. The steady state error for unit step input,

$$e_{ss} = \frac{1}{1 + K_p}$$

where $K_p = \lim_{s \to 0} G(s) \cdot H(s)$

for unity feedback system $H(s) = 1$

$$K_p = \lim_{s \to 0} G(s) = \lim_{s \to 0} \frac{10}{(s+1)\,(s+2)} = \frac{10}{3} = 3.33$$

$$e_{ss} = \frac{1}{1 + K_p} = \frac{1}{1 + 3.33} = \frac{1}{4.33}$$

Q.37 What are generalized error coefficients. They are the coefficients of generalized error series. The generalized error series is given by,

$$e(t) = C_0 r(t) + C_1 r'(t) + \frac{C_2}{2!} r''(t) + \frac{C_3}{3!} r'''(t) + \ldots + \frac{C_n}{n!} r^{n'}(t) \ldots$$

The coefficients C_0, C_1, C_2, ... C_n are called generalized error coefficients (or) dynamic error coefficients.

The nth coefficient, $C_n = \lim\limits_{s \to 0} \frac{d^n}{ds^n} F(s)$,

where $F(s) = \frac{1}{1 + G(s)\,H(s)}$.

Q.38 Give the relation between generalized and static error coefficients?

Ans. The following expression shows the relation between generalized and static error coefficient,

$$C_0 = \frac{1}{1+K_p}, C_1 = \frac{1}{K_v}, c_2 = \frac{1}{K_a}$$

Q.39 Mention the advantage of generalized error constants over static error constants?

Ans. Generalized error series gives error signal as a function of time.

Q.40 What is the effect on system performance when a proportional controller is introduced in a system?

Ans. The proportional controller improves the steady-state tracking accuracy, disturbance signal rejection and relative stability of the system. It also increases the loop gain of the system which results in reducing the sensitivity of the system to parameter variations.

Q.41 What is the disadvantage in proportional controller?

Ans. The disadvantage of proportional controller is that it produces a constant steady state error.

Q.42 What is the effect of PI controller on the system performance?

Ans. The PI controller increases the order of the system by one, which results in reducing the steady state error. But the system becomes less stable than the original system.

Q.43 What is the effectof PD controller on the system performance?

Ans. The effect of PD controller is to increase the damping ratio of the system and the peak overshoot is reduced.

Q.44 Why derivative controller is not used in control systems?

Ans. The derivative controller produces a control action based on rate of change of error signal and it does not produce corrective measures for any constant error. Hence derivative controller is not used in control systems.

EXERCISE

1. What is the unit-step response of the following system. Fig. (1.1).

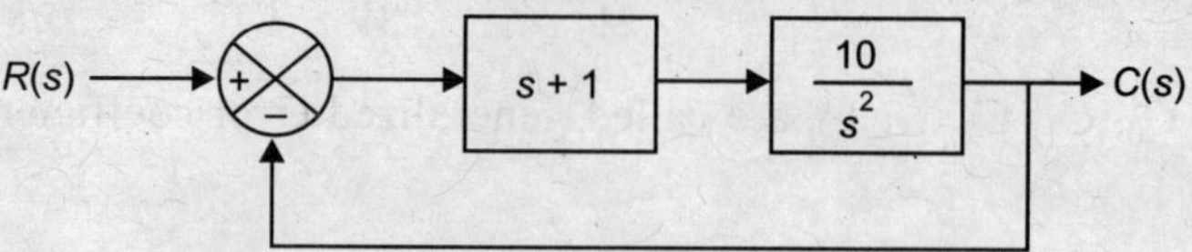

Ans. $C(t) = -1.4s\ 5e^{-8.8lt} + .145e^{-1.13t} + 1$

2. Obtain the unit-step response of a unity feedback system whose open-loop transfer function is

$$G(s) = \frac{5(s+20)}{s(s+4.59)(s^2+3.41s+16.35)}$$

Ans. $C(t) = 1 + \frac{3}{8}e^{-t}\cos 3t - \frac{17}{24}e^{-t}\sin 3t - \frac{11}{8}e^{-3t}\cos t - \frac{13}{8}e^{-3t}\sin t$

3. The open loop transfer function of an unity feedback control system is given by $G(s) = 100/s\ (s+2)(s+5)$. For unit step input, find time response of the closed loop system and determine % overshoot and the rise time.

Ans. $C(t) = \left[1 - 0.186\,e^{-7.45t} - 0.88\,e^{.225t}\cos(3.65t - 22°)\right]$

As t tends to infinity, $C(t)$ tends to infinity and so the system is unstable. Therefore % overshot and rise time are not defined.

4. A servomechanism has its moment of inertia $J = 10 \times 10^{-6}$ kg – m², retarding friction, B = 400 × 10^{-6} N – m/(rad/sec) and elasticity coefficient, $K = 0.004$ N – m/ rad. Find the natural frequency and damping factor of the system.

Ans. Natural frequency, $\omega_n = 20$ rad/sec

Damping factor, $\zeta = 1$

5. For a second order system whose open loop transfer function $G(s) = 4/s(s+2)$, determine the maximum overshoot, the time to reach the maximum overshoot when a step displacement of 18° is given to the system. Find the rise time, time constant and the settling time for an error of 7%.

Ans. Maximum overshoot = 0.16, when input is 18°, $M_p = 2.88°$

Peak time, $t_p = 1.181$ sec

Time constant, $T = 1$ sec

Rise time, $t_r = 1.21$ sec

Settling time for 7% error = 2.66 sec

6. Consider the unity feedback closed loop system where the forward transfer functino is $G(s) = 25/s(s+5)$. Obtain the rise time, peak time, maximum overshoot and the settling time when the system is subjected to a unit-step input.

Ans. Rise time, $t_r = 0.55$ sec

Peak time, $t_p = 0.785$ sec

% peak overshoot, $M_p = 9.5\%$

Settling time, $t_s = 1.33$ sec (for 2% error)

$= 1$ sec (for 5% error)

7. Consider the system given below (Fig. 7) where $\zeta = 0.6$ and $\omega_n = 0.5$ rad/sec. Determine the rise time, peak time, maximum overshoot and settling time, when the system is subjected to a unit step input.

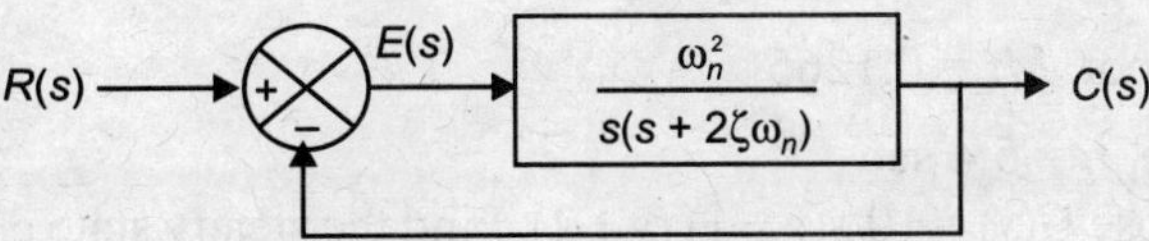

Ans. Rise time, $t_r = 0.55$ sec

Maximum overshoot, $M_p = 0.095$

Peak time, $t_p = 0.785$ sec

Settling time, $t_s = 1$ sec (for 5% criterian)

8. For the system (given below), determine the values of K and K_h. So that the maximum overshoot in the unit step response is 0.2 and the peak time is 1 sec with these values of K and K_h, obtain rise time and settling time.

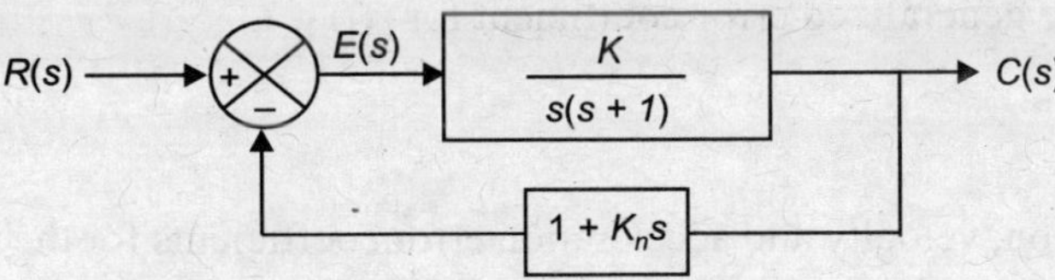

Ans. $K = 12.5$

$K_h = 0.178$

Rise time, $t_r = 0.65$ sec

Settling time, $t_s = 2.48$ sec (for 2% error)

$t_s = 1.86$ sec (for 5% error)

9. The system (given below) subjected to a unit-step input. Determine the values of K and T, where the maximum overshoot of the system is 2.5 4% corresponding to $\zeta = 0.4$.

Ans. $K = 1.42$

$T = 1.09$

10. Determine the values of K and T of the closed-loop system shown below. So that the maximum overshoot in unit-step response is 25% and the peak time is 2 sec. Answer that $J = 1$ kg – m^2.

Ans. $K = 2.95$ N – m

$T = 0.471$ sec

11. A unity-feedback system is characterized by the open-loop transferfunction, G(s) = $1/s(0.5s + 1)(0.2s + 1)$.

(a) Determine the steady-state errors to unit-step, unit-ramp and unit parabolic inputs.

(b) Determine rise time, peak time, peak overshoot and settling time of the unit-step response of the system.

Ans. (a) e_{ss} (unit step) = 0

e_{ss} (unit ramp) = 1

e_{ss} (unit parabola) = ∞

(b) Rise time, $t_r = 1.91$ sec

Peak time, $t_p = 2.79$ sec

Peak overshoot, $M_p = 0.1265$

Settling time, $t_s = 5.4$ sec

12. For a system whose $G(s) = 10/s(s + 1)(s + 2)$. Find the steady state error when it is subjected to the input, $r(t) = 1 + 2t + 1.5t^2$.

Ans. The total steady state error is ∞.

13. A unity feedback system has $G(s) = 1/s(1 + s)$. The input to the system is described by $r(t) = 4 + 6t + 2t^3$. Find the generalized error coefficients and steady state error.

Ans. $c_0 = 0$ $c_1 = 1$ $e_{ss} = \infty$

$c_2 = 0$ $c_3 = -6$

14. A unity feedback system has the forward path transfer function $G(s) = 10/(s + 1)$. Find the steady state error and the generalized error coefficient for $r(t) = t$.

Ans. $C_0 = 1/11$ $e_{ss} = \infty$

$C_1 = 10/121$

15. Find out the position, velocity and acceleration error coefficients for the following unity feedback systems having forward loop transfer function $G(s)$ as

(a) $100/(1 + 0.5s)(1 + 2s)$

(b) $K/s(1.01s)(1 + s)$

(c) $K/s^2(s^2 + 8s + 100)$

(d) $K(1 + s)(1 + 2s)/s^2(s^2 + 4s + 20)$

Ans. $K_p = 100$, $K_v = 0$ and $K_a = 0$ for the system $100/(1 + 0.5s)(1 + 2s)$

$K_p = \infty$, $K_v = K$ and $K_a = 0$ for the system $K/s(1 + 0.1s)(1 + s)$

$K_p = \infty$, $K_v = \infty$ and $K_a = K/100$ for the system $K/s^2(s^2 + 8s + 100)$

$K_p = \infty$, $K_v = \infty$ and $K_a = K/100$ for the system $K(1 + s)(1 + 2s)/s^2(s^2 + 4s + 20)$

16. The open loop transferfunction of a unity feedback control system is $G(s) = 9(s + 1)$, using the generalized error series determine the error signal and steady state error of the system when the system is excited by (i) $r(t) = 2$, (ii) $r(t) = t$, (iii) $r(t) = 3t^2$ (iv) $r(t) = 1 + 2t + 3t^2/2$

Ans. (i) $e(t) = 0.2$; $e_{ss} = 0.2$

(ii) $e(t) = 0.1t + 0.09$; $e_{ss} = \infty$

(iii) $e(t) = 0.15t^2 + 0.24$; $e_{ss} = \infty$

(iv) $e(t) = 0.15t^2 + 0.77t + 0.226$; $e_{ss} = \infty$

17. For unity feedback system having open loop transfer function as $G(s) = K(s + 2)/s^2(s^2 + 7s + 12)$. Determine (i) type of system, (ii) error constants K_p, K_v, and K_a, (ii) steady state error for parabolic input.

Ans. (i) It is type 2 system

(ii) $K_p = \infty$

$K_v = \infty$

$K_a = K/6$

(iii) $e_{ss} = 6/K$

18. Determine the values of K and T of the closed loop system. So that the maximum overshoot in unit step response is 25% and the peak time is 2 sec.

Ans. $K = 2.95$ N-m

$T = .471$ sec

SOME WORKED EXAMPLES

EXAMPLE 1: When the system shown in Fig. 1 in subjected to a unit step input, the system output responds such that it attains maximum peak overshoot of 25.4%. Determine the value of K and T. Assume $\zeta = .4$ $t_p = 3$.

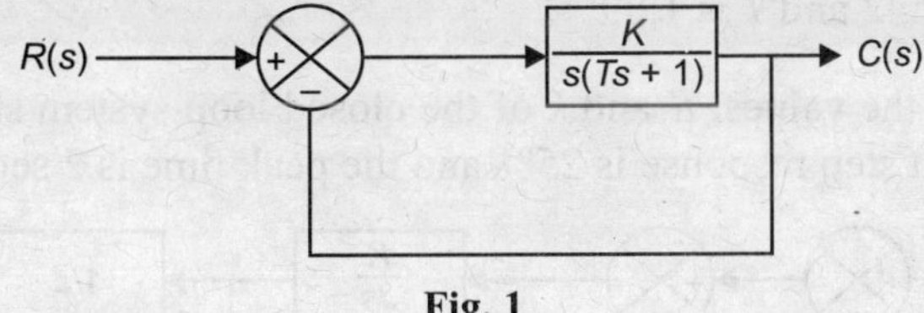

Fig. 1

Solution: We know

$$t_p = \frac{\pi}{\omega_d} = \frac{\pi}{\omega_n \sqrt{1-\zeta^2}}$$

$$\therefore \quad \frac{\pi}{\omega_n \sqrt{1-\zeta^2}} = 3$$

$$\frac{\pi}{\omega_n \sqrt{1-(.4)^2}} = 3$$

$$\therefore \quad \omega_n = \frac{\pi}{3\sqrt{1-(.4)^2}}$$

$$\boxed{\omega_n = 1.14}$$

From block diagram,

$$\frac{C(s)}{R(s)} = \frac{K}{Ts^2 + s + K}$$

This can be rewritten as (Divide by T in both numerator and denominator)

$$\frac{C(s)}{R(s)} = \frac{K/T}{s^2 + s/T + K/T}$$

Compare $s^2 + s/T + K/T$ with standard characteristic equation (i.e., $s^2 + 2\zeta\omega_n^s + \omega_n^2$)

$$\therefore \quad \omega_n^2 = K/T$$

$$\omega_n = \sqrt{K/T}$$

$$2\zeta\omega_n = 1/T$$

$$\therefore \quad T = \frac{1}{2\zeta\omega_n} = \frac{1}{2\times .4\times 1.14} = 1.09$$

$$\boxed{T = 1.09}$$

$$K = \omega_n^2 T \Rightarrow (1.14)^2\ (1.09)$$

$$\boxed{K = 1.42}$$

Result: The value of K is 1.42 and T is 1.09

EXAMPLE 2: Determine the values, K and k of the closed loop system shown in Fig. (2) so that the maximum overshoot in unit step response is 25% and the peak time is 2 sec. Assume that $J = 1$ kg-m^2.

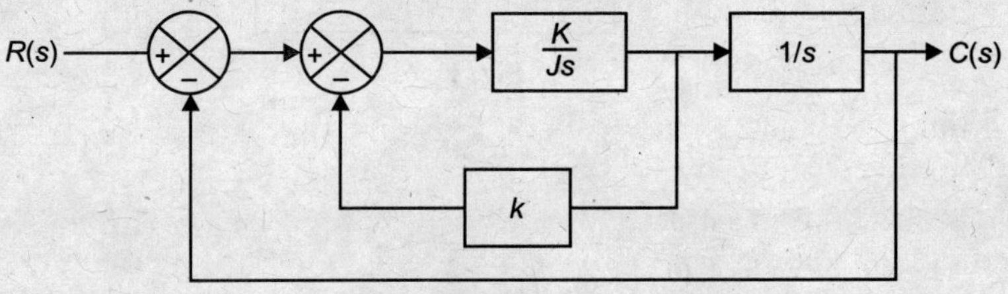

Fig. 2

Solution: The closed loop transfer function can be derived from Fig. (2), using the lock diagram reduction technique,

$$\frac{C(s)}{R(s)} = \frac{K}{Js^2 + Kks + K} \tag{1}$$

From given data, $J = 1$ kg-m^2, so substitute this value in equation (1),

$$\therefore \quad \frac{C(s)}{R(s)} = \frac{K}{s^2 + Kks + K}$$

Compare the characteristic equation $s^2 + Kks + K$ with standard characteristic equation of second order system ($s^2 + 2\zeta\omega_n + \omega_n^2$)

so, $$\omega_n^2 = K \Rightarrow \omega_n = \sqrt{K}$$

$$2\zeta\omega_n = Kk \Rightarrow k = \frac{2\zeta\omega_n}{K}$$

we know, $$\underset{\text{(Peak overshoot)}}{M_P} = e^{-\zeta\pi/\sqrt{1-\zeta^2}}$$

$$M_P = 25\% \text{ (given)}$$

$\therefore$ $$.25 = e^{-\zeta\pi/\sqrt{1-\zeta^2}}$$

from that, $$\zeta\pi/\sqrt{1-\zeta^2} = 1.386$$

$\therefore$ $$\boxed{\zeta = .404}$$

we know t_p is 2 sec which is given is the problem

$$t_p = \frac{\pi}{\omega_d} = 2$$

$\therefore$ $$\boxed{\omega_d = 1.57}$$

$$\omega_n = \frac{\omega_d}{\sqrt{1-\zeta^2}} = \frac{1.57}{\sqrt{1-.404^2}} = 1.72$$

$$\boxed{\omega_n = 1.72}$$

$\therefore$ $$K = \omega_n^2 = 2.95 \text{ N-m}$$

and $$k = \frac{2\zeta\omega_n}{K} = \frac{2\times.404\times1.72}{2.95} = .471 \text{ sec}$$

Result: The value of K is 2.96 N-m and the value of k is 4.71 sec.

EXAMPLE 3: The Fig. (3) shows a mechanical vibratory system. When step input is applied to the system, (where step input value = 2) the mass oscillates such that it has M_P of 9.5 and t_p = 2 sec. Determine M, B and K of the system. The displacement x is measured from the equilibrium position. (The final value is 0.1 and ζ = 0.6)

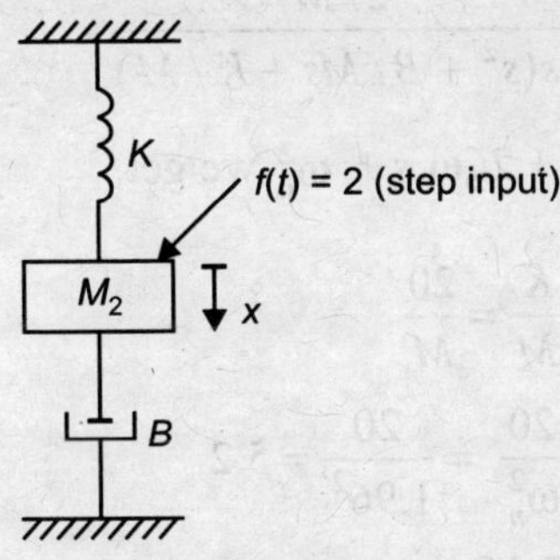

Fig. 3

CHAPTER 4

Solution: From the Fig. (3), we can derive the transfer function of the system,

$$f(t) = M\frac{d^2x}{dt^2} + B\frac{dx}{dt} + Kx$$

on taking Laplace transform

$$F(s) = X(s)[Ms^2 + Bs + K]$$

$$\therefore \quad \frac{X(s)}{F(s)} = \frac{1}{Ms^2 + Bs + K}$$

The input is ie $r(t) = 2$

$$\therefore \quad R(s) = 2/s$$

Hence,
$$X(s) = \frac{2}{s(Ms^2 + Bs + K)} \tag{1}$$

The steady state value of x is,

$$x(\infty) = \underset{s \to 0}{\text{Lt}}\ s\,X(s) = \underset{s \to 0}{\text{Lt}}\ s \cdot \frac{2}{s(Ms^2 + Bs + K)}$$

$$x(\infty) = \frac{2}{K} = 0.1 \text{ (Final value)}$$

$$\therefore \quad K = 20$$

Maximum peak over shoot is 9.5 and $\zeta = .6$ (given)

we know,
$$t_p = \frac{\pi}{\omega_d} = \frac{\pi}{\omega_n\sqrt{1-\zeta^2}} = \frac{\pi}{.8\omega_n}$$

$$t_p = 2 \text{ sec [given]}$$

$$\therefore \quad 2 = \frac{\pi}{.8\omega_n} \Rightarrow \omega_n = 1.96$$

$$\boxed{\omega_n = 1.96 \text{ rad/sec}}$$

The equation (1) can be written as by dividing by M in both numerator and denominator,

$$X(s) = \frac{2/M}{s(s^2 + B/Ms + K/M)} \tag{2}$$

Compare $s^2 + \frac{B}{M}s + \frac{K}{M}$ with $s^2 + 2\zeta\omega_n s + \omega_n^2$ we get,

$$\omega_n^2 = \frac{K}{M} = \frac{20}{M}$$

$$M = \frac{20}{\omega_n^2} = \frac{20}{1.96^2} = 5.2$$

$$2\zeta\omega_n = B/M$$

$$\therefore \qquad B = M \times 2 \times \zeta \times \omega_n = 2 \times .6 \times 1.96 \times 5.2$$

$$\boxed{B = 12.2}$$

Result: The value of

$$M = 5.2$$
$$B = 12.2$$
$$K = 20$$

EXAMPLE 4: For unity feedback system having open loop transfer function as $G(s) = K(s + 2)/s^2 (s^2 + 7s + 12)$. Determine (i) Type of system (iii) error constants K_P, K_v and K_a. (iii) Steady state error for step input.

Solution:

(i) Type of the system is 2

$$\text{(ii)}\quad K_P = \underset{s \to 0}{\text{Lt}}\ G(s)\,H(s) = \underset{s \to 0}{\text{Lt}}\ \frac{K(s+2)}{s^2(s^2+7s+12)}$$

$$\boxed{K_P = \infty}$$

$$K_V = \underset{s \to 0}{\text{Lt}}\ s \cdot G(s)\,H(s) = \underset{s \to 0}{\text{Lt}}\ s \cdot \frac{K(s+2)}{s^2(s^2+7s+12)}$$

$$\boxed{K_V = \infty}$$

$$K_a = \underset{s \to 0}{\text{Lt}}\ s^2\, G(s)H(s) = s^2 \cdot \frac{K(s+2)}{s^2(s^2+7s+12)}$$

$$\boxed{K_a = K/6}$$

(iii) The steady state error for unit step input

$$e_{ss} = \frac{1}{1+K_p} = \frac{1}{1+\infty} = 0$$

$$e_{ss} = 0 \quad \text{for unit step input}$$

Result: (i) Type of the system is 2

(ii) $K_p = \infty$, $K_v = \infty$, $K_a = K/6$

(iii) $e_{ss} = 0$ [For unit step]

EXAMPLE 5: Find out K_P, K_V and K_a for the unity feedback system $G(s) = 100/(1 + .5s)(1 + 2s)$.

Solution: The open loop transfer function is

$$G(s) = \frac{100}{(1+.5s)(1+2s)}$$

$$K_p = \underset{s \to 0}{\text{Lt}}\ G(s)\,H(s) = \underset{s \to 0}{Lt}\ \frac{100}{(1+.5s)(1+2s)}$$

$$\boxed{K_P = 100}$$

$$K_v = \underset{s \to 0}{Lt}\ s \cdot G(s)\, H(s)\ \underset{s \to 0}{Lt}\ s \cdot \frac{100}{(1+5s)\,(1+250}$$

$$\therefore \quad \boxed{K_V = 0}$$

$$K_a = \underset{s \to 0}{Lt}\ s^2 G(s)\, H(s) = \underset{s \to 0}{2t}\ s^2 - \frac{100}{(1+.5s)\,(1+2s)}$$

$$\therefore \quad \boxed{K_a = 0}$$

Result: $K_P = 100$, $K_v = 0$ and $K_a = 0$

EXAMPLE 6: Consider the unity feedback closed loop system where the transfer function is $G(s) = 25/s\,(s+5)$. Obtain t_p when subjected to unit step input.

Solution: The open loop transfer function is

$$\frac{C(s)}{R(s)} = \frac{\dfrac{25}{s(s+5)}}{1+\dfrac{25}{s(s+5)}} = \frac{25}{s(s+5)+25}$$

$$\frac{C(s)}{R(s)} = \frac{25}{s^2 + 5s + 25} \qquad (1)$$

Comparing the equation (1) with standard second order system characteristic equation we get,

$$\omega_n^2 = 25 \Rightarrow \boxed{\omega_n = 5}$$

$$2\zeta\omega_n = 5$$

$$2 \times \zeta(5) = 5$$

$$\zeta = \frac{5}{10} = .5$$

we know,

$$t_p = \frac{\pi}{\omega_n\sqrt{1-\zeta^2}} = \frac{\pi}{5\sqrt{1-.5^2}}$$

$$\boxed{t_p = .785} \text{ sec}$$

Result: $t_p = .785$ sec, when subjected to unit step input.

EXAMPLE 7: Consider the system given in Fig. (4) where $\zeta = .6$ and $\omega_n = .5$ rad/sec. Determine the rise time and peak time when the system is subjected to a unit step input.

Solution: We know,

$$t_r = \frac{\pi - \tan^{-1}\sqrt{\dfrac{1-\zeta^2}{\zeta}}}{\omega_n\sqrt{1-\zeta^2}}$$

$$t_r = \frac{\pi - \tan^{-1}\dfrac{\sqrt{1-.6^2}}{.6}}{.5\sqrt{1-.6^2}} = .55 \text{ sec}$$

$$\boxed{t_r = .55 \text{ sec}}$$

$$t_p = \frac{\pi}{\omega_d} = \frac{\pi}{\omega_n\sqrt{1-\zeta^2}} = \frac{\pi}{\omega_n\sqrt{1-.6^2}}$$

$$\boxed{t_p = .785} \text{ sec}$$

R(s) → (+/−) → $\frac{\omega_n^2}{s(s+2\zeta\omega_n)}$ → C(s)

Fig. 4

Result: $t_r = .55$ sec

$t_p = .785$ sec.

II. STABILITY

4.10 INTRODUCTION

The term stability refers to the stable working condition of a control system. Every working system is designed to be stable. In a stable system the response (or) output is predictable, finite and stable for a given input (or) changes in input (or) changes in system parameters.

The different definitions of the stability are the following

1. A system is stable if its output is bounded (finite) for any bounded (finite) input.
2. A system is asymptotically stable if in the absence of input, the output tends towards zero (or to the equilibrium state) irrespective of initial conditions.
3. A system is stable if for a bounded disturbing input signal the output vanishes ultimately as *t* approaches infinity.
4. A system is unstable if for a bounded disturbing input signal the output is of infinite amplitude (or) oscillatory.
5. For a bounded input signal, if the output has constant amplitude oscillations then the system may be stable (or) unstable under some limited constraints such a system is called limitedly stable.
6. If a system output is stable for all variations of its parameters then the system is called absolutely stable system.
7. If a system output is stable for a limited range of variations of its parameters, then the system is called conditionally stable system.

The concept of stability is very important to analyse and design the system. A system is said to be stable if its response cannot be made to increase indefinitely by the application of a bounded input excitation. If the output approaches towards infinite value for sufficiently large time, the system is said to be unstable.

A Linear Time Invariant (LTI) system is stable if

1. The system is excited by a bounded input, the output is bounded. (BIBO stability criteria)
2. In the absence of the input, the output tends towards zero (the equilibrium state of the system)

This is known as asymptotic stable.

Consider the transfer function.

$$\frac{C(s)}{R(s)} = \frac{a_m s^m + a_{m-1} s^{m-1} + \ldots a_1 s + a_0}{b_n s^n + b_{n-1} s^{n-1} + \ldots b_1 s + b_0} \tag{4.127}$$

The output is given by

$$C(t) = \int_0^{\infty} g(\tau)\, r(t - \tau)\, d\tau$$

where $g(\tau) = L^{-1}\, G(s)$ = Impulse response of the system. So, a system is said to be stable if the impulse response approaches zero for sufficiently large time. If the impulse response approaches infinity for sufficiently large time, the system is said to be unstable. If the impulse response approaches a constant value for sufficiently large time, the system is said to be marginally stable.

4.11 RESPONSE OF A SYSTEM

Let the closed loop transfer function (or overall transfer function)

$$\frac{C(s)}{R(s)} = M(s)$$

∴ The response (or) output in *s*-domain.

$$C(s) = M(s)\, R(s)$$

Let,

$$C(t) = L^{-1}\, [C(s)] \qquad \text{(Response in time domain)}$$

$$r(t) = L^{-1}\, [R(s)] \qquad \text{(Input in time domain)}$$

For an impulse input,

$$r(t) = \delta(t);$$

$$R(s) = L[\delta(t)] = 1$$

The impulse response of the system

$$L^{-1}\, [C(s)] = L^{-1}\, [M(S)\, R(s)]$$

$$= L^{-1}\, M(S) = m(t) \tag{4.128}$$

Hence the impulse response of a system is the inverse Laplace transform of the system transfer function. The advantage of impulse response is that the knowledge of impulse response can be used to determine the output of a system for any arbitrary input.

The output is *s*-domain is the product of transfer function and input is *s*-domain. The convolution theorem states that the inverse Laplace transform of the product of two s-domain functions is equal to the convolution of their time domain functions.

i.e., $$L^{-1}[M(s)R(s)] = \int_{-\infty}^{+\infty} m(\tau)r(t-\tau)d\tau$$

∴ The response of a system in time domain

$$C(t) = L^{-1}[M(s)\,R(s)] = \int_{-\infty}^{+\infty} m(\tau)\,r(t-\tau)\,d\tau \qquad (4.129)$$

From the Eq. (4.129) it can be concluded that the response for any arbitrary input is given by convolution of impulse response and the input.

4.12 BOUNDED-INPUT BOUNDED OUTPUT STABILITY (BIBO)

A linear relaxed system is said to have BIBO stability if every bounded (finite) input results in a bounded (finite) output. A test for BIBO stability can be obtained from convolution theorem.

The response of a system $C(t)$ for any input is given by convolution of the input and impulse response.

The response in time domain, $$C(t) = \int_{0}^{\infty} m(\tau)\,r(t-\tau)\,d\tau \qquad (4.310)$$

Note: A relaxed system is one in which the initial conditions are zero. Hence the limits of integration is from 0 to ∞.

If the input $r(t)$ is bounded then there exist a constant A, such that $|r(t)| \le A_1 < \infty$. The condition for bounded output for this bounded input condition can be derived as follows.

On taking the absolute value on both sides of Eq. (4.130), we get

$$|C(t)| = \left| \int_{0}^{\infty} m(\tau)\cdot r(t-\tau)\,d\tau \right| \qquad (4.131)$$

Since the absolute value of an integral is not greater then the integral of the absolute value of the integrand the Eq. (4.131) can be written as,

$$|C(t)| \le \int_{0}^{\infty} |\,m(\tau)\,r(t-\tau)\,d\tau\,|$$

$$\le \int_{0}^{\infty} |m(\tau)|\,r(t-\tau)\,d\tau$$

$$\le \int_{0}^{\infty} |m(\tau)|\,A_1\,d\tau \qquad [\because \text{For bounded input } |r(t-\tau) < A_1\,]$$

$$\le A_1 \int_{0}^{\infty} |\,m(\tau)\,|\,d\tau \qquad (4.132)$$

CHAPTER 4

It the output $C(t)$ is bounded then there exists a constant A_Q

Such that $$| C(t) | \leq A_2 < \infty$$

$$\therefore \quad A_1 \int_0^{\infty} | m(\tau) | \, d\tau \leq A_2 < \infty \tag{4.13.3}$$

The above condition is satisfied if

$$\int_0^{\infty} | m(\tau) | \, d\tau < \infty$$

Hence the bounded output,

$$\int_0^{\infty} | m(\tau) | \, d\tau < \infty$$

therefore we can conclude that a system with impulse response $m(\tau)$ is BIBO stable if and only if the impulse response is absolutely integrable i.e.,

$$\int_0^{\infty} | m(\tau) | \, d\tau \text{ is finite}$$

This means that area under the absolute value curve of the impulse response $m(\tau)$ evaluated from $t = 0$ to $t = \infty$ must be finite.

4.13 EFFECT OF LOCATION OF POLES ON STABILITY

The closed loop transfer function, $M(s)$ can be expressed as a ratio of two polynomials in s. The denominator polynomial of closed loop transfer function is called characteristic equation. The roots of characteristic equation are poles of closed loop transfer function.

For BIBO stability the integral of impulse response should be finite. Which implies that the impulse response should be finite as t tends to infinity. (The impulse response is the inverse Laplace transform of the transfer function). This requirement for stability can be linked to the location of roots of characteristic equation in the s-plane.

The closed loop transfer function $M(s)$ can be expressed as ratio of two polynomials in s as shown in equation (4.130)

$$M(s) = \frac{b_0 s^m + b_1 s^{m-1} + b_2 s^{m-2} + \ldots + b_{m-1} s + b_m}{a_0 s^n + a_1 s^{n-1} + a_2 s^{n-2} + \ldots a_{n-1} s + a_n} \tag{4.134}$$

$$= \frac{(s + Z_1)(s + Z_2)(s + Z_3) + \ldots (s + Z_m)}{(s + P_1)(s + P_2)(s + P_3) + \ldots (s + P_n)} \tag{4.135}$$

The roots of numerator polynomial $Z_1, Z_2 \ldots Z_n$ are zeros. The roots of denominator polynomial $P_1, P_2, \ldots P_n$ are poles. The denominator polynomial is the characteristic equation and so the poles are roots of characteristic equation.

By partial fraction expansion we can write,

$$M(s) = \frac{A_1}{s+P_1} + \frac{A_2}{s+P_2} + \frac{A_3}{s+P_3} + \ldots \frac{A_n}{s+P_n} \tag{4.136}$$

The poles, $P_1, P_2, P_3 \ldots P_m$ may be at origin or lying on imaginary axis or lying on right or left half s-plane. The impulse response is given by inverse Laplace transform of $M(s)$. The inverse Laplace transform of each term of $M(s)$ depends on the location of poles. The impulse response of various terms of $M(s)$ are shown in Table (4.2).

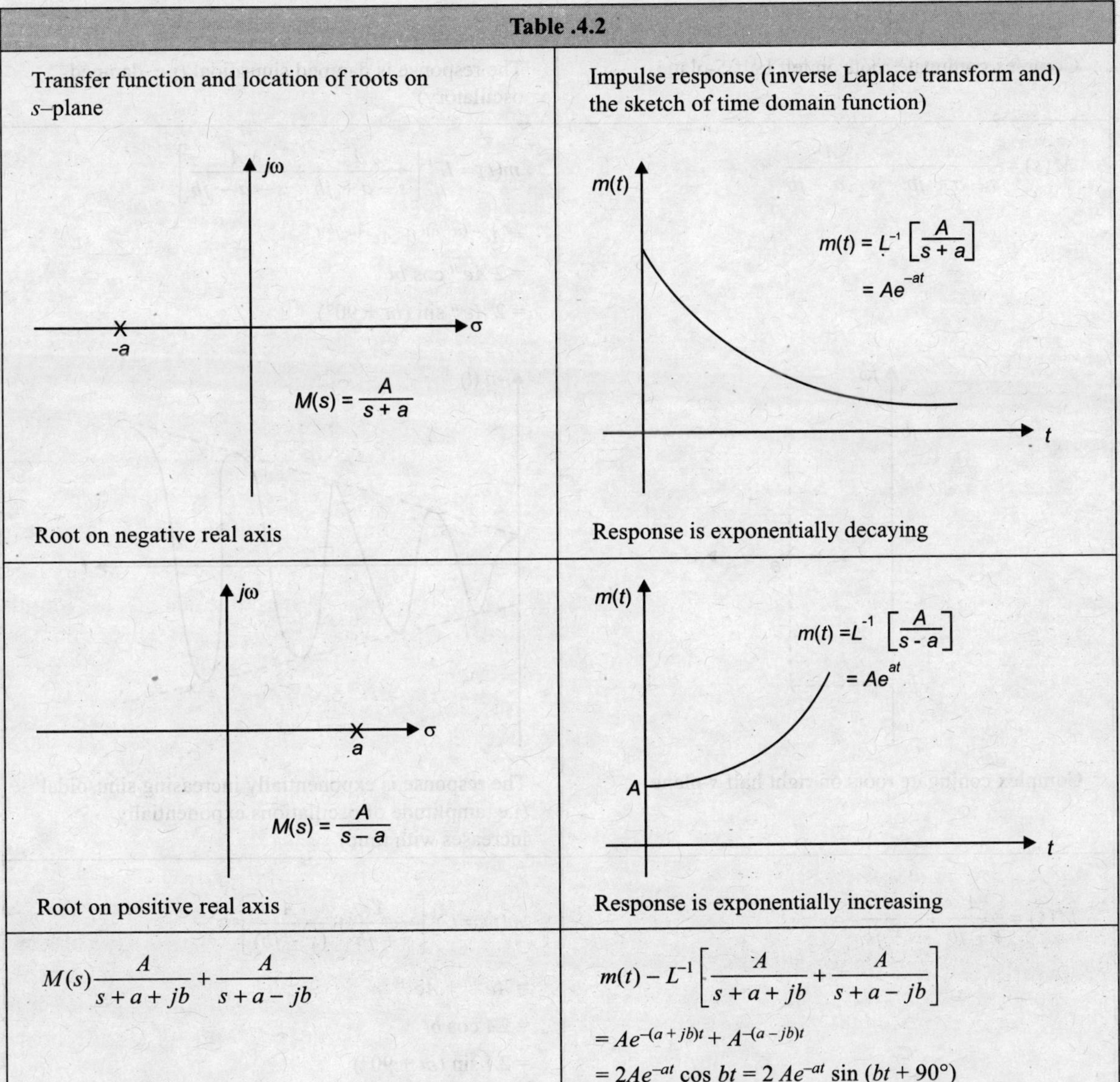

Table .4.2

Transfer function and location of roots on s–plane	Impulse response (inverse Laplace transform and) the sketch of time domain function)
jω, σ, -a; $M(s) = \frac{A}{s+a}$ Root on negative real axis	m(t); $m(t) = L^{-1}\left[\frac{A}{s+a}\right] = Ae^{-at}$; t Response is exponentially decaying
jω, σ, a; $M(s) = \frac{A}{s-a}$ Root on positive real axis	m(t); $m(t) = L^{-1}\left[\frac{A}{s-a}\right] = Ae^{at}$; A; t Response is exponentially increasing
$M(s) \frac{A}{s+a+jb} + \frac{A}{s+a-jb}$	$m(t) - L^{-1}\left[\frac{A}{s+a+jb} + \frac{A}{s+a-jb}\right]$ $= Ae^{-(a+jb)t} + A^{-(a-jb)t}$ $= 2Ae^{-at}\cos bt = 2\,Ae^{-at}\sin(bt + 90°)$

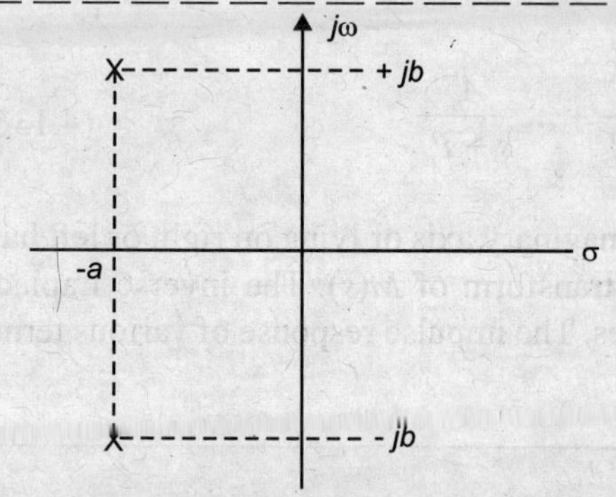

Complex conjugate roots on left half *S*-plane

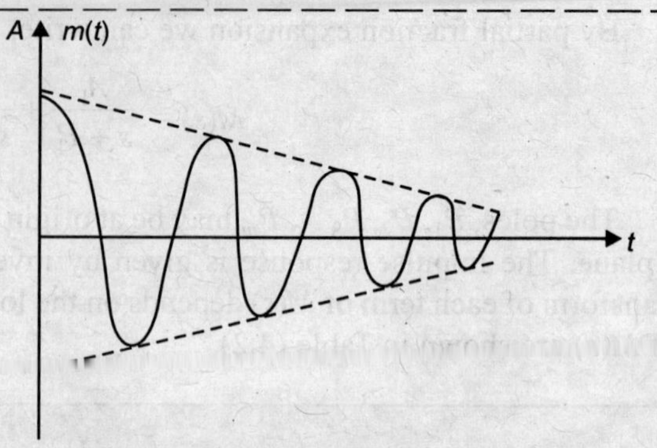

The response is damped sinusoidal (i.e. damped oscillatory)

$$M(s) = \frac{A}{s - a + jb} + \frac{A}{s - a - jb}$$

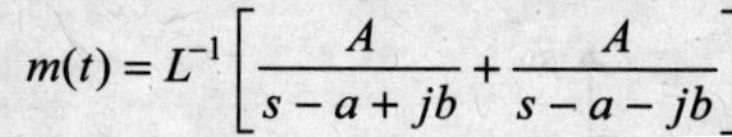

$$m(t) = L^{-1}\left[\frac{A}{s - a + jb} + \frac{A}{s - a - jb}\right]$$

$$= Ae^{-(a+jb)t} + Ae^{-(a-jb)t}$$

$$= 2\,Ae^{at}\cos bt$$

$$= 2\,Ae^{at}\sin(bt + 90°)$$

jω
+ jb
σ
a
- jb

Complex conjugate roots on right half *s*-plane

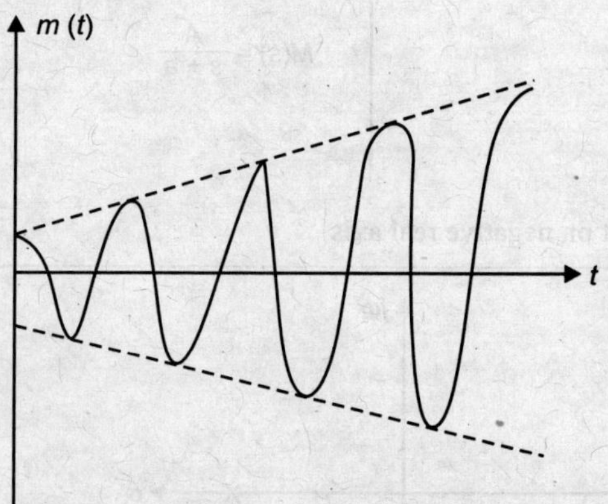

The response is exponentially increasing sinusoidal (i.e. amplitude of oscillations exponentially increases with time)

$$M(s) = \frac{A}{s + jb} + \frac{A}{s - jb}$$

$$m(t) = L^{-1}\left[\frac{A}{s + jb} + \frac{A}{(s - jb)}\right]$$

$$= Ae^{-jb} + Ae^{+jb}$$

$$= 2A\cos bt$$

$$= 2A\sin(bt + 90°)$$

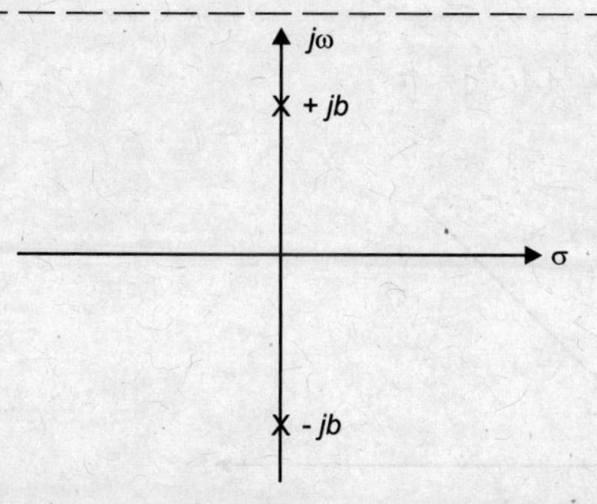 Single pair of roots on imaginary axis.	The response is oscillatory
$M(s) = \frac{A}{(s+jb)^2} + \frac{A}{(s-jb)^2}$ 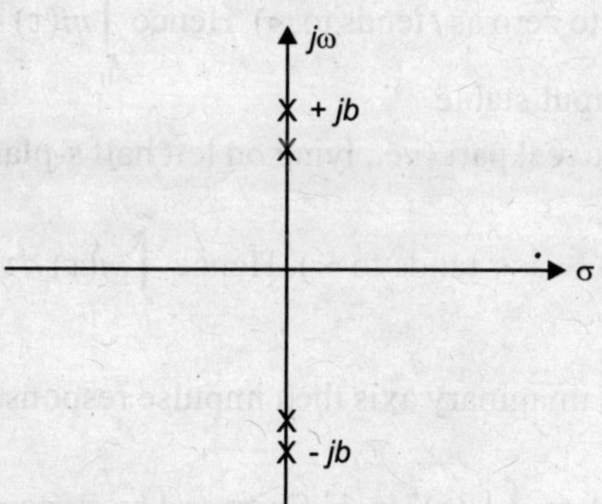 Double pair of roots on imaginary axis	$M(t) = L^{-1}\left[\frac{A}{(s+jb)^2} + \frac{A}{(s-jb)^2}\right]$ $= At\, e^{-jb} + At\, e^{jb}$ $= 2\,At \cos bt = 2At \sin(bt + 90°)$ 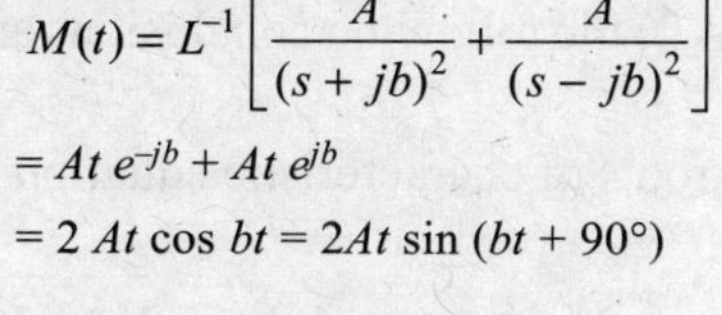 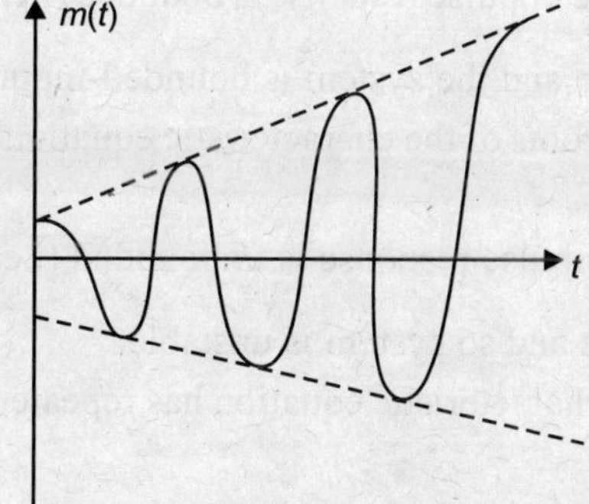 Response is Linearly increasing sinusoidal (i.e. amplitude of oscillations linearly increases with time)
$M(s) = \frac{A}{s}$ 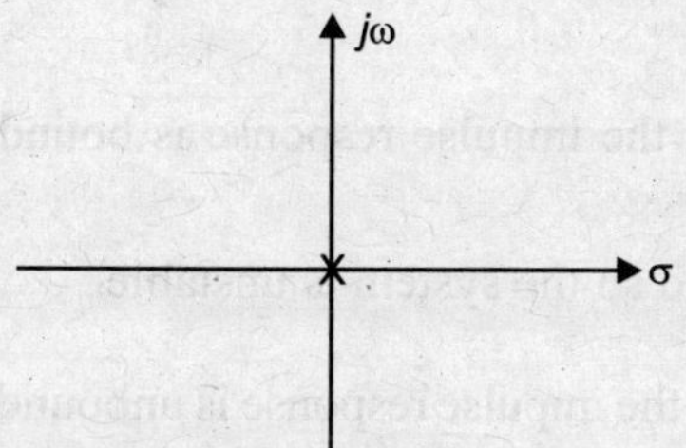 Single root at origin	$m(t) = L^{-1}[A/s] = A$ 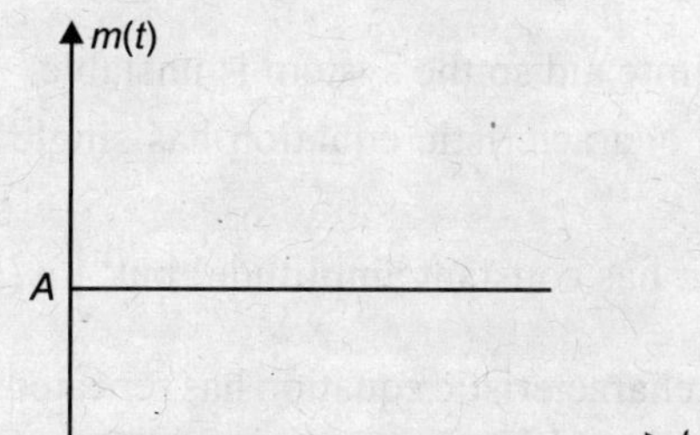 The response is constant

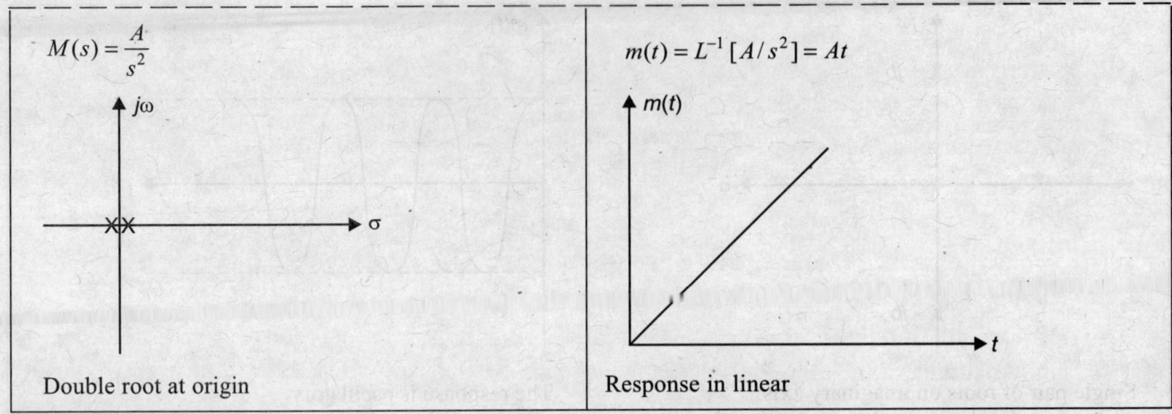

Double root at origin | Response in linear

From Table (4.2) the following conclusions are drawn based on the location of roots of characteristic equation.

1. If all the roots of characteristic equation have negative real parts (i.e. lying on left half s-plane) then the impulse response is bounded (i.e., it decreases to zero as t tends to ∞). Hence $\int_0^{\infty} m(\tau)\, d\tau$ is finite and the system is bounded-input bounded output stable.
2. If any roots of the characteristic equation has a positive real part (i.e., lying on left half s-plane) then impulse response is unbounded (i.e. increases to ∞ as t tends to ∞). Hence $\int_0^{\infty} m(\tau)\, d\tau$ is infinite and so system is unstable.
3. If the characteristic equation has repeated roots on the imaginary axis then impulse response is unbounded (i.e., it increases to ∞ as t tends to ∞). Hence $\int_0^{\infty} m(\tau)\, d\tau$ is finite and so system is unstable.
4. If one (or) more non repeated roots of the characteristic equation are lying on the imaginary axis, then impulse response is bounded (i.e., it has constant amplitude oscillations) but $\int_0^{\infty} m(\tau)\, d\tau$ is infinite aid so the system is unstable.
5. If the characteristic equation has single root at origin then the impulse response is bounded (i.e., it has constant amplitude) but $\int_0^{\infty} m(\tau)\, d\tau$ is infinite and so the system is unstable.
6. If the characteristic equation has repeated roots at origin then the impulse response is unbounded (i.e. it linearly increases to infinity as t tends to ∞) and sot the system is unstable.
7. In system with one (or) more repeated roots on imaginary axis (or) with single root at origin, the output is bounded for bounded inputs except for the inputs having poles matching the

system poles. These cases may be treated as acceptable (or) non-acceptable. Hence when the system has non repeated poles on imaginary axis (or) single pole at origin, it is referred as limitedly (or) marginally stable system.

In summary the following three points may be stated regarding the stability of the system depending on the location of roots of characteristic equation.

(i) If all the roots of characteristic equation has negative real parts, then the system is stable.

(ii) If any roots of the characteristic equation has a positive real part (or) if there is a repeated root on the imaginary axis then the system is unstable.

(iii) If the condition (i) is satisfied except for the presence of one (or) more non repeated roots on the imaginary axis, then the system is limitedly (or) marginally stable.

In order to ascertain the stability of a system, it is necessary to determine if any of the roots of the characteristic equation lie in the right half *s*-plane. The characteristic equation is given by the denominator polynomial of closed loop transfer function, [Eq. (3.4.1)]

Consider the *n*th order characteristic equation shown below.

$$a_0\, s^n + a_1\, s^{n-1} + a_2\, s^{n-2} + \ldots\, a_{n-1}\, s + a_n = 0 \qquad (4.137)$$

Let the roots of *n*th order characteristic equation [equation (3.4.3)] be $s = r, r_2 \ldots r_n$

These roots are functions of the coefficients $a_0, a_1, a_2 \ldots a_{n-1}, a_n$

Consider a second order polynomial,

$$\begin{aligned} a_0 s^2 + a_1 s + a_2 &= a_0 \left(s^2 \frac{a_1}{a_0} s + \frac{a_2}{a_0} \right) \\ &= a_0\,(s - r_1)\,(s - r_2) \\ &= a_0 s^2 - a_0\,(r_1 + r_2)s + a_0\, r_1 r_2 \end{aligned} \qquad (4.138)$$

Consider a third order polynomial

$$\begin{aligned} a_0 s^3 + a_1 s^2 + a_2 s + a_3 &= a_0 \left[s^3 + \frac{a_1}{a_0} s^2 + \frac{a_2}{a_0} s + \frac{a_3}{a_0} \right] \\ &= a_0(s - r_1)\,(s - r_2)\,(s - r_3) \\ &= a_0 s^3 - a_0(r_1 + r_2 + r_3)\, s^2 + a_0\,(r_1\, r_2 + r_1\, r_3 + r_2\, r_3)\, s - a_0\, r_1\, r_2\, r_3 \end{aligned} \qquad (4.139)$$

on extending this expansion to the *n*th order polynomial, we get,

$$\begin{aligned} a_0 s^n + a_1 s^{n-1} + \ldots\, a_{n-1}\, s + a_n = {} & a_0 s^n - a_0 \text{ (sum of all the roots) } s^{n-1} + \\ & a_0 \text{ (sum of the products of the roots taken 2 at a time) } s^{n-2} - \\ & a_0 \text{ (sum of the products of the roots taken 3 at a time) } s^{n-3} + \ldots + \\ & a_0\,(-1)^n \text{ (product of all the } n \text{ roots)} \end{aligned} \qquad (4.140)$$

If the roots of a polynomial are real and in the left half s-plane, then all r_1 in Eqs. (4.138) and (4.139) are real and negative. Therefore all polynomial coefficients are positive. This characteristic also applies to the general case of Eq. (4.140). If at least one root is in the right half of s-plane then some of the coefficients will be negative. Also, it can be observed that if all the roots are in the left half plane, no coefficient can be zero.

Since the characteristic polynomial coefficients are real, the complex roots should occur as conjugate pairs.

From Eq. (4.140) when polynomial coefficients are formed, the imaginary parts of roots products of roots will cancel.

Therefore, if all roots occur in the left half plane, (whether it is complex (or) real) then all coefficients of the general polynomial of Eq. (4.140) will be positive. Presence of a negative coefficient implies that there is atleast one root in the right half plane.

A zero coefficient indicates presence of complex–conjugate roots on the imaginary axis and/or one or more roots in the right half plane.

In summary the following conclusions can made about the coefficients of characteristic polynomial.

1. If all the coefficients are positive and if no coefficient is zero then all the roots are in the left half plane.
2. If any coefficient a_i is equal to zero then some of the roots may be one on the imaginary axis (or) on the right half plane.
3. If any coefficient a_i is negative then atleast one root is in the right half plane.

It can be concluded that the absence (or) negativeness of any of the coefficients of a characteristic polynomial indicates that the system is either unstable (or) at most marginally stable. Thus the necessary condition for stability of the system is that all the coefficients of its characteristic polynomial be positive. If any coefficient is zero/negative, we can immediately say that the system is unstable.

When all the coefficients are positive, the system is not necessarily stable, because there may be roots in the right half plane and/or on the imaginary axis. In order for all the roots to have negative real parts, it is necessary but not sufficient that all of the coefficients of characteristic equation be positive. [For example, consider the characteristic polynomial with all positive coefficients $s^3 + s^2 + 2s + 8 = 0$. The characteristic polynomial can be written as

$$(s^3 + s^2 + 2s + 8) = (s + 2)\left(s - \frac{1}{2} - j\frac{\sqrt{15}}{2}\right)\left(s - \frac{1}{2} + j\frac{\sqrt{15}}{2}\right) = 0$$

The coefficients of the polynomial are all positive but two roots are on right half plane and so the system is unstable.

4.14 ROUTH HURWITZ CRITERION

The closed loop transfer function of a system is a ratio of two polynomials in s. The denominator polynomial of closed loop transfer function is called the characteristic equation of the system.

The roots of the characteristic equation of a stable system should lie on the left of s-plane. Hence the roots should have negative real parts. The Routh–Hurwitz stability criterion is an analytical procedure for determining whether all the roots of a polynomial have negative real parts (or) not.

The first step in analysing the stability of a system is to examine its characteristic equation. The necessary condition for stability is that all the coefficients of the polynomial be positive. If some of the coefficients are zero (or) negative it can be concluded that the system is not stable.

When all the coefficients are positive, the system is not necessarily stable Even though the coefficients are positive, some of the roots may lie on the right half of s-plane (or) on the imaginary axis. In order for all the roots to have negative real parts, it is necessary but not sufficient that all of the coeffcients of the characteristic equation be positive. If all the coeffcients of the characteristic equation are positive, then the system may be stable and one should proceed further no examine the sufficient conditions of stability.

The Hurwitz criterion is in terms of determinants and Routh criterion is in terms of array formulation.

The Routh stability criterion is based on ordering the coefficients of the characteristic equation,

$$a_0 s^n + a_1 s^{n-1} + a_2 s^{n-2} + \ldots + a_{n-1} s + a_n = 0$$

where $a_0 > 0$ Routh array as shown in below.

$$\begin{array}{llllllll} s^n & : & & a_0 & a_2 & a_4 & a_6 & a_8 \quad \ldots \\ s^{n-1} & : & a_1 & a_3 & a_5 & as_7 & a_9 & \ldots \\ s^{n-2} & : & & b_0 & b_1 & b_2 & b_3 & b_4 \quad \ldots \\ s^{n-1} & : & & \cdots\cdots\cdots & & & & \\ \vdots & \vdots & & \vdots & \vdots & \vdots & \vdots & \vdots \\ s^1 & : & & \vdots & \vdots & \vdots & \vdots & \vdots \\ s^0 & & & \vdots & \vdots & \vdots & \vdots & \vdots \end{array}$$

Definiton of routh stability criterion **"The necessary and sufficient condition for stability is that all of the elements in the first column of the routh array be positive. If this condition is not met, the system is unstable and the number of sign changes in the elements of the first column of the Rough array corresponds to the number of roots of the characteristic equation in the Right half of the *s*-plane"**.

Note: If the order of sign column elements is +, +, –, + and +. Then + to – is considered as one sign change and – to + as another sign change.

4.14.1 Construction of Routh Array

Let the characteristic polynomial be

$$a_0 s^n + a_1 s^{n-1} + a_2 s^{n-2} + \ldots a_{n-1} s^1 + a_n s^0$$

The coefficients of the polynomial are in two rows as shown below

$$s^n : a_0 \; a_2 \; a_4 \; a_6 \; \ldots$$
$$s^{n-1} : a_1 \; a_3 \; a_5 \; a_7 \; \ldots$$

If n is even then s^n row is formed by coefficients of even order terms and s^{n-1} row is formed by coefficients of odd order terms (i.e. coefficients of odd power of s).

If n is odd, then s^n row is formed by coefficients of even order terms (i.e., coefficients of even powers of s).

The other rows of routh array upto s^0 row can be formed by the following procedure. Each row of routh array is constructed by using the elements of previous two rows.

CHAPTER 4

Consider two consecutive rows of routh array as shown below.

$$s^{n-x} : x_0\ x_1\ x_2\ x_3\ x_4 \ldots$$
$$s^{n-x-1} : y_0\ y_1\ y_2\ y_3\ y_4 \ldots$$

Let the next row be, $s^{n-x-2} : z_0\ z_1\ z_2\ z_3 \ldots$

The elements of s^{n-x-2} row are given by,

$$z_0 = (-1)\begin{vmatrix} x_0 & x_1 \\ y_0 & y_1 \end{vmatrix} = \frac{y_0x_1 - y_1x_0}{y_0}$$

$$z_1 = (-1)\begin{vmatrix} x_0 & x_2 \\ y_0 & y_2 \end{vmatrix} = \frac{y_0x_2 - x_0y_2}{y_0}$$

$$z_2 = (-1)\begin{vmatrix} x_0 & x_3 \\ y_0 & y_3 \end{vmatrix} = \frac{y_0x_3 - y_3x_0}{y_0}$$

The elements $z_0, z_1, z_2, z_3 \ldots$ computed until an element equals to zero for all possible computations as shown above.

In the process of constructing Routh array the missing terms are considered as zeros. Also, all the elements of any row can be multiplied or divided by a positive constant for simplification.

For constructing Routh array the following three cases are important.

Case I: Normal Routh array (non-zero elements in the first column of routh array)

Case II: A row of all zeros

Case III: First element of a row is zero but some or other elements are not zero

Case I: Normal routh array

In this case there is no difficulty in forming Routh array. The routh array can be constructed as explained above. The sign changes are not noted to find the number of roots lying on the right half of s-plane and the stability of the system can be estimated.

In this case if there is no sign change in the first column of routh array then all the roots are lying on left half of s-plane and the system is stable.

If there is sign change in the first column of routh array, then the system is unstable and the number of roots lying on the right half of *s*-plane is equal to number of sign changes the remaining roots are lying on the left half of *s*-plane.

Case II: A row of all zeros

An all zero row indicates the existence of an even polynomial as a factor of the given characteristic equation.

In an even polynomial the exponents of s are even integers (or) zero only. This even polynomial factor is called the auxiliary polynomial. The coefficients of the auxiliary polynomial will always be the elements of the row directly above the row of zeros in the array.

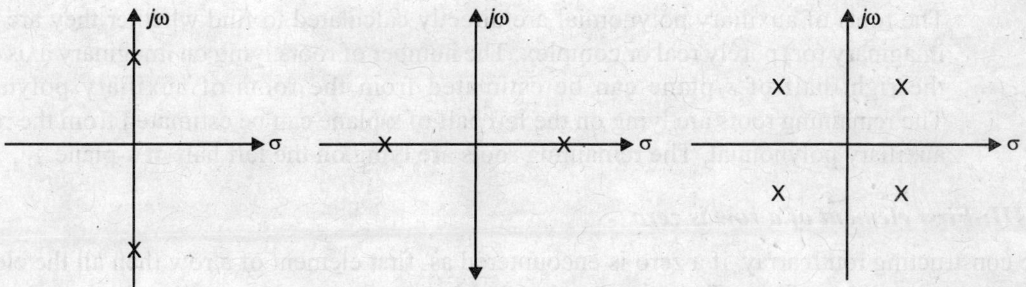

Fig. 4.19: The roots of an even polynomial.

Method 1

1. Determine the auxiliary polynomial, $A(s)$
2. Differentiate the auxiliary polynomial with respect to s, get d $A(s)$ ds.
3. The row of zeros is replaced with coefficients of $dA(s)/ds$.
4. Continue the construction of the array in the usual manner (as that of case–I) and the array is interpreted as follows.
   (a) If there are sign changes in the first column of routh array then the system is unstable. The number of roots lying on right half of s-plane is equal to number of sign charges. The number of roots on imaginary axis can be estimated from the roots of auxiliary polynomial. The remaining roots are lying on the left half of s-plane.
   (b) If there are sign changes in the first column of routh array then the all zeros row indicate the existence of purely imaginary roots and so the system is limitedly or marginally stable. The roots of auxiliary equation lies on imaginary axis and the remaining roots lies on left half of s-plane.

Method 2:

1. Determine the auxiliary polynomial, $A(s)$.
2. Divide the characteristic equation by auxiliary polynomial.
3. Construct routh array using the coefficients of quotient polynomial.
4. The array is interpreted as follows.
   (a) If there are sign changes in the first column of routh array of quotient polynomial then the system is unstable. The number of roots of quotient polynomial lying on right half of s-plane is given by number of sign changes in first column of routh array

   The roots of auxiliary polynomial are directly calculated to find whether they are purely imaginary or purely real or complex.

   The total number of roots on right half of s-plane is given by the sum of number of sign changes and the number of roots of auxiliary polynomial with positive real part. The number of roots on imaginary axis can be estimated from the roots of auxiliary polynomial. The remaining roots are lying on the left half of s-plane.

   (b) If there is no sign change in the first column of routh array of quotient polynomial then the system is limitedly (or) marginally stable. Since there is no sign change all the roots of quotient polynomial are lying on the left half of s-plane.

The roots of auxiliary polynomial are directly calculated to find whether they are purely imaginary (or) purely real or complex. The number of roots lying on imaginary axis and on the right half of s-plane can be estimated from the roots of auxiliary polynomial. The remaining roots are lying on the left half of s-plane can be estimated from the roots of auxiliary polynomial. The remaining roots are lying on the left half of s-plane.

Case III: First element of a row is zero

While constructing routh array, if a zero is encountered as first element of a row then all the elements of the next row will be infinite. To overcome this problem let $0 - \in$ and complete the construction of array in the usual way (as that of case)

Finally let $\in \to 0$ and determine the values of the elements of the array which are functions of $\in$. The resultant array is interpreted as follows.

Note: If all the elements of a row are zeros then the solution is attempted by considering the polynomial as case II polynomial. Even if there is a single element zero one's row. It is considered as a row of all zeros.

(a) If there is no sign change in first column of routh array and if there is no row with all zeros, then all the roots are lying on left half of s-plane and the system is stable.

(b) if there are sign changes in first column of routh array and there is no row with all zeros, then some of the roots are lying on the right half of s-plane is equal to number of sign changes and the remaining roots are lying on the left half of s-plane.

(c) If there is a row of all zeros after leting $\in \to 0$, then there is a possibility of roots on imaginary axis. Determine the auxiliary polynomial and divide the characteristic equation by auxiliary polynomial to eliminate the imaginary roots. The routh array is constructed using the coefficients of quotient polynomial and divide the characteristic equation by auxiliary polynomial to eliminate the imaginary roots. The routh array is constructed using the coefficients of quotient polynomial and the characteristic equation is interpreted as explained in method-2 of Case II: polynomial

Statement of Routh–Hurwitz Criterion: The system is stable if and only if all the elements in the first column have the same algebraic sign. If all elements are not of the same sign then the number of sign changes of the elements in first column equals the number of roots of the characteristic equation in the right of the s-plane.

EXAMPLE 4.25: Using Routh criterion, determine the stability of the system represented by the characteristic equation, $s^4 + 8s^3 + 18s^2 + 16s + 5 = 0$. Comment on the location of the roots of characteristic equation.

Solution: The characteristic equation of the system is $s^4 + 8s^3 + 18s^2 + 16s + 5 = 0$. The given characteristic equation is 4th order equation and so it has 4 roots. Since the highest power of s is even number, from the first row of routh array using the coefficients of even powers of s and from the second row using the coefficients of odd powers of s.

$$
\begin{array}{lllll}
s^4: & 1 & 18 & 5 & \ldots\text{Row–1} \\
s^3: & 8 & 16 & & \ldots\text{Row–2}
\end{array}
$$

The elements of s^3 row can be divided by 8 to simplify the computations.

$$\begin{array}{lllll} s^4: & 1 & 18 & 5 & \text{... Row 1} \\ s^3: & 1 & 2 & & \text{... Row 2} \\ s^2: & 16 & 5 & & \text{... Row 3} \\ s^1: & 1.7 & & & \text{... Row 4} \\ s^0: & 5 & & & \text{... Row 5} \end{array}$$

Column – 1

$s^2 : \dfrac{1\times 18 - 2\times 1}{1} \qquad \dfrac{1\times 5 - 0\times 1}{1}$

$s^2 : 16 \qquad 5$

$s^1 : \dfrac{16\times 2 - 5\times 1}{16}$

$s^1 : 1.6875 \approx 1.7$

$s^0 : 1.7\times 5 - 0\times 16/1.7$

$s^0 : 5$

on examining the elements of first column of routh array it is observed that all the elements are positive and there is no sign change. Hence all the roots are lying on the left half of s-plane and system is stable.

Result

1. Stable system
2. All the four roots are lying on the left of s-plane

EXAMPLE 4.26: Construct Routh array and determine the stability of the system whose characteristic equation is $s^6 + 2s^5 + 8s^4 + 12s^3 + 20s^2 + 16s + 16 = 0$. Also determine the number of roots lying on right half of s-plane, left half of s-plane and on imaginary axis.

Solution: The characteristic equation of the system is $s^6 + 2s^5 + 8s^4 + 12s^3 + 20s^2 + 16s + 16 = 0$. The given characteristic polynomial is 6th order equation and so it has 6 roots. Since the highest power of s is even number, form the first row of routh array using the coefficients of even powers of s and form the second row using the coefficient of odd powers of s.

$$\begin{array}{llllll} s^6: & 1 & 8 & 20 & 16 & \text{...Row 1} \\ s^5: & 2 & 12 & 16 & & \text{...Row 2} \end{array}$$

The elements s^5 can be divided by 2 to simplify the calculations.

$$\begin{array}{llllll} s^6: & 1 & 8 & 20 & 16 & \text{... Row 1} \\ s^5: & 1 & 6 & 8 & & \text{... Row 2} \\ s^4: & 1 & 6 & 8 & & \text{... Row 3} \\ s^3: & 0 & 0 & & & \text{... Row 4} \\ s^3: & 1 & 3 & & & \text{... Row 5} \\ s^2: & 3 & 8 & & & \text{... Row 6} \\ s^1: & .33 & & & & \text{... Row 7} \\ s^0: & 8 & & & & \text{... Row 8} \end{array}$$

Column – 1

s^4 :	$\frac{1\times 8-6\times 1}{1}=2$	$\frac{1\times 20-8\times 1}{1}=12$	$\frac{1\times 16-0\times 1}{1}=16$
Divided by 2			
s^4 :	1	6	8
s^3 :	$\frac{1\times 6-6\times 1}{1}$	$\frac{1\times 8-8\times 1}{1}$	
s^3 :	0	0	

The auxiliary equation is

$$A = s^4 + 6s^2 + 8 \quad \text{(This is } s^4 \text{ row)}$$

On differentiating A with respect to s we get

$$\frac{dA}{dS} = 4s^3 + 12s$$

The coefficients of $\frac{dA}{dS}$ are used to form s^3 row. $\left(\text{Form } s^3 \text{ row again using } \frac{dA}{dS}\right)$

s^3 :	4	12
Divide by 4		
s^3 :	1	3
s^2 :	$\frac{1\times 6-3\times 1}{1}$	$\frac{1\times 8-0\times 1}{1}$
s^2 :	3	8
s^1 :	$\frac{3\times 3-8\times 1}{3}$	
s^1:	0.33	
s^0 :	$\frac{0.33\times 8-0\times 3}{0.33}$	
s^0 :	8	

On examining the elements of 1st column of routh array it is observed that there is no sign change the row with all zeros indicate the possibility of roots on imaginary axis. Hence the system is limitedly (or) marginally stable.

The auxiliary polynomial is

$$s^4 + 6s^2 + 8 = 0$$

Let $\qquad s^2 = x$

$$\therefore \quad x^2 + 6x + 8 = 0$$

The roots of quadratic are,

$$x = \frac{-6 \pm \sqrt{6^2 - 4 \times 8}}{2} = -3 \pm 1 = -2 \text{ (or)} -4.$$

The roots of auxiliary polynomial is,

$$s = \pm\sqrt{x} = \pm\sqrt{-2} \quad \text{and} \quad \pm\sqrt{-4}$$

$$= +j\sqrt{2}, -j\sqrt{2}, +j2 \quad \text{and} \quad -j2$$

The roots of auxiliary polynomial are also roots of characteristic equation. Hence 4 roots are lying on imaginary axis and the remaining two roots are lying on the left half of s-plane.

Result:

1. The system is limitedly (or) marginally stable.
2. Four roots are lying on imaginary axis and the remaining two roots are lying on the left half of s-plane.

EXAMPLE 4.27: Construct Routh array and determine the stability of the system represented by the characteristic equation $s^5 + s^4 + 2s^3 + 2s^2 + 3s + 5 = 0$. comment on the location of the roots of characteristic equation.

Solution: The characteristic equation of the system is

$$s^5 + s^4 + 2s^3 + 2s^2 + 3s + 5 = 0$$

The given characteristic polynomial is 5th order equation and so it has to roots. Since the highest power of s is odd number, form the first row of routh array using the coefficients of odd powers of s and form the second row using the coefficients of even powers of s.

s^5 :	1	2	3	... Row 1
s^4 :	1	2	5	... Row 2
s^3 :	$\in$	-2		... Row 3
s^2 :	$\dfrac{2\in+2}{\varepsilon}$	5		... Row 4
s^1 :	$-\left(\dfrac{5\in^2+4\in+4}{2\in+2}\right)$			... Row 5
s^0 :	5			... Row 6

On letting $\in \to 0$, we get

s^5 :	1	2	3	... Row 1
s^4 :	1	2	5	... Row 2
s^3 :	0	-2		... Row 3
s^2 :	∞	5		... Row 4
s^1 :	-2			... Row 5
s^0 :	5			... Row 6
	Column – 1			

On observing the elements of first column of routh array, it is found that there are two sign changes. Hence two roots are lying on right half of s-plane and the system is unstable. The remaining three roots are lying on the left half of *s*-plane.

s^3 :	$\frac{1\times 2.2\times 1}{1}$	$\frac{1\times 3-5\times 1}{1}$
s^3 :	0	– 2
Replace 0 by ∈		
s^3 :	∈	–2
s^2 :	$\frac{\in\times 2-(-2\times 1)}{\in}$	$\frac{\in\times 5-0\times 1}{\in}$
s^2 :	$\frac{2\in+2}{\in}$	5
s^1 :	$\frac{\frac{2\in+2}{\in}\times(-2)-(5\times\in)}{2\frac{\in+2}{\in}}$	
s^1 :	$\frac{-(5\in^2+4\in+4)}{2\in+2}$	
s^0 :	$\frac{\frac{-(5\in^2+4\in+4)}{2\in+2}\times 5-0\times\frac{2\in+2}{\in}}{\frac{-(5\in^2+4\in+4)}{2\in+2}}$	
s^0 :	5	

Result:

1. The system is unstable.
2. Two roots are lying on the right of *s*-plane and three roots are lying on the left half of *s*-plane.

EXAMPLE 4.28: By routh stability criterion determine the stability of the system represented by the characteristic equation $9s^5 - 20s^4 + 10s^3 - s^2 - 9s - 10 = 0$. Comment on the location of roots of characteristic equation.

Solution: The characteristic polynomial of the system is $9s^5 - 20s^4 + 10s^3 - s^2 - 9s - 10 = 0$ characteristic polynomial, it is found that some of the coefficients are negative and so some roots will lie on the right of *s*-plane. Hence the system is unstable. The routh array can be constructed to find the number of roots lying on right half of *s*-plane.

$$9s^5 - 20s^4 + 10s^3 - s^2 - 9s - 10 = 0$$

The given characteristic polynomial is 5th order equation and so it has 5 roots. Since the highest power of s is odd number, form the first row of routh array using the coefficients of odd powers of s and form the second row using the coefficients of even powers of s.

	Column – 1			
s^5 :	9	10	– 9	... Row 1
s^4 :	– 20	– 1	– 10	... Row 2
s^3 :	9.55	– 13.5		... Row 3
s^2 :	– 29.3	– 10		... Row 4
s^1 :	– 16.8			... Row 5
s^0 :	– 10			... Row6

s^3 :	$\dfrac{-20\times 10-(-1\times 9)}{-20}$	$\dfrac{-20\times(-9)-(-10)\times 9}{-20}$
s^3 :	9.55	– 13.5
s^2 :	$\dfrac{9.55\times(-1)-(-13.5)\times(-20)}{9.55}$	$\dfrac{9.55\times(-10)}{9.55}$
s^2 :	– 29.3	– 10
s^1 :	$\dfrac{-29.3\times(-13.5)-(-10)\times 9.55}{-29.3}$	
s^1 :	– 16.8	
s^0 :	$\dfrac{-16.8\times(-10)}{-16.8}$	
s^0 :	– 10	

By examining the elements of 1st column of routh array it is observed that there are three sign changes and so three roots are lying on the right half of s-plane and the remaining two roots are lying on the left half of s-plane.

Result:

1. The system is unstable
2. Three roots are lying on the right half and two roots are lying on the left half of s-plane.

EXAMPLE 4.29: The characteristic polynomial of a system is $s^7 + 9s^6 + 24s^5 + 24s^4 + 24s^3 + 24s^2 + 23s + 15 = 0$. Determine the location of roots on s-plane and hence the stability of the system.

Solution: The characteristic equation is $s^7 + 9s^6 + 24s^5 + 24s^4 + 24s^3 + 24s^2 + 23s + 15 = 0$. The given characteristic polynomial is 7th order equation and so it has 7 roots. Since the highest power of s is odd number, form the first row of array using the coefficients of odd powers of s and form the second row using the coefficients of even power of s as shown below.

s^7 :	1	24	24	23	... Row 1
s^6 :	9	24	24	15	... Row 2

Divide s^6 row by 3 to simplify the computations.

s^7 :	1	24	24	23	... Row 1
s^6 :	3	8	8	5	... Row 2
s^5 :	1	1	1		... Row 3
s^4 :	1	1	1		... Row 4
s^3 :	0	0			... Row 5
s^3 :	2	1			... Row 5
s^2 :	0.5	1			... Row 6
s^1 :	− 3				... Row 7
s^0 :	1				... Row 8

Column – 1

s^5 :	$\frac{3\times24-8\times1}{3}$	$\frac{3\times24-8\times1}{3}$	$\frac{3\times23-5\times1}{3}$
s^5 :	21.33	21.33	21.33
Divide by 21.33			
s^5 :	1	1	1
s^4 :	$\frac{1\times8-1\times3}{1}$	$\frac{1\times8-1\times3}{1}$	$\frac{1\times5-0\times3}{1}$
s^4 :	5	5	5
Divide by 5			
s^4 :	1	1	1
s^3 :	$\frac{1\times1-1\times1}{1}$	$\frac{1\times1-1\times1}{1}$	
s^3 :	0	0	

The auxiliary polynomial is

$$A = s^4 + s^2 + 1$$

Differentiate A with respect to s.

$$\frac{dA}{dS} = 4s^3 + 2s$$

Divided by 2		
s^3 :	4	2
s^3 :	2	1
s^2 :	$\frac{2\times1-1\times1}{2}$	$\frac{2\times1-0\times1}{2}$
s^2 :	0.5	1
s^1 :	$\frac{0.5\times1-1\times2}{0.5}$	
s^1 :	– 3	
s^0 :	$\frac{-3\times1}{-3}$	
s^0 :	1	

On examining the first column elements of routh array it is found that there are two sign changes. Hence two roots are lying on the right half of s-plane and so the system is unstable the row of all zeros indicate the possibility of roots an imaginary axis. This can be tested by evaluating the roots of auxiliary polynomial.

The auxiliary equation is $s^4 + s^2 + 1 = 0$.

Put $s^2 = x$ in the auxiliary equation

$$\therefore \quad s^4 + s^2 + 1 = x^2 + x + 1 = 0$$

The roots of quadratic are,

$$x = \frac{-\pm\sqrt{1-4}}{2} = -\frac{1}{2} \pm j\frac{\sqrt{3}}{2}$$

$$= 1\angle 120° \text{ or } 1\angle -120°$$

But

$$s^2 = x \therefore s = \pm\sqrt{x} = \pm\sqrt{1\angle 120°} \text{ (or) } \pm\sqrt{1\angle -120°}$$

$$= \pm\sqrt{1\angle 120°/2} \text{ (or) } \pm\sqrt{1\angle -120°/2}$$

$$= \pm\angle 60° \text{ (or) } \pm -60°$$

$$= \pm(0.5 + j0.866) \text{ (or) } \pm(0.5 - j0.866)$$

The roots of auxiliary polynomial are lying on the right half of s-plane and the remaining two on the left half of s-plane the roots of auxiliary equation one also the roots of characteristic polynomial. The two roots lying on the right half of s-plane are indicated by two sign changes in the first column of routh array. The remaining five roots are lying on the left half of s-plane. No roots are lying on imaginary axis. So the system in unstable.

Result:

1. The system is unstable.
2. Two roots are lying on right half of s-plane and five roots are lying on left half of s-plane.

CHAPTER 4

EXAMPLE 4.30: The characteristic polynomial of a system is $s^7 + 5s^6 + 9s^4 + 4s^3 + 20s^2 + 36s + 36 = 0$. Determine the location of roots on the s-plane and hence the stability of the system.

Solution: The characteristic equation is

$$s^7 + 5s^6 + 9s^5 + 9s^4 + 4s^3 + 20s^2 + 36s + 36 = 0$$

The given characteristic polynomial is 7th order equation and so it has 7 roots. Since the highest power of is odd number, form the first row of array using the coefficients of odd powers of s and form the second row of array using the coefficients of even powers of s as shown below.

s^7 :	1	9	4	36	... Row 1
s^6 :	5	9	20	36	... Row 2

Divide s^6 row by 5 to simplify the computations.

s^7 :	1	9	4	36	... Row 1
s^6 :	1	1.8	4	7.2	... Row 2
s^5 :	1	0	4		... Row 3
s^4 :	1	0	4		... Row 4
s^3 :	0	0			... Row 5

The row of all zeros indicate the existence of even polynomial, which is also the auxiliary polynomial. the auxiliary polynomial is $s^4 + 4 = 0$. Divide the characteristic equation by auxiliary equation to get the quotient polynomial.

The characteristic equation can be expressed as a product of quotient polynomial and auxiliary equation.

$$\therefore \quad s^7 + 5s^6 + 9s^5 + 9s^4 + 4s^3 + 20s^2 + 36s + 36 = (s^4 + 4)\,(s^3 + 5s^2 + 9s + 9) = 0$$

The routh array is constructed for quotient polynomial as shown below.

	Column – 1	
s^3 :	1	9
s^2 :	5	9
s^1 :	7.2	
s^0 :	9	

s^5 :	$\frac{1\times 9 - 1.8\times 1}{1}$	$\frac{1\times 4 - 4\times 1}{1}$	$\frac{1\times 36 - 7.2\times 1}{1}$
s^5 :	7.2	0	28.8
Divide by 7.2			
s^5 :	1	0	4
s^4 :	$\frac{1\times 1.8 - 0\times 1}{1}$	$\frac{1\times 4 - 4\times 1}{1}$	$\frac{1\times 7.2 - 0\times 1}{1}$
s^4 :	1.8	0	7.2

Divide by 1.8			
s^4 :	1.8	0	7.2
s^4 :	1	0	4
s^3 :	$\dfrac{1\times0-0\times1}{1}$	$\dfrac{1\times4-4\times1}{1}$	
s^3 :	0	0	
s^1 :	$\dfrac{5\times9-9\times1}{5}$		
s^1 :	7.2		
s^0 :	$\dfrac{7.2\times9-0\times5}{7.2}$		
s^0 :	9		

There is no sign change in the elements of first column of routh array of quotient polynomial. Hence all the roots of quotient polynomial are lying on the left half of s-plane.

To determine the stability the roots of auxiliary polynomial should be evaluated.

The auxiliary polynomial is

$$s^4 + 4 = 0$$

Put $s^2 = x$ in the auxiliary equation.

$$\therefore \quad s^4 + 4 = x^2 + 4 = 0$$

$$\therefore \quad x^2 = -4$$

$$x = \pm\sqrt{-4} = \pm j2 - 2\angle 90^\circ \text{ (or) } 2\angle -90^\circ$$

But, $s = \pm\sqrt{x}$,

$$= \pm\sqrt{2\angle 90^\circ} \text{ (or) } \pm\sqrt{2\angle -90^\circ}$$

$$= \pm\sqrt{2}\angle 90^\circ/2 \text{ (or) } \pm\sqrt{2}\angle -90^\circ/2$$

$$= \pm\sqrt{2}\angle 45^\circ \text{ (or) } \pm\sqrt{2}\angle -45^\circ$$

$$= \pm(1 + j1) \text{ (or) } \pm(1 - j1)$$

The roots of auxiliary equation are complex and has quadrantal symmetry. Two roots of auxiliary equation are lying on the right half of s-plane and the other two on the left half of s-plane.

The roots of characteristic equation are given by roots of quotient polynomial and auxiliary polynomial. Hence we can conclude that two roots of characteristic equation are lying on the right half of s-plane and so the system is unstable. The remaining five roots are lying on the left half of s-plane

Result:

1. The system is unstable.

2. Two roots of s-plane and five roots are lying on the left of s-plane.

$$
\begin{array}{r|l}
 & s^3 + 5s^2 + 9s + 9 \\
\hline
s^2 + 4 & s^7 + 5s^6 + 9s^5 + 9s^4 + 4s^3 + 20s^2 + 36s + 36 \\
 & s^7 \qquad\qquad\qquad\quad + 4s^3 \\
\hline
 & 5s^6 + 9s^5 + 9s^4 \qquad + 20s^2 + 36s + 36 \\
 & 5s^6 \qquad\qquad\qquad + 20s^2 \\
\hline
 & 9s^5 + 9s^4 \qquad + 36s + 36 \\
 & 9s^5 \qquad\qquad + 36s \\
\hline
 & 9s^4 \qquad + 36 \\
 & 9s^4 \qquad + 36 \\
\hline
 & 0
\end{array}
$$

EXAMPLE 4.31: Use the routh stability criterion to determine the location of roots on the s-plane and hence the stability for the system represented by the characteristic equation $s^5 + 4s^4 + 8s^3 + 8s^2 + 7s + 4 = 0$.

Solution: The characteristic equation of the system is $s^5 + 4s^4 + 8s^3 + 8s^2 + 7s + 4 = 0$. The given characteristic polynomial is 5th order equation and so it has 5 roots. Since the highest power of s is odd number, form the first row of routh array using the coefficients of odd powers of s and form the second row using the coefficients of even power of s.

s^5 :	1	8	7	... Row 1
s^4 :	4	8	4	... Row 2

Divide s^4 row by 4 to simplify the calculation

s^5 :	1	8	7	... Row 1
s^4 :	1	2	1	... Row 2
s^3 :	1	1		... Row 3
s^2 :	1	1		... Row 4
s^1 :	$\in$			... Row 5
s^0 :	1			... Row 6
	Column – 1			

when $\in \to 0$, there is no sign change in the first column of routh array. But we have a row of all zeros (s^1 row (or) row 5) and so there is a possibility of roots on imaginary axis. This can be found from the roots of auxiliary polynomial. Here the auxiliary polynomial is given by s^2 row.

$$s = \pm\sqrt{-1} = \pm\, j1$$

s^3 :	$\dfrac{1\times 8 - 2\times 1}{1}$	$\dfrac{1\times 7 - 1\times 1}{1}$
s^3 :	6	6

Divide by 6		
s^3 :	1	1
s^2 :	$\frac{1\times 2-1\times 1}{1}$	$\frac{1\times 1-0\times 1}{1}$
s^2 :	1	1
s^1 :	$\frac{1\times 1-1\times 1}{1}$	
s^1 :	1	
s^0 :	$\frac{\in\times 1-0\times 1}{\in}$	
s^0 :	1	

The roots of auxiliary polynomial are $+j1$, $j1$, lying on imaginary axis. The roots of auxiliary polynomial are also roots of characteristic equation are lying on imaginary axis and so the system is limitedly (or) marginally stable. The remaining three roots of characteristic equation are lying on the left half of s-plane.

Results:

1. The system is limitedly (or) marginally stable.
2. The two roots are lying on imaginary axis and three roots are lying on left of s-plane.

EXAMPLE 4.32: Use the routh stability criterion to determine the location of roots on the s-plane and hence the stability for the system represented by the characteristic equation $s^6 + s^5 + 3s^4 + 3s^3 + 3s^2 + 2s + 1 = 0$

Solution: The characteristic polynomial of the system is

$$s^6 + s^5 + 3s^4 + 3s^3 + 3s^2 + 2s + 1 = 0$$

The given characteristic polynomial is 6th order equation and so it has 6 roots. Since the highest power of s is even number, form the first row of routh array using the coefficients of even powers of s and form the second row using the coefficients of odd powers of s as shown below.

s^4 :	$\frac{1\times 3-3\times 1}{1}$	$\frac{1\times 3-2\times 1}{1}$	$\frac{1\times 1-0\times 1}{1}$
s^4 :	0	1	1
Let $0\to\in$			
s^4 :	$\in$	1	1
s^3:	$\frac{\in\times 3-1\times 1}{\in}$	$\frac{\in\times 2-1\times 1}{\in}$	

CHAPTER 4

s^3 :	$\dfrac{3\epsilon-1}{\epsilon}$	$\dfrac{2\epsilon-1}{\epsilon}$
s^2 :	$\dfrac{\dfrac{3\epsilon-1}{\epsilon}-\dfrac{2\epsilon-1}{\epsilon}\times\epsilon}{\dfrac{3\epsilon-1}{\epsilon}}$	$\dfrac{\dfrac{3\epsilon-1}{\epsilon}\times 1-0\times\epsilon}{\dfrac{3\epsilon-1}{\epsilon}}$
s^2:	$-2\epsilon^2+4\epsilon-1/3\epsilon-1$	1
s^1 :	$\dfrac{\dfrac{-2\epsilon^2+4\epsilon-1}{3\epsilon-1}\times\dfrac{2\epsilon-1}{\epsilon}-\dfrac{3\epsilon-1}{\epsilon}\times 1}{\dfrac{-2\epsilon^2+4\epsilon-1}{3\epsilon-1}}$	
s^1 :	$\dfrac{(-2\epsilon^2+4\epsilon-1)(2\epsilon-1)-(3\epsilon-1)(3\epsilon-1)}{\epsilon(-2\epsilon^2+4\epsilon-1)}$	
s^1 :	$\dfrac{-4\epsilon^2+\epsilon^2}{s(-2\epsilon^2+4\epsilon-1)}=\dfrac{4\epsilon^2-\epsilon}{2\epsilon^2-4\epsilon+1}$	
s^0 :	$\dfrac{\dfrac{4\epsilon^2-\epsilon}{2\epsilon^2-4\epsilon+1}\times 1-0\times\dfrac{-2\epsilon^2+4\epsilon-1}{3\epsilon-1}}{(4\epsilon^2-\epsilon)/(2\epsilon^2-4\epsilon+1)}$	
s^0 :	1	

s^6 :	1	3	3	1	… Row 1
s^5 ;	1	3	2		… Row 2
s^4:	ϵ	1	1		… Row 3
s^3 :	$\dfrac{3\epsilon-1}{\epsilon}$	$\dfrac{2\epsilon-1}{\epsilon}$			… Row 4
s^2:	$\dfrac{-2\epsilon^2+4\epsilon-1}{3\epsilon-1}$	1			… Row 5
s^1 :	$\dfrac{4\epsilon^2-\epsilon}{2\epsilon^2-4\epsilon+1}$				… Row 6
s^0 :	1				… Row 7

On letting $\epsilon \to 0$, we get

s^6 :	1	3	3	1	…Row 1
s^5 :	1	3	2		…Row 2
s^4 :	0	1	1		…Row 3

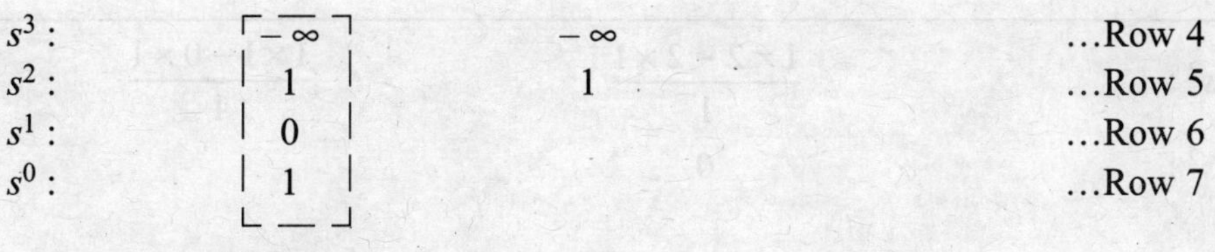

	Column 1		
s^3 :	$-\infty$	$-\infty$	...Row 4
s^2 :	1	1	...Row 5
s^1 :	0		...Row 6
s^0 :	1		...Row 7

Column 1

Since there is a row of all zeros (s^1 row) there is a possibility of roots on imaginary axis. The auxiliary polynomial is $s^2 + 1 = 0$.

The roots of auxiliary polynomial are

$$s = \pm\sqrt{-1} = \pm\, j1$$

The roots of auxiliary polynomial are also roots of characteristic equation. Hence roots are lying on imaginary axis. Therefore divide the characteristic polynomial by auxiliary equation and construct the routh array for quotient polynomial to find the roots lying on right half of s-plane.

The characteristic polynomial can be expressed as a product of auxiliary polynomial and quotient polynomial.

$$\therefore \quad s^6 + s^5 + 3s^4 + 3s^3 + 3s^2 + 2s + 1$$
$$= (s^2 + 1)\,(s^9 + s^3 + 2s^2 + 2s + 1) = 0$$

The routh array for quotient polynomial is constructed as shown below.

s^4 :	1	2	1	... Row 1
s^3 :	1	2		... Row 2
s^2 :	$\in$	1		... Row 3
s^1 :	$\dfrac{2\in -1}{\in}$			... Row 4
s^0 :	1			... Row 5

$$\begin{array}{r|l}
 & s^4 + s^3 + 2s^2 + 1 \\
\hline
s^2 + 1 & s^6 + s^5 + 3s^4 + 3s^3 + 3s^2 + 2s + 1 \\
 & s^6 \quad + s^4 \\
\hline
 & s^5 + 2s^4 + 3s^3 + 3s^2 + 3s^2 + 2s + 1 \\
 & s^5 \quad + s^3 \\
\hline
 & 2s^4 + 2s^3 + 3s^2 + + 2s + 1 \\
 & 2s^4 \quad + 2s^2 \\
\hline
 & 2s^3 + s^2 + 2s + 1 \\
 & 2s^3 \quad + 2s \\
\hline
 & s^2 + 1 \\
 & s^2 + 1 \\
\hline
 & 0 \\
\hline
\end{array}$$

CHAPTER 4

s^2 :	$\frac{1\times 2-2\times 1}{1}$	$\frac{1\times 1-0\times 1}{1}$
s^2 :	0	
Let $0 \to \in$		
s^2 :	$\in$	1
s^1 :	$\frac{\in\times 2-1\times 1}{\in}$	
s^1 :	$\frac{2\in-1}{\in}$	
s^0 :	$\frac{\frac{2\in-1}{\in}\times 1-0\times\in}{(2\in-1)/\in}$	
s^0 :	1	

On letting $\in \to 0$ we get,

s^4 :	1	2	1	...Row 1
s^3 :	1	2		...Row 2
s^2 :	0	1		...Row 3
s^1 :	$-\infty$			...Row 4
s^0 :	1			...Row 5

Column 1

On examining the first column of routh array of quotient polynomial, we found that there are two sign changes. Hence two roots are lying on imaginary axis, two roots are lying on right half of s-plane and the remaining two roots are lying on left half of s-plane. Hence the system is unstable.

Result:

1. The system is unstable.
2. The two roots are lying on imaginary axis, two roots are lying on right half s-plane and two roots are lying on left half s-plane.

EXAMPLE 4.33: Determine the range of K for stability of unity feedback system whose open loop transfer function is $G(s) = K/s(s + 1)(s + 2)$.

Solution: The closed loop transfer function,

$$\frac{C(s)}{R(s)} = \frac{G(s)}{1+G(s)} = \frac{K/s(s+1)(s+2)}{1+\frac{K}{s(s+1)(s+2)}}$$

$$= \frac{K}{s(s+1)(s+2)+K}$$

The characteristic equation is

$$s(s + 1)(s + 2) + K = 0$$

$$s(s^2 + 3s + 2) + K = 0$$

$$s^3 + 3s^2 + 2s + K = 0$$

The routh array is constructed as shown below.

The highest power of s in the characteristic polynomial is odd number. Hence form the first row using the coefficients of odd powers of s and form the second row using the coefficients of even powers of s.

$$\begin{array}{lcc} s^3: & 1 & 2 \\ s^2: & 3 & K \\ s^1: & \frac{6-K}{3} & \\ s^0: & K & \end{array}$$

Column 1

$$\begin{array}{ll} s^1: & \frac{2\times 2 - K\times 1}{3} \\ s^1: & \frac{6-K}{3} \\ s^0: & \frac{\frac{6-K}{3}\times K - 0\times 3}{(6-K)/3} \\ s^0: & K \end{array}$$

For the system to be stable there should not be any sign change in the elements of first column. Hence choose the value of K so that the first column elements are positive.

Form s^0 row, for the system to be stable,

$$K > 0$$

Form s^1 row, for the system to be stable,

$$\frac{6-K}{3} > 0$$

For $\frac{6-K}{3} > 0$, the value of K should be less than 6.

∴ The range of K for the system to be stable is $0 < K < 6$.

Result: The value of K is in the range $0 < K < 6$ for the system to be stable.

EXAMPLE 4.34: The open loop transfer function of a unity feedback control system is given by

$$G(s) = \frac{K}{(s+2)(s+4)(s^2 + 6s + 25)}$$

CHAPTER 4

By applying the routh criterion, discuss the stability of the closed-loop system as a function of K. Determine the value of K which will cause sustained oscillations in the closed loop system. What are the corresponding oscillating frequencies?

Solution: The closed loop transfer function

$$\frac{C(s)}{R(s)} = \frac{G(s)}{1+G(s)} = \frac{\dfrac{K}{(s+2)\,(s+4)\,(s^2+6s+25)}}{1+\dfrac{K}{(s+2)\,(s+4)\,(s^2+6s+25)}}$$

$$= \frac{K}{(s+2)\,(s+4)\,(s^2+6s+25)+K}$$

The characteristic equation is given by the denominator polynomial of closed loop transfer function. The characteristic equation is

$$(s+2)\,(s+4)\,(s^2+6s+25)+K = 0$$

$$(s^2+6s+8)\,(s^2+6s+25)+K = 0$$

$$s^4+12s^3+69s^2+198s+200+K = 0$$

The routh array is constructed as shown below. The highest power of s is the characteristic equation is even number. Hence form the first row using the coefficients of even powers of s and form the second new using the coefficients of odd powers of s.

$$\begin{array}{llll} s^4: & 1 & 69 & 200+K \quad \text{...Row 1} \\ s^3: & 12 & 198 & \quad \text{...Row 2} \end{array}$$

Divide s^3 row by 12 to simplify the calculations.

$$\begin{array}{llll} s^4: & 1 & 69 & 200+K \quad \text{...Row 1} \\ s^3: & 1 & 16.5 & \quad \text{...Row 2} \\ s^2: & 52.5 & 200+K & \quad \text{...Row 3} \\ s^1: & \dfrac{666.25-K}{52.5} & & \quad \text{...Row 4} \\ s^0: & 200+K & & \quad \text{...Row 5} \end{array}$$

Column – 1

$$\begin{array}{lll} s^2: & \dfrac{1\times 69-16.5\times 1}{1} & \dfrac{1\times(200+K)}{1} \\ s^2: & 52.5 & 200+K \\ \hline s^1: & \dfrac{52.5\times 16.5-(200+K)\times 1}{52.5} & \\ & & \end{array}$$

s^1 :	$\dfrac{666.25 - K}{52.5}$
s^0 :	$\dfrac{666.25 - K}{52.5} \times (200 + K)$
s^0 :	$200 + K$

For the system to be stable there should not be any sign change in the elements of first column. Hence choose the value of K so that the first column elements are positive.

From s^1 row, for the system to be stable, $(666.25 - K) > 0$

Since $(666.25 - K) > 0$, K should be less than 666.25.

From s^0 row, for the system to stable, $(200 + K) > 0$

Since $(200 + K) > 0$, K should be greater than – 200, but practical value of K-starts from 0. Hence K should be greater than 0.

∴ The range of K for the system to be stable is $0 < K < 666.25$.

When $K = 666.25$ the s^1 row becomes zero, which indicates the possibility of roots on imaginary axis. A system will oscillate if it has roots on imaginary axis and no roots on right half of s-plane.

When $K = 666.25$, the coefficients of auxiliary equation are given by the s^2 row.

∴ The auxiliary equation is

$$52.5\, s^2 + 200 + K = 0.$$

$$52.5\, s^2 + 200 + 666.25 = 0$$

$$s^2 = \frac{-200 - 666.25}{52.5} = -16.5$$

$$s = \pm j\, 4.06$$

When $K = 666.25$, the system has roots on imaginary axis and so it oscillates. Frequency of oscillation is given by the value of root on imaginary axis.

∴ The frequency of oscillation,

$$\omega = 4.06 \text{ rad/sec}$$

Result:

1. The range of k for stability is $0 < K < 666.25$.
2. The system oscillates when $K = 666.25$.
3. The frequency of oscillation, $\omega = 4.06$ rad/sec (when $K = 666.25$)

EXAMPLE 4.35: The open loop transfer function of a unity feedback system is given by $G(s) = \dfrac{K(s+1)}{s^3 + as^2 + 2s + 1}$. Determine the value of K and a so that the system oscillates at a frequency of 2 rad/sec.

Solution: The closed loop transfer function

$$\frac{C(s)}{R(s)} = \frac{G(s)}{1 + G(s)}.$$

CHAPTER 4

$$= \frac{\dfrac{K(s+1)}{s^3+as^2+2s+1}}{1+\dfrac{K(s+1)}{s^3+as^2+2s+1}} = \frac{K(s+1)}{s^3+as^2+1+K(s+1)}$$

$\therefore$ The characteristic equation is

$$s^3 + as^2 + 2s + 1 + K(s+1) = 0$$
$$s^3 + as^2 + 2s + 1 + Ks + K = 0$$
$$s^3 + as^2 + (2+K)s + 1 + K = 0$$

The routh array of characteristic polynomial is constructed as shown below. The maximum power of s is odd, hence the first row of routh array is formed using coefficients of odd powers of s and the second row of routh array is formed using coefficients of even powers of s.

If the elements of s row are all zeros then there exist an even polynomial (for auxiliary polynomial). If the roots of the auxiliary polynomial are purely imaginary then the roots are lying on imaginary axis and the system oscillates. The frequency of oscillation is the root of auxiliary polynomial.

From routh array , From s row.

$$\frac{a(2+K)-(1+K)}{a} = 0$$
$$a(2+K)-(1+K) = 0$$
$$2a + Ka - 1 - K = 0$$
$$2a - 1 + K(a-1) = 0$$

Put $K = 4a - 1$

$$\therefore 2a - 1 + (4a-1)(a-1) = 0$$
$$2a - 1 + 4a^2 - 4a - a + 1 = 0$$
$$4a^2 - 3a = 0 \text{ (or) } a(4a-3) = 0$$

since $a \neq 0$, $4a - 3 = 0$

$$a = 3/4$$

when $a = (3/4)$, $K = 4a - 1$

$$K = 2$$

Routh array

s^3 :	1	$2+K$
s^2 :	a	$1+K$
s^1 :	$\dfrac{a(2+K)-(1+K)}{a}$	
s^0 :	$1+K$	

The auxiliary polynomial is

$$as^2 + (1+K) = 0$$
$$as^2 = -(1+K)$$
$$s = \pm j\sqrt{\frac{1+K}{a}}$$

Given that $s = \pm j2$

$$\therefore \quad \sqrt{\frac{1+K}{a}} = 2$$

$$\frac{1+K}{a} = 4$$

$$K = 4a = 1$$

Result: When the system oscillates at a frequency of 2 rad/sec, $K = 2$ and $a = 3/4$.

4.15 NYQUIST STABILITY CRITERION

Let take closed loop transfer function

$$\frac{C(s)}{R(s)} = \frac{G(s)}{1+G(s)H(s)}$$

The characteristic equation of the system is given by the condition, $1 + G(s)\,H(s) = 0$

Let, $F(s) = 1 + G(s)\,H(s)$

The loop transfer function $G(s)\,H(s)$ can be written as

$$G(s)\,H(s) = \frac{K(s+z_1)\,(s+z_2)\ldots(s+z_m)}{(s+p_1)\,(s+p_n)\ldots(s+p_n)} \tag{4.141}$$

where $m \le n$

$$\therefore \quad F(s) = 1 + G(s)\,H(s) = 1 + \frac{K(s+z_1)\ldots(s+z_m)}{(s+p_1)\ldots(s+p_n)}$$

$$= \frac{(s+p_1)\,(s+p_2)\ldots(s+p_n) + K(s+z_1)\,(s+z_2)\ldots(s+z_m)}{(s+p_1)\,(s+p_2)\ldots+(s+p_n)} \tag{4.142}$$

$$= \frac{(s+z_1')\,(s+z_2')\ldots(s+z_n')}{(s+p_1)\,(s+p_2)\ldots(s+p_n)} \tag{4.143}$$

where, $z_1', z_2' \ldots z_n'$ are zeros of $F(s)$ which are obtained by combining the numerator and denominator polynomial of $G(s)\,H(s)$.

For the condition $F(s) = 0$, the numerator of $F(s)$ should be equal to zero

$$\therefore \quad (s+z_1')\,(s+z_2')\ldots.(s+z_n') = 0 \tag{4.144}$$

we can say that equation (4.144) is the characteristic equation of the system. For the stability of the system the roots of the characteristic equation should not lie on the right half s-plane. The roots of characteristic equation are zeros of $F(s)$ and also they are poles of closed loop transfer function.

Hence we can conclude that for the stability of closed loop system the zeros of $F(s)$ should not lie on the right half S-plane.

Note: For unity feedback system

$$\frac{C(s)}{R(s)} = \frac{K(s+z_1)\,(s+z_2)\ldots(s+z_m)}{(s+z_1')\,(s+z_2')\ldots(s+z_n')}$$

Let us choose an arbitrary contour in the s-plane which encircles the right half zeros and poles of $F(s)$ [equation (4.143)]. The principle of argument states that (explained in section 3.8.1) the corresponding contour in $F(s)$-plane, N times in the anticlock wise direction.

where $$N = p - z \tag{4.145}$$

Here,

p = Number of poles of $F(s)$ lying on right half s-plane

z = Number of zeros of $F(s)$ lying on right half s-plane

For the stability of the system the roots of characteristic equation and zeros of $F(s)$ should not lie on the right half of s-plane. Hence for a stable system $z = 0$. Hence from equation (4.145) when $z = 0$, we get

when $$z = 0, \quad N = p \tag{4.146}$$

$$z \neq 0, \quad N \neq p \tag{4.147}$$

From equation (4.141) and (4.142) we can say that the poles of $F(s)$ are also poles of open loop transfer function, hence for the stability of the system, (with reference to equation (4.146) and (4.147) number of poles of loop transfer function lying on right of s-plane should be equal to counter clock wise encirclement of the origin of $F(s)$ plane. If this condition is not met the system is unstable.

The loop transfer function, $G(s)\,H(s)$ can be expressed as

$$G(s)\,H(s) = [1 + G(s)\,H(s)] - 1 = F(s) - 1 \tag{4.148}$$

From equation (4.148) it can be concluded that the $F(s)$-plane contour drawn with respect to origin is same as the $F(s)$-plane contour of $F(s) - 1$ drawn with respect to $(-1 + j0)$ as shown in Fig. (4.20).

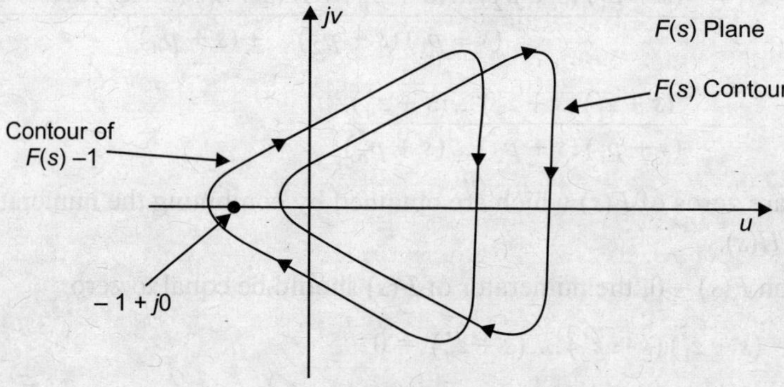

Fig. 4.20

Thus the encirclement of the origin of $F(s)$ plane by the $F(s)$ contour is equivalent to the encirclement of the pint $(-1 + j0)$ by the $G(s)\,H(s)$ contour. The Nyquist stability criterion have been proposed based on this concept.

In order to investigate the presence of poles of $G(s)\,H(s)$ on the right half s-plane a contour, C is chosen such that it encloses the entire right half s-plane as shown in Fig. (4.21) such contour C is called Nyquist contour.

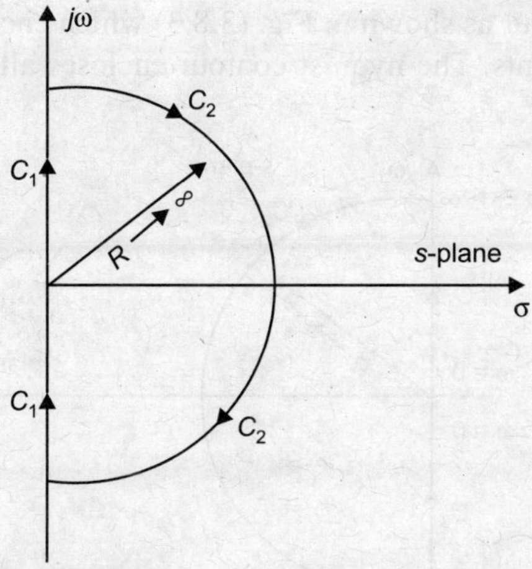

Fig. 4.21: Nyquist contour.

The nyquist contour is directed clockwise and comprise of two segments.

(i) An infinite line segment C_1 along the imaginary axis.

(ii) An arc, C_2 of ∞ radius.

along C_1, $s = j\omega$ with s varying form $-j\infty$ to $+j\infty$.

Along C_2, $s = \underset{K \to \infty}{\text{Lt}} \; Ke^{j\theta}$ with θ varying from $+\pi/2$ to $-\pi/2$.

Note: The s-plane is a complex plane. Any point on a complex plane can be expressed by the complex number in polar form, $Re^{j\theta}$, where R is the magnitude and θ is the argument (or phase).

Definition of Nyqist stability criterion:

The nyquist stability criterion can be stated follows. "If the $G(s)\,H(s)$ contour in the $G(s)\,H(s)$ plane corresponding to nyquist contour in the s-plane encircles the pint $(-1 + j0)$ in the anticlockwise direction as many times as the number of right half s-plane poles of $G(s)\,H(s)$, then the closed loop system is stable".

Important points to be noted.

1. There is no encirclement of $-1 + j0$ point. This implies that the system is stable if there are no poles of $G(s)\,H(s)$ in the right half s-plane. If there are poles on right half s-plane then the system is unstable.
2. An anticlockwise encirclement of $-1 + j0$ point. In this case the system is stable if the number of anticlockwise encirclements is same as the number of poles of $G(s)\,H(s)$ in the right half s-plane. If the number of encirclement is not equal to number of poles an right half *s-plane* then the system is unstable.
3. There is a clock-wise encirclement of the $(-1 + j0)$ point. In this case the system is always unstable.

Steps to Analyse the Stability using Nyquist Criterion

The following procedure can be followed to investigate the stability of closed loop system from the knowledge of open loop system, using Nyquist stability criterion.

1. Choose a nyquist contour as shown in Fig. (3.8.3) which encloses the entire right half s-plane except the singular points. The nyquist contour encloses all the right half s-plane poles and zeros of $G(s)\,H(s)$.

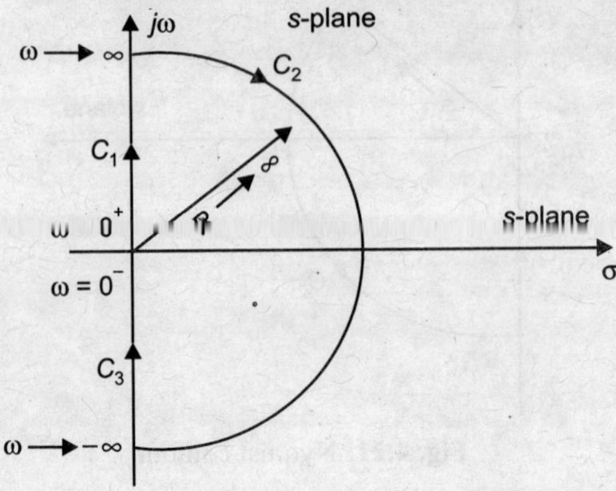

Fig. 4.22: (a) Nyquist contour when there is no pole on imaginary.

2. The nyquist contour should be mapped to the $G(s)\,H(s)$ plane using the function $G(s)\,H(s)$ to determine the encirclement $-1 + j0$ point in the $G(s)\,H(s)$ plane. The nyquist contour of Fig. (3.8.3b) can be divided into four sections C_1, C_2, C_3 and C_4. The mapping of the four sections in the $G(s)\,H(s)$ plane can be carried sectionwise and then combined together to get entire $G(s)\,H(s)$ contour.
3. In section C_1 the value of ω varies from 0 to ∞. The mapping of section C_1 is obtained by letting $s = j\omega$ in $G(s)\,H(s)$ and varying ω from 0 to $+\infty$.

i.e., $G(s)\,H(s)\big|_{\substack{s = j\omega \\ \omega = 0 \text{ to } \infty}} = G(j\omega)\,H(j\omega)\big|_{\omega = 0 \text{ to } \infty}$

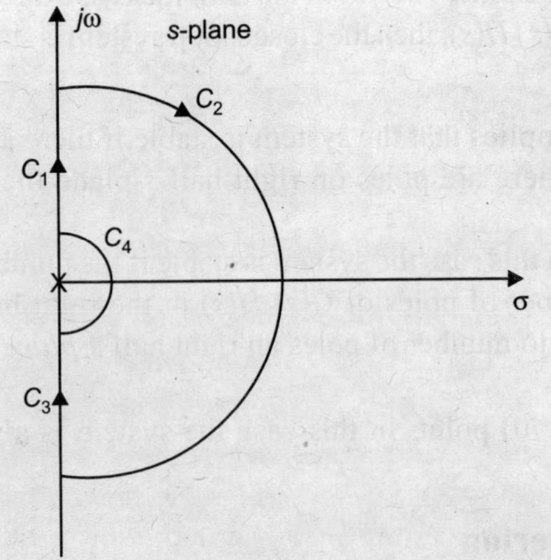

Fig. 4.22: (b) Nyquist contour when there are poles at origin.

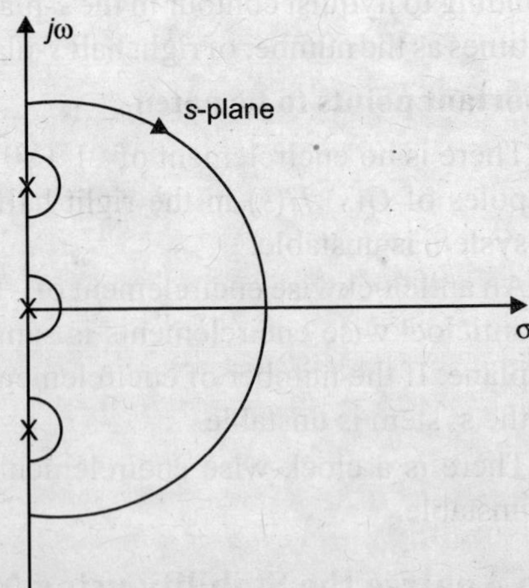

Fig. 4.22: (c) Nyquist contour when there are poles on imaginary axis and at origin.

Fig. 4.22: Nyquist contour.

The locus of $G(j\omega)\,H(j\omega)$ as ω is varies from 0 to ∞ will be the $G(s)\,H(s)$ contour in $G(s)\,H(s)$ plane corresponding to section C_1 in s plane. This locus is the polar plot of $G(j\omega)\,H(j\omega)$. There are three ways of mapping this section of $G(s)\,H(s)$ contour they are,

(i) Calculate the values of $G(j\omega)\,H(j\omega)$ for various values of ω and sketch the actual locus of $G(j\omega)\,H(j\omega)$.

(ii) Separate the real part and imaginary part of $G(j\omega)\,H(j\omega)$. Equate the imaginary part to zero, to find the frequency at which the $G(j\omega)\,H(j\omega)$ locus crosses real axis.Substitute this frequency on real part and find the crossing point of the locus on real axis. Sketch the approximate locus of $G(j\omega)\,H(j\omega)$ form the knowledge of type number and order of the system (or) from the value of $G(j\omega)\,H(j\omega)$ at $\omega = 0$ and $\omega = \infty$.

(iii) Separate the magnitude and phase of $G(j\omega)\,H(j\omega)$ to -180 and solve for ω. This value of ω is the phase crossover frequency and the magnitude at this frequency is the crossing point on real axis. Sketch the approximate root locus.

4. The section C_2 of nyquist contour has a semicircle of infinite radius. Therefore, every point on section C_2 has infinite magnitude but the argument varies from $+\pi/2$, to $-\pi/2$. Hence the mapping of section C_2 from s-plane to $G(s)\,H(s)$ plane can be obtained by letting $s = \underset{R \to \infty}{\text{Lt}}\; Re^{j\theta}$ in $G(s)\,H(s)$ and varying θ from $+\pi/2$ to $-\pi/2$.

Consider the loop transfer function in time constant form and with y number of poles at origin, as shown here,

$$G(s)\,H(s) = \frac{K(1+sT_1)\,(1+sT_2)}{s^y\,(1+sT_a)\,(1+sT_b)}$$

Let $G(s)\,H(s)$ has n zeros and n poles including poles at orign. For practical systems, $m > n$ since $s \to Re^{j\theta}$ and $R \to \infty$, the term $(1 + sT)$ can be approximated to sT [i.e., $(1 + sT)$ □ sT]

$$\therefore \qquad G(s)\,H(s) \;\square\; K \cdot \frac{sT_1 \cdot sT_2 \cdot sT_3 \;....}{s^y \cdot sT_a \cdot sT_b \cdot sT_c \;...}$$

$$= K_1 \frac{s^m}{s^n} = \frac{K_1}{s^{n-m}}$$

$$G(s)\,H(s) \approx K_1/s^n - s^m$$

on letting,

$$s = \underset{R \to \infty}{\text{Lt}}\; Re^{j\theta}$$

we get

$$\frac{G(s)\,H(s)}{s = \underset{R \to \infty}{\text{Lt}}\; Re^{j\theta}} = \frac{K_1}{\underset{R \to \infty}{\text{Lt}}\; (Re^{j\theta})^{n-m}} = 0e^{-j\theta(n-m)}$$

when $\theta = \pi/2$,

$$G(s)\,H(s) = 0\,e^{-j\,\pi/2(n-m)}$$

when $\theta = -\pi/2$,

$$G(s)\,H(s) = 0\,e^{+j\,\pi/2(n-m)}$$

From the above two equations we can conclude that the section C_2 of nyquist contour in s-plane is mapped as circles/circular are around is mapped as circles /circular are around origin with radius tending to zero in the $G(s)\,H(s)$ plane.

5. In section C_3 the value of ω varies from –∞ to 0. The mapping of section C_3 is obtained by letting $s = +j\omega$ in $G(s)\,H(s)$ and varying from –∞ to 0.

   i.e.,
$$G(s)\,H(s)\big|_{\substack{s=+j\omega \\ \omega=-\infty \text{ to } 0}} = G(j\omega)\,H(j\omega)\big|_{\omega=-\infty \text{ to } 0}$$

   The locus of $G(j\omega)\,H(j\omega)$ as ω is varies form –∞ to 0 will be the $G(s)\,H(s)$ contour in $G(s)\,H(s)$ plane corresponding section C_3 in s-plane. This locus is the inverse polar plot of $G(j\omega)\,H(j\omega)$. The inverse polar plot is given by the mirror image of polar plot with respect to real axis.

6. The section C_4 of nyquist contour has a semicircle of zero radius. Therefore every point on semicircle has zero magnitude but the argument varies from $-\pi/2$ to $+\pi/2$. Hence the mapping of section C_4 from s-plane to $G(s)\,H(s)$ plane can be obtained by letting $s = \underset{R\to 0}{\text{Lt}}\ Re^{j\theta}$ in $G(s)\,H(s)$ and varying θ from $-\pi/2$ to $+\pi/2$.

   Consider the loop transfer function in time constant from and with y number of poles at orign as shown below.

$$G(s)\,H(s) = \frac{K(1+sT_1)\,(1+sT_2)\,(1+sT_3)\ldots}{s^y\,(1+sT_a)\,(1+sT_b)\,(1+sT_c)\ldots}$$

   Let $G(s)\,H(s)$ has m zeros and n poles including poles at origin. For practical systems $m > n$. Since $s \to Re^{j\theta}$ and $R \to 0$ the term $1 + sT$ can be approximated to 1 [i.e., $(1 + sT) \approx 1$]

$$\therefore \qquad G(s)\,H(s) \approx K\frac{1}{sy}$$

   on letting $s = \underset{R\to 0}{\text{Lt}}\ Re^{j\theta}$ we get

$$G(s)\,H(s)\big|_{s=\underset{R\to 0}{Lt}\ Re^{j\theta}} = \frac{K}{\underset{R\to 0}{Lt}\ (Re^{j\theta})^y} = \infty e^{-j\theta y}$$

   when $\theta = -\pi/2$,

$$G(s)\,H(s) = \infty\ e^{j\pi/2y}$$

   when $\theta = \pi/2$,

$$G(s)\,H(s) = \infty\ e^{-j\pi/2y}$$

   From the above two equations we can conclude that the section of nyquist. Contour in s-plane is mapped as circles/circular are in $G(s)\,H(s)$ plane with origin as centre and infinite radius.

Note:

1. If there are no poles on the origin then the section C_4 of nyquist contour will be absent.
2. If there are poles on imaginary axis as shown below then the nyquist contour is divided into the following 8 sections and the mapping is performed section wise.

Section C_1: From $+jo^+$ to $+j\omega_1^-$
Section C_2: From $+j\omega_1^-$ to $+j\omega_1^+$
Section C_3: From $+j\omega_1^+$ to $+j\infty$
Section C_4: From $+j\infty$ to $-j\infty$
Section C_5: From $-j\infty$ to $-j\omega_1^-$
Section C_6: From $-j\omega_1^-$ to $-j\omega_1^+$
Section C_7: From $-j\omega_1^+$ to $-jo^-$
Section C_8: From $-jo^-$ to $+jo^+$

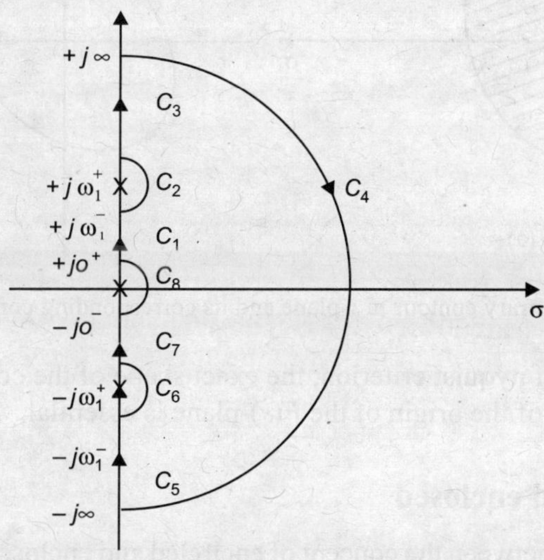

Fig. 4.23

4.15.1 Mathematical Preliminaries for Nyquist Stability Criterion

Let
$$F(s) = \frac{(s - z_1)(s - z_2) \dots (s - z_m)}{(s - p_1)(s - p_2) \dots (s - p_n)} \tag{4.149}$$

The roots of numerator polynomial are zeros and the roots of denominator polynomial are poles.

The function has m number of zeros and n number of poles.

Let s be a complex variable represented by $s = \sigma + j\omega$ on the complex s-plane. The function $F(s)$ is also complex and may be defined as $F(s) = u + jv$ and represented on the complex $F(s)$-plane with coordinates u and v.

The equation (4.149) indicates that, the function $F(s)$ maps the point in the s-plane into the $F(s)$ plane.

Note: A function is analytic in the s-plane provide the function and all its derivatives exist. The points in the s-plane where the function (or it derivatives) does not exist are called singular points.

Since any number of points of analycity in the s-plane can be mapped into the s-plane which does not go through any singular point, there exists a corresponding contour in the $F(s)$-plane as shown in Fig. (4.24).

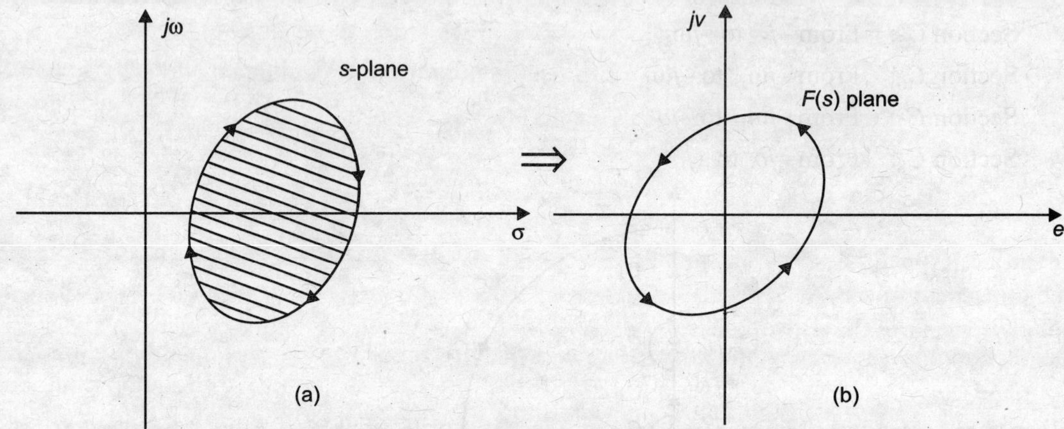

Fig. 4.24: An arbitrary contour in s-plane and its corresponding contour in $F(s)$ plane.

Note: For the development of nyquist criterion, the exact shape of the contour is not required but only the number of encirclements of the origin of the $F(s)$-plane is essential.

Concept of encircled and enclosed

It is important to distinguish between the concept of encircled and enclosed which are frequently used to apply nyquist criterion.

Encircled

A point is said to be encircled by a closed path if it is found inside the path with reference to Fig. (4.25a) the point A is encircled in the clockwise direction and the point B is not encircled.

Enclosed

Any point or region is said to be enclosed by a closed path, if it is found to the right of the path when the path is traversed in the prescribed direction. The shaded regions in Fig. (4.25b and c) are the region enclosed by the closed path the point A is enclosed by closed path and the point B is not enclosed in Fig. (4.25c), the point A is not enclosed by closed path but point B is enclosed.

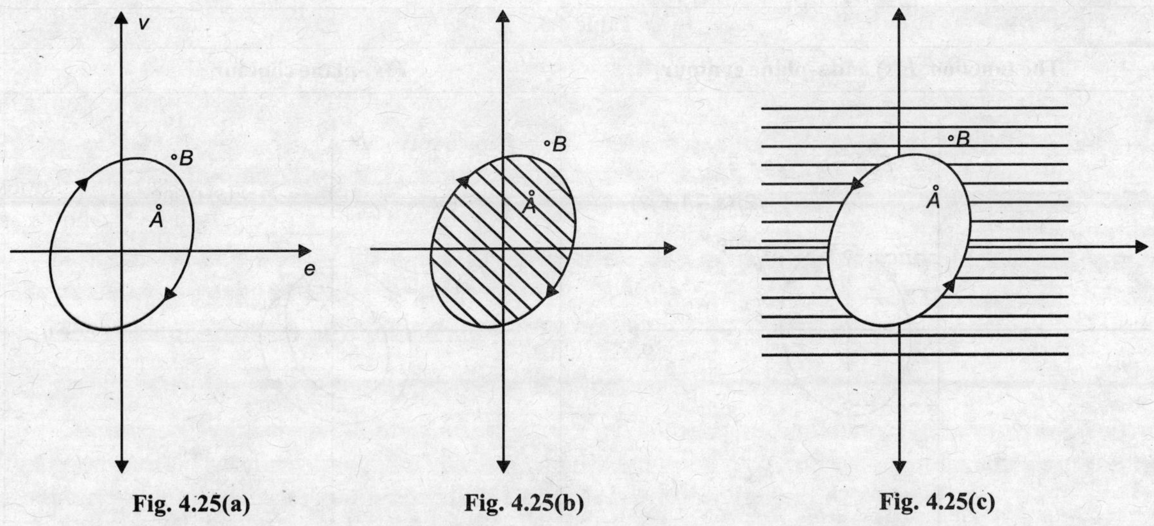

Fig. 4.25(a) Fig. 4.25(b) Fig. 4.25(c)

It can be proved that there exists a relationship between the enclosure of poles and zeros by the s-plane closed contour and number of encirclements of the origin of $F(s)$ plane by the corresponding $F(s)$-plane contour. The following points is the summary of this relationship.

1. If s-plane closed contour enclose z-number of zeros in the right half s-plane then the corresponding contour in $F(s)$ plane will encircle, the origin of $F(s)$ plane Z times in the clockwise direction.
2. If s-plane closed contour encloses P number of poles in the right half s-plane then the corresponding contour in $F(s)$ plane will encircle the origin of $F(s)$ plane p times in anticlockwise direction.
3. If the s plane closed contour encloses Z zeros and P poles in the right half s-plane and if $P > Z$, then the corresponding contour in $F(s)$-plane will encircle the origin of $F(s)$-plane $(P\text{-}Z)$ times in the anticlockwise direction.
4. If the s-plane closed contour enclosed Z zeros and P poles in the right half s-plane and if $P < Z$ then the corresponding contour in $F(s)$-plane will encircle the orign of $F(s)$-plane $(Z - P)$ times in the clockwise direction.
5. If the s-plane closed contour encloses Z zeros and P poles in the right half x-plane and if $P = Z$, then the corresponding contour in $F(s)$-plane will not encircle the origin of $F(s)$-plane.
6. If the s-plane closed contour does not enclose any pole (or) zero then the corresponding contour in $F(s)$-plane will not encircle the origin of $F(s)$-plane.

   The Table (4.3) shows examples of arbitrary s-plane contours and their corresponding $F(s)$-plane contours. Here zeros are marked by small circles (0) and poles by (X).

The relation between the enclosure of poles and zeros of $F(s)$ lying on the right half s-plane by the s-plane contour and the encirclements of the origin of $F(s)$-plane by the corresponding $F(s)$-plane contour is called principle of argument.

CHAPTER 4

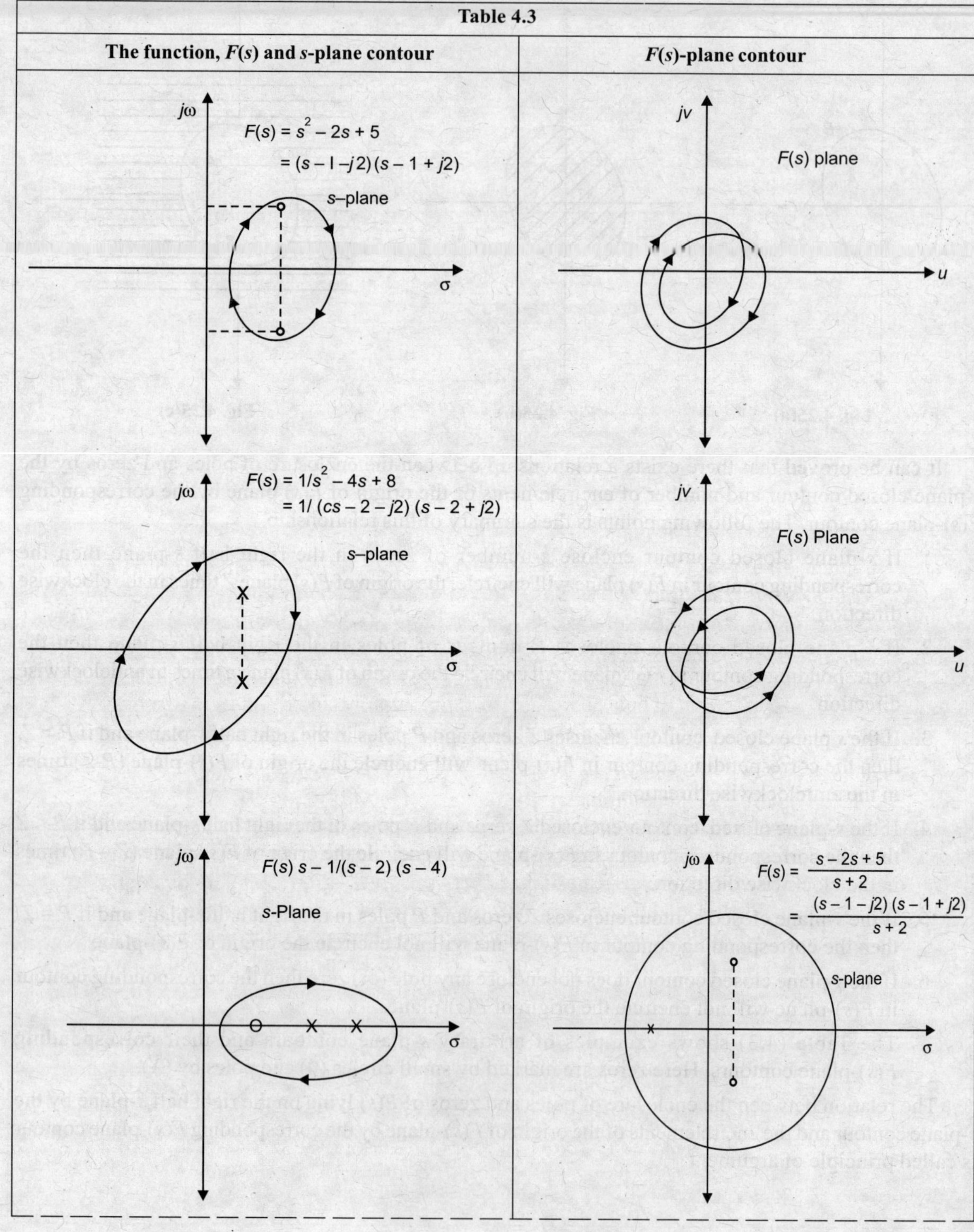
Table 4.3
The function, F(s) and s-plane contour
F(s)-plane contour
jω
F(s) = s² – 2s + 5
= (s – l –j2) (s – 1 + j2)
s–plane
σ
jv
F(s) plane
u
jω
F(s) = 1/s² – 4s + 8
= 1/ (cs – 2 – j2) (s – 2 + j2)
s–plane
σ
jv
F(s) Plane
u
jω
F(s) s – 1/(s – 2) (s – 4)
s-Plane
σ
jω
F(s) = s – 2s + 5 / s + 2
= (s – 1 – j2) (s – 1 + j2) / s + 2
s-plane
σ

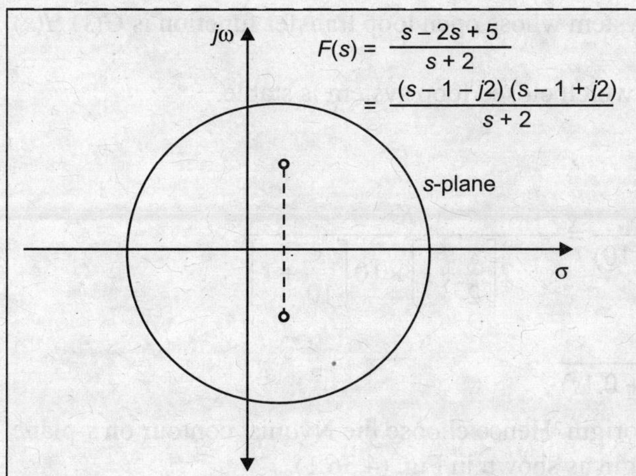
jω
F(s) = s – 2s + 5 / s + 2
= (s – 1 – j2) (s – 1 + j2) / s + 2
s-plane
σ

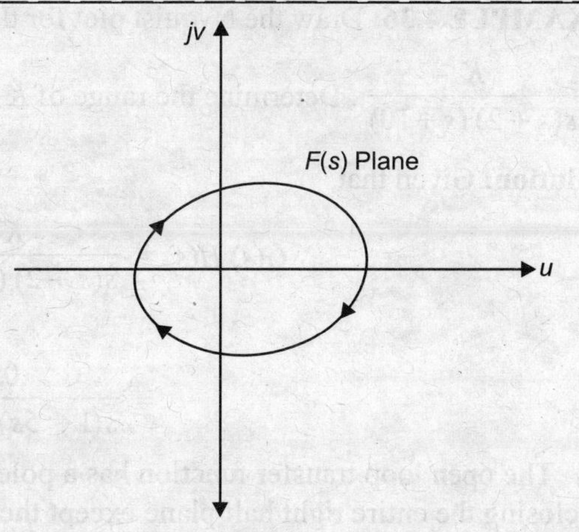
jv
F(s) Plane
u

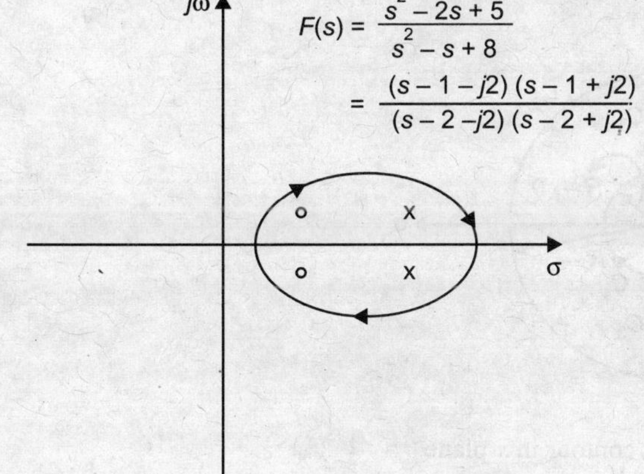
jω
F(s) = s² – 2s + 5 / s² – s + 8
= (s – 1 – j2) (s – 1 + j2) / (s – 2 –j2) (s – 2 + j2)
σ

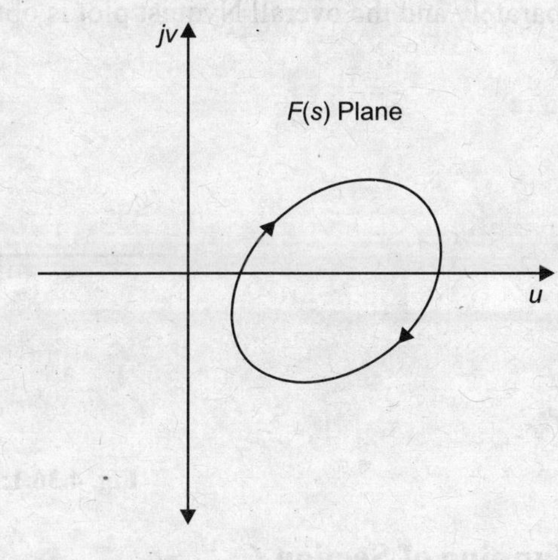
jv
F(s) Plane
u

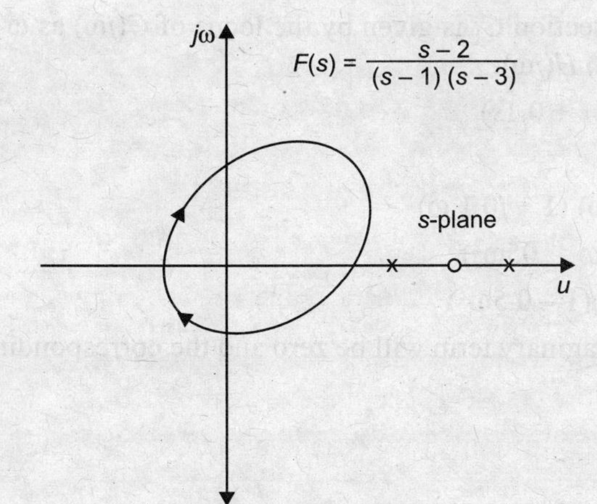
jω
F(s) = s – 2 / (s – 1) (s – 3)
s-plane
u

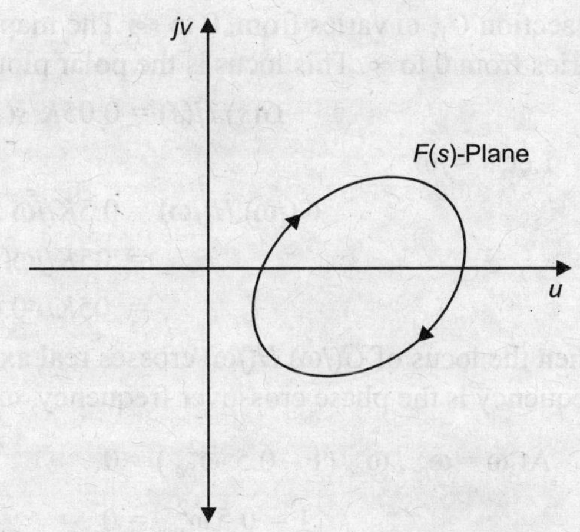
jv
F(s)-Plane
u

EXAMPLE 4.36: Draw the Nyquist plot for the system whose open loop transfer function is $G(s)\,H(s)$ $= \dfrac{K}{s(s+2)(s+10)}$. Determine the range of K for which closed loop system is stable.

Solution: Given that,

$$G(s)\,H(s) = \frac{K}{s(s+2)(s+10)} = \frac{K}{s \times 2\left[\dfrac{s}{2}+1\right] \times 10\left[\dfrac{s}{10}+1\right]}$$

$$= \frac{.05K}{s(1+.5s)(1+0.1s)}$$

The open loop transfer function has a pole at origin. Hence choose the Nyquist contour on s-plane enclosing the entire right half plane except the origin as shown in Fig. (4.36.1).

The Nyquist contour has 4 sections C_1, C_2, C_3 and C_4. The mapping of each section's performed separately and the overall Nyquist plot is obtained by combining the individual sections.

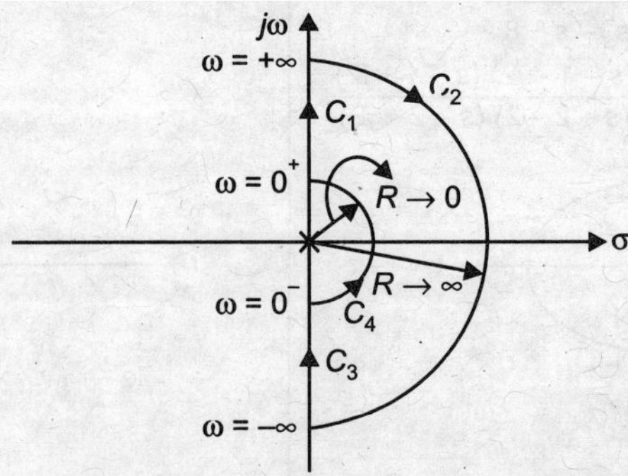

Fig. 4.36.1: Nyquist contour in s-plane.

Mapping of Section C_1

In section C_1, ω varies from, 0 to ∞. The mapping of section C_1 is given by the locus of $G(j\omega)$ as ω is varies from 0 to ∞. This locus is the polar plot of $G(j\omega)\,H(j\omega)$.

$$G(s)\,H(s) = 0.05K/s(1+.5s)(1+0.1s)$$

Let, $s = j\omega$,

$$\therefore \quad G(j\omega)\,H(j\omega) = 0.5K/j\omega\,(1+0.5\omega)\,(1+j0.1\,\omega)$$

$$= .05K/j\omega(1+j0.6\,\omega - .05\omega^2)$$

$$= .05K/{-0.6\omega^2} + j\omega(1-0.5\omega^2)$$

when the locus of $G(j\omega)\,H(j\omega)$ crosses real axis the imaginary term will be zero and the corresponding frequency is the phase crossover frequency, ω_{pc}.

$$\therefore \quad \text{At } \omega = \omega_{pc},\ \omega_{pc}(1-0.5\,\omega_{pc}^2) = 0$$

$$1 - 0.5\,\omega_{pc}^2 = 0$$

$$\omega_{pc} = \sqrt{\frac{1}{.05}} = 4.472 \text{ rad/sec}$$

At $\omega = \omega_{pc} = 4.472$ rad/sec,

$$G(j\omega)\,H(j\omega) = .05K/-0.6\omega^2 = \frac{-.05K}{0.6 \times (4.472)^2}$$

$$G(j\omega)\,H(j\omega) = -0.00417\,K$$

The open loop system is type-1 and third order system it is a minimum phase system with all poles. Hence the polar plot of $G(j\omega)\,H(j\omega)$ starts at –90º axis at infinity, crosses real axis at $-0.0417\,K$ and ends at origin in second quadrant. The section C_1 and its mapping is show in Fig. (4.36.2) and (4.36.3)

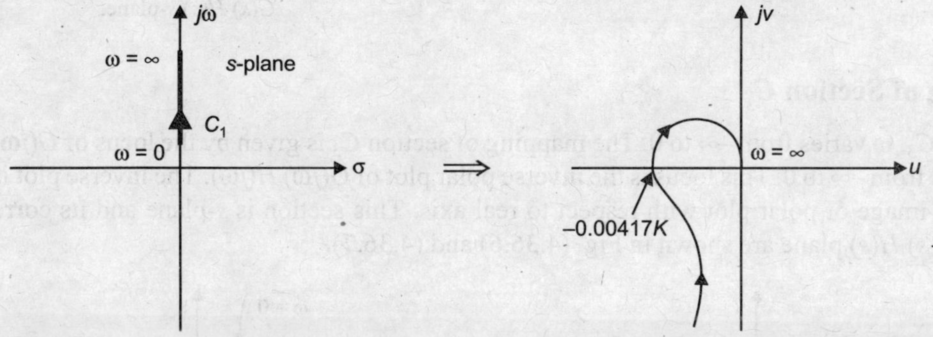

Fig. 4.36.2: Section C_1 in s-plane.

Fig. 4.36.3: Mapping of section C_1 in $G(s)\,H(s)$ s-plane.

Mapping of Section C_2

The mapping of section C_2 from s-plane to $G(s)\,H(s)$ s-plane is obtained by letting $s = \underset{R\to\infty}{Lt}\; Re^{j\theta}$ in $G(s)\,H(s)$ and varying θ from $+\pi/2$ to $-\pi/2$. Since $s \to Re^{j\theta}$ and $R \to \infty$, the $G(s)\,H(s)$ can approximated as shown below [i.e., $(1 + sT) \approx sT$]

$$G(s)\,H(s) = .05K/s(1 + .5s)\,(1 + .1s) \approx \frac{.05K}{s \times .5s \times 1.s} = \frac{K}{s^3}$$

Let, $s = \underset{R\to\infty}{Lt}\; Re^{j\theta}$

$$\therefore \quad G(s)\,H(s)\Big|_{s=\underset{R\to\infty}{Lt} Re^{j\theta}} = \frac{K}{s^3}\Bigg|_{s=\underset{R\to\infty}{Lt} Re^{j\theta}} = \frac{K}{\underset{R\to\infty}{Lt}\,(Re^{j\theta})^3} = 0e^{-j3\theta}$$

when $\theta = \pi/2$, $\quad G(s)\,H(s) = oe^{-j3\pi/2}$

when $\theta = -\pi/2$, $\quad G(s)\,H(s) = oe^{+j3\pi/2}$

From above two equations, we can say that section C_2 in s-plane is mapped as circular arc of zero radius around origin $G(s)\,H(s)$-plane with argument varying from $-3\pi/2$ to $+3\pi/2$ as shown in Fig. (4.36.4) and (4.36.5)

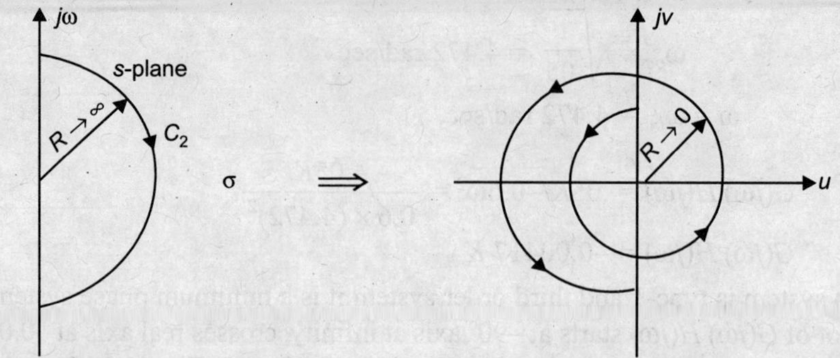

Fig. 4.36.4: Section C_2 in s-plane.

Fig. 4.36.5: Mapping of section C_2 in $G(s)\,H(s)$ s-plane.

Mapping of Section C_3

In section C_3, ω varies from $-\infty$ to 0. The mapping of section C_3 is given by the locus of $G(j\omega)\,H(j\omega)$ as ω is varied from $-\infty$ to 0. This locus is the inverse polar plot of $G(j\omega)\,H(j\omega)$. The inverse plot is given by the mirror image of polar plot with respect to real axis. This section is s-plane and its corresponding contour $G(s)\,H(s)$ plane are shown in Fig. (4.36.6) and (4.36.7).

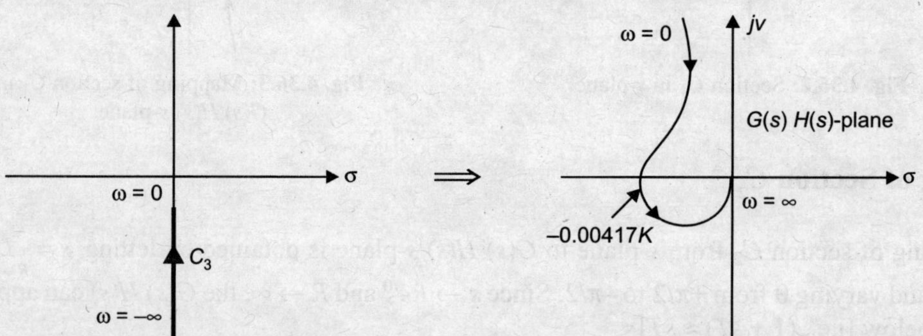

Fig. 4.36.6: Section C_3 in s-plane.

Fig. 4.36.7: Mapping of section C_3 in $G(s)\,H(s)$ s-plane.

Mapping of Section C_4

The mapping of section C_4 from s-plane to $G(s)\,H(s)$-plane is obtained by letting $s = \underset{R \to 0}{Lt}\ Re^{j\theta}$ in $G(s)\,H(s)$ and varying θ from $-\pi/2$ to $+\pi/2$. Since $s \to Re^{j\theta}$ and $R \to 0$ the $G(s)\,H(s)$ can be approximated $1 + sT \approx 1$ [$\because 1 \gg sT$]

$$G(s)\,H(s) = \frac{.05K}{s(1+.5s)\,(1+.1s)} \approx \frac{.05K}{s \times 1 \times 1} = \frac{.05K}{s}$$

Let $$s = \underset{R \to 0}{Lt}\ Re^{j\theta}$$

$$\therefore \quad G(s)\,H(s)\Big|_{\substack{Lt \\ R\to 0}\ \mathrm{Re}^{j\theta}} = \frac{.05K}{s}\Big|_{s=\ \substack{Lt \\ R\to 0}\ \mathrm{Re}^{j\theta}} = \frac{.05K}{\underset{R\to 0}{Lt}\ (\mathrm{Re}^{j\theta})} \infty e^{-j\theta}$$

when $\theta = -\pi/2$, $\quad G(s)\,H(s) = \infty e^{+j\pi/2}$

when $\theta = +\pi/2$, $\quad G(s)\,H(s) = \infty e^{-j\pi/2}$

From above two equations, we can say that section C_4 in s-plane is mapped as circular arc of infinite radius with argument varying from $+\pi/2$ to $-\pi/2$ as shown in Fig. (4.36.8) and (4.36.9).

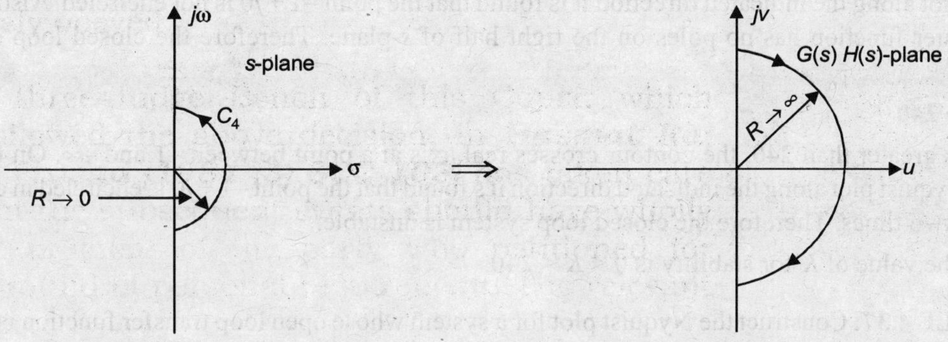

Fig. 4.36.8: Section C_4 in s-plane.

Fig. 4.36.9: Mapping of section C_4 in $G(s)\,H(s)$ s-plane.

Complete nyquist plot

The entire Nyquist plot in $G(s)\,H(s)$-plane can be obtained by combining the mappings of individual sections, as shown in Fig. (4.36.10).

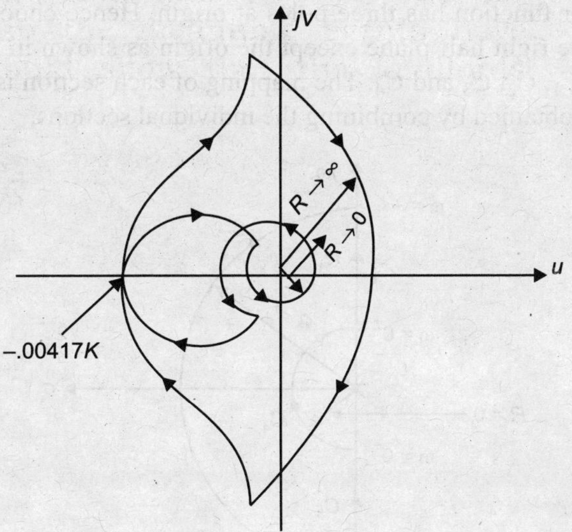

Fig. 4.36.10: Nyquist plot of $G(s)H(s) = K/s(s + 2)\,(s + 10)$.

Stability analysis

When, $-0.00417K = -1$, the contour passes through $(-1 + j0)$ point and corresponding value of K is the limiting value of K for stability.

$$\therefore \qquad \text{Limiting value of } K = \frac{1}{.00417} = 240$$

when K < 240

when K is less than 240, the contour crosses real axis at a point between 0 and -1. On travelling through Nyquist plot along the indicated direction it is found that the point $-1 + j0$ is not encircled. Also the open loop transfer function has no poles on the right half of s-plane. Therefore the closed loop system is stable.

when *K* > 240

when K is greater than 240, the contour crosses real axis at a point between -1 and $-\infty$. On travelling through Nyquist plot along the indicated direction it s found that the point $-1 + j0$ is encircled in clockwise direction two times. Therefore the closed loop system is unstable.

Result: The value of K for stability is $0 < K < 240$.

EXAMPLE 4.37: Construct the Nyquist plot for a system whose open loop transfer function is given by $G(s)\,H(s) = \dfrac{K(1+s)^2}{s^3}$. Find the range of K for stability.

Solution: Given that,

$$G(s)\,H(s) = \frac{K(1+s)^2}{s^3}$$

The open loop transfer function has three poles at origin. Hence choose the Nyquist contour on s-plane enclosing the entire right half plane except the origin as shown in Fig. (4.37.1). The Nyquist contour has four sections C_1, C_2, C_3 and C_4. The mapping of each section is performed separately and the overall Nyquist plot is obtained by combining the individual sections.

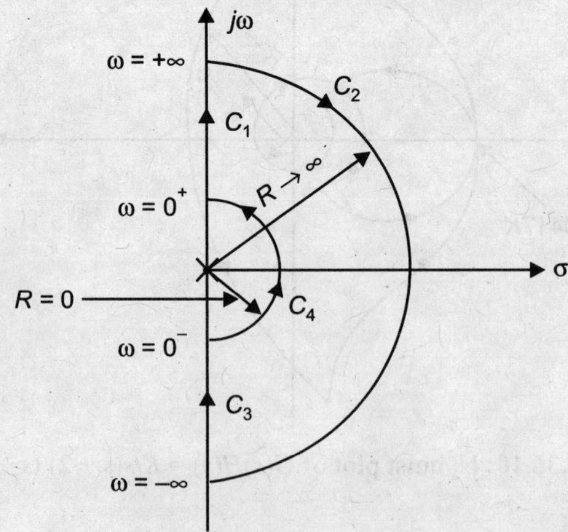

Fig. 4.37.1: Nyquist contour in s-plane.

Mapping of Section C_1

In section C_1, ω varies from, 0 to +∞. The mapping of section C_1 is given by the locus of $G(j\omega)\,H(j\omega)$ as ω is varied from 0 to ∞. This locus is the polar plot of $G(j\omega)\,H(j\omega)$.

$$G(s)\,H(s) = \frac{K(1+s)^2}{s^3}$$

Let, $s = j\omega$,

$$\therefore \quad G(j\omega)\,H(j\omega) = \frac{K(1+j\omega)^2}{(j\omega)^3} = \frac{K(1+\omega^2+2j\omega)}{-j\omega^3}$$

$$= \frac{K(1-\omega)^2}{-j\omega^3} + \frac{K\,2j\omega}{-j\omega^3} = \frac{-2K}{\omega^2} + j\frac{K(1-\omega^2)}{\omega^3}$$

when the locus of $G(j\omega)\,H(j\omega)$ crosses real axis the imaginary term will be zero and the corresponding frequency is the phase crossover frequency, ω_{pc}.

$$\therefore \quad \text{At } \omega = \omega_{pc},\ \omega_{pc} \quad K(1-\omega_{pc}^2) = 0$$

$$\therefore \quad 1-\omega_{pc}^2 = 0 \qquad \omega_{pc} = 1 \text{ rad/sec}$$

At $\omega = \omega_{pc} = 1$ rad/sec,

$$G(j\omega)\,H(j\omega) = \frac{2K}{\omega^2} = \frac{-2K}{1^2} = -2K$$

$$G(j\omega)\,H(j\omega) = K(1+j\omega)^2 / (j\omega)^3$$

$$= K\sqrt{1+\omega^2}\,\angle \tan^{-1}\omega \sqrt{1+\omega^2}\,\angle \tan^{-1}\omega \,/\, \omega^3\,\angle 270^\circ$$

$$= \frac{K(1+\omega^2)}{\omega^3}\angle(2\tan^{-1}\omega - 270^\circ)$$

As $\omega \to 0$, $\quad G(j\omega)\,H(j\omega) \to \infty \angle -270^\circ$

As $\omega \to \infty$, $\quad G(j\omega)\,H(j\omega) \to 0 \angle -90^\circ$

From above two equations we can say that the polar plots starts at –270° axis at infinity, crosses axis at $-2K$ and ends at origin in third quadrant. The section C_1 and its mapping are shown in Fig. (4.36.2) and (4.36.3).

Mapping of Section C_2

The mapping of section C_2 from s-plane to $G(s)\,H(s)$-plane is obtained by letting $s = \underset{R\to\infty}{Lt}\ Re^{j\theta}$ in $G(s)\,H(s)$ and varying θ from +π/2 to –π/2. Since $s \to Re^{j\theta}$ and $R \to \infty$, the $G(s)\,H(s)$ can approximated as $(1+sT) \approx sT]$

$$G(s)\,H(s) = \frac{K(1+s)^2}{s^3} = \frac{Ks^2}{s^3} = K/s$$

CHAPTER 4

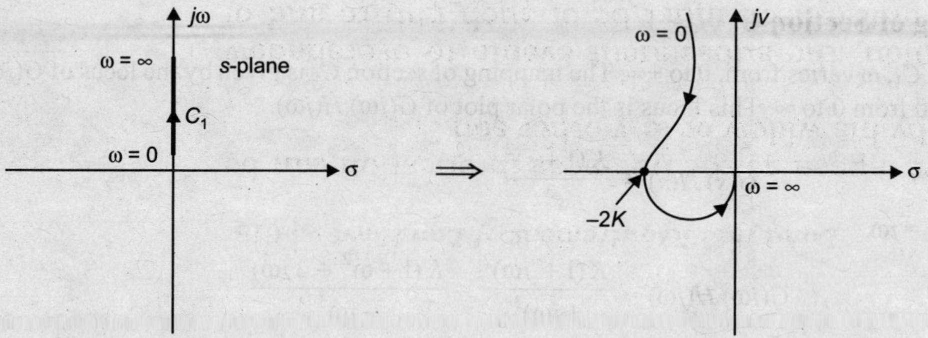

Fig. 4.37.2: Section C_1 in *s*-plane.

Fig. 4.37.3: Mapping of section C_1 in $G(s)\,H(s)$ *s*-plane.

Let, $s = \underset{R \to \infty}{Lt}\ Re^{j\theta}$.

$$\therefore \qquad G(s)\,H(s)\Big|_{s = \underset{R \to \theta}{Lt}\ Re^{j\theta}} = \frac{K}{s}\Big|_{s = \underset{R \to \infty}{Lt}\ Re^{j\theta}} = \frac{K}{\underset{R \to \infty}{Lt}\ Re^{j\theta}} = oe^{-j\theta}$$

when $\theta = \pi/2$, $\qquad G(s)\,H(s) = oe^{-j\,\pi/2}$

when $\theta = -\pi/2$, $\qquad G(s)\,H(s) = oe^{j\,\pi/2}$

From above two equations, we can say that section C_2 in *s*-plane is mapped as circular arc of zero radius around origin $G(s)\,H(s)$-plane with argument varying from $-\pi/2$ to $+3\pi/2$ as shown in Fig. (4.36.4) and (4.36.5).

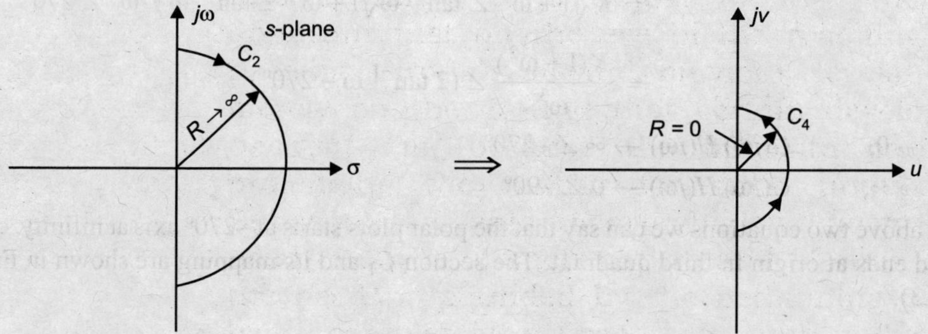

Fig. 4.37.4: Section C_2 in *s*-plane.

Fig. 4.37.5: Mapping of section C_2 in $G(s)\,H(s)$ *s*-plane.

Mapping of Section C_3

In section C_3, ω varies from $-\infty$ to 0. The mapping of section C_3 is given by the locus of $G(j\omega)\,H(j\omega)$ as ω is varied from $-\infty$ to 0. This locus is the inverse polar plot of $G(j\omega)\,H(j\omega)$.

The inverse plot is given by the mirror image of polar plot with respect to real axis. The section is C_3 in *s*-plane and its corresponding contour $G(s)\,H(s)$ plane are shown in Fig. (4.37.6) and (4.37.7).

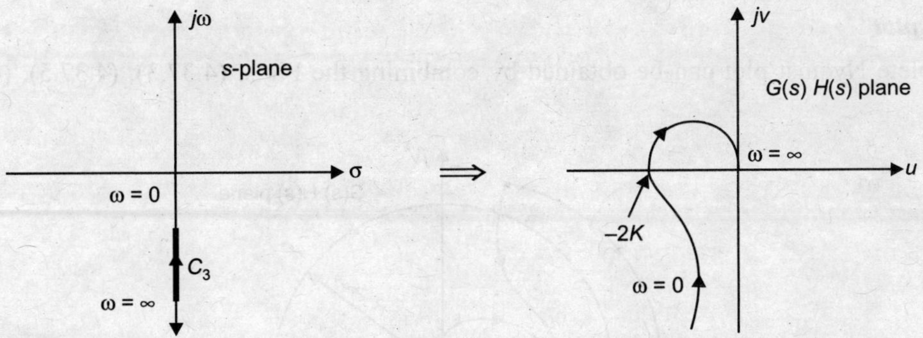

Fig. 4..37.6: Section C_3 in s-plane.

Fig. 4.37.7: Mapping of section C_3 in $G(s)\,H(s)$ s-plane.

Mapping of Section C_4

The mapping of section C_4 from s-plane to $G(s)\,H(s)$-plane is obtained by letting $s = \underset{R \to 0}{Lt}\ \text{Re}^{j\theta}$ in $G(s)$ $H(s)$ and varying θ from $-\pi/2$ to $+\pi/2$. Since $s \to \text{Re}^{j\theta}$ and $R \to 0$ the $G(s)\,H(s)$ can be approximated $1 + sT \approx 1$ [$\because sT \lll 1$]

$$G(s)\,H(s) = \frac{K(1+s)^2}{s^3} \approx \frac{K \times 1}{s^3} = K/s^3$$

Let $$s = \underset{R \to 0}{Lt}\ \text{Re}^{j\theta}$$

$$\therefore \qquad G(s)\,H(s)\Big|_{\underset{R \to 0}{Lt}\ \text{Re}^{j\theta}} = \frac{K}{s^3}\Bigg|_{s = \underset{R \to 0}{Lt}\ \text{Re}^{+j\theta}} = \frac{K}{\underset{R \to 0}{Lt}\ (\text{Re}^{+j\theta})^3} \infty e^{-j3\theta}$$

when $\theta = -\pi/2$, $\qquad G(s)\,H(s) = \infty e^{+j3\pi/2}$

when $\theta = \pi/2$, $\qquad G(s)\,H(s) = \infty e^{-j3\pi/2}$

From above two equations, we can say that section C_4 in s-plane is mapped as circular arc of finite radius varying from $+3\pi/2$ to $-3\pi/2$ as shown in Fig. (4.37.8) and (4.37.9).

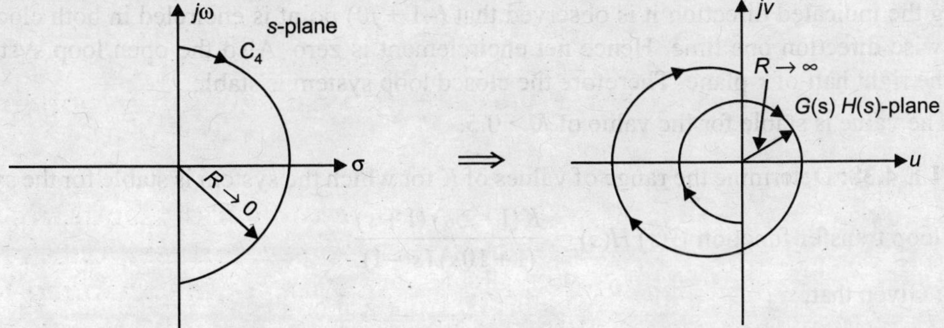

Fig. 4.37.8: Section C_4 in s-plane.

Fig. 4.37.9: Mapping of section C_4 in $G(s)\,H(s)$ s-plane.

Complete plot

The complete Nyquist plot can be obtained by combining the Fig's. (4.37.3), (4.37.5), (4.37.7) and (4.37.9).

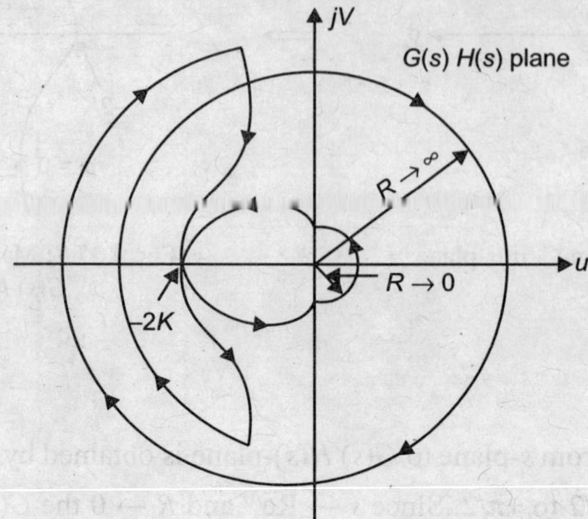

Fig. 4.37.10: Nyquist plot of $G(s)\,H(s) = \dfrac{K(1+s)^2}{s^3}$.

Stability analysis

When, $-2K = -1$, the contour passes through $(-1 + j0)$ point and corresponding value of K is the limiting value of K for stability.

$\therefore$ Limiting value of $K = 1/2 = .5$

when $K < 0.5$

when K is less than .5, the contour crosses real axis at a point between 0 and –1. On travelling through Nyquist plot along the indicated direction it is observed that the $-1 + j0$ point is encircled in clockwise direction one time therefore the system is unstable.

when $K > .5$

when $K > 0.5$, the contour crosses real axis at a point between –1 and $-\infty$. On travelling through Nyquist plot along the indicated direction it is observed that $(-1 + j0)$ point is encircled in both clockwise and anticloskwise direction one time. Hence net encirclement is zero. Also the open loop system has no poles at the right half of s-plane. Therefore the closed loop system is stable.

Result: The value is stable for the value of $K > 0.5$.

EXAMPLE 4.38: Determine the range of values of K for which the system is stable for the system with the open loop transfer function $G(s)\,H(s) = \dfrac{K(1+.5s)\,(1+s)}{(1+10s)\,(s-1)}$.

Solution: Given that,

$$G(s)\,H(s) = \frac{K(1+0.5s)\,(1+s)}{(1+10s)\,(s-1)}$$

The open loop transfer function does not have a pole at origin. Hence choose the Nyquist contour on s-plane enclosing the entire half plane as shown in Fig. (4.38.1).

The Nyquist contour has 3 sections. C_1, C_2 and C_3. The mapping of each section is performed separately and the overall Nyquist plot is obtained by combining the individual sections.

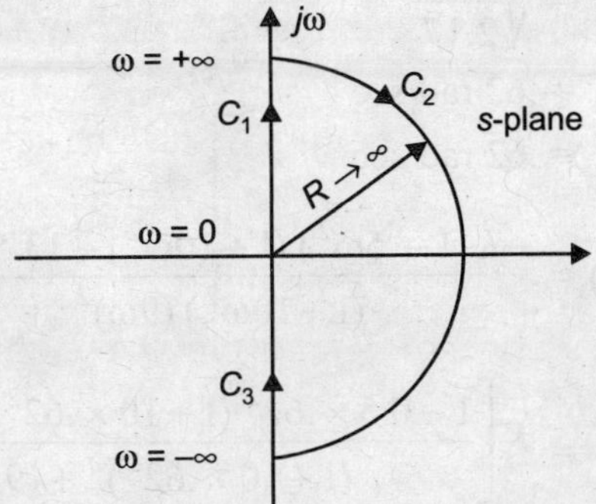

Fig. 4.38.1: Nyquist contour in s-plane.

Mapping of Section C_1

In section C_1, ω varies from 0 to $+\infty$. The mapping of section C_1 is given by the locus of $G(j\omega)\,H(j\omega)$ as w is varied from 0 to ∞. This locus is the polar plot of $G(j\omega)\,H(j\omega)$

$$G(s)\,H(s) = \frac{K(1+.5s)\,(1+s)}{(1+10s)\,(s-1)}$$

Let $s = j\omega$

$$\therefore \quad G(j\omega)\,H(j\omega) = \frac{K(1+j0.5\omega)\,(1+j\omega)}{(1+j\,10\omega)\,(-1+j\omega)} = \frac{K(1+j1.5\omega-0.5\omega^2)}{-1-j9\omega-10\omega^2}$$

$$= \frac{K(1-.5\omega^2)+j1.5\omega K}{-(1+10\omega^2)-j9\omega}$$

$$G(j\omega)\,H(j\omega) = \frac{K(1-0.5^2\,\omega)+j1.5\omega K}{-(1+10\omega^2)-j9\omega} \times \frac{-1(1+10\omega^2)+j9\omega}{-(1+10\omega^2)+j9\omega}$$

$$= \frac{-K(1-0.5\omega^2)\,(1+10\omega^2)-13.5\omega^2 K + j[9\omega K(1-0.5\omega^2)-1.5\omega K(1+10\omega^2)}{(1+10\omega^2)^2+(9\omega)^2}$$

when the$G(j\omega)\,H(j\omega)$ locus crosses real axis the imaginary term in zero and the corresponding frequency is the phase crossover frequency.

$$\therefore \quad \text{At } \omega = \omega_{pc},\ 9\omega_{pc}\,K(1-0.5\,\omega_{pc}^2)-1.5\,\omega_{pc}\,K\,(1+10\,\omega_{pc}^2)=0$$

$$9\omega_{pc}\;K\left(1-0.5\,\omega_{pc}^2\right) = 1.5\,\omega_{pc}K\left(1+10\omega_{pc}^2\right)$$

$$1 - .5\omega_{pc}^2 = .167 + 1.67\omega_{pc}^2$$

$$2.17\omega_{pc}^2 = .833$$

$$\omega_{pc} = \sqrt{\frac{.833}{2.17}}$$

$$\omega_{pc} = .62 \text{ rad/sec}$$

At $\omega = \omega_{pc} = .62$ rad/sec

$$G(j\omega)\,H(j\omega) = \frac{-K(1-.5\omega^2)(1+10\omega^2) - 13.5\omega^2 K}{(1+10\omega^2)(9\omega)^2}$$

$$= K\left[\frac{1-0.5\times.62^2\,(1+10\times.62^2)+13.5\times.62^2}{(1+10\times.62^2)^2+(9\times.62)^2}\right]$$

$$G(j\omega)\,H(j\omega) = -.167\,K$$

The $G(j\omega)\,H(j\omega)$ locus crosses real axis at a point –0.167.

The shape of $G(j\omega)\,H(j\omega)$ locus is determined by calculating the magnitude and phase of $G(j\omega)\,H(j\omega)$ for various values of ω.

$$G(j\omega)\,H(j\omega) = \frac{K(1+j.5\omega)(1+j\omega)}{(1+j10\omega)(-1+j\omega)}$$

$$= \frac{K\sqrt{1+(.5\omega)^2}\,\angle\tan^{-1}.5\omega\sqrt{1+\omega^2}\,\angle\tan^{-1}\omega}{\sqrt{1+(10\omega)^2}\,\angle\tan^{-1}10\omega\sqrt{1+\omega^2}\,\angle(180^\circ-\tan^{-1}\omega)}$$

$$= \frac{K\sqrt{1+.25\omega^2}}{\sqrt{1+100\omega^2}}\angle(\tan^{-1}.5\omega + 2\tan^{-1}\omega - \tan^{-1}10\omega - 180^\circ)$$

$$|G(j\omega)\,H(j\omega)| = K\sqrt{1+.25\omega^2}\,/\sqrt{1+100\omega^2}$$

$$\angle G(j\omega)\,H(j\omega) = \tan^{-1}.5\omega + 2\tan^{-1}\omega - \tan^{-1}10\omega - 180^\circ$$

ω, rad/sec	0	0.1	0.5	1.5	2.0	5.0	∞
$\lvert G(j\omega)\,H(j\omega)\rvert$	K	$.707K$	$.202K$	$.083K$	$.07K$	$.05K$	0
$\angle G(j\omega)\,H(j\omega)$, deg	–180	–210	–191	–116	–95	–43	0

It is understood from the above table is that

1. The locus of $G(j\omega)\,H(j\omega)$ starts at $K\angle -180^\circ$ when $\omega = 0$ and travels in second quadrant,
2. Locus crosses axis at $-.167\,K$ and enters third quadrant.
3. Then the locus corsses negative imaginary axis and enters 4th quadrant.
4. Finally the locus ends at origin when $\omega = \infty$.

The section C_1 in s-plane end its corresponding mappings are shown in below Fig. (4.38.2) and (4.38.3).

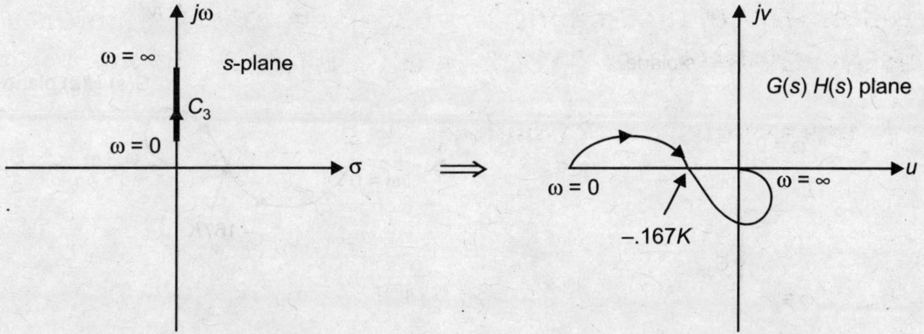

Fig. 4.38.2: Section C_1 in s-plane.

Fig. 4.38.3: Mapping of section C_1 in $G(s)\,H(s)$ s-plane.

Mapping of Section C_2

The mapping of section C_2 from s-plane to $G(s)\,H(s)$-plane is obtained by letting $s = \underset{R \to \infty}{Lt}\ Re^{j\theta}$ in $G(s)\,H(s)$ and varying θ from $-\pi/2$ to $+\pi/2$. Since $s \to Re^{j\theta}$ and $R \to \infty$, the $G(s)\,H(s)$ can be approximated as $(1 + sT) \approx sT]\ [\because sT \gg 1]$

$$G(s)\,H(s) = \frac{K(1+.5s)\,(1+s)}{(1+10s)\,(s-1)} \approx \frac{K.5s \times s}{10s \times s} = .05K$$

The approximate $G(s)\,H(s)$ is independent of s and so the contour of section C_2 in s-plane is mapped as a point at $.05K$ in $G(s)\,H(s)$ plane.

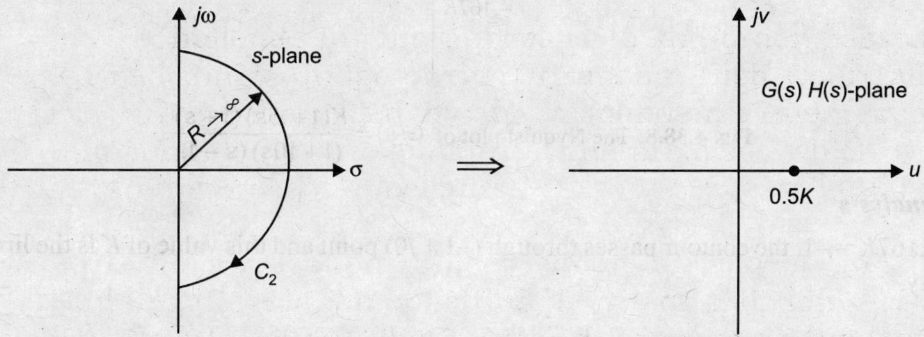

Fig. 4.38.4: Section C_2 in s-plane.

Fig. 4.38.5: Mapping of section C_2 in $G(s)\,H(s)$ s-plane.

Mapping of Section C_3

In section C_3, ω varies from $-\infty$ to 0. The mapping of section C_3 is given by the locus of $G(j\omega)\,H(j\omega)$ as ω is varied from $-\infty$ to 0. This locus is the inverse polar plot of $G(j\omega)\,H(j\omega)$.

CHAPTER 4

The inverse plot is given by the mirror image of polar plot with respect to real axis. The section is C_3 in s-plane and its corresponding contour $G(s)\ H(s)$ plane are shown in below Fig. (4.38.6) and (4.38.7).

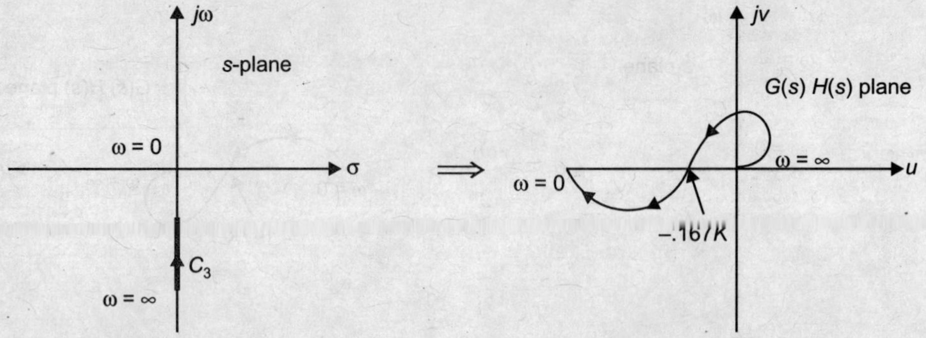

Fig. 4.38.6: Section C_3 in s-plane.

Fig. 4.38.7: Mapping of section C_3 in $G(s)\ H(s)$ s-plane.

Complete nyquist plot

The entire nyquist plot in $G(s)\ H(s)$ is obtained by combining Fig. (4.38.3), (4.38.5) and (4.38.7). The complete plot is shown in Fig. (4.38.8).

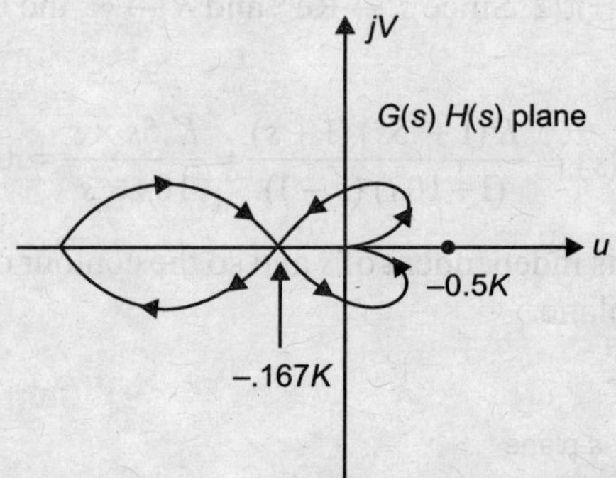

Fig. 4.38.8: The Nyquist plot of $G(s) = \dfrac{K(1+.5s)\,(1+s)}{(1+10s)\,(s-1)}$.

Stability analysis

When, $-0.167K = -1$, the contour passes through $(-1 + j0)$ point and this value of K is the limiting value for stability.

$$\text{The limiting value of } K = \frac{1}{.167} = 6$$

when K < 6

when $K < 6$, the locus crosses red axis between 0 and -1. On travelling through the locus it is observed that the $-1 + j0$ point is encircled clockwise and so the closed loop system is unstable.

when K > 6

when $K > 6$, the locus crosses real axis between -1 and $-\infty$. On travelling through the locus is observed that the $-1 + j0$ point is encircled anticlockwise and one time. Hence net encirclement is zero. Also the open loop system has one pole at the right half of *s*-plane. Hence the system is stable.

Result:

1. The open loop system is unstable.
2. For stability of the closed loop system the value of *K* should be greater them six ($K > 6$).

EXAMPLE 4.38: Construct Nyquist plot for a feedback control system whose open loop transfer function is given by, $G(s)\,H(s) = \dfrac{5}{s(1-s)}$. Commenton the stability of open loop and closed loop system.

Solution: Given that,

$$G(s)\,H(s) = \frac{5}{s(1-s)}$$

The open loop transfer function has a pole at origin. Hence choose the Nyquist contour on *s*-plane, enclosing the entire right half *s*-plane except the origin as shown in Fig. (4.38.1).

The Nyquist contour has 4 sections, C_1, C_2, C_3 and C_4. The mapping of each section is performed separately and the overall plot is obtained by combining all individual sections.

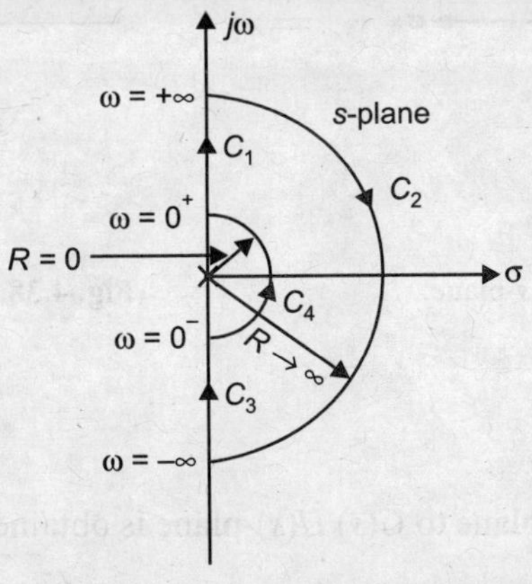

Fig. 4.38.1: Nyquist contour in *s*-plane.

Mapping of Section C_1

In section C_1, ω varies from 0 to $+\infty$. The mapping of section C_1 is given by the locus of $G(j\omega)\,H(j\omega)$ as ω is varied from 0 to ∞. This locus is the polar plot of $G(j\omega)\,H(j\omega)$

$$G(s)\,H(s) = 5/s\,(1-s)$$

Let, $s = j\omega$

$$\therefore \qquad G(j\omega)\,H(j\omega) = \frac{5}{j\omega(1-j\omega)} = \frac{5}{\omega\,\angle 90^{\circ}\,\sqrt{1+\omega^2}\,\angle -\tan^{-1}\omega}$$

$$= \frac{5}{\omega\sqrt{1+\omega^2}} \angle(-90^\circ + \tan^{-1}\omega)$$

$$\therefore \quad |G(j\omega)\,H(j\omega)| = 5/\omega\sqrt{1+\omega^2}$$

$$\angle G(j\omega)\,H(j\omega) = -90^\circ + \tan^{-1}\omega$$

The exact shape of $G(j\omega)\,H(j\omega)$ locus is determined by calculating the magnitude and phase of $G(j\omega)\,H(j\omega)$ for various values of ω.

ω, rad/sec	0	0.6	1.0	2.0	10.0	∞
$\lvert G(j\omega)\,H(j\omega)\rvert$	∞	7.15	3.53	1.12	.05	0
$\angle G(j\omega)\,H(j\omega)$, deg	–90	–59	–45	–26	–5	0

$G(j\omega)\,H(j\omega)$ locus in section C_1 starts at –90° axis at infinity for $\omega = 0$ and meets the origin along 0°axis when $\omega = \infty$. The mapping of section C_1 is shown in Fig. (4.38.2) and (4.38.3).

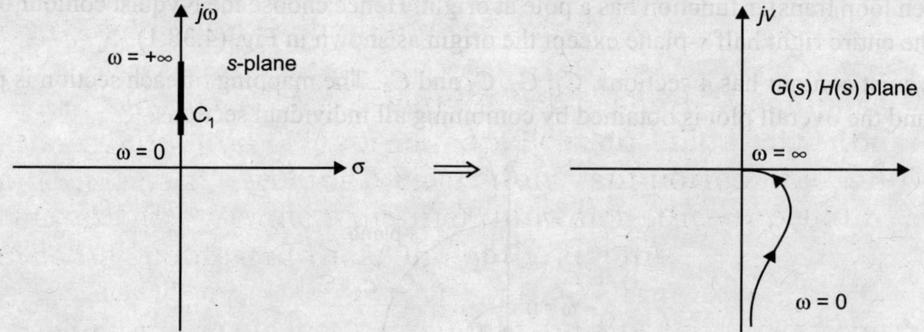

Fig. 4.38.2: Section C_1 in s-plane.

Fig. 4.38.3: Mapping of section C_1 in $G(s)\,H(s)$ plane.

Mapping of Section C_2

The mapping of section C_2 from s-plane to $G(s)\,H(s)$-plane is obtained by letting $s = \underset{R\to\infty}{Lt}\ Re^{j\theta}$ in $G(s)\,H(s)$ and varying θ from $+\pi/2$ to $-\pi/2$. Since $s \to Re^{j\theta}$ and $R \to \infty$, the $G(s)\,H(s)$ can be approximated as $[(1-s) \approx s]$

$$\therefore \quad G(s)\,H(s) = \frac{5}{s(1-s)} \approx \frac{5}{s(-s)} = \frac{5}{s^2 e^{j\pi}}$$

Let, $\quad s = \underset{R\to\infty}{Lt}\ \text{Re}^{j\theta}$

$$\therefore \quad G(s)\,H(s)\Big|_{s = \underset{R\to\theta}{Lt}\ \text{Re}^{j\theta}} = \frac{5}{\underset{R\to\infty}{Lt}\ (\text{Re}^{j\theta})^2 e^{j\pi}} = oe^{-j(2\theta+\pi)}$$

when $\theta = \pi/2$, $\quad G(s)\, H(s) = oe^{-j2\pi}$

when $\theta = -\pi/2$, $\quad G(s)\, H(s) = oe^{jo}$

From above two equations, we can say that section C_2 in s-plane is mapped as circular arc of zero radius around origin in $G(s)\ H(s)$-plane with argument varying from -2π to $+0$ as shown in Fig. (4.38.4) and (4.38.5).

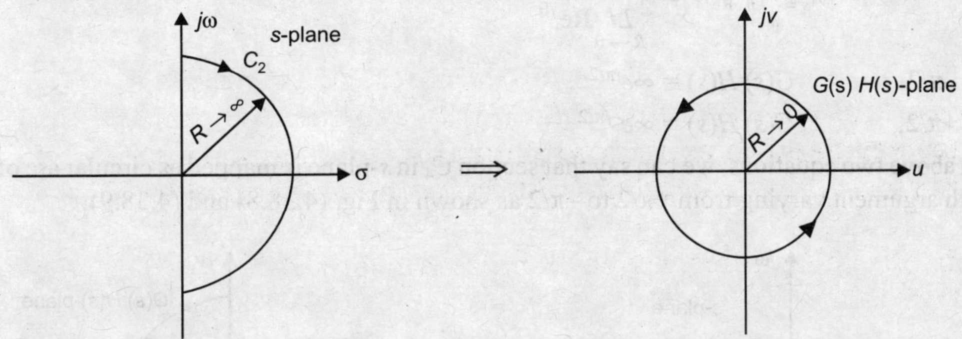

Fig. 4.38.4: Section C_2 in s-plane.

Fig. 4.38.5: Mapping of section C_2 in $G(s)\ H(s)$ s-plane.

Mapping of Section C_3

In section C_3, ω varies from $-\infty$ to 0. The mapping of section C_3 is given by the locus of $G(j\omega)\ H(j\omega)$ as ω is varied from $-\infty$ to 0. This locus is the inverse polar plot of $G(j\omega)\ H(j\omega)$.

The inverse polar plot is with respect to real axis. The section is C_3 in s-plane and its corresponding contour $G(s)\ H(s)$ plane are shown in below Fig. (4.38.6) and (4.38.7).

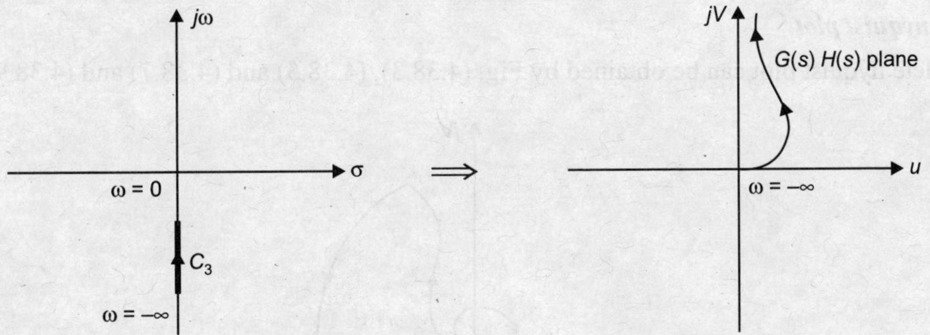

Fig. 4.38.6: Section C_3 in s-plane.

Fig. 4.38.7: Mapping of section C_3 in $G(s)\ H(s)$ s-plane.

Mapping of Section C_4

The mapping of section C_4 from s-plane to $G(s)\ H(s)$-plane is obtained by letting $s = \underset{R \to 0}{Lt}\ \text{Re}^{j\theta}$ in $G(s)$ $H(s)$ and varying θ from $-\pi/2$ to $+\pi/2$. Since $s \to \text{Re}^{j\theta}$ and $R \to 0$, $G(s)\ H(s)$ can be approximated $1 - s \approx 1$ [$\because s \ll 1$]

$$\therefore \qquad G(s)\,H(s) = \frac{5}{s(1-s)} \approx \frac{5}{s\times 1} = 5/s$$

$$\text{Let} \qquad s = \underset{R\to 0}{Lt}\ \mathrm{Re}^{j\theta}$$

$$\therefore \qquad G(s)\,H(s)\Big|_{s=\underset{R\to 0}{Lt}\ \mathrm{Re}^{j\theta}} = \frac{5}{\underset{R\to 0}{Lt}\ \mathrm{Re}^{j\theta}} = \infty e^{-j\theta}$$

when $\theta = -\pi/2$, $\quad G(s)\,H(s) = \infty e^{j\pi/2}$

when $\theta = +\pi/2$, $\quad G(s)\,H(s) = \infty e^{-j\pi/2}$

From above two equations, we can say that section C_4 in s-plane is mapped as circular arc of infinite radius with argument varying from $+\pi/2$ to $-\pi/2$ as shown in Fig. (4.38.8) and (4.38.9).

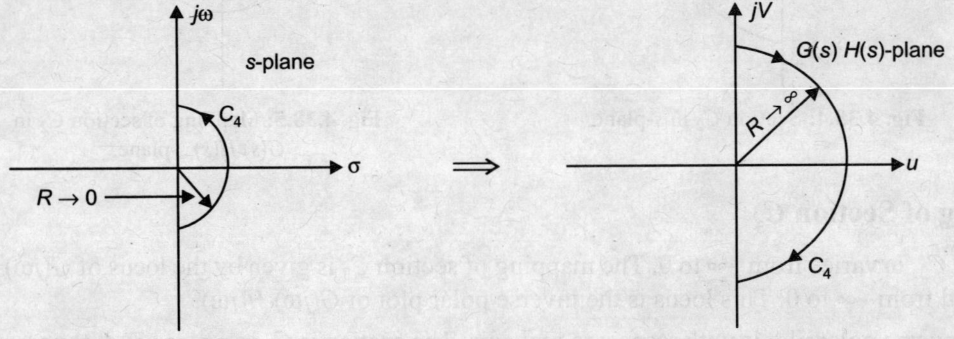

Fig. 4.38.8: Section C_4 in s-plane.

Fig. 4.38.9: Mapping of section C_4 in $G(s)\,H(s)$ s-plane.

Complete nyquist plot

The complete nyquist plot can be obtained by Fig. (4.38.3), (4.38.5) and (4.38.7) and (4.38.9).

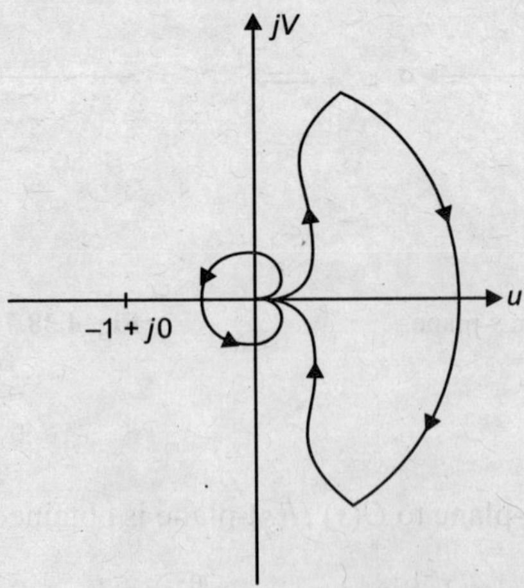

Fig. 4.38.10: Complete Nyquist plot of $G(s) = 5s(1 - s)$.

Stability analysis

The nyquist contour in $G(s)\,H(s)$ place does not encircle the point $(-1 + j0)$ but the open loop transfer function has one pole on the right half s-plane. ∴ The system is unstable.

Result: Both open and closed loop systems are unstable.

EXAMPLE 4.39: Determine the stability of closed loop system, whose open loop transfer function in $G(s)\,H(s) = (s + 2)/(s + 1)\,(s - 1)$, use nyquist stability criterion. Comment on the stability of open-loop and closed loop system.

Solution: Given that

$$G(s)\,H(s) = (s + 2)\big/(s + 1)\,(s - 1)$$

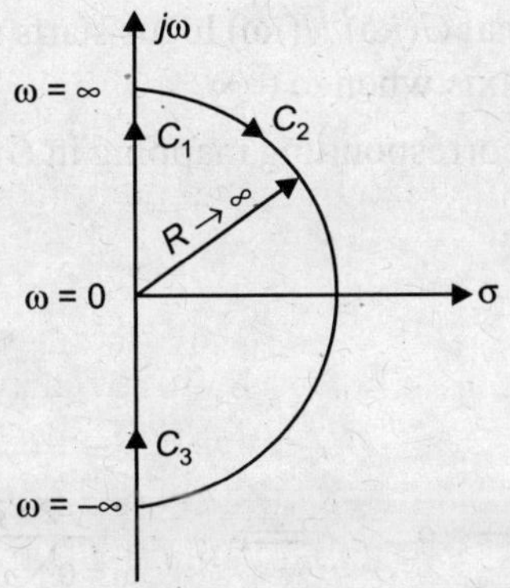

jω
ω = ∞
C₂
C₁
R → ∞
ω = 0
σ
C₃
ω = −∞

Fig. 4.39.1: Nyquist contour in s-plane.

The open loop transfer function does not have a pole at origin. Hence choose the Nyquist contour on s-plane enclosing the entire right half place as shown in Fig. (4.39.1).

The Nyquist contour has 3 sections C_1, C_2, C_3. The mapping of each section is performed separately and the overall Nyquist plot is obtained by combining them.

Mapping of Section C_1

In section C_1, ω varies from 0 to +∞. The mapping of section C_1 is given by the locus of $G(j\omega)\,H(j\omega)$ as ω is varied from 0 to ∞. This locus is the polar plot of $G(j\omega)\,H(j\omega)$

$$G(s)\,H(s) = (s+2)\big/(s+1)\,(s-1) = \frac{2(1+.5s)}{(1+s)\,(-1+s)}$$

Let, $s = j\omega$

$$\therefore \qquad G(j\omega)\,H(j\omega) = \frac{2(1+j0.5\omega)}{(1+j\omega)\,(-1+j\omega)} = \frac{2\sqrt{1+2.5\omega^2}\;\angle\tan^{-1}.5\omega}{\sqrt{1+\omega^2}\;\angle\tan^{-1}\omega\sqrt{1+\omega^2}\;\angle 180^\circ - \tan^{-1}\omega}$$

$$= \frac{2\sqrt{1+2.5\omega^2}}{1+\omega^2}\angle(180^\circ + \tan^{-1}.5\omega)$$

$$\therefore \qquad |G(j\omega)\,H(j\omega)| = \frac{2\sqrt{1+.25\omega^2}}{1+\omega^2}$$

$$\angle G(j\omega)\,H(j\omega) = -180^\circ + \tan^{-1}.5\omega$$

The exact shape of $G(j\omega)\ H(j\omega)$ locus as determined by calculating the magnitude and phase of $G(j\omega)\ H(j\omega)$ for various of ω values.

ω, rad/sec	0	0.4	1	2	10	∞
$\lvert G(j\omega)\,H(j\omega)\rvert$	2	1.76	1.12	.57	0.1	0
$\angle G(j\omega)\,H(j\omega)$, deg	–180	–168	–153	–135	–101	-90

From above values, we can say that $G(j\omega)\ H(j\omega)$ locus starts at two axis at a magnitude for –2 for $\omega = 0$ and meets the origin along –90° axis when $\omega = \infty$.

The section C_1 in s-plane and its corresponding mapping in $G(s)\ H(s)$ plane is shown in Fig. (4.39.2) and (4.39.3).

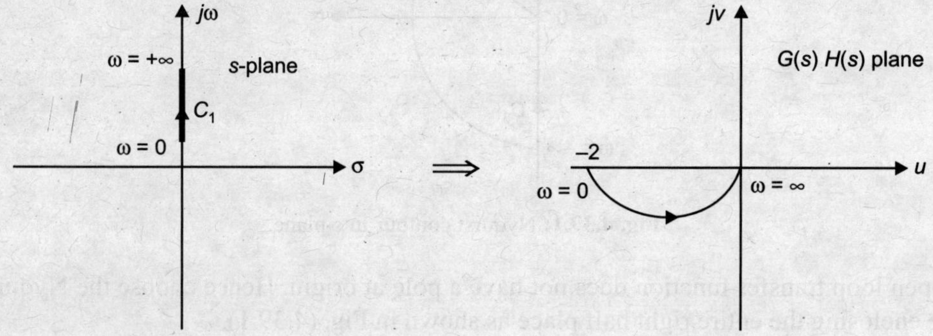

Fig. 4.39.2: Section C_1 in s-plane.

Fig. 4.39.3: Mapping of section C_1 in $G(s)\ H(s)$ plane.

Mapping of Section C_2

The mapping of section C_2 from s-plane to $G(s)\ H(s)$-plane is obtained by letting $s = \underset{R\to\infty}{Lt}\ Re^{j\theta}$ in $G(s)\ H(s)$ and varying θ from $+\pi/2$ to $-\pi/2$. Since $s \to Re^{j\theta}$ and $R \to \infty$, the $G(s)\ H(s)$ can be approximated as $1 + sT \approx sT$

$$\therefore \qquad G(s)\,H(s) = \frac{2(1+.5s)}{(1+s)\,(1-s)} \approx \frac{2\times .5s}{s\times s} = \frac{1}{s}$$

Let, $$s = \underset{R\to\infty}{Lt}\ \mathrm{R}e^{j\theta}$$

$$\therefore \qquad G(s)\,H(s)\Big|_{s=\underset{R\to\infty}{Lt}\ \mathrm{R}e^{j\theta}} = \frac{1}{\underset{R\to\infty}{Lt}\ \mathrm{R}e^{j\theta}} = oe^{-j\theta}$$

when $\theta = \pi/2$, $\quad G(s)\,H(s) = oe^{-j\pi/2}$

when $\theta = -\pi/2$, $\quad G(s)\,H(s) = oe^{j\pi/2}$

From above two equations, we can say that section C_2 in s-plane is mapped as circular arc of zero radius around origin in $G(s)\,H(s)$-plane with argument varying from $-\pi/2$ to $+\pi/2$ as shown in Fig. (4.39.4) and (4.39.5).

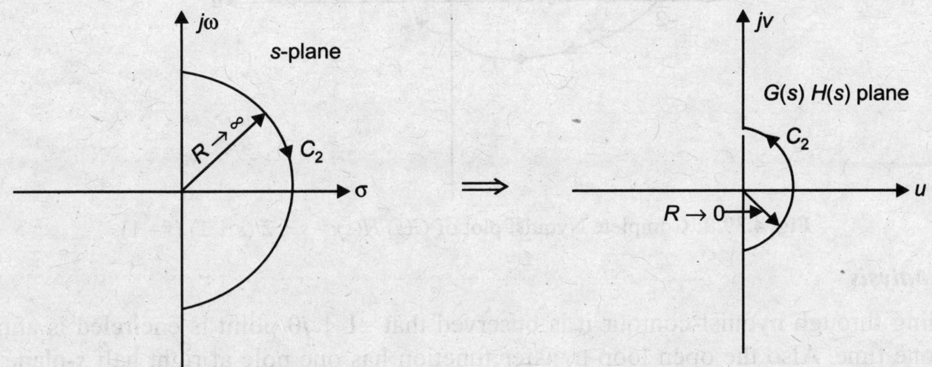

Fig. 4.39.4: Section C_2 in s-plane.

Fig. 4.39.5: Mapping of section C_2 in $G(s)\,H(s)$ s-plane.

Mapping of Section C_3

In section C_3, ω varies from -0 to 0. The mapping of section C_3 is given by the locus of $G(j\omega)\,H(j\omega)$ as ω is varied from $-\infty$ to 0. This locus is the inverse polar plot of $G(j\omega)\,H(j\omega)$.

The inverse polar plot is by the mirror image of polar plot with respect to real axis. The section is C_3 in s-plane and its corresponding contour $G(s)\,H(s)$ plane are shown in below Fig. (4.39.6) and (4.39.2).

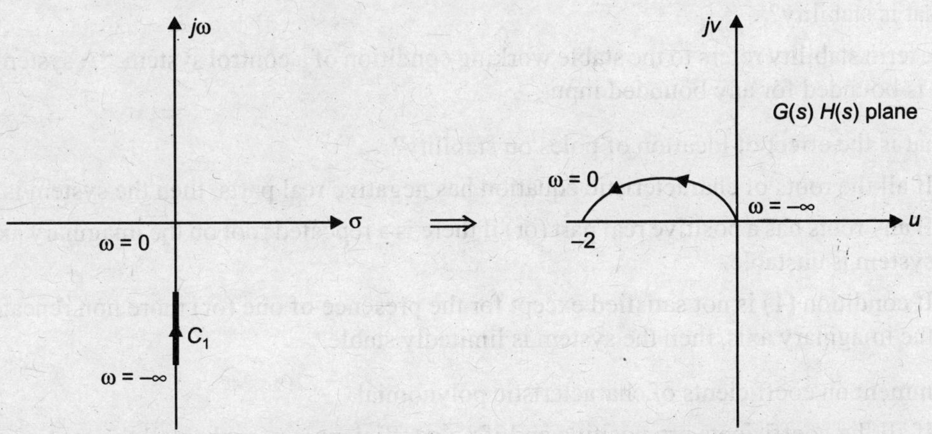

Fig. 4.39.6: Section C_3 in s-plane.

Fig. 4.39.7: Mapping of section C_3 in $G(s)\,H(s)$ s-plane.

Complete nyquist plot

The complete nyquist plot can be obtained by combining Fig. (4.39.3), (4.39.5) and (4.39.7).

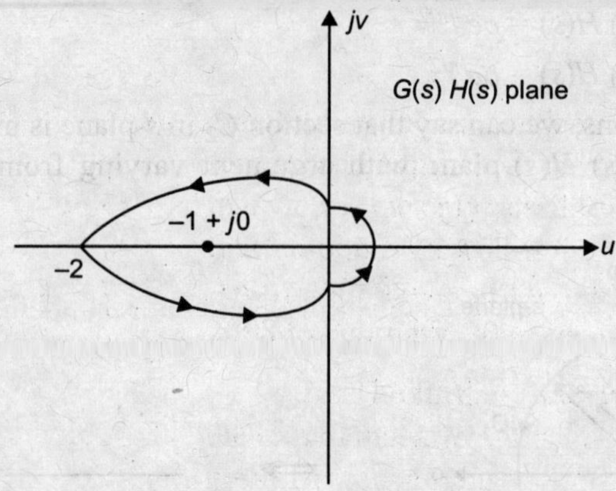

Fig. 4.39.8: Complete Nyquist plot of $G(s)\ H(s) = s + 2/(s + 1)\ (s - 1)$.

Stability analysis

On travelling through nyquist contour it is observed that $-1 + j0$ point is encircled is anticlockwise direction one time. Also the open loop transfer function has one pole at right half *s*-plane. Since the number of anticlockwise encirclement is equal to number of open loop poles on right half *s*-plane, the closed loop system is stable.

Result:

- Open loop system is unstable.
- Closed loop system is stable.

SOME QUESTION AND ANSWERS

Q.1 What is stability?

Ans. The term stability refers to the stable working condition of a control system. "A system is stable if its output is bounded for any bounded input.

Q.2 What is the effect of location of poles on stability?

Ans.
- If all the roots of characteristic equation has negative real parts, then the system is stable.
- If any roots has a positive real part (or) if there is a repeated root on the imaginary axis, then the system is unstable.
- If condition (1) is not satisfied except for the presence of one (or) more non repeated roots on the imaginary axis, then the system is limitedly stable.

Q.3 Comment on coefficients of characteristic polynomial.

Ans.
- If all the coefficients are positive and if no coefficient is zero then all the roots are in the left half plane.
- If any coefficient is equal to zero then some of the roots may be one on the imaginary axis (or) on the right half plane.
- If any coefficient is negative then atleast one root is in the right half plane.

Q.4 What is Routh-Hurwitz criterion?

Ans. Routh-Hurwitz stability criterion is an analytical procedure for determining whether all the roots of a polynomial have negative real parts (or) not.

Q.5 Differentiate the different cases in routh array.

Ans. 1. Normal Routh array

2. A row of all zeros

3. First element of a row is zero but some (or) other elements are not zero.

Q.6 What is Nyquist stability criterion?

Ans. The closed loop system is stable, If $G(s)\,H(s)$ contour in the $G(s)\,H(s)$ plane corresponding to nyquist contour in the s-plane encircles the point $(-1 + j0)$ in the anti-clockwise direction as any times as the number of right half s plane poles of $G(s)\,H(s)$.

Q.7 At what condition the system will be always unstable in Nyquist stability criterion?

Ans. When there is a clock-wise encirclement of the $(-1 + j0)$ point. In this case the system is always unstable.

EXERCISE

1. Check weather the give equations are stable (or) not using Routh Hurwitz criterion.
   (a) $s^7 + 3s^6 + 4s^5 + 3s^4 + 2s^3 + 20s^2 + 72s + 72 = 0$
   (b) $s^6 + 2s^5 + 3s^4 + 2s^3 + 2s^3 + 3s + 5 = 0$
   (c) $s^5 + 3s^4 + 7s^3 + 8s^2 + 9s + 8 = 0$
2. Comment on stability for the equations given below using Nyquist stability criterion
   (a) $G(s)\,H(s) = 20/s(s-1)\,(s+5)$
   (b) $G(s)\,H(s) = (s+1)/(s+1)\,(s+2)$
   (c) $G(s)\,H(s) = (s+3)/(s-2)\,(s+2)$
3. Using Routh criterion determine the locations of the roots of the following characteristic equations and comment on stability of the system.
   (a) $3s^4 + 10s^3 + 5s^2 + 5s + 3 = 0$
   (b) $s^5 + 2s^4 + 24s^3 + 48s^2 - 25s - 50 = 0$
   (c) $2s^6 + 4s^5 + s^4 - 32s^3 + 51s^2 + 3s + 15 = 0$
   (d) $s^6 + 13s^5 + 5s^4 + 9s^3 + 8s^2 + 6s + 4 = 0$
4. The characteristic equations for certain feedback control systems are given below. In each case, determine the range of values of k, for which the system is stable.
   (a) $s^4 + 3s^3 + 3s^2 + s + k = 0$
   (b) $s^4 + s^3 + s^2 + s + k = 0$

5. Open loop transfer functions of certain system using feedback are given below. Determine location of closed loop poles in the s-plane, using Routh criterion. Comment on stability.
   (a) $G(s) = 200(1 + s)/s(1 + 0.1s)\,(1 + 0.2s)\,(1 + 0.5s)$
   (b) $G(s) = 10/(s + 2)\,(s + 1)\,(s^2 + 6s + 25)$
6. Open loop transfer functions of certain unity feedback systems are given below. In each case determine the range of values of k for which the system is stable.
   (a) $G(s) = k(s + 13)/s(s + 3)\,(s + 7)$
   (b) $G(s) = k(s - 1)/s(s + 2)$
   (c) $G(s) = k(s + 1))/s(s - 1)\,(s + 6)$

CHAPTER 5

Root Locus Technique

5.1 ROOT LOCUS TECHNIQUE

Root locus is a simple technique which is used to analyse the systems. The root locus technique is a powerful tool for adjusting the location of closed loop poles to achieve the desired system performance by varying one (or) more system parameters.

Consider the open loop transfer function of system,

$$G(s) = K/s(s + P_1)\,(s + P_2)$$

The closed loop transfer function of the system with unity feedback is given by

$$\frac{C(s)}{R(s)} = \frac{G(s)}{1+G(s)} = \frac{\dfrac{K}{s(s+p_1)\,(s+p_2)}}{1+\dfrac{K}{s(s+p_1)\,(s+p_2)}}$$

$$= \frac{K}{s(s+p_1)\,(s+p_2)+K}$$

The denominator polynomial of $C(s)/R(s)$ is the characteristic Eq. of the system. The characteristic Eq. is given by

$$s(s+p_1)\,(s+p_2)+K=0$$

The roots of characteristic Eq. is a function of open loop gain K. When the gain K is varied from 0 to ∞, the roots of characteristic Eq. will take different values. When $K = 0$, the roots are given by open loop poles. When $K \to \infty$, the roots will take the value of open loop zeroes.

The path taken by the roots of characteristic Eq. when open loop gain K is varied. from 0 to ∞ are called root loci.

Note: In general the roots of characteristic Eq. can be varied by varying any other system parameter other than gain.

In general the closded loop transfer function of the system with multiple loops is obtained from the signal flow graph of the system using Mason's gain formula.

$$\frac{C(s)}{R(s)} = T(s) = \frac{1}{\Delta}\sum P_K \ \Delta_K$$

The determinant, Δ is the denominator polynomial of $C(s)/R(s)$. The characteristic Eq. of the system is given by, $\Delta = 0$. For the single loop system shown in Fig. (3.6.1).

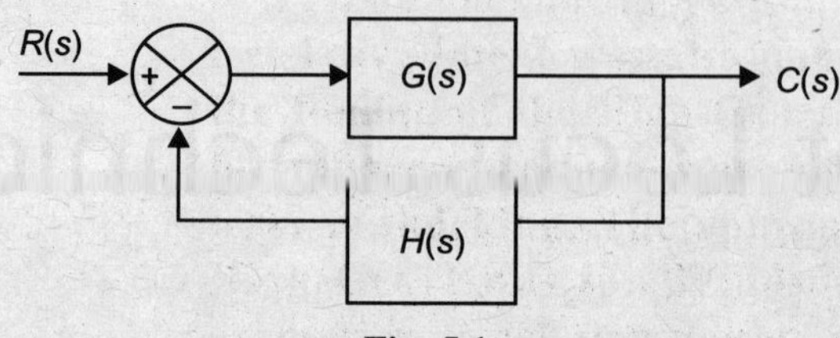

Fig. 5.1

$$\frac{C(s)}{R(s)} = \frac{G(s)}{1 + G(s)\,H(s)}$$

The characteristic Eq., is

$$1 + G(s)\,H(s) = 0$$

Let, $$D(s) = G(s)\,H(s)$$

$$1 + G(s)\,H(s) = 1 + D(s) = 0$$

or $$D(s) = -1 \qquad (5.1)$$

From Eq. (3.6.1) it can be concluded that the roots of the characteristic Eq. occur only for those values of s for which, $D(s) = -1$

The Eq. (3.6.1) can be converted to two Even s conditions given below,

$$|\,D(s)\,| = 1$$

$$\angle D(s) = \pm\, 180°\ (2q + 1) \qquad (5.2)$$

where $$q = 0, 1, 2, 3\ldots$$

The Eq. (3.6.2) is called magnitude criterion and Eq. (3.6.3) is called angle criterion. The magnitude criterion states that $s = s_a$ will be a point on root Locus if for that value of s, $|D(s)| = |G(s)\ H(s)| = 1$. The angle criterion states that $s = s_a$ will be a point on root Locus if for that value of s, $\angle D(s) = \angle G(s)\ H(s)$ is equal to an odd multiply of 180°.

The function $D(s)$ can be expressed as a ratio of two polynomials in s as shown in below.

$$D(s) = G(s)\,H(s) = K\,\frac{(s + z_1)\,(s + z_2)\,(s + z_3)\ldots}{(s + p_1)\,(s + p_2)\,(s + p_3)\ldots} \qquad (5.3)$$

$$\therefore \qquad D(s) = K\,\frac{|\,s + z_1\,|\,|\,s + z_2\,|\,|\,s + z_3\,|\ldots}{|\,s + p_1\,|\,|\,s + p_2\,|\,|\,s + p_3\,|\ldots} = K\,\frac{\prod_{i=1}^{m}|s + z_i|}{\prod_{i=1}^{n}|s + p_i|}$$

where, m = Number of zeros

n = Number of poles

The magnitude criterion states that

$$|D(s)| = 1$$

$$\therefore \qquad K\frac{\sum_{i=1}^{m}|s+z_i|}{\sum_{i=1}^{n}|s+p_i|} = 1$$

or
$$K = \frac{\sum_{i=1}^{n}|s+p_i|}{\sum_{i=1}^{m}|s+z_i|} \tag{5.4}$$

The open-loop gain K corresponding to a point $s = s_a$ on root Locus can be calculated using Eq. (3.6.5). If can be shown that $|s + p_i|$ is equal to the length of vector drawn from $s = p_i$ to $s = s_a$ and $|s + z_i|$ is equal to the length of vector drawn from $s = z_i$ to $s = s_a$. Hence the Eq. for K can be written as

$$K = \frac{\text{Product of length of vector from open loop poles to the point } s = s_a}{\text{Product of length of vectors from open loop zeros of the point } s = s_a}$$

From Eq. (3.6.4)

$$\angle D(s) = \angle(s+z_1) + \angle(s+z_2) + \angle(s+z_3) \ldots - \angle(s+p_1) - \angle(s+p_2) - \angle(s+p_3) \ldots$$

$$= \sum_{j=1}^{m}\angle(s+z_i) - \sum_{i=1}^{n}\angle(s+p_i)$$

The angle criterion states that

$$\angle D(s) = \pm\, 180° \, (2q+1)$$

$$\therefore \qquad \sum_{i=1}^{m}\angle(s+z_i) - \sum_{i=1}^{n}\angle(s+p_i) = \pm\, 180° \, (2q+1) \tag{5.5}$$

The Eq. (3.6.6) can be used to check whether a point $s = s_a$ is a point on root Locus or not. It can be shown that $\angle(s + p_i)$ is equal to the angle of vector drawn from $s = p_i$ to $s = s_a$ and $\angle\, s + z_i$ is equal to the angle of vector drawn from $s = z_i$ to $s = s_a$. Hence Eq. (3.6.6) can be written as

$$\begin{bmatrix}\text{Sum of angles vectors from} \\ \text{open loop zeros to the point} \\ s = s_a\end{bmatrix} - \begin{bmatrix}\text{Sum of angles of vectors from} \\ \text{open loop poles to the point} \\ s = s_a\end{bmatrix}$$

$$= \pm\, 180° \, (2q+1)$$

Construction of Root Locus

The sketching of root locus is difficult task, so we always go for trial and error method. In this method, the poles and zeros of $G(s)\ H(s)$ are located on the s-plane on a graph sheet and a trial point $s = s_a$ is selected. Determine the angles of vectors drawn from poles and zeros to the trial point. From the angle criterion, determine the angle to be contributed by these vectors to make the trial point as a point on root Locus. Shift the trial point suitably so that the angle criterion is satisfied.

A number of points are determined using the above procedure. Join the points by a smooth curve which is the root Locus. The value of K for a particular root can be obtained from the magnitude criterion.

Since, the trial and error method is difficult, we have developed set of rules to reduce the difficulties involved in sketching root locus and to develop a quick approximate sketch. From the approximate sketch a more accurate root locus can be obtained.

Rules for Construction of Root Locus

Rule 1: The root locus is a symmetrial about the real axis.

Rule 2: Each Branch of the root locus originates from an open-loop pole corresponding to $K = 0$ and terminates at either on a finite open loop zero corresponding to $K = \infty$. The number of branches of the root locus terminating on infinity is equal to $(n - m)$,

where $m \rightarrow$ Number of zeros

$n \rightarrow$ Number of poles

Rule 3: Segments of the real axis having an odd number of real axis open-loop poles plus zeros to their right side is a part of the root locus.

Rule 4: Asymptotes making angles with the real axis is given by,

$$\phi_A = \frac{\pm 180° \, [2q + 1]}{(n - m)}; \; q = 0, 1, 2, \ldots (n - m)$$

Rules 5: The point of intersection of the asymptotes with the real axis is at $s = \sigma_A$

where $$\sigma_A = \frac{\text{Sum of poles} - \text{Sum of zeros}}{n - m}$$

Rule 6: The breakaway and break in points of the root locus are determined from the roots of the Eq. $dK/ds = 0$. If r numbers of branches of root locus meet at a point, then they away at angle of $\pm 180°/r$.

Rule 7: The angle of departure from a complex open-loop pole is given by

$\phi_p = \pm 180° \, (2q + 1) + \phi; \; q = 0, 1, 2, \ldots$

where ϕ is the net angle contribution at the pole by all other open loop poles and zeros.

Rule 8: The intersection of root Locus branches with the imaginary axis can be determined by use of the routh criterion, or by letting $s = j\omega$ in the characteristic Eq. and equating the real part and imaginary part to zero, to solve for ω and K. the value of ω is the intersection point on imaginary axis and K is the value of gain at the intersection point.

Rule 9: The open-loop gain K at any point $s = s_a$ on the root locus is given by

$$K = \frac{\text{Product of vector lengths from open loop poles to the point } s_a}{\text{Product of vector lengths from open zeros to the point } s_a}$$

Note: The length of vector should be measured to scale. If there is no finite zero then the product of vector lengths from zeros is equal to 1.

Procedure for Constructing Root Locus

Step 1: Locate the poles and zeros of $G(s)$ $H(s)$ on the s-plane. The root locus branch start from open loop poles and terminate at zeros.

Step 2: Determine the root locus on real axis.

Step 3: Determine the asymptotes of root locus branches and meeting point of asymptotes with in real axis.

Step 4: Find the break away and breaking points.

Step 5: Determine the angle of departure from the complex pole if there is a complex pole. Determine the angle of arrival at the complex zero, if there is a complex zero.

Step 6: Find the points where the root loci crosses the imaginary axis.

Step 7: The value of gain K at any point on the locus can be determined from magnitude condition. The magnitude conditon is, given by

$$\text{Gain } K \text{ at a point } (s = s_a) = \frac{\text{Product of length of vectors from poles to the point } (s = s_a)}{\text{Product of length of vectors from finite zeros to the point } (s = s_a)}$$

Note: When there is no finite zero, the denominator is taken as unity. The length of vectors should be measured to scale.

Brief Explanation for the various steps in the Procedure for Constructing Root Locus

Step 1: Location of poles and zeros

Draw the real and imaginary axis on an ordinary graph sheet and choose same scales both on real and imaginary axis. The poles are marked by cross "X" and zeros and marked by small circle "O". The number of root locus branches is equal to number of poles of open loop transfer function. The origin of a root locus is at a pole and the end is at a zero.

If n = Number of poles and

m = Number of finite zeros,

then n root locus branches ends at finite zeros. The remaining $(n - m)$ root locus branches will end at zeros at infinity.

Step 2: Root locus on real axis:

To decide the part of root locus on real axis, take a test point on real axis. If the total number of poles and zeros on the real axis to the right of the test point is odd number then the test point lies on the root locus. if it is even then the test point does not lie on the root locus.

Step 3: Angles of asymptotes and centroid:

If n = Number of poles and

m = Number of zeros,

then $(n - m)$ root locus branches will terminate at zeros at infinity. These root locus branches will go along an asymptotic path and meets the asymptotes at infinity. Hence number of asymptotes is equal to number of root locus branches going to infinity. The formula for angles of asymptotes and the centroid is,

$$\text{Angles of asymptotes} = \frac{\pm 180\,(2q+1)}{n-m}$$

where, $q = 0, 1, 2, 3\ldots (n-m)$

$$\text{centroid} = \frac{\text{Sum of poles} - \text{Sum of zeros}}{n-m}$$

Step 4: Breakaway and break in points:

The breakaway (or) breaking points either lie on real axis (or) exists as complex conjugate pairs. If there is a root locus on real axis between two poles then there exists a breakaway point.

If there is root locus on real axis between two zeros then there exist a breakin point. If there is a not locus on real axis between pole and zero then there may be (or) may not be breakaway (or) breakin point.

Let the characteristic Eq. be in the form

$$B(s) + K\,A(s) = 0$$

$$\therefore \qquad K = -\frac{B(s)}{A(s)}$$

The breakaway and breakin point is given by roots of the Eq. $\frac{dK}{ds} = 0$. The roots of $\frac{dK}{ds} = 0$ are actual breakaway (or) breakin point provided for this value of root the gain K should be positive and real.

Step 5: Angle of departure and angle of arrival

$$\text{Angle of departure (from a complex pole } A) = 180^\circ - \begin{bmatrix}\text{Sum of angles of vector to} \\ \text{the complex pole A from} \\ \text{other poles}\end{bmatrix} + \begin{bmatrix}\text{Sum of angles of vectors} \\ \text{to the complex pole A from} \\ \text{zeros}\end{bmatrix}$$

If poles are complex then they exist only as conjugate pair. Consider the two complex conjugate roles A and A* shown in Fig. (5.2).

$$\text{Angle of departure at pole } A = 180° - (\theta_1 + \theta_3 + \theta_5) + (\theta_2 + \theta_4)$$

$$\text{Angle of departure at pole } A^* = [\text{Angle of departure at pole } A]$$

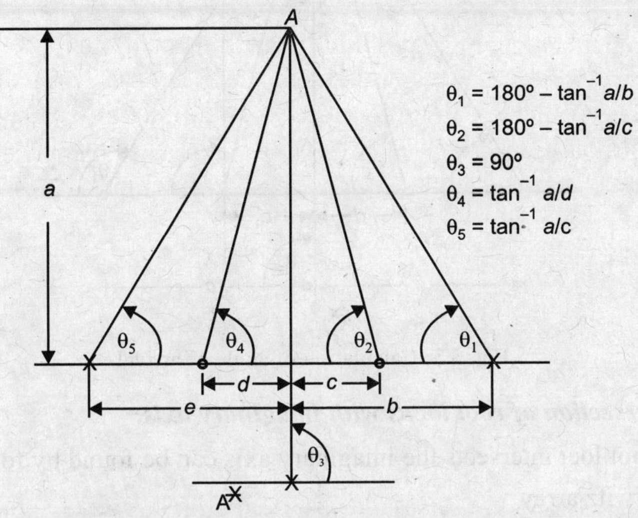

Fig. 5.2: Calculation of angles of departure.

Note: The angles can be calculated as shown in Fig. (5.2) or they can be measured using protractor.

$$\text{Angle of arrival at a complex zero } A = 180° - \begin{bmatrix}\text{Sum of angles of vectors} \\ \text{to the complex zero A from} \\ \text{all other zeros}\end{bmatrix} + \begin{bmatrix}\text{Sum of angles of vectors} \\ \text{to the complex zero } A \text{ from} \\ \text{poles}\end{bmatrix}$$

From Fig. (5.3)

$$\theta_1 = -180° - \tan^{-1}(a/c)$$

$$\theta_2 = 180° - \tan^{-1}(a/b)$$

$$\theta_3 = 90°$$

$$\theta_4 = \tan^{-1}(a/d)$$

$$\theta_5 = \tan^{-1}(a/e)$$

If zeros are complex then they exist only as conjugate pairs. Consider the two complex conjugate zeros A and A^* as shown in Fig. (5.3)

$$\text{Angle of arrival at zero } A = 180° - (\theta_1 + \theta_3) + (\theta_2 + \theta_4 + \theta_5)$$

$$\text{Angle of arrival at zero } A^* = -[\text{Angle of arrive at zero } A]$$

Note: The angles can be calculated as shown in Fig. (5.3) or they can be measured using protractor.

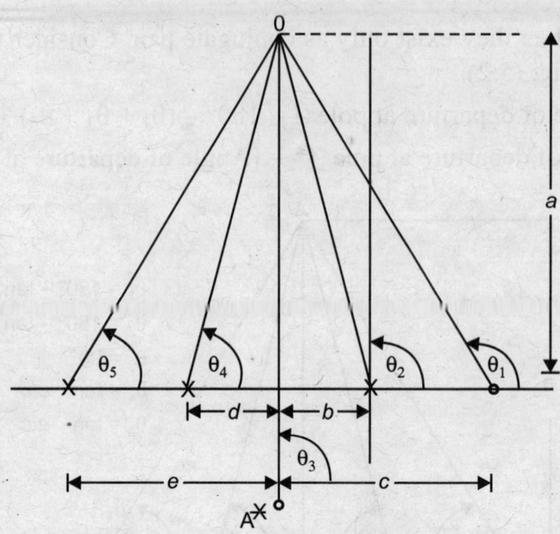

Fig. 5.3: Calculation of angle of arrival.

Step 6: Point of intersection of root locus with imaginary axis:

The point where the root loci intersects the imaginary axis can be found by following three methods.

1. By Routh Hurwitz array.
2. By trial and error approach.
3. Letting $s = j\omega$ in the characteristic Eq. and separate the real part and imaginary part. Two Eq.s are obtained-one by equating real part to zero and the other by equating imaginary part to zero. Solve the two Eq.s for ω and K. The value of ω gives the point where the root locus crosses imaginary axis. The value of K gives the value of gain K at this crossing point. Also this value of K is the limiting value of K for stability of the system.

Step 7: Test points and root locus:

Choose a test point. Using a protractor roughly estimate the angles of vectors drawn to this point and adjust the point to satisfy angle criterion. Repet the procedure for few more test points. Sketch the root locus from the knowledge of typical sketches and the informations obtained in step 1 through 6.

Determination of Open Loop Gain for a specified Damping of the Dominant Roots

The dominant pole is a pair of complex conjugate pole which decides the transient response of the system. In higher order systems the dominant poles are given by the poles which are very close to origin, provide all other poles are lying far away from the dominant poles. The poles which are far away from the origin will have less effect on the transient response of the system.

The transfer function of higher order systems can be approximated to a second order transfer function, whose standard form of closed loop transfer function is

$$\frac{C(s)}{R(s)} = \frac{\omega_n^2}{s^2 + 2\zeta\omega_n s + \omega_n^2}$$

The dominant poles (*sd* and *sd**) are given by the roots of quadratic factor,

$$(s^2 + 2\zeta\omega_n s + \omega_n^2) = 0$$

$$\therefore \quad s_d = \frac{-2\zeta\,\omega_n \pm \sqrt{4\zeta^2\omega_{n^2} - 4\omega_{n^2}}}{2}$$

$$= -\zeta\omega_n \pm j\omega_n\sqrt{1-\zeta^2}$$

The dominant pole can be plotted on the *s*-plane as shown in Fig. (5.4)

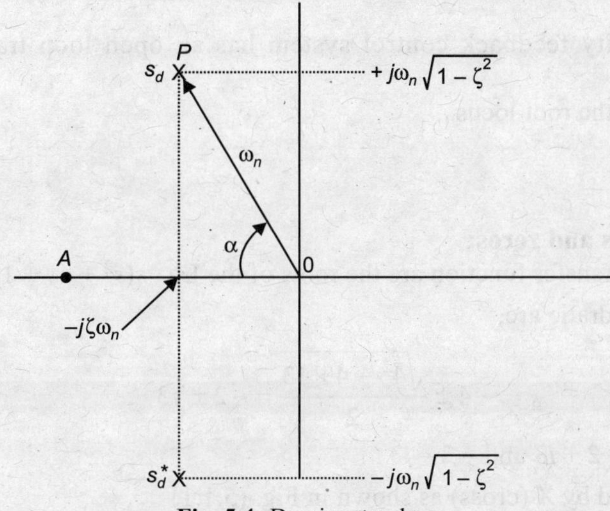

Fig. 5.4: Dominant pole, s_d.

In the right angle triangle, *OAP*

$$\cos\alpha = \frac{\zeta\omega_n}{\omega_n} = \zeta$$

$$\therefore \quad \alpha = \cos^{-1}\zeta$$

To fix a dominant pole on root locus, draw a line at an angle of $\cos^{-1}\zeta$ with respect to negative real axis. The meeting point of this line with root locus will give the location of dominant pole. The value of *K* corresponding to dominant pole can be obtained from magnitude condition.

The gain *K* corresponding to dominant pole,

$$s_d = \frac{\text{Product of length of vectors from open loop poles to dominant pole}}{\text{Product of length of vectors from open loop zeros to dominant pole}}$$

Advantages of Root Locus

- The root locus technique is an important tool in designing control systems with desired performance characteristics. The desired performance of the system can be achieved by adjusting the location of its closed-loop poles in the *s*-plane by varying one (or) more system parameters.

- The root locus can be plotted in the s-plane by varying a system parameter (usually gain, K) over the complete range of values. The roots corresponding to a particular value of the system parameter can then be located on the locus (or) the value of the parameter for a desired root location an be determined from the locus.
- The root locus technique is also used for stability analysis. Using root locus the range of values of K, for a stable system can be determined. It is also easier to study the relative stability of the system from the knowledge of location of closed loop poles. The dominant roots are used to estimate the damping ratio and natural frequency of oscillation of the system. From ζ and ω_n the time domain specifications can be calculated.

EXAMPLE 5.1: A unity feedback control system has an open loop transfer function $G(s) = \dfrac{K}{s(s^2+4s+13)}$. Sketch the root locus.

Solution:

Step 1: To locate poles and zeros:

The poles of open loop transfer function are the roots of the Eq., $s(s^2 + 4s + 13) = 0$

The roots of the quadratic are,

$$s = -4 \pm \frac{\sqrt{4^2 - 4\times 13}}{2} = -2 \pm j3$$

$\therefore$ The poles are 0, $-2 + j3$ and $-2 - j3$

The poles are marked by X (cross) as shown in Fig. (5.1.1)

Step 2: To find the root locus on real axis:

There is only one pole on real axis at the origin. Hence if we choose any test point on the negative real axis then to the right of that point the total number of real poles and zeros is one, which is an odd number. Hence the entire negative real axis will be part of root locus. The root locus on real axis shown as a bold live in Fig. (5.1.1).

Note: For the given transfer function one root locus branch will start at the pole at the origin and travel through the negative real axis to meet the zero at infinity.

Step 3: To find angles of asymptotes and centroid:

Since there are s poles, the number of root locus branches are three. There is no finite zero. Hence all the three root locus branches ends at zeros at infinity. The number of asymptotes required are three.

$$\text{Angles of asymptotes} = \frac{\pm 180° (2q+1)}{n-m}$$

Here $n = 3$, and $m = 0$

$$\text{If } q = 0, \text{ Angles} = \frac{\pm 180°}{3} = \pm 60°$$

$$\text{If } q = 1, \text{ Angles} = \frac{180° \times 3}{3} = \pm 180°$$

If $q = 2$, Angles= $\dfrac{\pm 180° \times 5}{3} = \pm 300° = \pm 60°$

If $q = 3$, Angles = $\dfrac{\pm 180° \times 7}{3} = \pm 420° = \pm 60°$

Since ±60º is repeating, the angle of asymptotes = ±60º.

$$\text{Centroid} = \frac{\text{Sum of poles} - \text{Sum of zeros}}{n - m}$$

$$= \frac{0 - 2 + j3 - 2 - j3 - 0}{3} = \frac{-4}{3} = -1.33$$

The centroid is marked on real axis and from the centroid the angles of asymptotes are marked using a protractor. The asymptotes are drawn as dotted lines as shown in Fig. (5.1.1)

Step 4: To find the break away and breakin points:

The closed loop transfer function,

$$\frac{C(s)}{R(s)} = \frac{G(s)}{1 + G(s)} \qquad [\because H(s) = 1]$$

$$\frac{C(s)}{R(s)} = \frac{K}{s(s^2 + 4s + 13) + K}$$

The characteristic Eq. is

$$s\,(s^2 + 4s + 13) + K = 0$$

$$s^3 + 4s^2 + 13s + K = 0$$

$$\therefore \quad K = -\,s^3 - 4s^2 - 13s$$

on differentiating the Eq. of K w.r.to s we get

$$\frac{dK}{ds} = -\,(3s^2 + 8s + 13)$$

Put $$\frac{dK}{ds} = 0$$

$$\therefore \quad -\,(3s^2 + 8s + 13) = 0$$

$$3s^2 + 8s + 13 = 0$$

$$\therefore \quad s = \frac{-8 \pm \sqrt{8^2 - 4 \times 13 \times 3}}{2 \times 3}$$

$$= -\,1.33 \pm j\,1.6$$

check for K

when $s = (-1.33 + j\,1.6)$, the value of K is given by

$$K = -(s^3 + 4s^2 + 13s)$$

$$= -[-1.33 + j(1.6)^3 + 4(-1.33 + j1.6)^2 + 13(-1.33 + j1.6)]$$

$$\neq \text{Positive and real}$$

when $s = -1.33 - j1.6$. The value of K is not equal to real and positive.

Since the value of K for $s = (-1.33 \pm j1.6)$ are not real and positive, the points are not on actual break away (or) break in points. The root locus has neither breakaway nor breakin point.

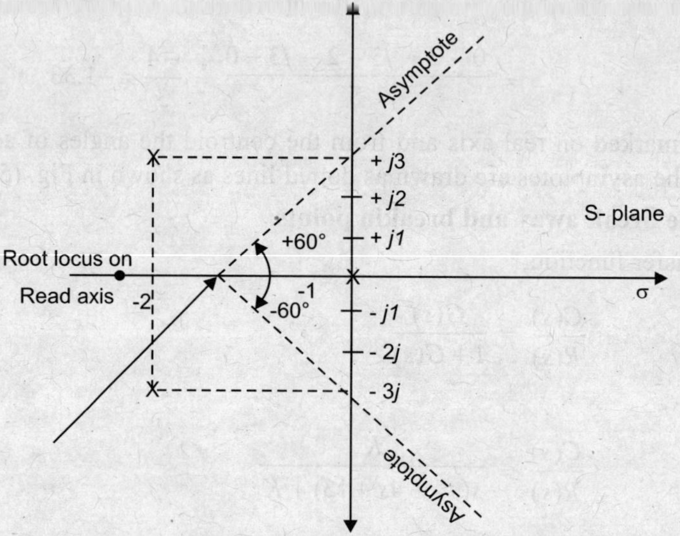

Fig. 5.1.1: Figure showing the asymptotes, root locus on real axis and location of poles and centroid.

Step 5: To find angle of departure

Let us consider the complex pole A as shown in Fig. (5.1.2). Draw vectors from all other poles to the Pole A as shown in Fig. (5.1.2). Let the angles of these vectors be θ_1 and θ_2

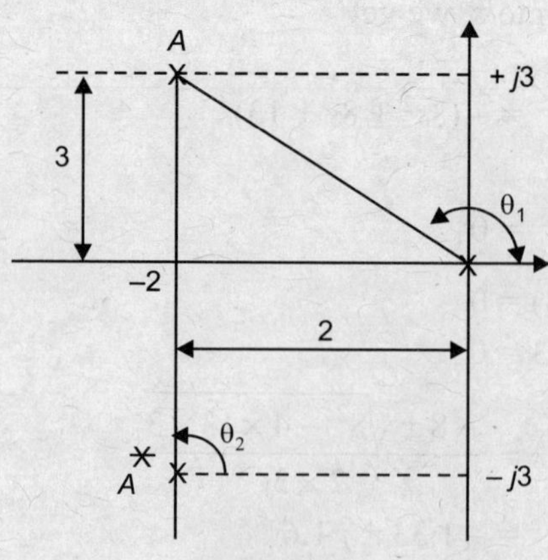

Fig. 5.1.2

Here, $\theta_1 = 180° - \tan^{-1}(3/2) = 123.7°$

$\theta_2 = 90°$

Angle of departure from the complex pole A

$$= 180° - (\theta_1 + \theta_2)$$

$$= 180° - (123.7° + 90°) = -33.7°$$

The angle of departure at complex pole A^* is negative of the angle of departure at complex pole A.

∴ Angle of departure at pole $A^* = +33.7$

Make the angles of departure at complex pole using protractor.

Step 6: To find the crossing point on imaginary axis:

The characteristic Eq. is given by $s^3 + 4s^2 + 13s + K = 0$

Put $s = j\omega$

$$(j\omega)^3 + 4(j\omega)^2 + 13(j\omega) + K = 0$$

$$-j\omega^3 - 4\omega^2 + 13\,j\omega + K = 0$$

on equating imaginary part to zero,

we get, $-\omega^3 + 13\omega = 0$

$$-\omega^3 = -13\omega$$

$$\omega = \pm\sqrt{13}$$

$$\omega = \pm 3.6$$

on equating real part to zero, we get,

$$-4\omega^2 + K = 0$$

$$K = 4\omega^2$$

$$K = 4(13) = 52$$

$$K = 52$$

The crossing point of root locus is $\pm j3.6$. The value of the K at this crossing point is $K = 52$.

The complete root locus sketch is shown in Figure (3.6.1.3). The root locus has three branches one branch starts at the pole at origin and travel through negative real axis to meet the zero at infinity. The other two root locus branches starts at complex poles (along the angle of departure) crosses the imaginary axis at $\pm j3.6$ and travel parallel to asymptotes to meet the zeros at infinity.

EXAMPLE 5.2: Sketch the root locus of the system whose open loop transfer function is $G(s) = \dfrac{K}{s(s+2)(s+4)}$. Find the value of K so that the damping ratio of the closed loop system is 0.5.

Solution:

Step 1: To locate poles and zeros

The poles of open loop transfer function are the roots of the Eq., $s(s + 2)(s + 4) = 0$.

∴ The poles are, $s = 0, -2, -4$.

The poles one marked by X (cross) as shown in Figure (5.2.1)

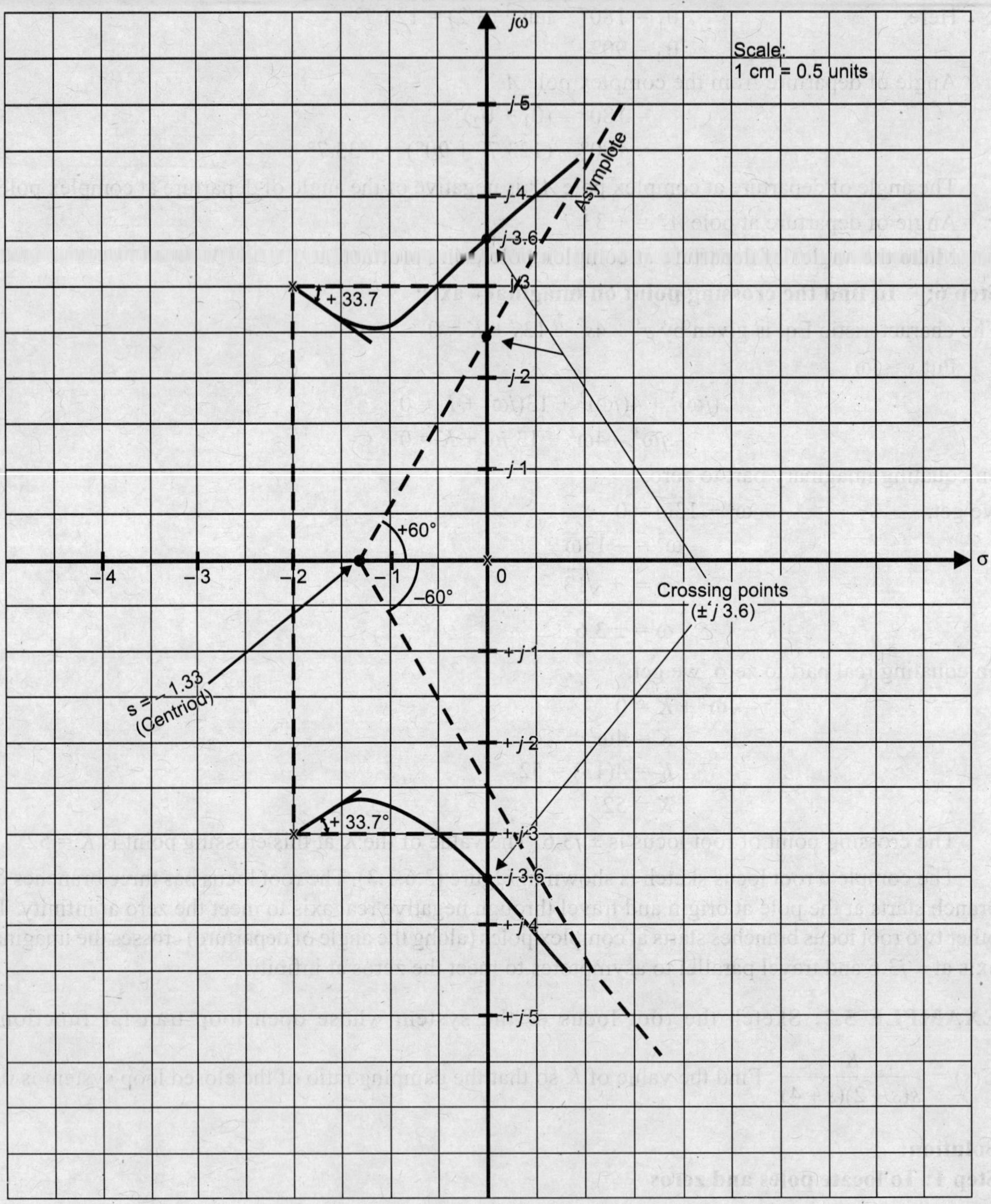

Fig. 5.1.3: Root locus of $1 + G(s) = 1 + \dfrac{K}{s(s^2 + 4s + 13)}$.

Step 2: To find root locus on real axis

There are 3 poles on the real axis. Choose test point on real axis between $s = 0$ and $s = -2$. To the right of this point the total number of real poles and zeros is one, which is an odd number. Hence the real axis between $s = 0$ and $s = -2$ will be part of root locus. Chose a test point on real axis between $s = -2$ and $s = -4$. to the right of this point the total number of real poles and zeros is two which is an even number. Hence the real axis between $s = -2$ and $s = -4$ will not be a part of root locus.

Choose a test point on real axis to the lest of $s = -4$. To the right of this point, the total number of real poles and zeros is three, which is an odd number. Hence the entire negative real axis from $s = -4$ to $-\infty$ will be a part of root locus.

Step 3: To find asymptotes and centroid

Since there are three poles the number of root locus branches are three. There is no finite zero. Hence all the three root locus branches ends at zeros at infinity. The number of asymptotes required are three.

$$\text{Angles of asymptotes} = \pm\frac{180(2q+1)}{n-m}$$

Hence $n = 3$ and $m = 0$

$$\text{If } q = 0, \text{ angle} = \pm\frac{180}{3} = \pm 60$$

$$\text{If } q = 1, \text{ angles} = \pm\frac{180\times 3}{3} = \pm 180°$$

$$\text{If } q = 2, \text{ angles,} = \pm\frac{180\times 5}{3} = \pm 300° = \mp 60°$$

Note: It is enough if you calculate the required number of angles. Here it is given by first three values of angles. The remaining values will be repetitions of the previous values.

$$\text{Centroid} = \frac{\text{Sum of poles} - \text{Sum of zeros}}{n-m} = \frac{0-2-4-0}{3} = -2.$$

The centroid is marked on real axis and from the centroid the angle of asymptotes are marked using a protractor. The asymptotes are drawn as dotted lines as shown in Figure (5.2.1.)

Figure (5.2.1) showing the asymptotes, root locus on real axis and location of poles, centroid and breakaway points.

Step 4: To find the breakaway and breakin point

The closed loop transfer function,

$$\frac{C(s)}{R(s)} = \frac{G(s)}{1+G(s)} \qquad [\because H(s) = 1]$$

$$\therefore \quad \frac{C(s)}{R(s)} = \frac{\dfrac{K}{s(s+2)(s+4)}}{1+\dfrac{K}{s(s+2)(s+4)}} = \frac{K}{s(s+2)(s+4)+K}$$

CHAPTER 5

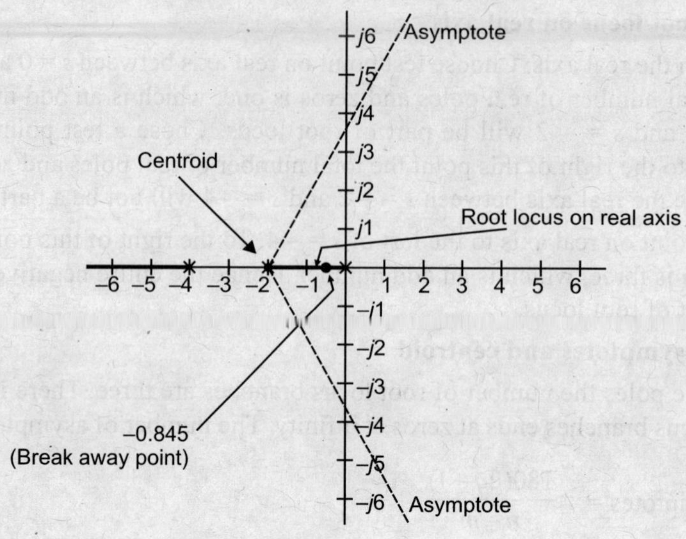

Fig. 5.2.1

The characteristic Eq. is given by

$$s(s + 2)\,(s + 4) + K = 0$$

$$s(s^2 + 6s + 8) + K = 0$$

$$s^3 + 6s^2 + 8s + K = 0$$

$$\therefore \qquad K = - s^3 - 6s^2 - 8s$$

On differentiating the Eq. of K with respect to s we get,

$$\frac{dK}{ds} = - (3s^2 + 12s + 8)$$

Put

$$\frac{dK}{ds} = 0$$

$$\therefore \qquad -(3s^2 + 12s + 8s) = 0$$

$$(3s^2 + 12s + 8s) = 0$$

$$s = \frac{-12 \pm \sqrt{12^2 - 4 \times 3 \times 8}}{2 \times 3} = -0.815 \text{ (or)} - 3.154$$

Check for K

when $s = - 0.845$, the value of K is given by

$$K = [-(-0.845)^3 + 6(-0.0845)^2 + 8(-0.845)] = 3.08$$

Since K is positive and real for $s = - 0.845$, this point is actual breakaway point,

when $s = - 3.154$, the value of K is given by

$$K = [(-3.154)^3 + 6(-3.154)^2 + 8(-3.154)] = - 3.08$$

Since K is negative for $s = - 3.154$, this is not a actual breakaway point. The breakaway point is marked on the negative real axis as shown in Figure (5.2.1).

Step 5: To find angle of departure

Since there are no complex pole or zero, we need not find angle of departure or arrival.

Step 6: To find the crossing point of imaginary axis

The characteristic Eq. is given by

$$s^3 + 6s^2 + 8s + K = 0$$

Put

$$s = j\omega$$

$$(j\omega)^3 + 6(j\omega)^2 + 8(j\omega) + K = 0$$

$$-j\omega^3 - 6\omega^2 + j8\omega + K = 0$$

Equating imaginary part of zero

$$-j\omega^3 + j8\omega = 0$$

$$-j\omega^3 = j8\omega$$

$$\omega^2 = 8$$

$$\omega = \pm\sqrt{8} = \pm 2.8$$

$$\omega = \pm 2.8$$

Equating real part to zero

$$-6\omega^2 + K = 0$$

$$K = 6\omega^2 = 6 \times 8 = 48$$

$$K = 48$$

The crossing point of root locus ± 2.8. The value of K corresponding to this point is $K = 48$. (This is two limiting value of K for the stability of the system.

The complete root locus sketch is shown in Figure (5.2.2). The root locus has three branches. One branch starts at the pole at meet the zero at infinity. The other two root locus branches starts at $s = 0$ and $s = -2$ and travel through negative real axis, breakaway from real axis at $s = -0.845$, then crosses imaginary axis at $s = \pm j2.8$ and travel parallel to asymptotes to meet the zeros at infinity.

To find the value of K corresponding to $\zeta = 0.5$

Given that $\zeta = 0.5$

Let $\alpha = \cos^{-1}\zeta = \cos^{-1} 0.5 = 60°$.

Draw a loine OP, such that the angle between line OP and negative real axis is 60° ($\alpha = 60°$) as shown in Figure (5.2.2). The meeting point of the line OP and root locus gives the dominant pole, *sd*.

The value of K corresponding to the point, $s = s_d$

$$= \frac{\text{Product of length of vector from all poles to the points, } s = s_d}{\text{Product of length of vectos from all zeros to the point, } s = s_d}$$

$$= \frac{l_1.l_2.l_3}{1}$$

$$= 1.3 \times 1.75 \times 3.5 = 7.96 \approx 8$$

Result: $K = 8$ at $\zeta = 0.5$

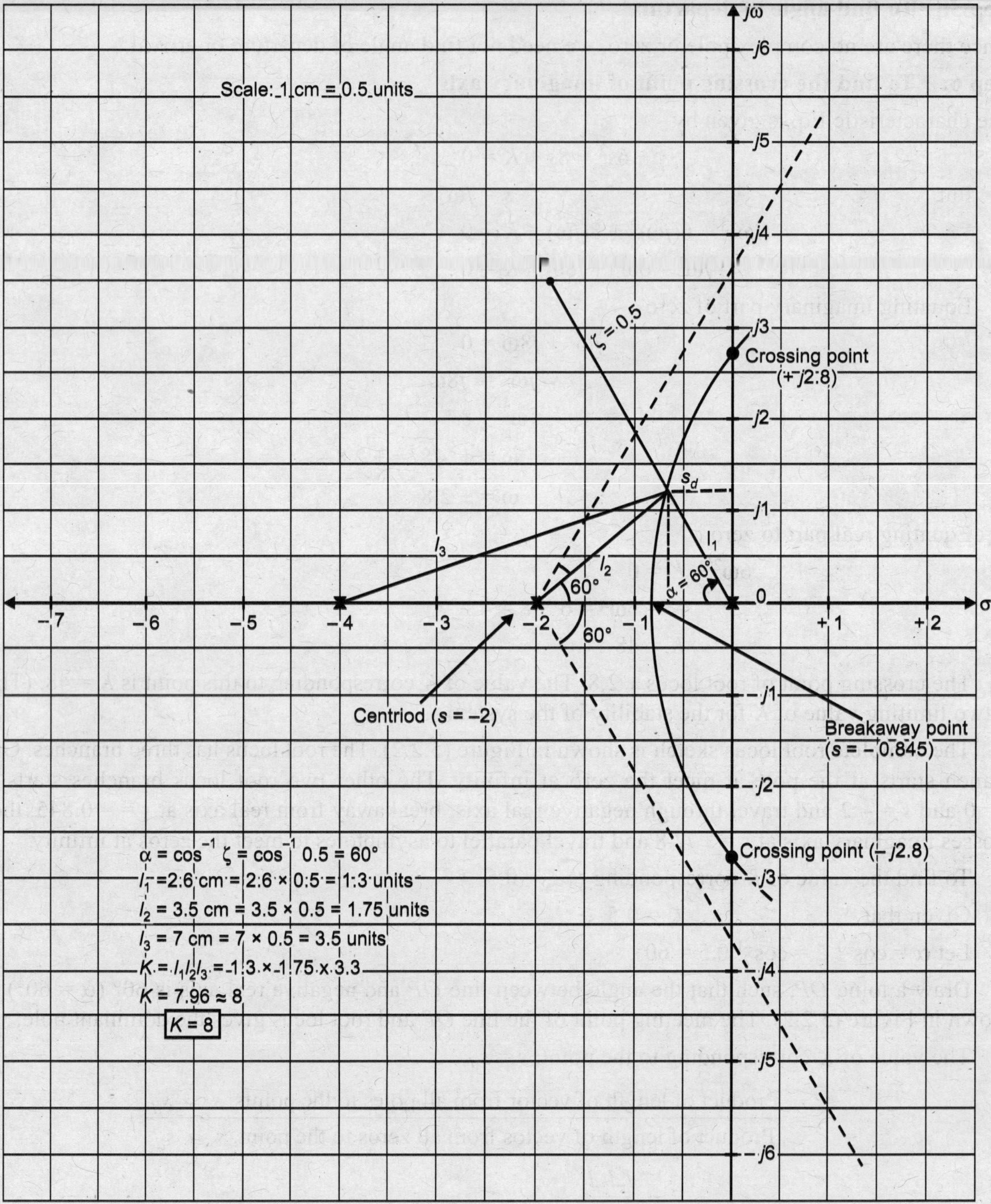

Fig. 5.2.2: Root locus of $1 + G(s) = 1 + \dfrac{1}{s(s+2)(s+4)}$.

EXAMPLE 5.3: The open loop transfer function of a unity feedback system is given by $G(s) = \frac{K(s+9)}{s(s^2+4s+11)}$. Sketch the root locus of the system.

Solution:

Step 1: To locate poles and zeros

The poles of open loop transfer function are the root locus of the Eq.

$$s(s^2 + 4s + 11) = 0$$

The roots of the quadratic are,

$$s = \frac{-4 \pm \sqrt{4^2 - 4 \times 11}}{2} = -2 \pm j\,2.64$$

∴ The poles are lying at s = 0, – 2 + j2.64, – 2 – j2.64

The zeros are lying at s = – 9 and infinity

The poles are marked by X(cross) and zeros by "0" (circle) as shown in Figure (5.3.1).

Step 2: To find root locus on real axis

One pole and one zero lines on real axis choose a test point to the left of s = 0, then to the right of this point, the total number of poles and zeros is one which is an odd numbers. Hence the portion of real axis from s = 0 to s = – 9 will be a part of root locus. If we choose a test point of the left of s = –9 then to the right of this point, the total number of pole and zeros is two, which is an even number. Hence the real axis from s = – 9 to – ∞ will not be a part of root locus. The root locus on real axis is shown as bold line in Figure (5.3.1).

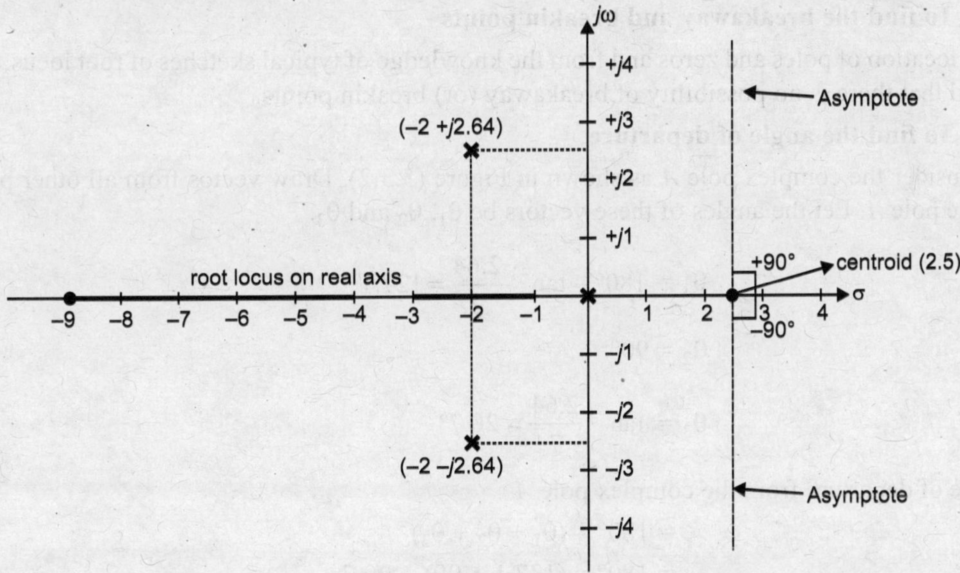

Fig. 5.3.1: Showing the asymptotes root locus on real axis and location of poles, zeros and centroid.

CHAPTER 5

Step 3: To find angles of asymptotes and centroid

Since there are 3 poles, the number of root locus branches are three one root locus branch start at the pole at origin and travel along negative real axis to meet the zero at $s = -9$. The other tow root locus branches meet the zeros at infinity, The number of asymptotes required are 2.

$$\text{Angle of asymptotes} = \pm\frac{180°(2q+1)}{n-m}$$

Here $n = 3$ and $m = 1$

$$\text{If } q = 0, \text{ angles} = \pm\frac{180}{2} = \pm 90°$$

$$\text{If } q = 2, \text{ angles} = \pm\frac{180\times 3}{2} = \mp 90°$$

$$\text{If } q = 2, \text{ angles} = \pm\frac{180\times 5}{2} = \pm 90°$$

Note: It is enough if you calculate the required number of angles. Here it is given by first two value of angles. The remaining values will be repetitions of the previous values.

$$\text{Centroid} = \frac{\text{Sum of poles} - \text{sum of zeros}}{n-m}$$

$$= 0 - 2 + j2.64 - 2 - j\,2.64 - (-9)/2 = 2.5$$

The centroid is marked and from the centroid, the angles of asymptotes are makrked using a protractor. The asymptotes are drawn as dotted lines as shown in Figure (5.3.1).

Step 4: To find the breakaway and breakin points

From the location of poles and zeros and from the knowledge of typical sketches of root locus, it can be concluded that there is no possibility of breakaway (or) breakin points.

Step 5: To find the angle of departure

Let us consider the complex pole A as shown in Figure (5.3.2). Draw vectos from all other poles and zero to the pole A. Let the angles of these vectors be θ_1, θ_2 and θ_3.

Here
$$\theta_1 = 180° - \tan^{-1}\frac{2.64}{2} = 127.1°$$

$$\theta_2 = 90°$$

$$\theta_3 = \tan^{-1}\frac{2.64}{7} = 20.7°$$

Angle of departure from the complex pole A

$$= 180° - (\theta_1 + \theta_2 + \theta_3)$$

$$= 180° - (127.1 + 90) + 20.7$$

$$= -16.4°$$

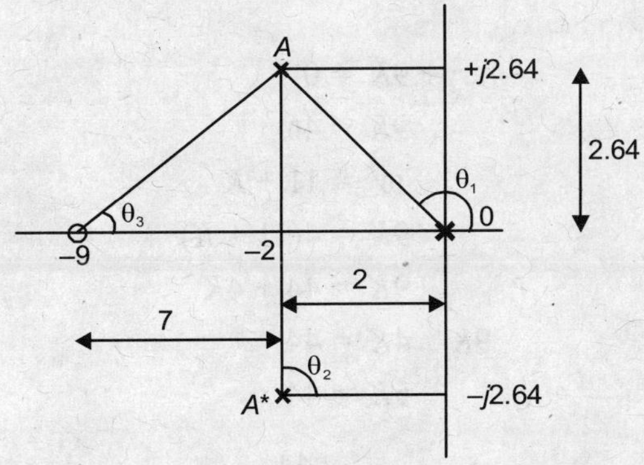

Fig. 5.3.2

The angle of departure at the complex pole A^* is negative of the angle of departure at complex pole A,

$\therefore$ Angle of departure at pole $A^* = -(-16.4) = +16.4°$

Mark the angles of departure at complex poles using protractor.

Step 6: To find the crossing point of imaginary axis

The closed loop transfer fucntion, $\dfrac{C(s)}{R(s)} = \dfrac{G(s)}{1+G(s)} = \dfrac{\dfrac{K(s+9)}{s(s^2+4s+11)}}{1+\dfrac{K(s+9)}{s(s^2+4s+11)}}$

$$= \frac{K(s+9)}{s(s^2+4s+11)+K(s+9)}$$

The characteistic Eq. is the denominator polynomial of $\dfrac{C(s)}{R(s)}$

$\therefore$ The characteristic Eq.s,

$$s(s^2 + 4s + 11) + K(s + 9) = 0$$

$$(s^3 + 4s^2 + 11s) + Ks + 9K = 0$$

Put $s = j\omega$

$$(j\omega)^3 + 4(j\omega)^2 + 11(j\omega) + K(j\omega) + 9K = 0$$

$$-j\omega^3 - 4\omega^2 + j11\omega + 9K = 0$$

on equating imaginary part to zero,

$$-j\omega^3 + j\,11\,\omega + jK\omega = 0$$

$$-j\omega^3 = -j\,11\,\omega - jK\omega$$

$$\omega^2 = 11 + K$$

Put $K = 8.8$

$\therefore$ $\omega^2 = 11 + 8.8 = 19.8$

$\omega = \pm\sqrt{19.8}$

$\omega = \pm 4.4$

CHAPTER 5

on equating real part to zero

$$-4\omega^2 + 9K = 0$$

$$9K = 4\omega^2$$

Put $\omega^2 = 11 + K$

$$\therefore \quad 9K = 4(11 + K)$$

$$9K = 44 + 4K$$

$$9K - 4K = 44$$

$$5K = 44$$

$$K = \frac{44}{5} = 8.8$$

$$K = 8.8$$

The crossing point of root locus is $\pm j\,4.4$. The value of K at this crossing point is $K = 8.8$ (This is limiting value of K for the stabiltiy of the system).

The complete root locus sketch is shown in Figure (5.3.3). The root locus has three branches one branch starts at pole at origin and travel through neagtive real axis to meet the zero at s = – 9. The other two root locus branches starts at complex poles (along the angle of departure) crosses the imaginary axis at $\pm j\,4.4$ and travel parallel t asymptotes to meet the zeros at infinity.

EXAMPLE 5.4: Sketch the root locus for the unity feedback system whose open loop transfer funciton is $G(s)\,H(s) = \dfrac{K}{s(s+4)(s^2+4s+20)}$.

Solution :

Step 1: To locate poles and zeros

The poles of open loop transfer function are the roots of the Eq. $s(s + 4)\,(s^2 + 4s + 20) = 0$.

The roots of the quadratic are,

$$s = \frac{-4 \pm \sqrt{4^2 - 4 \times 1 \times 20}}{2}$$

$$= -2 \pm j4$$

∴ The poles are lying at $s = 0, -4, -2 + j4$ and $-2 - j4$.

The zeros are lying at infinity.

The poles are marked by x(cross) as shown in Figure (5.4.1).

Step 2: To find root locus on real axis

There are two polesa on the real axis. Choose a test point on real axis between $s = 0$ and $s = -4$. To the right of this point, the total number of real poles is one which is an odd number. hence the real axdis between $s = 0$ and $s = -4$ will be a part of root locus. Choose a test point to the left of $s = -4$, now to the right of this test point the total number of poles and zeros is two which is even number. Hence the

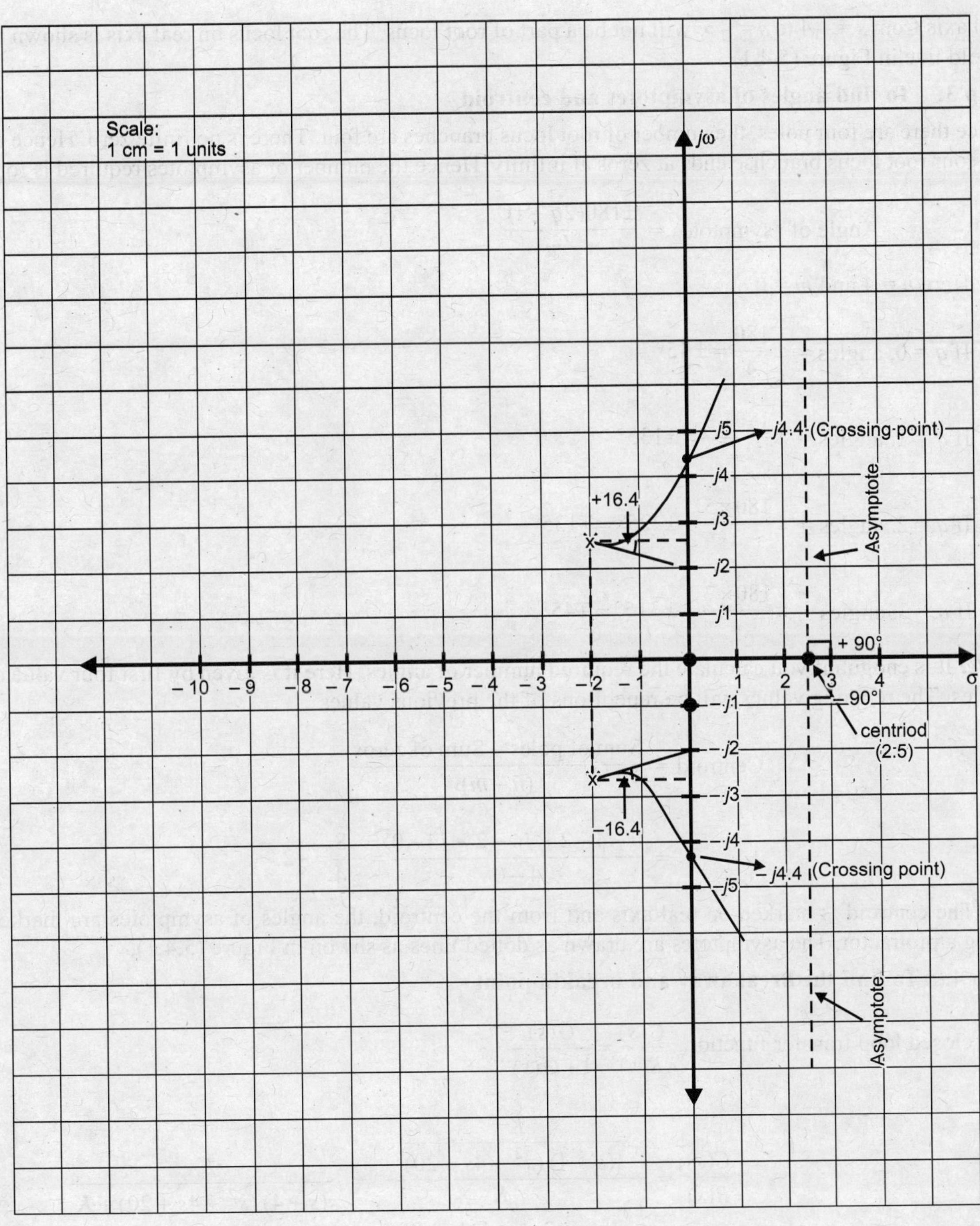

Fig. 5.3.3: Root locus sketch of $1 + G(s) = 1 + \dfrac{K(s+9)}{s(s^2+4s+11)}$.

real axis from $s = -4$ to $s = -\infty$ will not be a part of root locus. The root locus on real axis is shown as a bold line in Figure (5.4.1).

Step 3: To find angles of asymptotes and centroid

Since there are four poles, the number of root locus branches are four. There is no finite zero. Hence all the four root locus branchse ends at zeros at infinity. Hence the number of asymptotes required is four.

$$\text{Angle of asymptotes} = \frac{\pm 180(2q+1)}{n-m}$$

Here $n = 4$ and $m = 0$

If $q = 0$, angles $= \pm\frac{180}{4} = \pm 45°$

If $q = 1$, angles $= \pm\frac{180\times 3}{4} = \pm 135°$

If $q = 2$, angles $= \pm\frac{180\times 5}{4} = \pm 225° = \mp 135°$

If $q = 3$, angles $= \pm\frac{180\times 7}{4} = \pm 315° = \mp 45°$

Note: It is enoguh if you calculate the required number of angles. Here it is given by first four value of angles. The remaing values will be repetitions of the previous values.

$$\text{Centroid} = \frac{\text{Sum of poles} - \text{Sum of zeros}}{(n-m)}$$

$$= \frac{0-4-2+j4-2-j4-0}{4-0} = -\frac{8}{4} = -2.$$

The centroid is marked on real axis and from the centroid, the angles of asymptotes are marked using a protractor. The asymptotes are drawn as dotted lines as shwon in Figure (5.4.1).

Step 4: To find the breakaway and breakin point

The closed loop transfer function, $\frac{C(s)}{R(s)} = \frac{G(s)}{1+G(s)}$

$$\therefore \quad \frac{C(s)}{R(s)} = \frac{\dfrac{K}{s(s+4)(s^2+4s+20)}}{1+\dfrac{K}{s(s+2)(s^2+4s+20)}} = \frac{K}{s(s+4)(s^2+4s+20)+K}$$

The charactertistic Eq. is given by

$s(s + 4)(s^2 + 4s + 20) + K = 0$

$$\therefore \quad K = -s(s+4)(s^2+4s+20) = -(s^2+4s)(s^2+4s+20)$$

$$K = -(s^4+8s^3+36s^2+80s)$$

on differentiating the Eq. of K with respect to s we get,

$$\frac{dK}{ds} = -(4s^3+24s^2+72s+80)$$

Put $$\frac{dK}{ds} = 0$$

$$\therefore \quad -(4s^3+24s^2+72s+80) = 0$$

$$4s^3+24s^2+72s+80 = 0$$

on dividing by 4 we get,

$$s^3+6s^2+18s+20 = 0$$

The Eq. $s^3+6s^2+18s+20 = 0$ will have atleast one real root. By trial and error, the real root is found to be s = – 2 **(Refer Lin's method)**

The polynomial $(s^3+6s^2+18s+20) = 0$ can be expressed as

$$s^3+6s^2+18s+20 = (s+2)(s^2+4s+10) = 0$$

The root of the quadratic $(s^2+4s+10)$ are given by,

$$s = \frac{-4 \pm \sqrt{4^2 - 4\times 10}}{2} = -2.45$$

Check for K

When $s = -2$, $$K = -(s^4+8s^3+36s^2+80s)$$

$$= -[(-2)^4 + 8(-2)^3 + 36(-2)^2 + 80(-2)]$$

$$= -[-64] = 64$$

When $s = -2 \pm j\,2.45$,

$$K = -[s^4+8s^3+36s^2+80s]$$

$$= -[3.16\angle\pm129°)^4 + 8(3.16\angle\pm129°)^3 + 36(3.16\angle\pm129°)^2 + 80(3.16\angle\pm129°)$$

$$= -[99.7\angle\pm156° + 252.4\angle\pm27° + 359.5\angle\pm258° + 252.8\angle\pm129°]$$

For postive value of angles,

$$K = -[-91 + j40° + 225 + j115 - 75 - j351 - 159 + j196]$$

$$= -[-100] = 100$$

For negative values of angles,

$$K = -[-91 + j140 + 225 + j115 - 75 - j351 - 159 - j196]$$

$$= -[100] = 100$$

For all the roots of the Eq. , the value of K is positive and real. Hence all the three roots are actual breakaway points. The breakaway points are shown in Figure (5.4.1).

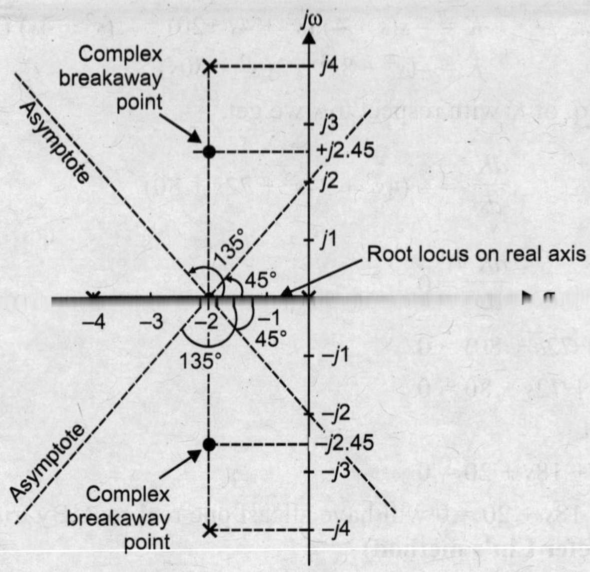

Fig. 5.4.1: Showing the asymptotes, root locus on real axis and location of poles, centroid and breakaway.

Step 5 : To find angle of departure

Let us consider the complex pole *A* shown is Figure (5.4.2). Draw vectos from all other poles to the pole A as shwon is Figure (5.4.2). Let angles of these vector be θ_1, θ_2 and θ_3.

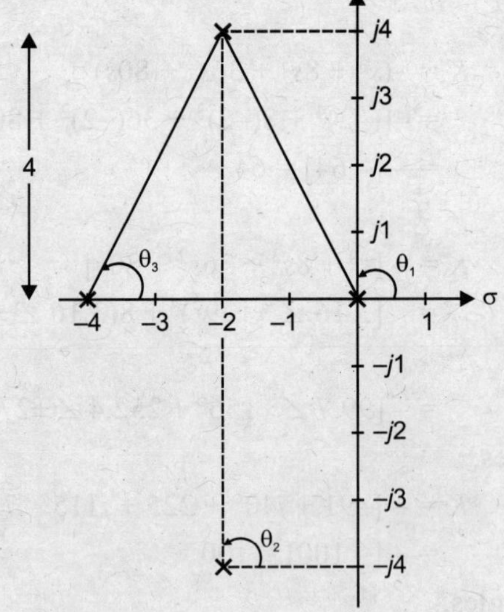

Fig. 5.4.2

Hence,

$$\theta_1 = 180° - \tan^{-1}\frac{4}{2} = 117°$$

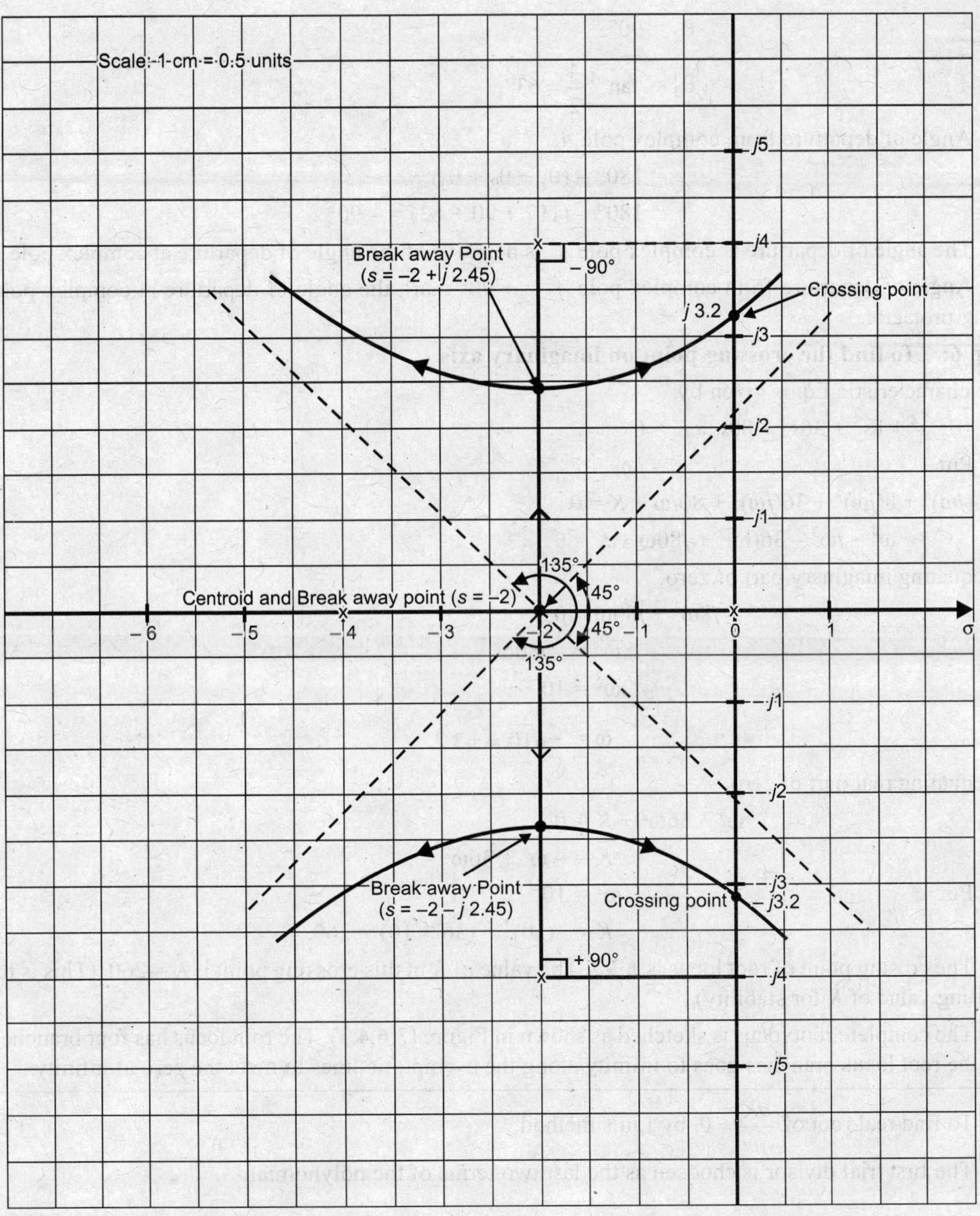

Fig. 5.4.3: Root locus sketch of $1 + G(s) = 1 + \dfrac{K}{s(s+4)(s^2+4s+20)}$.

$$\theta_2 = 90°$$

$$\theta_3 = \tan^{-1}\frac{4}{2} = 63°$$

Angle of departure from complex pole A,

$$= 180° - (\theta_1 + \theta_2 + \theta_3)$$

$$= 180° - (117 + 90 + 63) = -90°$$

The angle of departure at complex pole A^* is negative of the angle of departure at complex pole A.

∴ Angle of departure from complex pole $A^* = +90°$ mark the angle of departure at complex poles using protactor.

Step 6: To find the crossing point on imaginary axis

The characteristic Eq. is given by,

$$s^4 + 8s^3 + 36s^2 + 80s + K = 0$$

Put $s = j\omega$,

$$(j\omega)^4 + 8(j\omega)^3 + 36(j\omega)^2 + 80j\omega + K = 0$$

$$\omega^4 - j\omega^3 - 36(\omega)^2 + j80\omega + K = 0$$

on equating imaginary part of zero,

$$-j8\omega^3 + j80\omega = 0$$

$$-j8\omega^3 = -j80\omega$$

$$\omega^2 = 10$$

$$\omega = \pm\sqrt{10} = \pm 3.2$$

on equating real part of zero,

$$\omega^4 - 36\omega^2 + K = 0$$

$$K = -\omega^4 + 36\omega^2$$

Put $\omega^2 = 10$

∴ $K = -(10)^2 + (36 \times 10) = 260.$

The crosing point of root locus is ± 3.2. The value of K at this crossing point is $K = 260$. (This is the limiting value of K for stability).

The complete root locus is sketched as shown in Figure (3.6.4.3). The root locus has four branches. All the root locus branches goes to infinity along the asymptotic lines to meet the zero at infinity.

To find real root of $\frac{dK}{ds} = 0$ by Lin's method,

The first trial divisor is choosen as the last two terms of the polynomial.

Ist Trial:

$$\text{Trial divisor} = 18s + 20 = s + \frac{20}{18} = s + 1.11$$

$$
\begin{array}{r|l}
 & s^2 + 4.89s + 12.57 \\
\hline
s + 1.11 & s^3 + 6s^2 + 18s + 20 \\
 & s^3 + 1.11s^2 \\
\hline
 & 4.89s^2 + 18s \\
 & 4.89s^2 + 5.43s \\
\hline
 & 12.57s + 20 \quad \leftarrow \text{Next Trial Divisor} \\
 & 12.57s + 13.95 \\
\hline
 & \underline{6.05}
\end{array}
$$

IInd Trial:

Trial divisor = 12.57 s + 20

$$= s + \frac{20}{12.57} = s + 1.59$$

$$
\begin{array}{r|l}
 & s^2 + 4.41s + 11 \\
\hline
s + 1.59 & s^3 + 6s^2 + 18s + 20 \\
 & s^3 + 1.59s^2 \\
\hline
 & 4.41s^2 + 18s \\
 & 4.41s^2 + 7s \\
\hline
 & 11s + 20 \quad \leftarrow \text{Next Trial Divisor} \\
 & 11s + 17.49 \\
\hline
 & \underline{2.51}
\end{array}
$$

IIIrd Trial:

Trial divisor = 11s + 20 $= s + \frac{20}{11} = s + 1.82$

$$
\begin{array}{r|l}
 & s^2 + 4.18s + 10.4 \\
\hline
s + 1.89 & s^3 + 6s^2 + 18s + 20 \\
 & s^3 + 1.82s^2 \\
\hline
 & 4.18s^2 + 18s + 20 \\
 & 4.18s^2 + 7.6s \\
\hline
 & 10.4s + 20 \\
 & 10.4s + 18.9 \\
\hline
 & \underline{1.1}
\end{array}
$$

Since the remainder converge for every trial, let us approximate the root to $s = -2$. On dividing the polynomial by $s + 2$, we found that $(s + 2)$ is a divisor of the polynomial.

$$\begin{array}{r|l} & s^2 + 4s + 10 \\ \hline s+2 & s^3 + 6s^2 + 18s + 20 \\ & s^3 + 2s^2 \\ \hline & \quad 4s^2 + 18s \\ & \quad 4s^2 + 8s \\ \hline & \qquad 10s + 20 \\ & \qquad 10s + 20 \\ \hline & \qquad\quad 0 \end{array}$$

The real root is to be $s = -2$.

EXAMPLE 5.5: Sketch the root locus for the unity feedback system whose open loop transfer function is $G(s)H(s) = \dfrac{K(s+1.5)}{s(s+1)(s+5)}$.

Solution:

Step 1: To locate poles and zeros

The poles of open loop transfer function are the roots of the Eq., $s(s + 1)(s + 5) = 0$ and the zeros are the roots of the Eq. $(s + 1.5) = 0$.

The poles are lying at $s = 0, -1, -5$.

The zeros and lying at $s = -1.5$ and infinity. The poles are marked by X (cross) and zeros by "0" (circle) as shown in Figure (5.5.1).

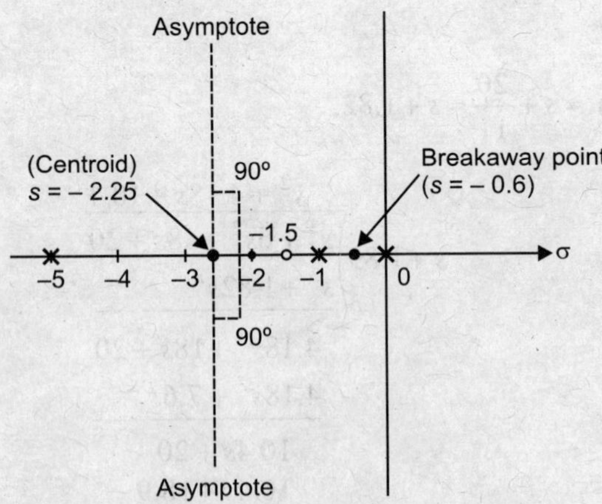

Fig. 5.5.1: Showing the asymptote, root locus on real axis and location of poles, zeros, centroid and breakaway point.

Step 2: To find root locus on real axis

The segment of real axis between $s = 0$ and $s = -1$ and the segment of real axdis between $s = -1.5$ and $s = -5$ will be part of root locus. Before if we choose a test point in this segment to the right of this point we have odd number of real poles and zeros. The root locus on real axis are shwon as bold lines is Figure (5.5.1).

Step 3: To find angles of asymptotes and centriod

Since there are three poles, the number of root locus branches are three. There is no finite zero, so one root locus branch will end at finite zero. The other two branches will meet the zero at infinity. Hence the number of asymptotes required is two.

$$\text{Angle of asymptotes} = \pm\frac{180(2q+1)}{n-m}$$

Here $n = 3$ and $m = 1$,

$$\text{If } q = 0, \text{ angles} = \pm\frac{180}{2} = \pm 90$$

$$\text{Centroid} = \frac{\text{Sum of poles} - \text{Sum of zeros}}{n-m}$$

$$= \frac{0-1-5-(-1.5)}{2} = -2.25$$

The centroid is marked on real axis and from the centroid the angles of asymptotes are marked using a protractor. The asymptotes are drawn as dotted lines as shown in Figure (5.5.1).

Step 4: To find the breakaway and breakin points

The closed loop transfer function,

$$\frac{C(s)}{R(s)} = \frac{G(s)}{1+G(s)} = \frac{\dfrac{K(s+1.5)}{s(s+1)(s+5)}}{1+\dfrac{K(s+1.5)}{s(s+1)(s+5)}}$$

$$= \frac{K(s+1.5)}{s(s+1)(s+5)+K(s+1.5)}$$

The characteristic Eq. is, $s(s+1)(s+5) + K(s+1.5) = 0$

$$\therefore \quad K = \frac{-(s+1)(s+5)}{s+1.5} = \frac{-s(s^2+6s+5)}{s+1.5}$$

$$= \frac{-(s^3+6s^2+5s)}{s+1.5}$$

on differentiating K with respect to s we get,

$$\frac{dK}{ds} = \frac{-(3s^2+12s+5)(s+1.5)-[s^3+6s^2+5s](1)}{(s+1.5)^2}$$

$$= \frac{-3s^2 - 4.5s^2 - 12s^2 - 18s - 5s - 7.5 + s^3 + 6s^2 + 9s}{(s+1.5)^2}$$

$$= \frac{-2s^3 - 10.5s^2 - 18s - 7.5}{(s+1.5)^2} = \frac{-2(s^3 + 5.25s^2 + 9s}{(5+1.5)^2}$$

For $\frac{dK}{ds} = 0$, the numerator should be zero.

$\therefore$ $s^3 + 5.25s^2 + 9s + 3.75 = 0$

The third order polynomial will have one real root. The real root of the above polynomial can be determined by Lin's method (Refer Lin's Method)

The polynomial cane bexpressed $s^3 + 5.25s^2 + 98 + 3.75 = (s + 0.6)(s^2 + 4.65s + 6.2) = 0$.

The roots of the quadratic $(s^2 + 4.65s + 6.2)$ are $\dfrac{s = -4.65 \pm \sqrt{(4.65)^2 - 4 \times 6.2}}{2}$ =-2.3 1 j 0.89.

Check for K

When $s = -0.6$. $K = -(s^3 + 6s^2 + 5s)/s + 1.5$

$$= \frac{-\left[(-0.6)^3 + 6(-0.6)^2 + 5(-0.6)\right]}{-0.6 + 1.5} = 1.17$$

For $s = -0.6$, the value of K is positve and real and so it is actual breakaway point. It can be shown that for $s = -2.3 \pm j\,0.86$ the value of K is not positve and real and so they cannot be breakaway point. The actual breakaway point is shown in Figure (5.5.1)

To find the real root of $s^3 + 5.25\,s^2 + 3.75 = 0$ by Lin's method.

The least two terms of the polynomial are choosen as Ist trial divisor.

Ist trial

Ist trial divisor = $9s + 3.75$

$s + 3.75/9$

$s = s + 0.42$

$$\begin{array}{r|l} & s^2 + 4.83s + 6.97 \\ \hline s + 0.42 & s^3 + 5.25s^2 + 9s + 3.75 \\ & s^3 + 0.42s^2 \\ \hline & 4.83s^2 + 9s \\ & 4.83s^2 + 2.03s \\ \hline \text{IInd trial divisor} \longrightarrow & 6.97s + 3.75 \\ & 6.97s + 2.93 \\ \hline & 0.82 \end{array}$$

IInd trial

IInd divisor = $6.97s + 3.75$

$= s + \dfrac{3.75}{6.97}$

$= s + .54$

$$\begin{array}{r|l} & s^2 + 9.713s + 6.46 \\ \hline s + .54 & s^3 + 5.25s^2 + 9s + 3.75 \\ & s^3 + .54s^2 \\ \hline & 4.71s^2 + 9s \\ & 4.71s^2 + 2.54s \\ \hline \text{IIIrd trial divisor} \longrightarrow & 6.46s + 3.75 \\ & 6.46s + 3.49 \\ \hline & .26 \end{array}$$

IIIrd trial:

IIIrd trial divisor = 6.46*s* + 3.75 $= s + \frac{3.75}{6.46}$

$$
\begin{array}{r|l}
 & s^2 + 4.67s + 6.3 \\
\hline
s + .58 & s^3 + 5.25s^2 + 9s + 3.75 \\
 & s^3 + .58s^2 \\
\hline
 & 4.67s^2 + 9s \\
 & 4.67s^2 + 2.7s \\
\hline
\text{IV}^{\text{th}} \text{ trial divisor} \longrightarrow & 6.3s + 3.75 \\
 & 6.3s + 3.65 \\
\hline
 & 0.1
\end{array}
$$

IVth trial divisor = 6.3 *s* + 3.75 $= s + \frac{3.75}{6.3} = s + 0.6$

$$
\begin{array}{r|l}
 & s^2 + 4.65s + 6.2 \\
\hline
s + 0.6 & s^3 + 5.25s^2 + 9s + 3.75 \\
 & s^3 + 0.6s^2 \\
\hline
 & 4.65s^2 + 9s \\
 & 4.65s^2 + 2.8s \\
\hline
 & 6.2s + 3.75 \\
 & 6.2s + 3.72 \\
\hline
 & .03
\end{array}
$$

On neglecting the small value of 0.03, one of the root of the polynomial is, $s = -0.6$

Step 5: To find angle of departure

Since there are no complex pole or zero we need not find angle of departure or arrival.

Step 6: To find crossing point of imaginary axis

The characteristic Eq. is

$$s(s + 1)(s + 5) + K(s + 1.5) = 0$$

$$s(s^2 + 6s + 5) + Ks + 1.5K = 0$$

$$s^3 + 6s^2 + 5s + Ks + 1.5K = 0$$

Put $s = j\omega$

$$(j\omega)^3 + 6(j\omega)^2 + 5(j\omega) + K(j\omega) + 1.5K = 0$$

$$-j\omega^3 - 6\omega^2 + j5\omega + jK\omega + 1.5K = 0$$

On equating imaginary part to zero we get,

$$-j5\omega^3 = jK\omega = 0$$

$$-j\omega^3 = -j5\omega - jK\omega$$

$$\omega^2 = 5 + K$$

On equating real part to zero we get,

$$-6\omega^2 + 1.5K = 0$$

Put $\omega^2 = 5 + K$

$$-6(5 + K) + 1.5K = 0$$

$$-30 - 4.5K = 0$$

$$K = \frac{30}{4.5} = -6.67$$

CHAPTER 5

Since the value of K is negative, there is no crossing point on imaginary axis, or for any positive values of K the root locus will not cross imaginary axis.

The complete root locus sketch is shown in Fig.(5.5.2). The root locus has three branches.

One branch starts at $s = -5$ and ends at finite zero at $s = -1.5$. The other two root locus starts at $s =$ 0 and $s =$- -1 and breakaway from real axis at $s = -0.6$, then travel parallel to asymptotes to meet the zeros get infinity.

EXAMPLE 5.6: Sketch the root locus for the unity feedback system whose open loop transfer function is $G(s) = \dfrac{K(s^2 + 6s + 25)}{s(s+1)(s+2)}$.

Solution:

Step 1: To locate poles and zeros

The poles of open loop transfer function are roots of the Eq. $s(s + 1)(s + 2) = 0$ and the zerros are the roots of the Eq. $(s^2 + 6s + 25) = 0$.

The roots of quadratic are,

$$s = \frac{-6 \pm \sqrt{6^2 - 4 \times 25}}{2}$$

$$= -3 \pm j4$$

The poles are lying at $s = 0, -1, -2$.

The zeros are lying at $s = -3 + j4, -3 - j4$.

The poles are marked by X(cross) and zeros by "o"(circle as shown in Fig. (5.6.1).

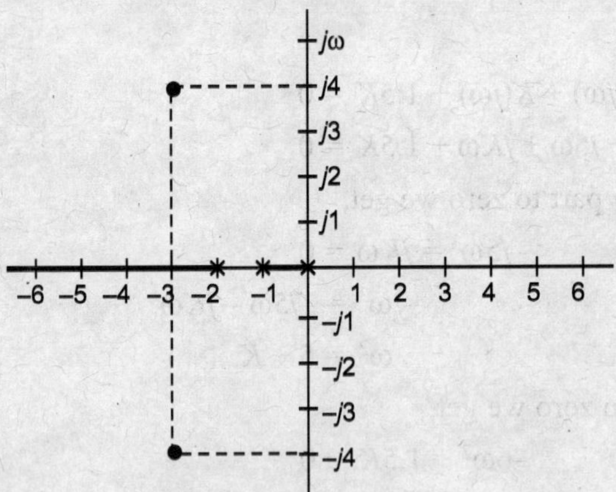

Fig. 5.6.1: Showing the root locus on real àxis, location fo poles & zeros.

Step 2: To find root locus on real axis

The segment of real axis between $s = 0$ and $s = -1$ and the entire negative real axis from $s = -2$ will be part of root locus. Because if we choose a test point in this segment then to the rigth of this point we

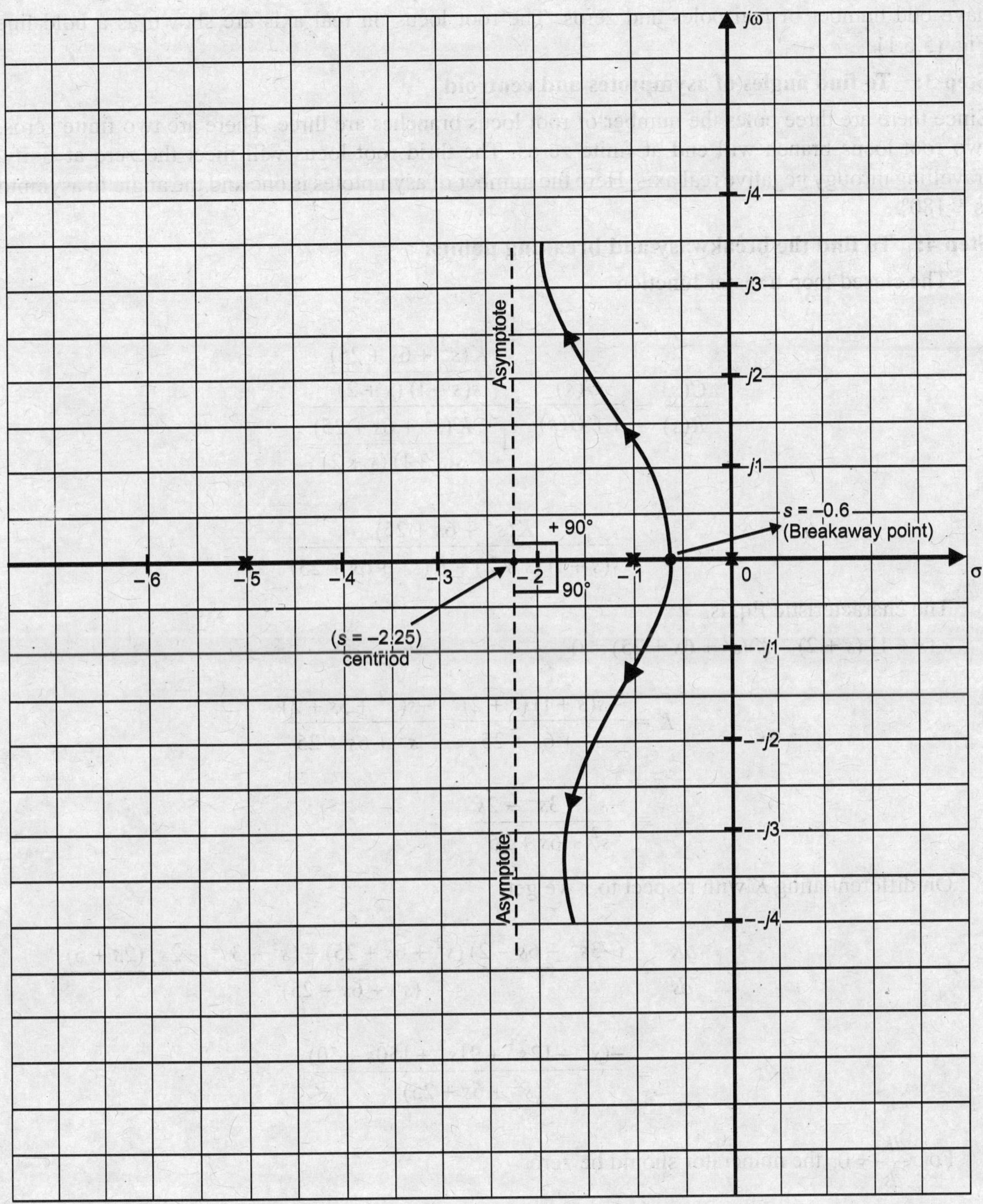

Fig. 5.5.2: Root locus sketch of $1 + G(s) = 1 + K(s + 1.5)/s(s + 1)(s + 5)$.

have odd number of real poles and zeros. The root locus on real axis are shown as a bold line in Fig. (5.6.1).

Step 3: To find angles of asymptotes and centroid

Since there are three poles the number of root locus branches are three. There are two finite zeros, so two root locus branch will end at finite zeros. The thrid root locus will meet the zero at finity by travelling througy negative real axis. Here the number of asymptotes is one and the angle fo asymptores is ± 180º.

Step 4: To find the breakwasy and breaking points:

The closed loop transfer function

$$\frac{C(s)}{R(s)} = \frac{G(s)}{1+G(s)} = \frac{\dfrac{K(s^2+6s+25)}{s(s+1)(s+2)}}{1+\dfrac{K(s^2+6s+25)}{s(s+1)(s+2)}}$$

$$= \frac{K(s^2+6s+25)}{s(s+1)(s+2)+K(s^2+6s+25)}$$

The characteristic Eq. is,

$s(s+1)(s+2) + K(s^2+6s+25) = 0$

$$\therefore \quad K = \frac{-s(s+1)(s+2)}{s^2+6s+25} = \frac{-s(s^2+3s+2)}{s^2+6s+25}$$

$$= \frac{-s^3-3s^2-2s}{s^2+6s+25}$$

On differentiating K with respect to s we get,

$$\frac{dK}{ds} = \frac{(-3s^2-6s-2)(s^2+6s+25)-(s^3-3s^2-2s)(2s+6)}{(s^2+6s+25)^2}$$

$$= \frac{-(s^4+12s^3+91s^2+150s+50)}{(s^2+6s+25)^2}$$

For $\dfrac{dK}{ds} = 0$, the numerator should be zero.

$\therefore \quad s^4 + 12s^3 + 91s^2 + 150s + 50 = 0$

The fourth order polynomial can be split into two quadratic Eq.s. The two quadratic factors can be obtained by lin's method. (Refer Lin's method).

To find quadratic factors by Lin's method.

The first trial divisor can be last three terms.

Ist trial:

$$\text{Ist trial divisor} = 91s^2 + 150s + 50$$

$$= s^2 + \frac{150}{91}s + \frac{50}{91} = s^2 + 1.65s + .55$$

$$\begin{array}{r|l}
 & s^2 + 10.35s + 73.37 \\
s^2 + 1.65s + .55 & s^4 + 12s^3 + 91s^2 + 150s + 50 \\
 & s^4 + 1.65s^3 + 0.55s^2 \\
\hline
 & 10.35s^3 + 90.45s^2 + 150s \\
 & 10.35s^3 + 17.08s^2 + 5.7s \\
\hline
\text{II}^{\text{nd}} \text{ trial divisor} \longrightarrow & 73.37s^2 + 144.3s + 50 \\
 & 73.37s^2 + 121.1s + 40.35 \\
\hline
 & 23.2s + 9.65 \\
\hline
\end{array}$$

IInd trial

$$\text{IInd trial divisor} = 73.37s^2 + 144.4s + 50$$

$$= s^2 + \frac{1443}{73.37}s + \frac{50}{73.37} = s^2 + 2s + .7$$

$$\begin{array}{r|l}
 & s^2 + 10s + 70.3 \\
s^2 + 2s + .7 & s^4 + 12s^3 + 91s^2 + 150s + 50 \\
 & s^4 + 2s^3 + .7s^2 \\
\hline
 & 10s^3 + 90.3s^2 + 150s \\
 & 10s^3 + 20s^2 + 7s \\
\hline
 & 70.3s^3 + 143s + 50 \\
 & 73.3s^2 + 140.6s + 49.2 \\
\hline
 & 2.4s + 0.8 \\
\hline
\end{array}$$

On neglecting the small remainder we can write

$s^4 + 12s^3 + 91s^2 + 150s + 50 \approx (s^2 + 2s + .7)\,(s^2 + 10s + 70.3)$

The roots of the quadratic $(s^2 + 2s + .7)$ are

$$s = \frac{-2 \pm \sqrt{2^2 - 4 \times .7}}{2} = -0.45, -1.55$$

The roots of the quadratic $(s^2 + 10s + 70.3)$ are

$$s = \frac{-10 \pm \sqrt{10^2 - 4 \times 70.3}}{2} = -5 \pm j6.73$$

Here, $s = 1.55$ is not a point on root locus, hence it cannot be a breakaway point.

Check the other three values for actual breakway point

when $$s = 0.45,\ K = \frac{-s^3 - 3s^2 - 2s}{s^2 + 6s + 25}$$

$$= \frac{-(-4.5)^3 - 3(-.45)^2 - 2(-.45)}{(-.45)^2 + 6(-.45) + 25}$$

$$= 0.017$$

For $s = -0.45$, the value of K is positive and real and so it is actual breakaway point. It can be shown that for $s = -5 \pm j6.73$ the value of K is not positive and real and so thay cannot be breakwasy points. The actual breakwasy point is shown in Fig. (5.6.1).

Step 5: To find angle of arival.

Let us consider the complex zero. A shown in Fig. (3.6.6.2). Draw vectors from all other poles to the A as shown in Fig. (3.6.6.2). Let the angles of these vectors be θ_1, θ_2, θ_3 and θ_4.

Hence, $$\theta_1 = 180^\circ \tan^{-1}\frac{4}{3} = 126.9^\circ$$

$$\theta_2 = 180^\circ \tan^{-1}\frac{4}{2} = 116.6^\circ$$

$$\theta_3 = 180^\circ \tan^{-1}\frac{4}{1} = 104^\circ$$

$$\theta_4 = 90^\circ$$

Angle of arrival at complex zero A

$$= 180^\circ - (\theta_4) + (\theta_1 + \theta_2 + \theta_3)$$

$$= 180^\circ - 90^\circ + 126.9^\circ + 116.6^\circ + 10.4^\circ$$

$$= 437.5^\circ = 77.5^\circ$$

Angle of arrival at complex zero A^* is negative of the angle of arrival at complex zero A.

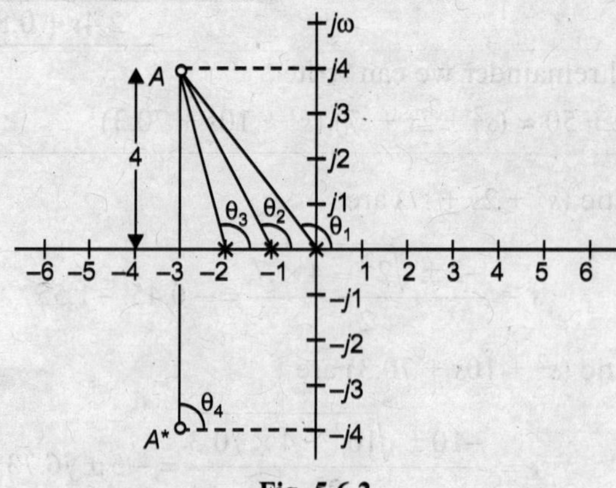

Fig. 5.6.2

∴ Angle of arrival at complex zero $A^* = -77.5°$

Mark the angles of arrival at complex zeros using protractor.

Step 6: To find the crossing point on imaginary axis:

The characteristic Eq. is

$$s(s + 1)(s + 2) + K(s^2 + 6s + 25) = 0$$

$$s(s^2 + 3s + 2) + Ks^2 + 6Ks + 25K = 0$$

$$s^3 + 3s^2 + 2s + Ks^2 + 6Ks + 25K = 0$$

$$s^3 + (3 + K)s^2 + (2 + 6K)s + 25K = 0$$

Put $s = j\omega$

$$(j\omega)^3 + (3 + K)(j\omega)^2 + (2 + 6K)(j\omega) + 25K = 0$$

$$-j\omega^3 - (3 + K)\omega^2 + j(2 + 6K)\omega + 25K = 0$$

On equating imaginary part to zero

$$-j\omega^3 + j(2 + 6K)\omega = 0$$

$$-j\omega^3 = -j(2 + 6K)\omega$$

$$\omega^2 = (2 + 6K)$$

On equating real part to zero,

$$-(3 + K)\omega^2 + 25K = 0$$

Put $\omega^2 = 2 + 6K$

$$-(3 + K)(2 + 6K) + 25K = 0$$

$$-(6 + 18K + 2K + 6K^2) + 25K = 0$$

$$-6K^2 + 5K - 6 = 0$$

$$K = \frac{-5 \pm \sqrt{5^2 - 4 \times (-6)(-6)}}{2 \times (-6)}$$

$$K = 0.4 \pm j\,0.9$$

Since the value of K is not real and positive, there is no crossing point on imaginary axis, or for any positive values of K the root locus will not cross imaginary axis.

Step 7: To find points on root locus

Choose test points a, b, c, d on the s plane and adjust the test points to satisfy angle criterion. The test points are shown in Fig. (5.6.3).

On the upper half of splane the root locus in sketched through the test points a, b, c and d. The root locus on the lower half of S-plane is the mirror image of the root locus on the upper half of splane.

The complete root locus sheten is shown in Fig. (5.6.3). the root locus has three branches, one branch starts at $s = -2$ and goes to infinity along negative real axis. The other two root locus branches starts at $s = 0$ and $s = -1$ and breaks from real axis at $s = -0.45$, then meets the complex zeros.

EXAMPLE 5.7: Sketch the root locus for the unity feedback system whose open loop transfer function is $G(s) = K/s(s^2 + 6s + 10)$.

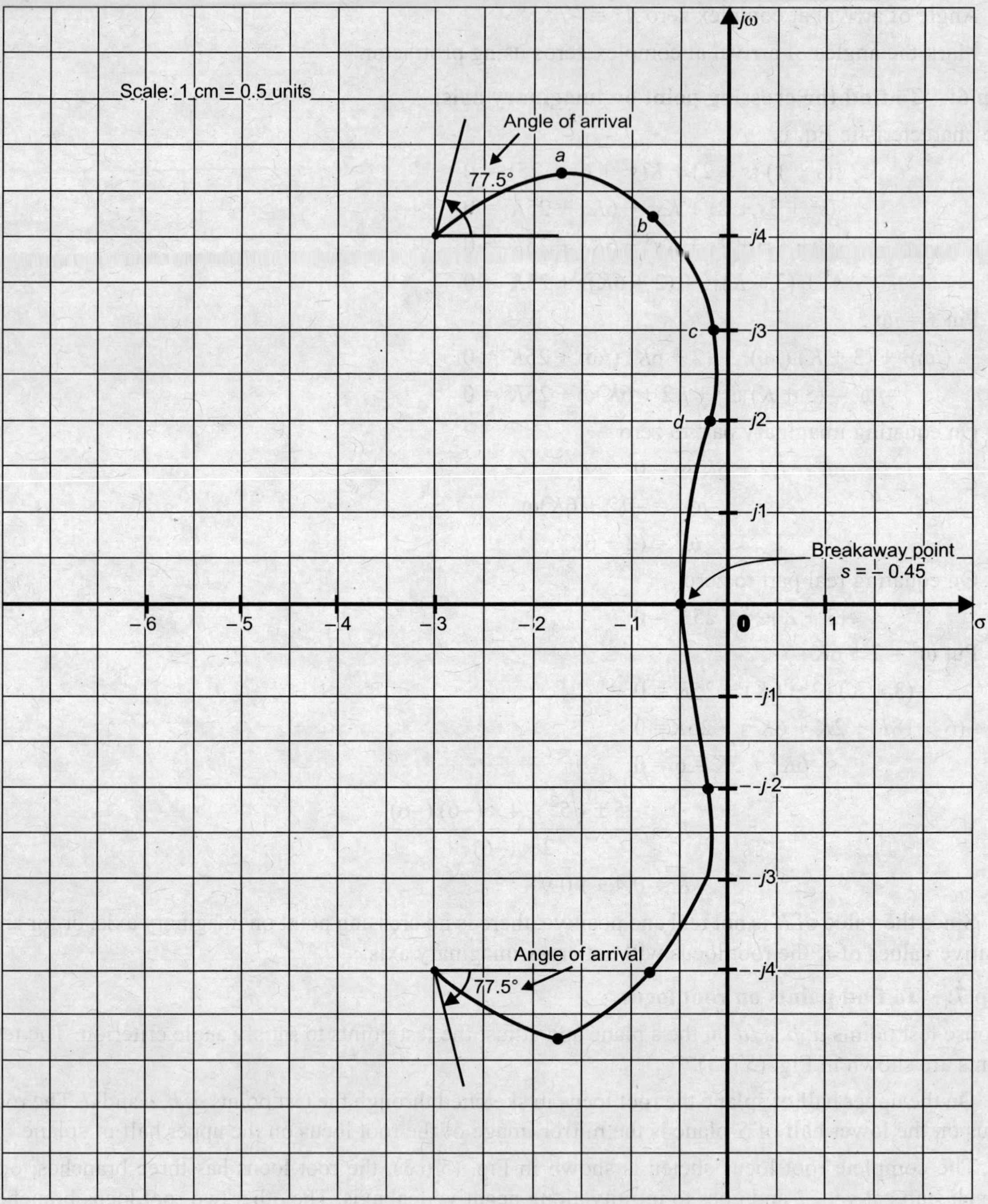

Fig. 5.6.3: Root locus sketch of $1 + G(s) = 1 + \dfrac{K(s^2 + 6s + 2s)}{s(s+1)(s+2)}$.

Solution:

Step 1: To locate poles and zeros

The poles of open loop transfer function are the roots of the Eq. $s(s^2 + 6s + 10) = 0$.

The roots of the quadratic are,

$$s = \frac{-6 \pm \sqrt{6^2 - 4 \times 10}}{2}$$

The poles are 0, $-3 + j1$ and $-3 - j1$.

The poles are marked by ×(cross) as shown in Fig. (5.7.1).

Step 2: To find the root locus on real axis.

There is only one pole on real axis at the origin. Hence if we choose any test point on the negative real axis then to the right of that point the total number of real poles and zero sis one, which is san odd number. hence the entire negtaive real axis will be part of root locus. The root locus on real axis is shown as bold line is Fig. (5.7.1).

Note: For the given transfer function are root locus branch will start at the pole at the origin and meet the zero at infinity through the negative real axis.

Step 3: To find angles of asymptotes and centroid:

Since there are 3 poles the number of root locus branches are three. There is no finite zero. hence all the three root locus branches ends at zeros at infinity. The number of asymptores required are three.

$$\text{Angle of asymptotes} = \frac{\pm 180^{\circ}(2q+1)}{n-m}$$

Here, $n = 3$ and $n = 0$,

If $q = 0$, $\quad \text{Angles} = \pm\frac{180}{3} = \pm 60^{\circ}$

If $q = 1$, $\quad \text{Angles} = \pm\frac{180 \times 3}{3} = \pm 180^{\circ}$

$$\text{Centroid} = \frac{\text{Sum of poles} - \text{Sum of zeros}}{n-m}$$

$$\text{Centroid} = \frac{-3 + j1 - 3 - j1}{3} = -2$$

The centroid is marked on real axis and from the centroid the angles of asymptotes are marked using or protractor. The asymptotes are drawn as dotted lines as shown in Fig. (5.7.1).

Step 4: To find the breakaway and breaking points

The closed loop transfer function, $\dfrac{C(s)}{R(s)} = \dfrac{G(s)}{1+G(s)}$

$$\therefore \quad \frac{C(s)}{R(s)} = \frac{\dfrac{K}{s(s^2+6s+10)}}{1+\dfrac{K}{s(s^2+6s+10)}} = \frac{K}{s(s^2+6s+10)+K}$$

The characteristic Eq. is $s(s^2 + 6s + 10) + K = 0$

$$\therefore \quad K = -s(s^2 + 6s + 10) = -s^3 - 6s^2 - 10s$$

on differentiating the Eq. of K with respect to s we get,

$$\frac{dK}{ds} = -3s^2 - 12s - 10$$

Put $\dfrac{dK}{ds} = 0$

$$-3s^2 - 12s - 10 = 0$$

$$3s + 12s + 10 = 0$$

$$\therefore \quad s = \frac{-12 \pm \sqrt{12^2 - 4\times 3\times 10}}{2\times 3} = -1.18 \text{ (or)} - 2.82$$

Check for K

when, $s = -1.18$, $K = -s^3 - 6s^2 - 10s$

$$= -(-1.18)^3 - 6(-1.18)^2 - 10\,(-1.18) = 5.09$$

when, $s = -2.82$, $K = -s^3 - 6s^2 - 10s$

$$= -(-2.82)^3 - 6(-2.82)^2 - 10\,(-2.82) = 2.91$$

Since the values of K for $s = -1.18$ and -2.82 and positive and real, both the points are actual breakaway or break in points. It can be proved that $s = -2.82$ in breaking point and $s = -1.18$ is a breakaway points. The breakin and breakaway points are shown in Fig. (5.7.1)

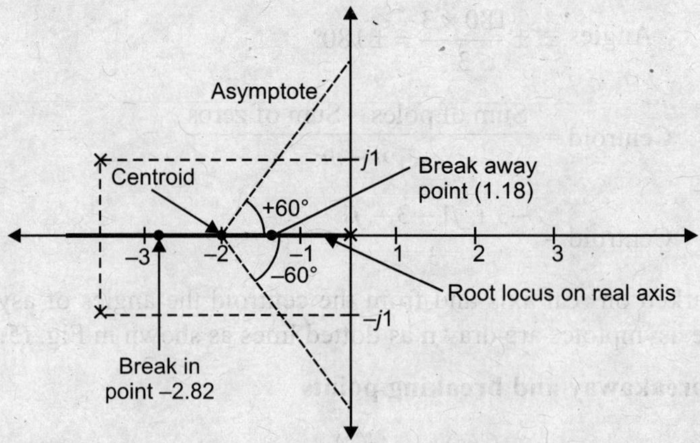

Fig. 5.7.1: Figure showing the asymptote, root locus on real axis and location of poles, centroid, breakin and breakaway points.

Step 5: To find the angle of departure

Consider the complex pole A shown in Fig. (5.7.2) draw vectors from all other poles to the pole A as shown in Fig. (5.7.2). Let the angle of these vectors be θ_1 and θ_2.

Here, $\theta_1 = -180° - \tan^{-1}(1/3) = 161.6°$

$\theta_2 = 90°$

Angle of departure from the complex pole $A = 180° - (\theta_1 + \theta_2)$

$= 180° - (61.6° + 90°) = -71.6° \approx -72°$

The angle of departure at complex pole A^* is negative of the angle of departure at complex pole A.

$\therefore$ Angle of departure at pole $A^* = +72$

Mark the angles of departure at complex poles using protractors.

Step 6: To find the corssing point on imaginary axis.

The characteristics Eq.is given by

$$s(s^2 + 6s + 10) + K = 0$$

$$s^3 + 6s^2 + 10s + K = 0$$

Put $s = j\omega$

$$(j\omega)^3 + 6(j\omega)^2 + 10(j\omega) + K = 0$$

$$-j\omega^3 - 6\omega^2 + j\,10\omega + K = 0$$

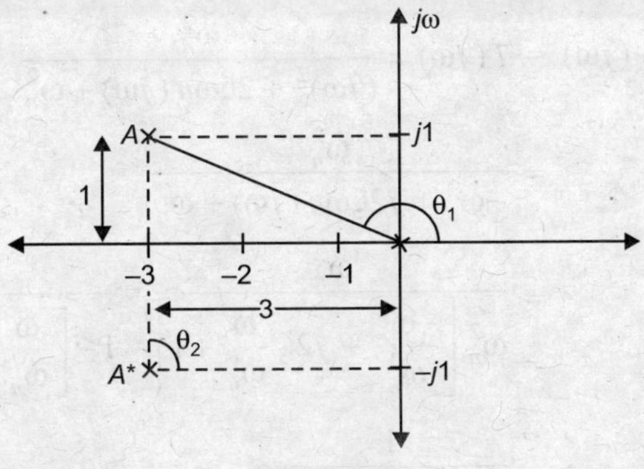

Fig. 5.7.2

On equating imaginary part to zero we get,

$$-\omega^3 + 10\omega = 0$$

$$\omega^3 = 10\omega$$

$$\omega^2 = 10$$

$$\omega = \pm\sqrt{10} = \pm 3.16 \approx \pm 3.2$$

On equating real part to zero we get,

$$-6\omega^2 + K = 0$$

$$K = 6\omega^2 = 6 \times 10 = 60 \Rightarrow K = 60$$

CHAPTER 5

The root locus crossing imaginary axis at $\pm j32$ and the gain K corresponding to this point is 60. This is the limiting value of K for the stability of the system.

The complete root locus sketch is shown in Fig. (5.7.3). The root locus has 3 branches. One branch starts at $s = 0$ and goes to infinity along negative real axis. The other two root locus branches starts at $s = -3 + j1$ and enter the real axis at $s = -2.82$ and then breakaway from real axis at $s = -1.18$. Finally theytravel paralllel to asymptotes to meet the zeros at infinity.

5.2 CORRELATION BETWEEN TIME AND FREQUENCY RESPONSE

The Correlation between time and frequency response has an explict form only for first and second order systems. The correlation for second order system is discussed here.

Consider the standard form of transfer function of second order system.

$$\frac{C(s)}{R(s)} = \frac{\omega_n^2}{s^2 + 2\zeta\omega_n\, s + \omega_n^2}$$

where,

ζ = Damping factor

ω_n = undamped natural frequency the sinusoidal transfer function of the system is obtained by letting $s = j\omega$,

$$\frac{C}{R}(j\omega) = T(j\omega) = \frac{\omega_n^2}{(j\omega)^2 + 2\zeta\omega n\,(j\omega) + \omega_n^2}$$

$$= \frac{\omega_n^2}{-\omega^2 + j2\zeta\omega n\,(j\omega) + \omega_n^2}$$

$$= \frac{\omega_n^2}{\omega_n^2\left[\frac{-\omega^2}{\omega_n^2} + j2\zeta\frac{\omega}{\omega_n} + 1\right]} = \frac{1}{1 - \left[\frac{\omega}{\omega_n}\right]^2 + j\,2\zeta\left[\frac{\omega}{\omega_n}\right]}$$

$$= \frac{1}{(1-u^2) + j\,2\zeta\,u}$$

where, $u = \left[\frac{\omega}{\omega_n}\right]$ is the normalised frequency magnitude of closed loop system,

$$M = |\,T(j\omega)\,| = \frac{1}{\sqrt{(1-u^2) + (2\zeta u)^2}} \tag{5.6}$$

Phase of closed loop system

$$\alpha = \angle T(j\omega) = -\tan^{-1}\left[\frac{2\zeta u}{1-u^2}\right] \tag{5.7}$$

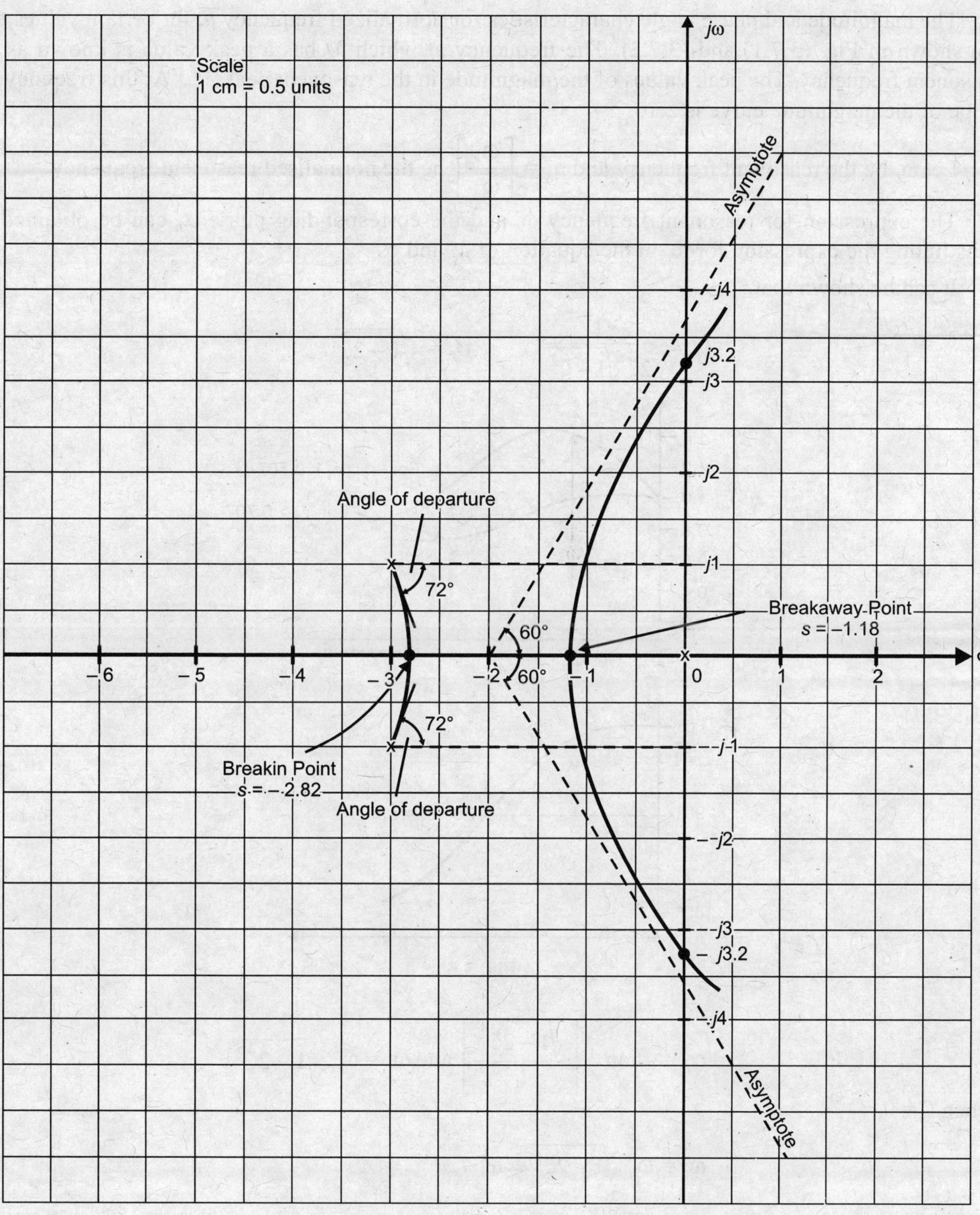

Fig. 5.7.3: Root 10 cm sketch of $= 1 + G(s) = \dfrac{K}{s(s^2 + 6s + 10)}$.

The magnitude and phase angle characteristics for normalized frequency u, for certain values of ζ are shown in Fig. (3.7.1) and (3.7.2). The frequency at which M has a peak value is known as the reasonent frequency. The peak values of the magnitude in the reasonent peak M_r. At this frequncy the slope of the magnitude curve is zero.

Let ω_r be the reasonant frequency and $u_r = \left[\frac{\omega_r}{\omega_n}\right]$ be the normalised reasonent frequency.

The expression for reasonant frequency ω_r and the corresponding phase α_r can be obtained by substituting the expression for ω_r in the equation of M and α.

It can be shown that,

$$M_r = \frac{1}{2\zeta\sqrt{1-\zeta^2}}; \qquad M_r = \sqrt{1-2\zeta^2}$$

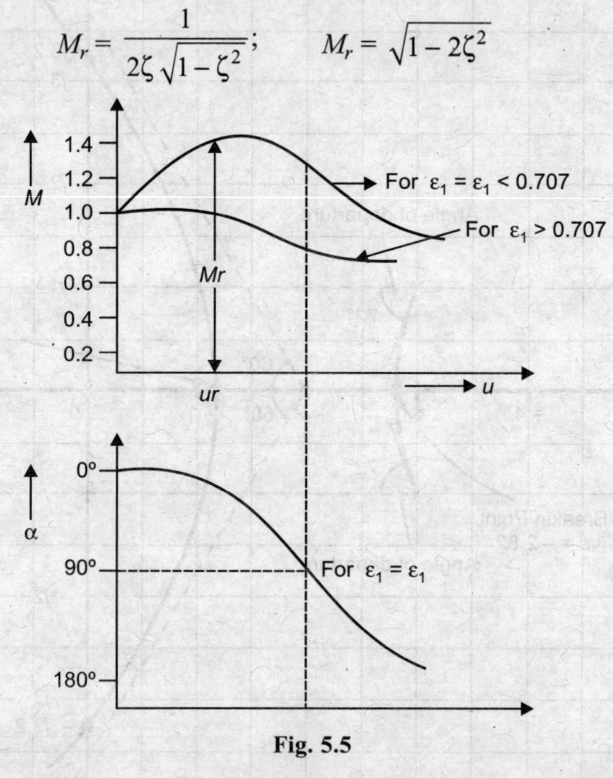

Fig. 5.5

$$\alpha_r = -\tan^{-1}\left[\frac{\sqrt{1-2\zeta^2}}{\zeta}\right] \text{ and } \omega_r = \omega_n\sqrt{1-2\zeta^2}$$

when $\zeta = 0$

$$\omega_r = \omega_n\sqrt{1-2\zeta^2} = \omega_n$$

$$M_r = \frac{1}{2\zeta\sqrt{1-\zeta^2}} = \infty$$

Hence, it is clear that as s tends to zero, ω_r approaches ω_n and M_r approaches infinity.

For $0 < \zeta < \frac{1}{\sqrt{2}}$, the reasonent frequency always hasa value less than ω_n and the reasonent peak has a value greater then one.

For $\zeta < \frac{1}{\sqrt{2}}$, the condition $\left(\frac{dM}{dc}\right) = 0$, will not be satisfied for any real value of ω.

Hence when $\zeta < \frac{1}{\sqrt{2}}$, the magnitude decreases motrotonically from $M = 1$ at $u = 0$ with increasing u.

It follows that for $\zeta > \frac{1}{\sqrt{2}}$, there is no reasonent peak and the greatest value of M equals one.

The frequency at which M has a value of $\frac{1}{\sqrt{2}}$ is specialsignificance and is called the cut-off frequency ω_c. the signal frequencies above cut-off are greatly alternated on passing through a system.

For feedback control system, the range of frequencies over which $M \geq \frac{1}{\sqrt{2}}$ is defined as Bandwidth ω_b. Control system being loco-pass filter, the bandwidth ω_b is equal to cut-off frequency ω_c.

In general the bandwidth of a control system indicates the noise-filtering characteristics of the system. Also, band width gives a measure of the transiont response.

The normalised bandwidth, $u_b = \left[\frac{\omega_b}{\omega_n}\right]$

$$u_b = \left[1 - 2\zeta^2 + \sqrt{2 - 4\zeta^2 + \zeta^4}\right]$$

from the equation of u_b it is clear that u_b is a function of ζ is shown in Fig. (5.6)

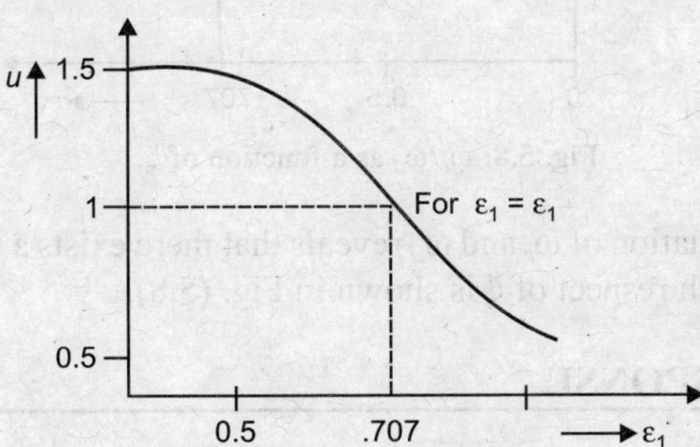

Fig. 5.6: Normalized bandwidth as a function of ζ.

The expression for the dumped frequency of oscillation ω_b and peak overshoot Mp of the step response for $0 \leq \zeta \leq 1$ are

Damped frequency, $\omega_d = \omega_n\sqrt{1 - \zeta^2}$

Peak overshoot, $M_p = \frac{-\zeta\pi}{\sqrt{1 - \zeta^2}}$

CHAPTER 5

Comparison of the equation of M_r and M_p reveals that both are functions of only ζ.

The sketch of M_r and M_p for various value of ζ are shown in Fig. (5.7) and (5.8). The sketch reveals that a system with a given value of M_r must exhibit a corresponding value of M_p if subjected to a step input. For $\zeta > \frac{1}{\sqrt{2}}$ the reasonent peak M_r does not exist and the correlation breaks down. This is not a serious problem as for this range of ζ, the step response oscillations are well dampled and M_p is negligible.

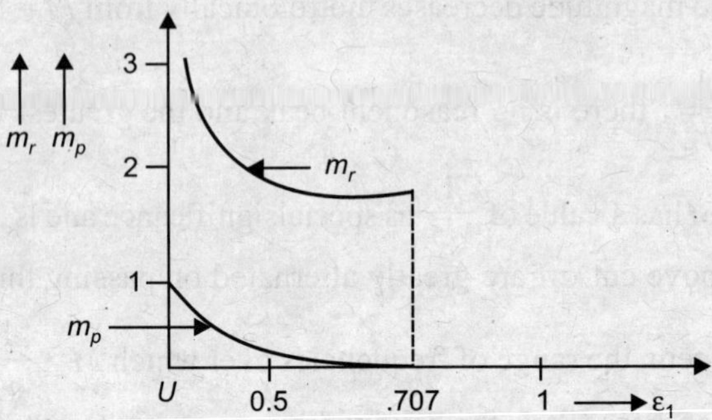

Fig. 5.7: M_r and M_p as a function ζ.

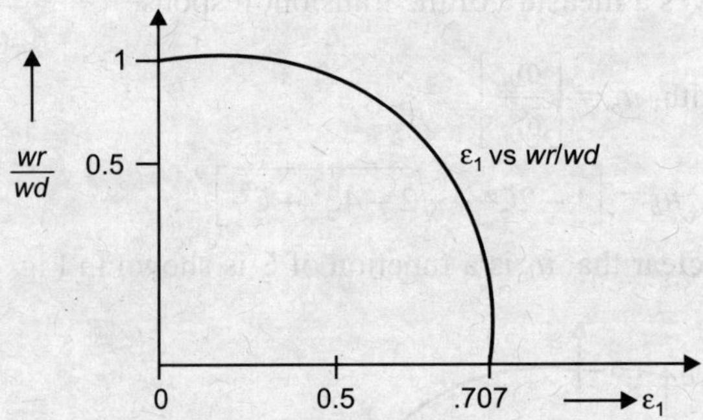

Fig. 5.8: ω_r/ω_d as a function of ζ.

The comparison of the equation of ω_r and ω_d reveals that there exists a definite correlation between them. The sketch of ω_r/ω_d with respect of ζ is shown in Fig. (5.8).

5.3 FREQUENCY RESPONSE

The frequency response is the steady state response of a system when the input to the system is a sinusoidal signal. Consider a linear time invariant (LTI) system, H shown in Fig. (5.9). Let $x(t)$ be the input. The output is $y(t)$ is also a sinusoidal signal of the same frequency but magnitude and phase angle are not same.

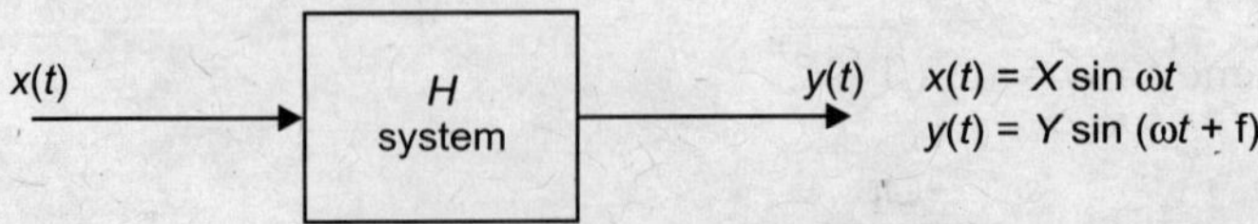

Fig. 5.9: LTI system.

The magnitude and phase relationship between the sinusoidal input and the steady state output of a system is termed as frequency response. In LTI systems the frequency respnse is independent of amplitude and phase of input. The freuqency response of a system is obtained by varying the frequency of the input signal by keeping the magnitude of the input signal at a constant value.

In they system transfer function $T(s)$, if s is replaced by $j\omega$, then the transfer function is called sinusoidal transfer function. The frequency response can be obtained by letting S by $j\omega$ i.e., obtained directly from $T(j\omega)$. $T(j\omega)$ is complex function of frequency.

The frequency response can be evaluated for both open loop system and closed loop system.

The advantages of frequency response analysis:

1. The absolute and relative stability of the closed loop system can be estimated from the knowledge of their open loop frequency response.
2. The practical testing of systems can be easily carried with available sinusoidal signal generators and precise measurement equipments.
3. The transfer function of complicated systems can be determined experimentally by frequency response tests.
4. The design and parameter adjustment of the open loop transfer function of a system for specified closed loop performance is carried out more easily in frequency domain.
5. When the system is designed by use of frequency analysis, the effects of noise disturbance and parameter variations are relatively easy to visualize and incorporate corrective measure.
6. The frequency response analysis and designs can be extended to certain non linear control system.

Frequency domain specifications

The performance and characteristics of a system in frequency domain are measured in terms of frequency domain specifications. The requirements of a system to be designed are usually specified in terms of these specifications.

The frequency domain specifications are:

1. Resonant peak, M_r
2. Resonant frequency, ω_r
3. Band width
4. Cut-off rate
5. Gain margin (Kg)
6. Phase margin (γ)

1. **Resonant peak (M_r):** The maximum value of the magnitude of closed loop transfer function is called the resonant peak. A large resonant peak corresponds to a large overshoot is transient response.
2. **Resonant frequency (ω_r):** The frequency at which the resonant peak occurs is called resonant frequency ω_r. This is related to the frequency of oscillation in the step response and thus its indicative of the speed of transient responses.

3. **Bandwidth (ω_b):** The bandwidth is the range of frequencies for which the system gain is more than –3 *db* the frequency at which the gain is –3 *db* is called cut of frequency. Bandwidth is usually defined for closed loop system and it transmits the signals whose frequencies are less than cut off frequency. The bandwidth is a measure of the ability of a feedback system to produced the input signal noise rejection characteristics and rise time. A large bandwidth corresponds to a small rise time (or) fast response.
4. **Cut off rate:** The slope of the log magnitude curve near the cut off frequency is called the cut off rate. The cut off rate indicates the ability of the system to distinguish the signal from noise.
5. **Gain margin, K_g:** The gain margin K_g is defined as the reciprocal of the magnitude of open loop transfer function at phase cross over frequency. The frequency at which the phase of open loop transfer function is 180° is called the phase cross – over frequency ω_{pc}.

$$\text{Gain margin, } K_g = \omega_{Pc}\frac{1}{|G(j\omega_{Pc})|}$$

The gain margin in db can be expressed as,

$$K_g \text{ in } db = 20\log K_g = 20\log\frac{1}{|G(j\omega_{Pc})|} = -20\log|G(j\omega_{Pc})|$$

Note: $|G(j\omega_{Pc})|$ is the magnitude of $G(j\omega)$ at $\omega = \omega_{Pc}$. (Gain margin is *db* is defined as the negative of the *db* magnitude of $G(j\omega)$ at phase cross over frequency)

The gain margin indicated the amount by which the gain of the system can be increased with out affecting the stability of the system.

6. **Phase Margin (γ):** The phase margin γ, is that amount of additional pahse Lag at the gain cross over frequency required to bring the system to the verge of instability. The gain cross over frequency ω_{gc} is the frequency at which the magnitude of the open loop transfer function is unity (or is frequency at which the db magnitude is zero).

The phase margin γ, is obtained by adding 180° to the phase angle ϕ of the open loop transfer function at the gain cross over frequency.

Phase margin, $\gamma = 180° + \phi_{gc}$, where,

$$\phi_{gc} = \angle G(j\omega_{gc})$$

Note: $\angle G(\omega_{gc})$ is the phase angle of $G(j\omega)$ at $\omega = \omega_{gc}$.

Estimation of Frequency Domain Specifications for a Second Order System

Resonant Peak (M_r)

Consider the closed loop transfer function of second order system

$$\frac{C(s)}{R(s)} = \frac{\omega_n^2}{s^2 + 2\zeta\omega_n s + \omega_n^2}$$

The sinusoidal transfer function $T(j\omega)$ is obtained by letting $s = j\omega$

$$\therefore \quad T(j\omega) = \frac{C(s)}{R(s)} = \frac{\omega_n^2}{(j\omega)^2 + 2\zeta\omega_n(j\omega) + \omega_n^2}$$

$$= \frac{\omega_n^2}{-\omega^2 + j2\zeta\omega_n\omega + \omega_n^2}$$

$$= \frac{\omega_n^2}{\omega_n^2\left[-\frac{\omega^2}{\omega_n^2} + j2\zeta\frac{\omega}{\omega_n} + 1\right]}$$

$$= \frac{1}{1 - \left(\frac{\omega}{\omega_n}\right)^2 + j2\zeta\left(\frac{\omega}{\omega_n}\right)}$$

Let, normalised frequency, $u = \left(\frac{\omega}{\omega_n}\right)$

$$\therefore \quad T(j\omega) = \frac{1}{\left(1-u^2\right) + j2\zeta u} = \frac{1}{\sqrt{\left(1-u^2\right)^2 + (2\zeta u)^2} \angle \tan^{-1}\frac{2\zeta u}{1-u^2}}$$

Let, $M \to$ Magnitude of closed loop transfer function

$\alpha \to$ phase of closed loop transfer function

$$M = |T(j\omega)| = \left[\frac{1}{\left(1-u^2\right)^2 + (2\zeta u)^2}\right]^{\frac{1}{2}} = \left[\left(1-u^2\right)^2 + 4\zeta^2u^2\right]^{\frac{1}{2}}$$

$$\alpha = \angle T(j\omega) = -\tan^{-1}\left(\frac{2\zeta u}{1-u^2}\right)$$

The resonant peak is the maximum value of M. The condition for maximum value of M can be obtained by differentiating the equation of M with respect to u and letting $\frac{dM}{du} = 0$ when $u = M_r$, where,

$u_r = \frac{\omega_r}{\omega}$ = Normalised resonant frequency

on differentiating equation of M with respect to u we get,

$$\frac{dM}{du} = \frac{d}{du}\left[\left(1-u^2\right)^2 + 4\zeta^2u^2\right]^{\frac{1}{2}}$$

$$= -\frac{1}{2}\left[\left(1-u^2\right)^2 + 4\zeta^2u^2\right]^{\frac{3}{2}}\left[2(1-u^2)(-2u) + 8\zeta^2u\right]$$

$$= \frac{-\left[-4u(1-u^2)+8\zeta^2 u\right]}{2\left[(1-u^2)^2+4\zeta^2u^2\right]^{3/2}} = \frac{4u(1-u^2)-8\zeta^2 u}{2\left[(1-u^2)^2+4\zeta^2u^2\right]^{3/2}}$$

Replace u by u_r in above equation

$$\frac{4u_r(1-u_r^2)-8\zeta^2 u_r}{2\left[(1-u_r^2)^2+4\zeta^2u_r^2\right]^{3/2}} = 0$$

Equating numerator to zero in above equation we get,

$$4u_r(1-u_r^2)-8\zeta^2u_r = 0$$

$$4u_r - 4u_r^3 - 8\zeta^2u_r = 0$$

$$4u_r^3 = 4u_r - 8\zeta^2 u_r$$

$$u_r^2 = 1-2\zeta^2$$

$$u_r = \sqrt{1-2\zeta^2}$$

The resonant peak occurs when $u_r = \sqrt{1-2\zeta^2}$ put this condition is equation for M and solve for M_r

$$\therefore \quad M_r = \left.\frac{1}{\left[\left(1-u^2\right)^2+4\zeta^2u^2\right]^{1/2}}\right|_{u=u_r}$$

$$\therefore \quad M_r = \frac{1}{\left[\left(1-u_r^2\right)^2+4\zeta^2u_r^2\right]^{1/2}}$$

$$M_r = \frac{1}{\left[1-\left(1-2\zeta^2\right)\right]^2+4\zeta^2(1-2\zeta^2)^{1/2}}$$

$$M_r = \frac{1}{\left(4\zeta^4+4\zeta^2-8\zeta^4\right)^{1/2}} = \frac{1}{\left(4\zeta^2-4\zeta^4\right)^{1/2}} = \frac{1}{\left[4\zeta^2(1-\zeta^2)\right]^{1/2}}$$

$$= \frac{1}{2\zeta\sqrt{1+\zeta^2}}$$

$\therefore$ The resonant peak, $$M_r = \frac{1}{2\zeta\sqrt{1-\zeta^2}}$$

Resonant frequency (ω_r)

The normalised resonant frequency,

$$u_r = \frac{\omega_r}{\omega_n} = \sqrt{1-2\zeta^2}$$

The resonant frequency $\omega_r = \omega_n\sqrt{1-2\zeta^2}$

Bandwidth (ω_b)

Let, normalized bandwidth, $u_b = \dfrac{\omega_b}{\omega_n}$

when $u = u_b$, the magnitude M, of the closed loop system is $\dfrac{1}{\sqrt{2}}$(or $-3db$).

Hence in the equation of M, put $u = u_b$ and equate to $\dfrac{1}{\sqrt{2}}$.

$$M = \frac{1}{\sqrt{2}} = \frac{1}{\left[(1-u_b^2)^2 + 4\zeta^2 u_b^2\right]^{1/2}}$$

on squaring and cross multiplying,

$$(1 - u_b^2)^2\ 4\zeta^2\ u_b^2 = 2$$

$$1 + u_b^4 + 2u_b^2 + 4\zeta^2\ u_b^2 = 2$$

$$u_b^4 - 2u_b^2\ (1 - 2\zeta^2) - 1 = 0$$

Let, $x = u_b^2$

$$\therefore \qquad x^2 - 2(1 - 2\zeta^2)x - 1 = 0$$

$$\therefore \qquad x = 2(1-2\zeta^2) \pm \frac{\sqrt{4(1-2\zeta^2)^2 + 4}}{2}$$

By taking only the positive sign,

$$x = 1 = -2\zeta^2 + \sqrt{2-4\zeta^2+4\zeta^4}$$

But, $\qquad u_b = \sqrt{x} = 1\left[1-2\zeta^2 + \sqrt{2-4\zeta^2+4\zeta^4}\right]^{1/2}$

Also, $\qquad u_b = \dfrac{\omega_b}{\omega_n}$

$\therefore$ bandwidth, $\omega_b = \omega_n\ ,\ u_b = \omega_n \left[1-2\zeta^2 + \sqrt{2-4\zeta^2+4\zeta^4}\right]^{1/2}$

Phase margin (γ)

The open loop transfer function of second order system,

$$G(s) = \frac{\omega_n^2}{s(s+2\zeta\omega_n)}$$

The sinusoidal transfer function $G(j\omega)$ is obtained by letting $s = j\omega$

$$G(j\omega) = \frac{\omega_n^2}{j\omega(j\omega + 2\zeta\omega_n)} = \frac{\omega_n^2}{-\omega^2 + j2\zeta\omega\omega_n}$$

$$= \frac{\omega_n^2}{\omega_n^2\left[-\frac{\omega^2}{\omega_n^2} + j2\zeta\frac{\omega}{\omega_n}\right]}$$

Let normalised frequency, $u = \frac{\omega}{\omega_n}$ on substituting $u = \frac{\omega}{\omega_n}$ is above equation we get,

$$G(j\omega) = \frac{1}{u^2 + j2\zeta u}$$

Magnitude of $\quad G(j\omega) = |G(j\omega)| = \frac{1}{\sqrt{u^4 + 4\zeta^2 u^2}}$

Phase of $\quad G(j\omega) = \angle G(j\omega) = -\tan^{-1}\frac{2\zeta u}{-u^2} = -\tan^{-1}\left(-\frac{2\zeta}{u}\right)$

$$= \tan^{-1}\frac{2\zeta}{u}$$

At the gain cross-over frequency code, the magnitude of $G(j\omega)$ is unity

Let normalised gain cross over frequency,

$$u_{gc} = \frac{\omega_{gc}}{\omega_n}$$

substitute u by u_{gc} in the equation for magnitude of $G(j\omega)$ and equates to unity

$\therefore$ At $u = u_{gc}$, $\quad |G(j\omega)| = 1 = \frac{1}{\sqrt{u_{gc}^4 + 4\zeta^2 u_{gc}^2}}$

$$\therefore \quad {u_{gc}}^4 + 4\zeta^2 u_{gc}^2 = 1$$

$${u_{gc}}^4 + 4\zeta^2 u_{gc}^2 - 1 = 0$$

Let $\quad x = {u_{gc}}^2$

$$\therefore \quad x^2 + 4\zeta^2 x - 1 = 0$$

$$x = \frac{-4\zeta^2 \pm \sqrt{16\zeta^4 + 4}}{2}$$

$$x = -2\zeta^2 \pm \sqrt{4\zeta^4 + 1}$$

on taking only the +ve sign

$$u_{gc} = \sqrt{x} = \left[-2\zeta^2 + \sqrt{4\zeta^4 + 1}\right]^{1/2}$$

The phase margin, $\gamma = 180 + \angle G(j\omega)/\ \omega = \omega_{gc},\ u = u_{gc}$

Substuituting for $\angle G(j\omega)$ from equations phase of $G(j\omega)$ and phase marging (γ) we get,

$$\gamma = 180^o + \tan^{-1}\frac{2\zeta}{u_{gc}} = 180^o + \tan^{-1}\frac{2\zeta}{\left[-2\zeta^2 + \sqrt{4\zeta^4 + 1}\right]^{-1/2}}$$

Note: Gain margin of 2nd order system is infinite.

5.4 FREQUENCY RESPONSE PLOTS

Frequency response analysis of control system can be carried either mathematically (or) graphically. The various graphical techniques to Frequency response is

1. Bode plot
2. Polar plot (or Nyauist Plot)
3. Nichols Plot
4. M and N circles
5. Nichols Chart

Bode plot, polar plot and Nichols plot are usually drawn for open loops system. The M and N circle and Nichols chart are used to graphically determine the frequency response of unity feedback closed loop system from the knowledge of open loop response.

The Frequency response plots are used to determine the Frequency domain specifications, to study the stability of the system and to adjust the gain of the system to satisfy the designed specifications.

5.5 BODE PLOTS

A bode plot consists of two graphs one is magnitude plot of sinusoidal transfer function versus log ω. The other is a plot of the phase angle of a sinusoidal transfer function versus log ω.

The bode plot can be drawn for both open loop and closed loop transfer function usually the bode plot is drawn for open loop system The standard representation the logarithmic magnitude of open loop transfer function is $G(j\omega)$ is $20\log|G(j\omega)|$ where base of the logarithm is 10.

The unit used is this representation of the magnitude is the decibel usually abbreviated as *db*. The curves are drawn on semilog paper, using the log scale for frequency and the linear scale for either magnitude (or) phase angle (in degree).

The main advantage of the bode plot is that multiplication of magnitude can be converted into addition. Also a simple method for sketching an approximate log-magnitude curve is available.

Consider the open loop transfer function,

$$G(s) = K(1 + sT_1)/s(1 + sT_2)\,(1 + sT_3)$$

$$G(j\omega) = K(1 + j\omega T_1)/j\omega(1 + j\omega T_2)\,(1 + j\omega T_3)$$

$$= \frac{K\angle 0^\circ \sqrt{1+\omega^2 T_1^2}\ \angle \tan^{-1}\omega T_1}{\omega\angle 90^\circ \sqrt{1+\omega^2 T_2^2}\angle \tan^{-1}\omega T_2 \sqrt{1+\omega^2 T_3^2}\ \angle \tan^{-1}\omega T_3}$$

The magnitude of $G(j\omega) = |G(j\omega)| = \dfrac{K\sqrt{1+\omega^2 T_1^2}}{\omega\sqrt{1+\omega^2 T_2^{\,2}}\sqrt{1+\omega^2 T_3^{\,2}}}$

The phase angle of the $G(j\omega) = \angle G(j\omega)$

$$= \tan^{-1}\omega T_1 - 90^\circ - \tan^{-1}\omega T_2 - \tan^{-1}\omega T_3$$

The magnitude or $G(j\omega)$ can be expressed in decibels as shown below —

$|G(j\omega)|$ in db = 20 log $|G(j\omega)|$

$$= 20\log\left[\frac{K\sqrt{1+\omega^2 T_1^2}}{\omega\sqrt{1+\omega^2 T_2^{\,2}}\sqrt{1+\omega^2 T_3^{\,2}}}\right]$$

$$= 20\log\left[\frac{K}{\omega}\times\sqrt{1+\omega^2 T_1^{\,2}}\times\frac{1}{\sqrt{1+\omega^2 T_2^{\,2}}}\times\frac{1}{\sqrt{1+\omega^2 T_3^{\,2}}}\right]$$

$$= 20\log\frac{K}{\omega}\times 20\log\sqrt{1+\omega^2 T_1^{\,2}} + 20\log\frac{1}{\sqrt{1+\omega^2 T_2^{\,2}}} + 20\log\frac{1}{\sqrt{1+\omega^2 T_3^{\,2}}}$$

$$= 20\log\frac{K}{\omega}\times 20\log\sqrt{1+\omega^2 T_1^{\,2}} - 20\log\sqrt{1+\omega^2 T_2^{\,2}} - 20\log\sqrt{1+\omega^2 T_3^{\,2}} \qquad (5.6)$$

From the equation (5.6) it si clear that, when the magnitude is expressed is *db*, the multiplication is converted to addition. Hence in magnitude plot, the *db* magnitude of individual factors of $G(j\omega)$ can be added.

Basic factors of G(jω)

1. Constant gain, K
2. Integral factor, $\dfrac{K}{j\omega}$ or $\dfrac{K}{(j\omega)^n}$
3. Derivative factor, $K(j\omega)$ or $K(j\omega)^n$
4. Fist order factor in denominator, $\dfrac{1}{1+j\omega T}$ or $\dfrac{1}{(1+j\omega T)^m}$

5. First order factor in numerator, $(1 + j\omega T)$ or $(1 + j\omega T)^m$

6. Second order factor in denominator, $\left[\dfrac{1}{1+2\zeta(j\omega/\omega_n)+(j\omega/\omega_n)^2}\right]$

7. Quadratic factor in numerator, $\left[1+2\zeta\left(\dfrac{j\omega}{\omega_n}\right)+\left(\dfrac{j\omega}{\omega_n}\right)^2\right]$

Constant Gain, K

Let $\qquad G(s) = K$

$$\therefore \qquad G(j\omega) = K = K \angle 0°$$

$$A = |G(j\omega)|\ db = 20 \log K$$

$$\phi = \angle G(j\omega) = 0°$$

The magnitude plot for a constant gain K is a horizontal straight line at the magnitude of 20 log K db. The phase plot is straight line at 0°.

when $K > 1$, 20 log K is positive

when $0 < K < 1$, 20 log K is negative

when $K = 1$, 20 log K is zero

Integral factor

Let $\qquad G(s) = \dfrac{K}{s}$

$$G(j\omega) = \left(\frac{K}{j\omega}\right) = \frac{K}{\omega} \angle -90°$$

$$A = |G(j\omega)| \text{ in } db = 20 \log \left(\frac{K}{\omega}\right)$$

and $\qquad \phi = \angle G(j\omega) = -90°$

when $\omega = 0.1\ K$, $A = 20 \log \left(\dfrac{1}{0.1}\right) = 20\ db$

when $\omega = K$, $A = 20 \log 1 = 0\ db$

when $\omega = 10\ K$, $A = 20 \log \dfrac{1}{10}\ db = -20\ db$

From the above analysis it is evident that the magnitude plot of the integral factor is a straight line with a slope of – 20 db/dec and passing through zero db, when $\omega = K$. Since the $\angle G(j\omega)$ is constant and independent of ω the phase plot is a straight line at – 90°.

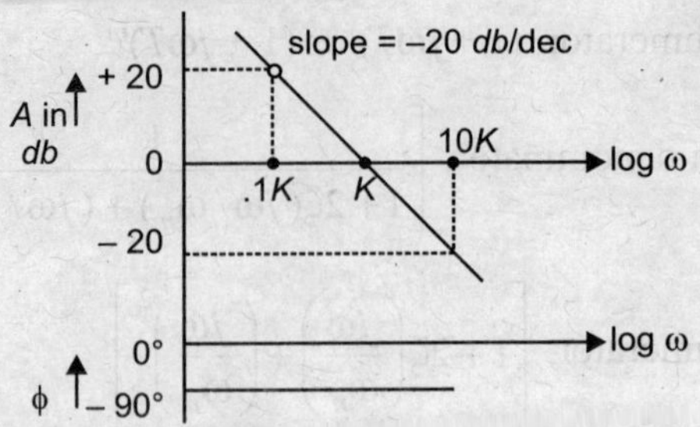

Fig. 5.10: Bode plot of integral factor $\left(\frac{k}{s}\right)$.

When an integral factor has multiplicity of *n*, then

$$G(s) = \frac{K}{s^n}$$

$$G(j\omega) = \frac{K}{(j\omega)^n} = \frac{K}{\omega^n}\angle -90n^\circ$$

$$A = |G(j\omega)| \text{ in } db = 20 \log \frac{K}{\omega^n}$$

$$= 20\log\left[\frac{K^{1/n}}{\omega}\right]^n = 20n\log\left[\frac{K^{1/n}}{\omega}\right]$$

$$\phi = \angle G(j\omega) = -90n^\circ$$

Now the magnitude plot of the integral factor in straight line with a slope of –20 *n db*/dec and passing through zero *db* with $\omega = K^{1/n}$. The phase plot is a straight line at – 90*n*°

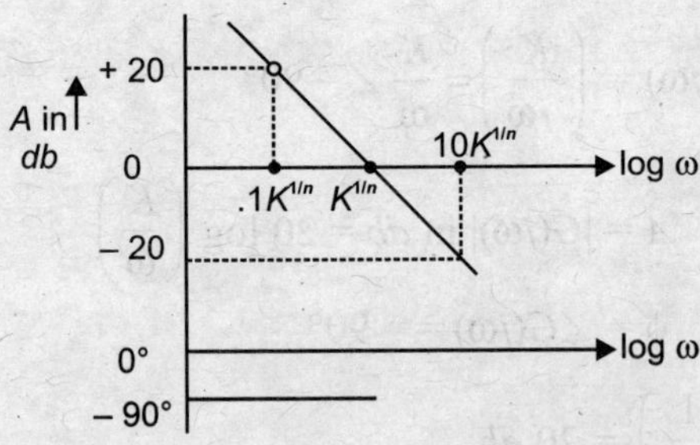

Fig. 5.11: Bode plot of integral factor $\frac{k}{(j\omega)^n}$.

Derivative Factor

Let $G(s) = Ks$

$$G(j\omega) = K(j\omega) = (K\omega)\angle 90^\circ$$

$$A = |G(j\omega)| \text{ in } db = 20 \log (K\omega)$$

and $\phi = \angle G(j\omega) = +90°$

when $\omega = 0.1/K$, $A = 20 \log (0.1) = -20\ db$

when $\omega = \dfrac{1}{K}$, $A = 20 \log 1 = 0\ db$

when $\omega = \dfrac{10}{K}$, $A = 20 \log 10 = +20\ db.$

Fig. 5.12

From above analysis it is evident that the magnitude plot of the derivative factor is a straight line with a slope of + 20 *db*/dec and passing through zero *db* when $\omega = \dfrac{1}{K}$. Since the $\angle G(j\omega)$ is a constant and impedance of ω, the phase plot is a straight line at +90°.

When the derivative factor has multiplicity of *n* then

$$G(s) = Ks^n$$

$$\therefore \quad G(j\omega) = K(j\omega)^n = K\omega^n \angle 90n°$$

$A = |G(j\omega)|$ in $db = 20 \log (K\omega^n) = 20 \log (K^{1/n}\,\omega) = 20 \log (K^{1/n}\,\omega)$ and $\phi = \angle G(j\omega) = 90n°$

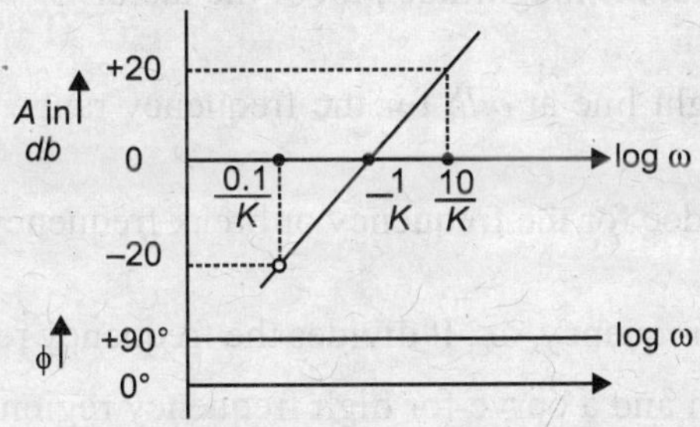

Fig. 5.13: Bode plot of derivative factor $K(j\omega)$.

Now the magnitude plot of the derivative factor is a straight line with a slope of + 20 db/dec and passing through zero db when $\omega = 1/K^{1/n}$. The phase plot is a straight line at $90n°$.

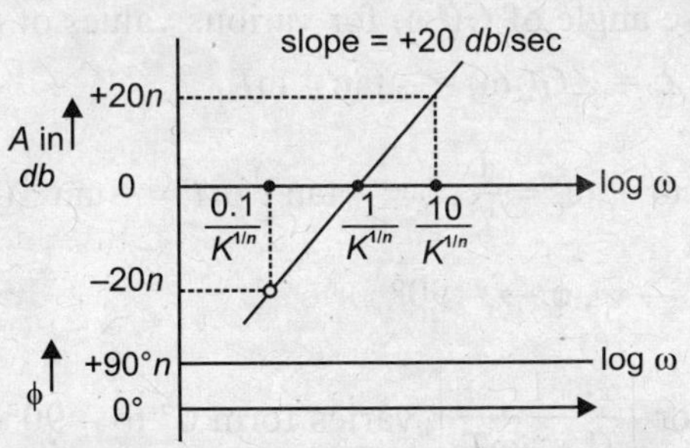

Fig. 5.14: Bode plot of derivative factor $K(j\omega)$.

First order factor in denominator

$$G(s) = \frac{1}{1+sT}$$

$$G(j\omega) = \frac{1}{1+j\omega T} = \frac{1}{\sqrt{1+\omega^2T^2}}\angle -\tan^{-1}\omega T$$

$$A = |G(j\omega)\text{in } db = 20\log\frac{1}{\sqrt{1+\omega^2T^2}} = -20\log\sqrt{1+\omega^2T^2}$$

At very low frequency $\omega T << 1$

$\therefore$ $$A = -20\log\sqrt{1+\omega^2T^2} = 20\log 1 = 0$$

At very high frequencies $\omega T >> 1$

$\therefore$ $$A = -20\log\sqrt{1+\omega^2T^2} \approx -20\sqrt{\omega^2T^2} = -20\log\omega T$$

at $$\omega = \frac{1}{T}, A = -20\log 1 = 0$$

at $$\omega = \frac{10}{T}, A = -20\log 10 = -20 \text{ db}$$

The above analysis shows that the magnitude plot of the factor $\frac{1}{(1+j\omega T)}$ can be approximated by two straight lines, one is *s* straight line at *odb* for the frequency range $0 < \omega < \frac{1}{T}$ and the other to a straight line with slope – 20 *db*/dec for the frequency or break frequency. For the factor $\frac{1}{(1+j\omega T)}$ the frequency, $\omega = \frac{1}{T}$ is the corner frequency, ω_c. It divides the frequency response curve into two regions, a curve for low frequency region and a curve for high frequency region.

The actual magnitude at the corner frequency, $\omega_c = 1/T$ is

$$A = -20\log\sqrt{1+1} = -3\,db.$$

Hence by this approximation the loss in *db* at the corner frequency is – 3*db*. The phase plot is obtained by calculating the phase angle of $G(j\omega)$ for various values of ω

phase angle $$\phi = \angle G(j\omega) = -\tan^{-1}\omega T$$

At the corner frequency, $\omega = \omega_c = \frac{1}{T}, \phi = -\tan^{-1}\omega T = -\tan^{-1}(1) - 45°$

As $\omega \to 0, \phi \to 0$ and As $\omega \to \infty, \phi \to -90°$

The phase angle of the factor $\left[\frac{1}{(1+j\omega T}\right]$ varies form 0° to – 90° as ω is varied from zero to ∞.
The phase plot is a curve passing through – 45° at ω_c.

When the first order factor is the denominator has a multiplicity of *m* then,

$$G(s) = \frac{1}{(1+sT)^m}$$

and $$G(j\omega) = \frac{1}{(1+j\omega T)^m} = \frac{1}{\left[\sqrt{(1+j\omega^2T^2)}\right]^m} \angle m \tan^{-1}\omega T$$

$$A = |G(j\omega)| \text{ in } db = 20\log\frac{1}{\sqrt{(1+\omega^2T^2)}^m} = -\,20\log\sqrt{1+\omega^2T^2}$$

$$\phi = \angle G(j\omega) = -\,m\tan^{-1}\omega T$$

Now the magnitude plot of the factor $\frac{1}{(1+j\omega T)^m}$ can be approximated by two straight lines, one is a straight line at zero *db* for the frequency range $0 < \omega < \frac{1}{T}$ and the other is a straight line with slope –20 m *db*/dec for the frequency range $\frac{1}{T} < \omega < \infty$. The corner frequency, $\omega_c = \frac{1}{T}$ and the loss in db at the corner frequency is – 3 *mdb*.

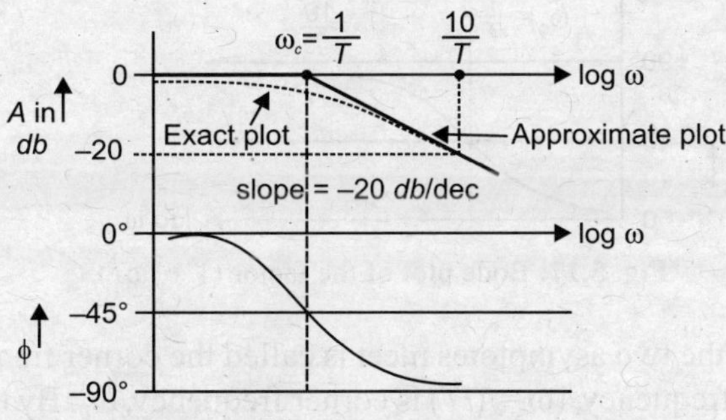

Fig. 5.15: Bode plot of the factor $\frac{1}{1+j\omega T}$.

The phase angle of the factor $\frac{1}{(1+j\omega T)^m}$ various from 0° to – 90° as ω is varied from zero to ∞. The phase plot is a curve through – 45 m° at ω_c.

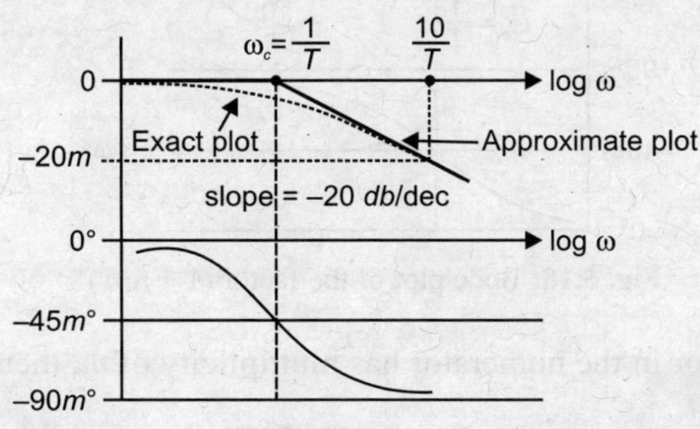

Fig. 5.16: Bode plot of the factor $\frac{1}{(1+j\omega)^m}$.

CHAPTER 5

First order factor in the numerator

$$G(s) = 1 + sT$$

$$G(j\omega) = 1 + j\omega T = \sqrt{1+\omega^2T^2} \angle -\tan^{-1}\omega T$$

$$A = |G(j\omega) \text{ in } db = 20\log\sqrt{1+\omega^2T^2} \text{ and } \phi = < G(j\omega) = \tan^{-1}\omega T$$

By analysis similar to that of previous section if can be shown that the magnitude plot of the factor $(1 + j\omega T)$ can be approximated by two starigh lines, one is a straight line at zero *db* for the frequency range $0 < \omega < \frac{1}{T}$ and the other is a straight line with slope + 20 *db*/dec for the frequency range $\frac{1}{T} < \omega < \infty$. The two straight lines are asymptotes of the exact curve.

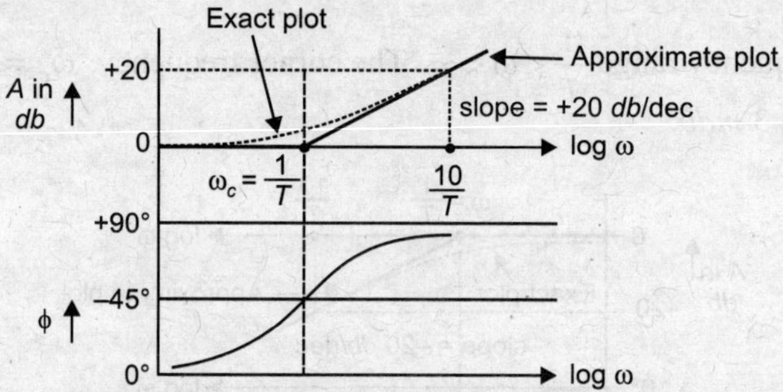

Fig. 5.17: Bode plot of the factor $(1 + j\omega T)$.

The frequency at which the two asymptotes meet is called the corner frequency or break frequency. For the factor $(1 + j\omega T)$, the frequency, $(\omega = 1/T)$ is corner frequency, ω_c. By this approximation the loss in *db* at the corner frequency is + 3 *db*. The phase angle of the factor $(1 + j\omega T)$ varies from zero to +90° as ω is varied from 0 to ∞. The phase plot in a curve passing through + 45° at ω_c.

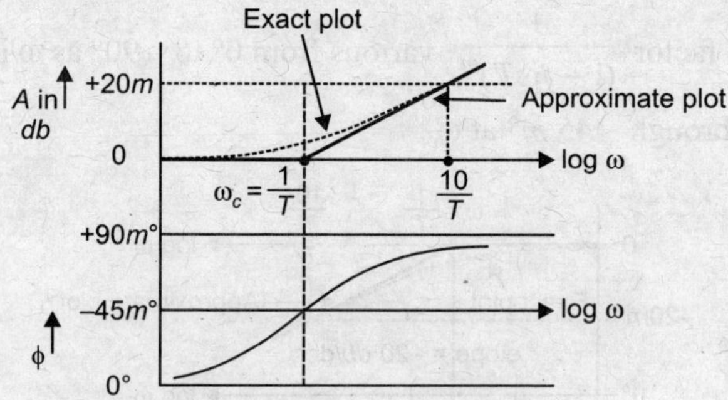

Fig. 5.18: Bode plot of the factor $(1 + j\omega T)^m$

When the first order factor in the numerator has multiplicity of m, then $G(s) = (1 + sT)^m$

$$G(j\omega) = (1 + sT)^m = \left(\sqrt{1+\omega^2T^2}\right)^m \angle m \tan^{-1}\omega T$$

$$A = |G(j\omega)| \text{ in } db = 20\log\left(\sqrt{1+\omega^2T^2}\right)^m$$

$$= 20m \log\sqrt{1+\omega^2T^2}$$

$$\phi = \angle G(j\omega) = m \tan^{-1} \omega T$$

Now the magnitude plot of the factor $(1 + j\omega T)^m$ can eb approximated by two straight lines, one is a straight line at zero *db* for the frequency range $0 < \omega < \frac{1}{T}$ and the other in a straight line with a slope of + 20 *m db*/dec for the frequency $\frac{1}{T} < \omega < \infty$. The corner frequency $\omega_c = \frac{1}{T}$ and the loss in *db* at this comes frequency is + 3 *m db*.

The phase angle of the factor $(1 + j\omega T)^m$ varies from zero to + 90 $m°$ as ω is varied form zero intinity. The phase plot is a curve passing through + 45 $m°$ at ω_c.

Quadratic factor in the denominator

$$G(s) = \frac{1}{1+2\zeta\frac{s}{\omega_n}+\left[\frac{s}{\omega_n}\right]^2}$$

$$G(j\omega) = \frac{1}{1+j\frac{2\zeta\omega}{\omega_n}+\left[\frac{j\omega}{\omega_n}\right]^2} = \frac{1}{1-\left[\frac{\omega}{\omega_n}\right]^2+j2\zeta\frac{\omega}{\omega_n}}$$

$$= \frac{1}{\sqrt{\left[1-\frac{\omega^2}{\omega_n^2}\right]^2+4\zeta^2\frac{\omega^2}{\omega_n^2}}}\angle\tan^{-1}\frac{2\zeta\frac{\omega}{\omega_n}}{1-\frac{\omega^2}{\omega_n^2}}$$

Let $A = |G(j\omega)|$ in *db*

$$A = 20\log\frac{1}{\sqrt{\left(1-\frac{\omega^2}{\omega_n^2}\right)^2+4\zeta^2\frac{\omega^2}{\omega_n^2}}} = -20\log\sqrt{\left(1-\frac{\omega^2}{\omega_n^2}\right)^2+4\zeta^2\frac{\omega^2}{\omega_n^2}}$$

$$= -20\log\frac{1}{1-\frac{\omega^4}{\omega_n^4}-2\frac{\omega^2}{\omega_n^2}+4\zeta^2\frac{\omega^2}{\omega_n^2}} - 20\log\sqrt{1-\frac{\omega^2}{\omega_n^2}+(2-4\zeta^2)+\frac{\omega^4}{\omega_n^4}}$$

CHAPTER 5

At very low frequencies when $\omega << \omega_n$ the magnitude is

$$A = -20\log\sqrt{1-\frac{\omega^2}{\omega_n^2}\left(2-4\zeta^2\right)+\frac{\omega^4}{\omega_n^4}} \approx -20\log 1 = 0$$

At very high frequencies when $\omega >> \omega_n$ the magnitude is

$$A = -20\log\sqrt{1-\frac{\omega^2}{\omega_n^2}\left(2-4\zeta^2\right)+\frac{\omega^4}{\omega_n^4}} \approx -20\log\sqrt{\frac{\omega^4}{\omega_n^4}} = -20\log\frac{\omega^2}{\omega_n^2}$$

$$\therefore \quad A = -40\log\frac{\omega}{\omega_n}$$

At $\omega = \omega_n$, $A = -40 \log 1 = 0$ *db*
At $\omega = 10\,\omega_n$, $A = -40 \log 10 = -40$ *db*

From the above analysis it is evident that the magnitude plot of the quadratic factor in the denominator can be approximated by two straight line, one is a straight line at 0 *db* for the frequency range $0 < \omega < \omega_n$ and the other is a straight line with slope -40 *db*/bec for the frequency range $\omega_n < \omega < \infty$. The two straight lines ar asymptotes of the exact curve. The frequency at which the two asymptotes meet is called the corner frequency. For the quadratic factor, the frequency ω_n is the corner frequency, ω_c.

The two asymptotes of the exact curve are independent of the damping ratio, ζ. In the exact magnitude plot resonant peak occurs near the corner frequency and the magnitude of resonant peak depends on ζ. Lower the value of ζ, larger will be resonent peak. Hence by this approximation the error at the corner frequency depends on damping ratio ζ. The phase plot is obtained by calculating the phase angle of $G(j\omega)$ for various value of ω.

$$\phi = \angle G(j\omega) = \tan^{-1}\left[\frac{2\zeta\dfrac{\omega}{\omega_n}}{1-\dfrac{\omega^2}{\omega_n^2}}\right]$$

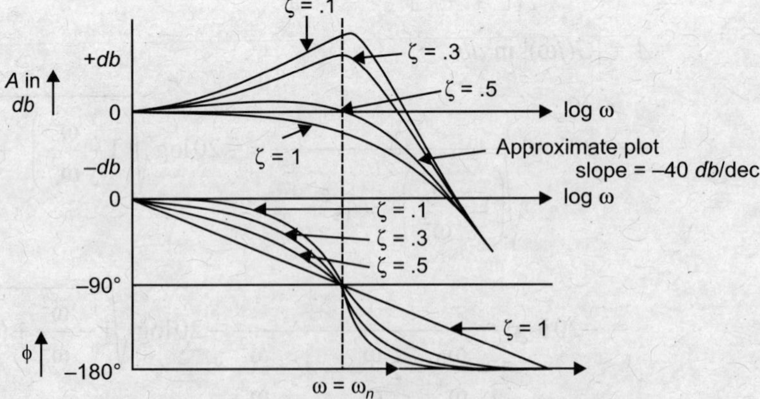

Fig. 5.19: Bode plot of qualratic factor in denominator.

At $\omega = \omega_n$, $\phi = \tan^{-1}\dfrac{2\zeta}{0} = \tan^{-1}\infty = -90°$

At $\omega = 0$, $\phi \to 0$ and as $\omega \to \infty$, $\phi = -180°$

The phase angle of the quadratic factor varies from 0 to – 180° as ω is varied from 0 to ∞. The phase plot is a curve passing through –90° at ω_c. At the corner frequency phase angle is – 90° and independent of ζ, but at all other frequency depends on ζ.

Quadratic factor in the numerator

$$G(s) = 1 + 2\zeta\left(\frac{s}{\omega_n}\right) + \left(\frac{s}{\omega_n}\right)^2$$

$$G(j\omega) = 1 + 2\zeta\frac{\omega}{\omega_n} + \left(\frac{j\omega}{\omega_n}\right)^2 = \sqrt{\left[1-\left(\frac{\omega^2}{\omega_n^2}\right)\right]^2 + 4\zeta\frac{\omega^2}{\omega_n^2}}\angle\tan^{-1}\frac{2\zeta\frac{\omega}{\omega_n}}{1-\frac{\omega^2}{\omega_n^2}}$$

The magnitude plot of the quadratic factor is the numerator can be approximated by two straight lines, one is a straight line at 0 *db* for the frequency range $0 < \omega < \omega_n$ and the other is a straight line with slope + 40 *db*/dec for the frequency range $\omega_n < \omega < \infty$. The corner frequency is ω_n. Due to this approximation the error at the corner frequency depends on ζ.

The phase angle varies from 0 to + 180°, as ω is varied from 0 to ∞. At the corner frequency the phase angle is + 90° and independent of ζ, but at all other frequency it depends on ζ.

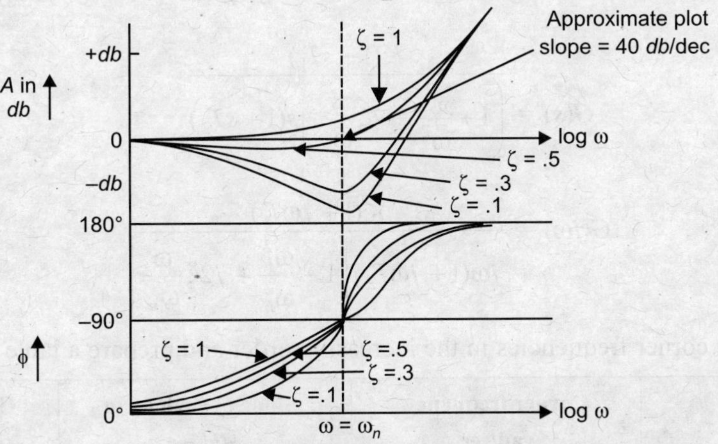

Fig. 5.20: Bode plot of quadratic factor in numerator.

Tite step by step procedure for plotting the magnitude plot is given below

Step 1: Convert the transfer function into bode form (or) time constant form. The bode form of the transfer function is

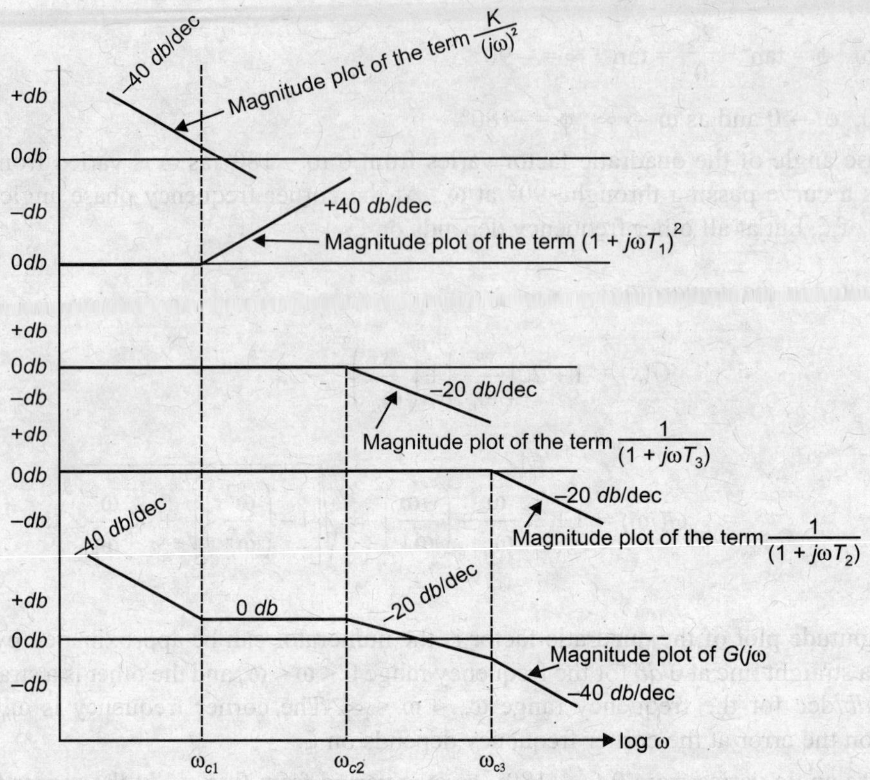

Fig. 5.21: Magnitude bode plot of $G(j\omega) = \dfrac{K(1+j\omega T_1)^2}{(j\omega)^2\,(1+j\omega T_2)\,(1+j\omega T_3)}$.

$$G(s) = \frac{K(1+sT_1)}{\left(1+\dfrac{s^2}{\omega_n^2}+2\zeta\dfrac{s}{\omega_n}\right)s(1+sT_2)}$$

$$G(j\omega) = \frac{K(1+j\omega T_1)}{j\omega(1+j\omega T_2)\left[1-\dfrac{\omega^2}{\omega_n^2}+j2\zeta\dfrac{\omega}{\omega_n}\right]}$$

Step 2: List the corner frequencies in the increasing order and prepare a table as shown below :

Term	Corner frequency rad/sec	Slope of each term *db*/dec	Change in slope from term to another term *db*/dec

In the above table enter K or $\dfrac{K}{(j\omega)^n}$ or $K(j\omega)^n$ as the first term and the other terms in the increasing order of corner frequencies. Then enter the corner frequency, slope contributed by each term and change in slope at every corner frequency.

Step 3: Choose an arbitrary frequency ω_l which is lesser than the lowest corner frequency. Calculate the *db* magnitude of K or $\dfrac{K}{(j\omega)^n}$ or $K(j\omega)^n$ at use and at the lowest corner frequency.

Step 4: Then calculate the gain (*db* magnitude) at every corner frequency are by one by using the formula, for first two terms the formual will be,

$$A_1 \text{ (gain of first term)} = 20 \log |\text{first term}|$$

$$A_2 \text{ (gain of second term)} = 20 \log |\text{first term}|$$

$$A_3 \text{ (gain of third term)} = \text{change in slope from } \omega_{c1} \text{ to } \omega_{c2} \times \log \left|\frac{\text{Present frequency}}{\text{Previous frequency}}\right| + \text{Previous gain } (A_2)$$

The same formula should be used till n^{th} term.

Step 5: Choose an arbitrary frequency ω_n. Which is greater than the highest corner frequency. Calculate the gain at ω_n by using the formula in step 4.

Step 6: In a semilog graph sheet mark the required range of frequency on *x*-axis (log scale) and the range of *db* magnitude on *y*-axis (ordinary scale).

Step 7: Mark all the points obtained in step 4 and 5 on the graph and join the points by straight lines. Mark the slope at every part of the graph.

Note: The magnitude plot obtained above is an approximate plot. If an exact plot is needed then appropriate correction should be made at every corner frequencies.

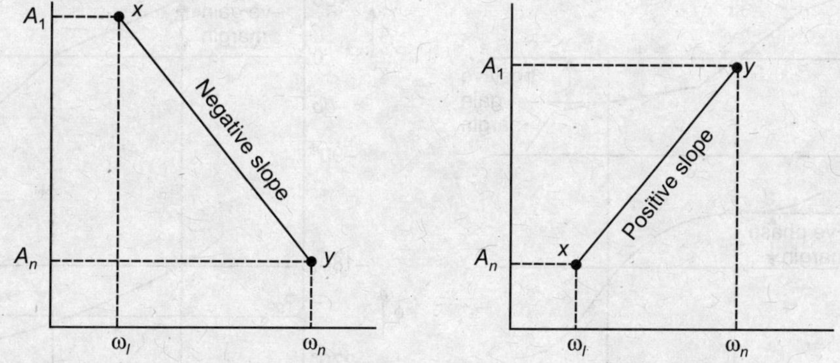

Fig. 5.22: Representation of magnitude plot for both +ve and –ve slopes.

Procedure for phase plot of bode plot

The phase point is an exact plot and no approximations are made while drawing the phase plot (i.e., plot can be drawn without using scale whereas magnitude plot should be drawn using scale. The exact phase angles of $G(j\omega)$ are computed for various values of ω and tabulated.

CHAPTER 5

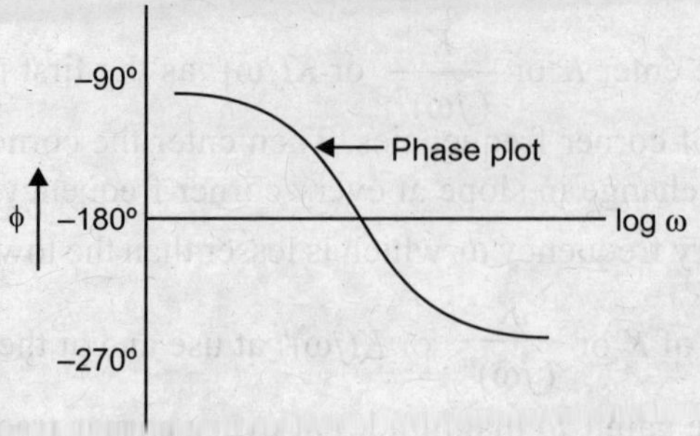

Fig. 5.23: Representation of phase plot.

The choice of frequencies are same as magnitude plot. The magnitude plot and phase plot are drawn in a single semilong sheet on a common frequency scale.

Take another y-axis in the graph where the magnitude plot is drawn and in this y-axis mark the range of phase angles. From the tabulated values of ω and phase angles, mark all the points on the graph. Join the points by a smooth curve.

Determination of Gain margin and phase margin from bode plot

The gain margin in *db* is given by the negative of *db* magnitude of $G(j\omega)$ at the phase cross-over frequency, ω_{pc}. The ω_{pc} is the frequency at which phase of $G(j\omega)$ is – 180°. If the *db* magnitude of $G(j\omega)$ at ω_{pc} is negative then gain margin is positive and vice versa.

Let ϕ_{gc} be the phase angle of $G(j\omega)$ at gain cross over frequency ω_{gc}. The ω_{gc} is the frequency at which the *db* magnitude of $G(j\omega)$ is zero. Now the phase margin γ is given by, $\gamma = 180° + \phi_{gc}$. If ϕ_{gc} is less then –180° than phase margin is positive and vice versa. The positive and negative gain margins are shown in Figure (5.24).

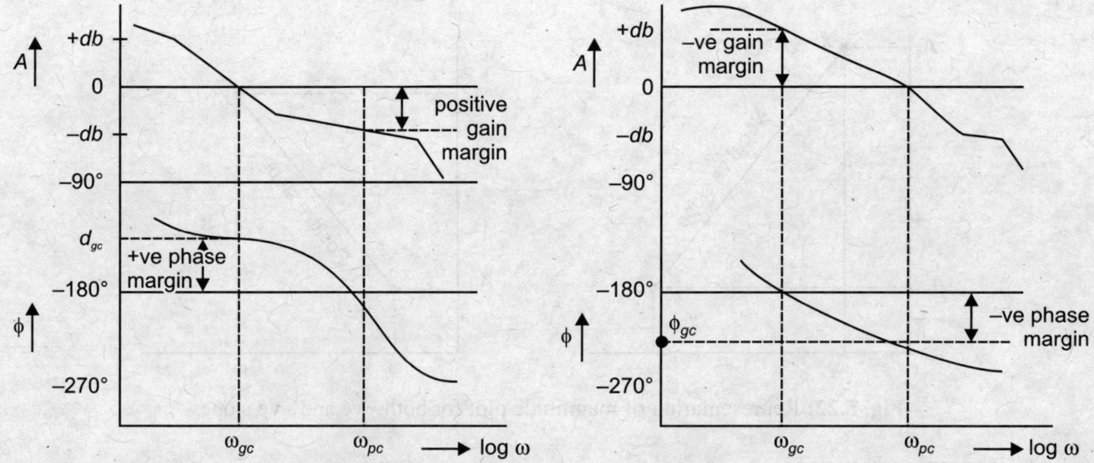

Fig. 5.24: Bode plot showing (i) Phase margin (ii) Gain margin (kg).

Gain Adjustment using bode plot: In the open loop transfer function $G(j\omega)$ the constant K contributes only magnitude. Hence by charging the values of K the system can be adjusted to meet the desired specifications. The desired specification are gain margin, phase margin, ω_{pc} and ω_{gc}. In a system transfer function if the value of K required to be estimated to satisfy a desired specification then draw the bode plot of the system with $K = 1$. The constant K can add 20 log K to very point of the magnitude plot and due to this addition the magnitude plot will shift vertically, up or down. Hence shift the magnitude plot vertically up or down to meet the desired specification. Equate the vertical distance by which the magnitude plot is shifted to 20 log K and solve for K.

Let x = change in db (The required db on which the plot have to be shifted)

Now $20 \log K = x$; $\log K = \frac{x}{20}$ $\therefore$ $K = 10^{\frac{x}{20}}$

EXAMPLE 5.8: Sketch bode plot for the following tarnster function and determine the system gain K for the gain cross over frequency to be 5 rad/sec $G(s) = \dfrac{Ks^2}{(1+0.2s)(1+.02s)}$.

Solution:

Step 1: Replace s by $j\omega$ in given transfer function

The sinusoidal transfer function $G(j\omega)$ is obtained by replacing s by $j\omega$ in the given s-domain transfer function

$$\therefore \qquad G(j\omega) = \frac{K(j\omega)^2}{(1+0.2j\omega)(1+.02j\omega)}$$

Let $K = 1$,

$$\therefore \qquad G(j\omega) = \frac{(j\omega)^2}{(1+0.2j\omega)(1+.02j\omega)}$$

Step 2: Find the required informations to draw magnitude plot

The corner frequencies are

$$\omega_{c1} = \frac{1}{0.2} = 5 \text{ rad/sec}$$

$$\omega_{c2} = \frac{1}{.02} = 50 \text{ rad/sec}$$

The various terms of $G(j\omega)$ are listed in Table 3.11.1 is the increasing order of their corner frequency. Also the table shows the slope contributed by each term and the change is slop at the corner frequency.

Choose a low frequency ωl such that $\omega l < \omega_{c1}$ and choose a high frequency ω_h such that $\omega_h > \omega_{c2}$

Let $\qquad \omega_l = .5$ rad/sec and $\omega_h = 100$ rad/sec

Let $\qquad A = |G(j\omega)|$ in db

Let us calculate A at ω_l, ω_{c1}, ω_{c2} and ω_h.

At $\omega = \omega_l$, $A_1 = 20 \log |(j\omega)^2| = 20 \log (\omega^2) = 20 \log (.5)^2 = -12\ db \Rightarrow A_1 = -12\ db$

At $\omega = \omega_{c1}$, $A_2 = 20 \log |(j\omega)^2| = 20 \log (\omega^2) = 20 \log (5)^2 = 28\ db \Rightarrow A_2 = 28$ db

Table 5.8.1

Term	Corner frequency rad/sec	Slope of each term *db*/dec	Change in slope from one term to another term *db*/dec
$(j\omega)^2$	–	+40	
$\dfrac{1}{1+j0.2\omega}$	$\omega_{c1} = \dfrac{1}{.2} = 5$	– 20	+40 – 20 = 20
$\dfrac{1}{1+j.02\omega}$	$\omega_{c2} = \dfrac{1}{.02} = 50$	– 20	20 – 20 = 0

At $\omega = \omega_{c2}$. A_3 = [slope from one term to another term) × log $\dfrac{\text{Present frequency}}{\text{Previous frequency}}$ + Previous gain (A_2)

$$\text{i.e., } A_3 = \left[\text{slope from } \omega_{c1} \text{ to } \omega_{c2} \times \log\frac{\omega_{c2}}{\omega_{c1}}\right] + A_2 \,(\text{at } \omega = \omega_{c1}) = 20 \log\frac{50}{5} + 28 = 48\,db.$$

At $\omega = \omega_h$, A_4 = [slope from one term to another term) × log $\dfrac{\text{Present frequency}}{\text{Previous frequency}}$ + Previous gain (A_3)

$$\text{i.e., } A_4 = \left[\text{slope from } \omega_{c2} \text{ to } \omega_h \times \log\frac{\omega_h}{\omega_{c2}}\right] + A\,(\text{at } \omega = \omega_{c2}) = 0 \times \log\frac{100}{50} + 48 = 48\; db.$$

Let the points *a*, *b*, *c* and d be the points corresponding to frequencies ω_l, ω_{c1}, ω_{c2} and ω_h respectively on the magnitude plot.

Phase plot

The phase angle of $G(j\omega)$ as a function of ω is given by $\phi = \angle G(j\omega) = 180° - \tan^{-1} 0.2\,\omega - \tan^{-1} .02\omega$.

The phase angle of $G(j\omega)$ are calculated for various value for ω and listed Table (5.8.2)

Table 5.8.2

ω, rad/sec	$\phi = \angle G(j\omega)$ deg
0.5	173.7 ≈ 174
1	167.6 ≈ 168
5	129.3 ≈ 130
10	105.3 ≈ 106
50	50.7 ≈ 50
100	29.5 ≈ 30

when $\omega = .5, \quad \phi = 180° - \tan^{-1} .2\,(.5) - \tan^{-1} .02\,(.5) \approx 174°$

$$\omega = 1, \quad \phi = 180° - \tan^{-1} .2\,(1) - \tan^{-1} .02\,(1) \approx 168°$$

$$\omega = 5, \quad \phi = 180° - \tan^{-1} .2\,(5) - \tan^{-1} .02\,(1) \approx 130°$$

$$\omega = 10, \quad \phi = 180° - \tan^{-1} .2\,(10) - \tan^{-1} .02\,(10) \approx 106$$

$$\omega = 50, \quad \phi = 180° - \tan^{-1} .2\,(50) - \tan^{-1} .02\,(50) \approx 50$$

$$\omega = 100, \quad \phi = 180° - \tan^{-1} .2\,(100) - \tan^{-1} .02\,(100) \approx 30$$

On the same semi log sheet choose scale on the y-axis on the right side of semilog sheet. Mark the calculated phase angle on the graph sheet. Join the points by smooth curve.

Calculation of K

We know that the gain cross over frequency 5 rad/sec (given). At $\omega = 5$ rad/sec the gain is 28 db. If gain cross over frequency is 5 rad/sec then at that frequency the *db* gain should be zero. Hence to every point of magnitude plot a *db* gain of –28 *db* should be added. The addition of –28 *db* shifts the plot downwards. The magnitude correction is independent of frequency. Hence the magnitude of –28 *db* is contributed by the term K. The value of K is calculated by equating 20 log k to –28 *db*.

$$20 \log K = -28\ db$$

$$\log K = -\frac{28}{20}$$

$$\therefore \quad K = 10^{-(28/20)} = .0398$$

The magnitude plot with $K = 1$ and .0398 and the phase plot are shown in Figure (5.8.1)

Note: The frequency $\omega = 5$ rad/sec is a corner frequency. Hence in the exact plot the *db* gain at $\omega = 5$ rad/sec will be 3 *db* less than the approximate plot. Therefore for exact plot the 20 log K will contribute a gain of – 25 *db*.

$$\therefore \quad 20 \log K = -25\ db$$

$$\log K = -\frac{25}{20}$$

$$\therefore \quad K = 10^{-\frac{25}{20}} = .0562$$

$$K = .0562$$

Results: The value of K at 5 rad/sec is .0562.

$$\boxed{K = .0562}$$

5.5.1 Relationship between System Type and Log Magnitude Curve

Consider the unity - feed back control system. The static position, velocity and acceleration error constant describe the low frequency behaviour of type 0, type 1, type 2 systems, respectively. For a given system, only one of the static error constant is finite and significant. (The larger value of the finite static error constant, the higher the loop gain; as ω approaches zero).

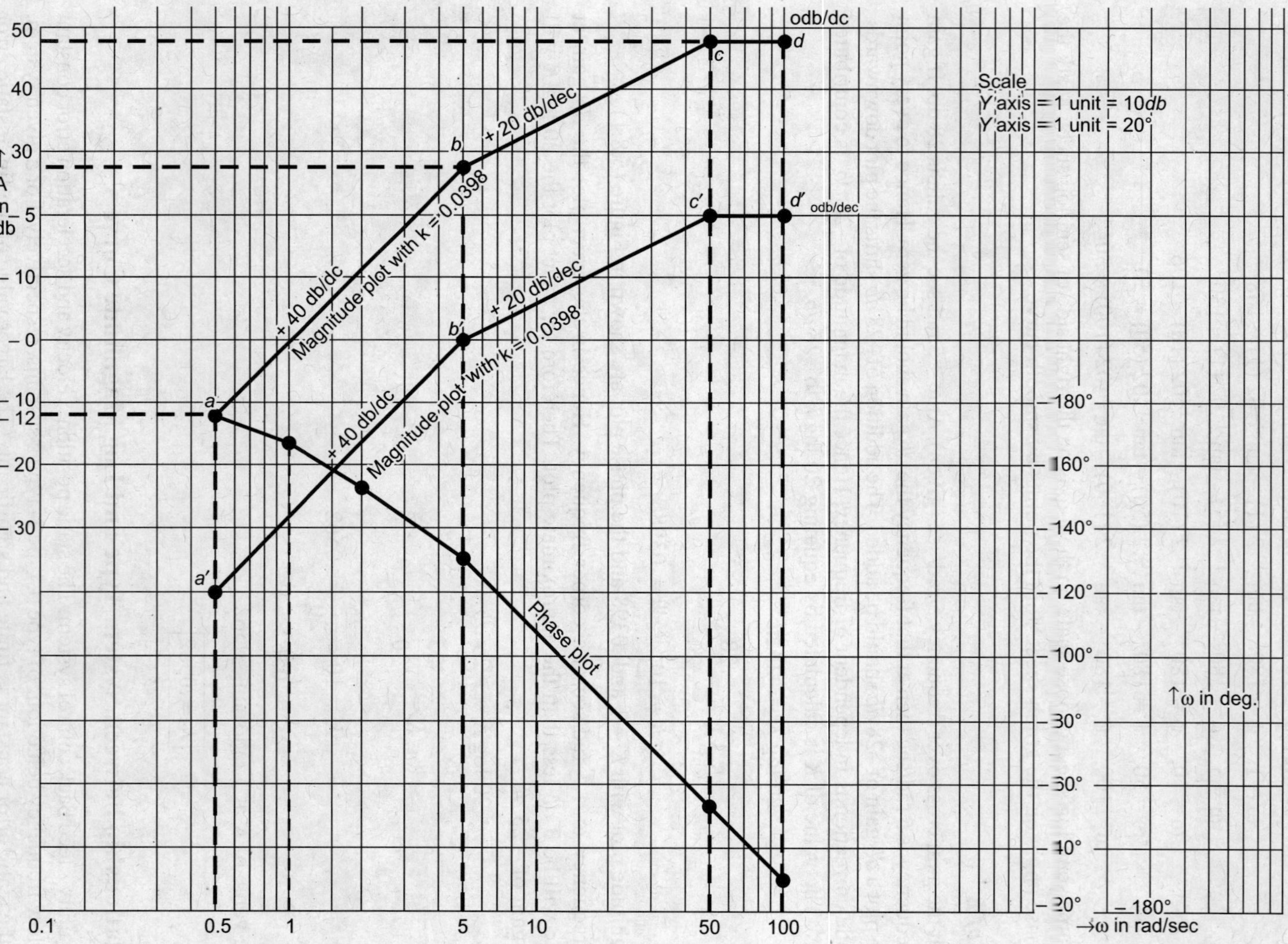

Fig. 5.8.1: Bode plot of transfer function $G(s0 = ks2/(1 + 0.2s)\,(1 + .2s)$.

The type of the system determines the slope of the log magnitude curve at low frequencies. Thus, information concerning the existence and magnitude of the steady state error of a control system to a given input can be determined from the observation of the low frequency region of the log-magnitude curve.

Determination of static position error constants: Consider the unity feedback control system shown in Figure (5.25). Assume that the open - loop transfer function is given by Figure (5.26).

$$G(s) = \frac{K(T_a s+1)(T_b s+1).....(T_m s+1)}{s^n (T_1 s+1)(T_2 s+1).....(T_p s+1)}$$

Figure (5.25) shows an example of the log-magnitude plot of a type 0 system. In such a system the magnitude of $G(j\omega)$ equals K_p at low frequencies, or

$$\underset{\omega \to 0}{Lt}\ G(j\omega) = K = K_p$$

If follows that the low-frequency asymptotes is a horizontal line at 20 log K_p *db*.

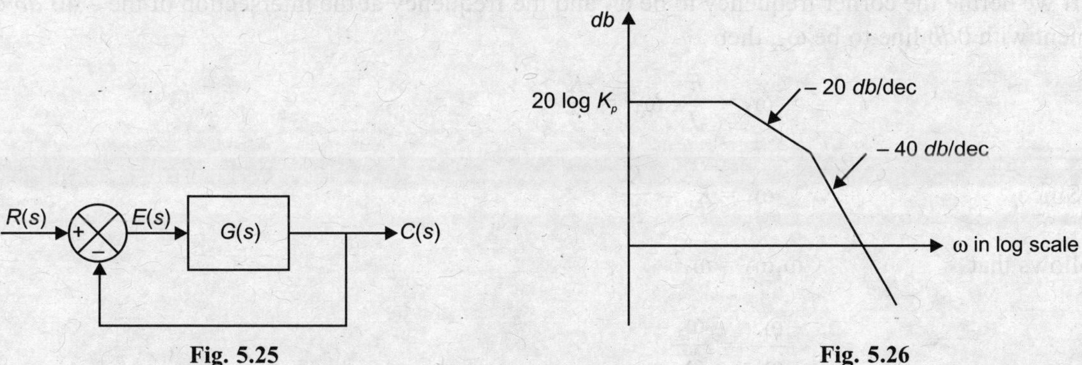

Fig. 5.25

Fig. 5.26

Determination of static velocity error constants: Consider the unity feedback control system shown in Figure (5.25) and (5.27) shows an example of the log magnitude plot of a type 1 system. The intersection of the initial – 20 *db*/dec segment with the line ω = 1 has the magnitude 20 log K_v. This can be explained as below.

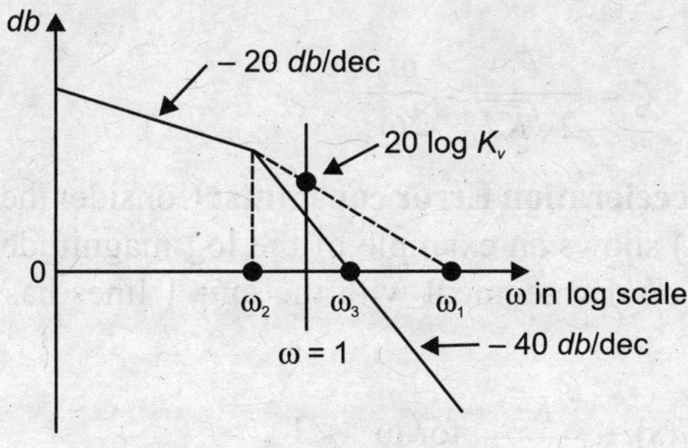

Fig. 5.27

CHAPTER 5

In a type 1 system, $G(j\omega) = \dfrac{K_v}{j\omega}$, for $\omega << 1$

thus $$20\log\left|\frac{K_v}{j\omega}\right|_{\omega=1} = 20 \log K_v$$

The intersection of the initial – 20 log *db*/dec segment with the 0 *dB* line has a frequency equal to K_v. To see this,

$$\left|\frac{K_v}{j\omega_1}\right|_{\omega=1} = 1 \text{ or } K_v = \omega_1$$

As an example consider the type system with unity feedback whose open loop transfer function is

$$G(s) = \frac{K}{s}\ (Js + F)$$

If we define the corner frequency to be ω_2 and the frequency at the intersection of the – 40 *db*/dec segment with 0*db* line to be ω_3, then

$$\omega_2 = \frac{F}{J},\ \omega_3^2 = \frac{K}{J}$$

Since $$\omega_1 = K_v = \frac{K}{F}$$

is follows that $$\omega_1\omega_2 = \omega_3^2$$

$$\frac{\omega_1}{\omega_3} = \frac{\omega_3}{\omega_2}$$

on the bode diagram

$$\log \omega_1 - \log \omega_3 = \log \omega_3 - \log \omega_2$$

Thus the ω_3 point is just midway between the ω_2 and ω_1 points. The damping ratio ζ of the system is then

$$\zeta = \frac{F}{2\sqrt{KJ}} = \frac{\omega_2}{2\omega_3}$$

Determination of Static Acceleration Error constants: Consider the unity feedback system shown Figure (5.25) and Figure (5.28) shows an example of the log magnitude plot of a type 2 system. The intersection of the initial – 40 *db*/dec segment with the $\omega = 1$ lines has the magnitude of 20 log K_a. Since at low frequencies

$$G(s) = \frac{K_a}{(j\omega)^2} \text{ for } \omega << 1$$

it follows that $20\log\left|\frac{K_a}{(j\omega)^2}\right|_{\omega=1} = 20\log K_a$

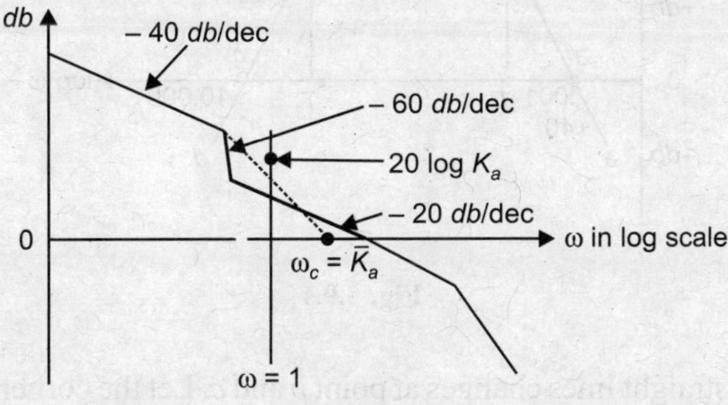

Fig. 5.28

The frequency ω_a at the intersection of the initial – 40 *db*/dec segment with the 0*db* line gives he square root of K_a. This can be seen from the following

$$20\log\left|\frac{K_a}{(j\omega)^2}\right| = 20\log 1 = 0$$

which yields $\omega_a = \sqrt{K_a}$

5.5.2 Determination of Transfer Function from Bode Magnitude Plot

If one is able to draw the bode magnitude plot for given transfer function then it should not be difficult to get the transfer function from given if the bode magnitude plot. However the method described below is valid only for transfer function of minimum phase.

The following steps are followed to get the transfer function from bode plot.

Step 1: Identify low frequency response portion (start) of the plot. 0 db/dec. (horizontal plot) corresponds to K, +20 db/dec. corresponds to Ks, 40 db/dec. Corresponds to Ks^2 etc. and – 20 db/sec. Corresponds to $\frac{K}{s}$, – 40 db/dec. corresponds to $\frac{K}{s^2}$ etc.

Step 2: Whenever there is a change of slope, assign the corner frequencies T_1, T_2 etc.

Step 3: Change of slope (for eg) from say + 20 db/dec to 0 db/dec at corner frequency T, corresponds to a pole, from + 20 db/dec. to –20 db/dec. corresponds to two repeated poles, from 0 db/dec. to + 20 db/dec corresponds to a zero, form – 20 db/dec to + 20 db/dec. corresponds to repeated zeros. Thus the poles and zero and the time constant of the transfer function are determined.

(i.e.) Identify change of slope at every corner frequencies.

Step 4: From the given bode plot, obtain the decibel. The gain and the corner frequencies are obtained by writing appropriate equation and then the complete transfer function is obtained.

EXAMPLE 5.9: For the bode magnitude plot shown in Figure (5.9.1.) obtain the transfer function.

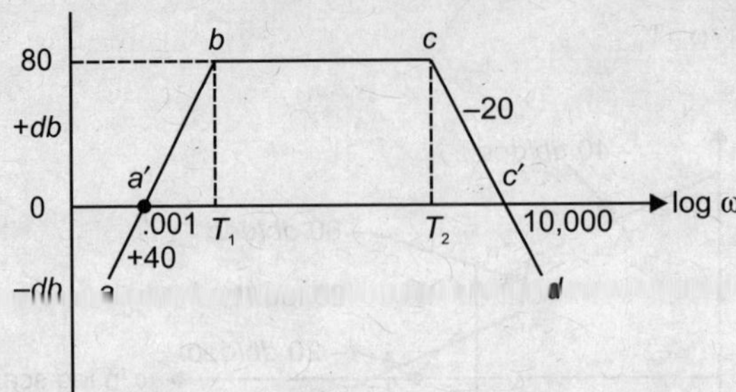

Fig. 5.9.1

Solution:

Step 1: The slope of the straight lines changes at point b and c. Let the corner frequencies corresponding to these points be T_1 and T_2. Since the low frequency lines $a\,a'b$ starts with $+40$ db/dec, the corresponding transfer function will be $G(s) = Ks^2$. At point b there should be two repeated poles with corner frequency T_1, to nulify the + 40 db/dec lines. Hence

$$G_2(s) = \frac{1}{\left(1+\frac{s}{T_1}\right)^2}$$

At point c there should be another pole so that the straight line cd goes with – 20 db/dec.

Hence $$G_3(s) = \frac{1}{(1+sT_1)}$$

the complete transfer function is

$$G(s) = G_1(s)\ G_2(s)\ G_3(s) = \frac{Ks^2}{\left(1+\frac{s}{T_1}\right)^2\left(1+\frac{s}{T_2}\right)}$$

Now it is necessary to determine K_1, T_1 and T_2 from Figure (3.11.2).

The straight line $aa'b$ is due to the function Ks^2. Hence

$$|G_1(j\omega)| = |K \cdot \omega^2| = 1 \text{ at } \omega = .001$$

$$K(.001)^2 = 1$$

$$\therefore \quad K = 10^6$$

Also from the triangle, $a'bT_1$

$$40 \log \frac{T_1}{.001} = 80$$

From which, $$T_1 = .1$$

$$20 \log \frac{10,000}{T_2} = 80$$

$$T_2 = 1$$

Hence $$G(s) = \frac{10^6 s^2}{(1+10s)^2(1+s)}$$

Result: The transfer function for the given bode magnitude plot is $\boxed{G(s) = \frac{10^6 s^2}{(1+10s)^2(1+s)}}$.

5.5.3 Frequency Response Specification

The gain margin

Consider the $G_1 + (j\omega)$ plot shown in Figure (5.29) where the magnitude of $(G_1 + j\omega)$ is crosses the negative real axis is d, then $\frac{1}{d}$ is called the gain margin and it is $-20 \log d$ is decibel. The frequency at which the gain is unity is called the gain cross–over frequency. The gain margin is defined as the reciprocal of the gain of open lop sinusoidal transfer function evaluated at the frequency ω_g at which the phase shift is 180°.

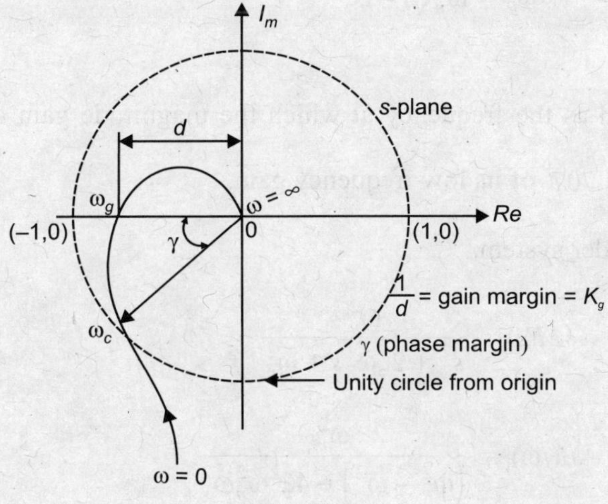

Fig. 5.29

The phase margin

The phase margin is defined as the angle through which $G_1 + (j\omega)$ plot is rotated in order that unity magnitude point on the locus passes through the critical point (–1, 0). In other words the phase margin ϕ_m as in Figure (5.29) is – 180° plus the phase angle at unity gain.

CHAPTER 5

Damping factor

There is an approximate relationship relating phase margin (γ) and damping factor ζ of a second order system. This is given by

$$\zeta = \frac{\pi}{360°}\gamma$$

Where γ is in degrees. For larger values of phase margin and gain margin, the stability is higher. For a maximum phase transfer function both gain margin and phase margin should be positive for the system to be stable. Negative phase margin indicates that the system is unstable.

Resonant peak

The resonant peak is the peak value of $M(j\omega) = \frac{C}{R}(j\omega)$ for a second order system.

$$M_P = \frac{1}{2\zeta\sqrt{1-\zeta_1^2}}$$

Resonant frequency

It is the frequency at which the resonance peak M_P occurs. For a second order system.

$$\omega_p = \omega_n\sqrt{1-2\zeta^2}$$

Bandwidth

The bandwidth is defined as the frequency at which the magnitude gain of the frequency response curve is reduced to $\frac{1}{\sqrt{2}} = .707$ of its low frequency gain.

Consider a second order system.

$$M(s) = \frac{\omega_n^2}{s^2 + 2\zeta\omega_n s + \omega_n^2}$$

$$|M(j\omega)| = \frac{\omega_n^2}{\left(\omega_n^2 - \omega^2\right) + 4\zeta^2\omega_n^2\omega^2}$$

Let ω_b be the frequency at which the gain is reduced to $\frac{1}{\sqrt{2}}$ times its low frequency value.

$$\frac{1}{\sqrt{2}} = \frac{\omega_n^2}{\sqrt{\left(\omega_n^2 - \omega_b^2\right) + 4\zeta^2\omega_b^2\omega_n^2}}$$

Solving the equation we get

$$\omega_b = \omega_n \left[1 - 2\zeta^2 + \sqrt{4\zeta^4 - 4\zeta^2 + 2}\right]^{1/2}$$

Hence the bandwidth of a second order system,

$$\text{B.W} = \omega_n \left[1 - 2\zeta^2 + \sqrt{4\zeta^4 - 4\zeta^2 + 2}\right]^{1/2}$$

Note: It is to be noted that for some system, large band width is necessary for the system to respond to a wide range of input signal. For certain other systems small band width is necessary so that it provides attenuation to the high frequency noise in the input signal. Large bandwidth corresponds to a small rise time and fasters of response. However large bandwidth requires high cost components.

EXAMPLE 5.10: For the bode magnitude plot shown in Figure (5.10.1) determine the transfer function.

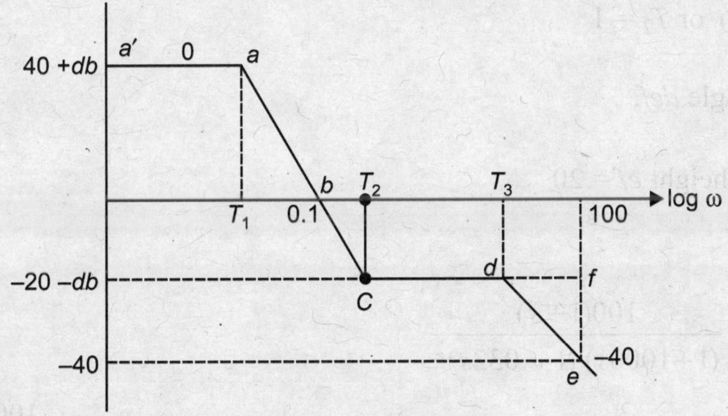

Fig. 5.10.1

Solution: Let the corner frequencies where the slope changes be T_1, T_2 and T_3

For the horizontal line $a'a$, the transfer function is

1. $G_1(s) = K$

2. For the straight line *abc*, $G_2(s) = \dfrac{1}{\left(1 + \dfrac{s}{T_1}\right)}$

3. For the horizontal line *cd*, $G_3(s) = \left(1 + \dfrac{s}{T_2}\right)$

4. For the straight line *de*, $G_4(s) = \left(1 + \dfrac{s}{T_3}\right)^2$

CHAPTER 5

5. Hence the resultant transfer function is $G(s) = \dfrac{K\left(1+\dfrac{s}{T_2}\right)}{\left(1+\dfrac{s}{T_1}\right)\left(1+\dfrac{s}{T_3}\right)^2}$

6. Now, K, T_1, T_2 and T_3 are to be determined. For the horizontal line $a'a$, only k is contributing *db* upto the point a.

   Hence $20 \log K = 40$ or $K = 100$.

7. From the triangle ab T_1,

$$20 \log \frac{1}{T_1} = 40 \text{ or } T_1 = \frac{1}{100} = .001$$

8. From the triangle *bc* T_2,

$$20 \log \frac{T_2}{.1} = 20 \text{ or } T_2 = 1$$

9. From the triangle *def*,

$$40 \log \frac{100}{T_3} = \text{height } ef = 20$$

$$T_3 = 31.6$$

10. Hence $G(s) = \dfrac{100(1+s)}{(1+1000s)\,(1+.032s)^2}$

Result: The transfer function of given magnitude bode plot is $G(s) = \dfrac{100(1+s)}{(1+1000s)\,(1+.032s)^2}$

EXAMPLE 5.11: For the log magnitude plot shown in Figure (5.11.1), obtain the transfer function.

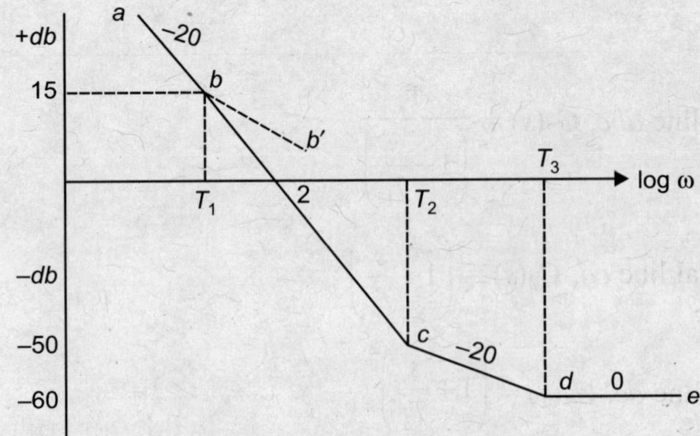

Fig. 5.11.1

Solution:

1. Let T_1, T_2, T_3 be the corner frequencies where the slopes change.

2. For the line *ab*, $G_1(s) = \frac{K}{s}$

3. For the line *bc*, $G_2(s) = \left(\frac{1}{1+\frac{s}{T_1}}\right)$

4. For the line *cd*, $G_3(s) = \left(1+\frac{s}{T_2}\right)$

5. For the line *de*, $G_4(s) = \left(1+\frac{s}{T_3}\right)$

6. The resultant transfer function is $G(s) = \frac{K\left(1+\frac{s}{T_2}\right)\left(1+\frac{s}{T_3}\right)}{s\left(1+\frac{s}{T_1}\right)}$

7. Now we have to determine K, T_1, T_2 and T_3

   From straight line *abb′*

   $\frac{K}{\omega} = 1$ at $\omega = 2$ hence $K = 2$.

8. At point *b*, the magnitude contribution is due to $\frac{K}{\omega}$.

   Hence at $\omega = T_1$, $20 \log \frac{K}{T_1} = 15$ or $T_1 = .36$.

9. From the straight line *bc*, $40 \log \frac{T_2}{T_1} = 65$ or $T_2 = 15$.

10. From the straight line *cd*, $20 \log \frac{T_3}{T_2} = (60 - 50) = 10$ or $T_3 = 47.4$.

11. Hence the transfer function is, $G(s) = \frac{2(1+.067s)(1+.021s)}{s(1+2.78s)}$.

Result: The transfer function of given bode magnitude is $G(s) = \frac{2(1+.067s)(1+.021s)}{s(1+2.78s)}$.

EXAMPLE 5.12: Determine analytically, the gain margin and phase margin for the system with the following transfer function $G(s)\, H(s) = \dfrac{5}{s\left(1+\dfrac{s}{2}\right)\left(1+\dfrac{s}{6}\right)}$.

Solution:

$$GH(j\omega) = \frac{\dfrac{5}{-90° - \tan^{-1}\dfrac{\omega}{2} - \tan^{-1}\dfrac{\omega}{6}}}{\omega\sqrt{\left(1+\dfrac{\omega^2}{4}\right)\left(1+\dfrac{\omega^2}{36}\right)}} \quad (5.12.1)$$

Let ω_g be the frequency at which the polar plot crosses the –ve real axis. Then,

$$-90° - \tan^{-1}\frac{\omega_g}{2} - \tan^{-1}\frac{\omega_g}{6} = -180°$$

$$\frac{\tan^{-1}\omega_g\left(\dfrac{1}{2}+\dfrac{1}{6}\right)}{1-\dfrac{\omega_g^2}{12}} = +90°$$

$$\frac{1-\omega_g^2}{12} = 0 \text{ form which } \omega_g = 3.46 \text{ rod/sec}$$

Substituting the value of ω_g in equation (3.11.5) for magnitude

$$d = \frac{5}{3.46\sqrt{(1+3)\left(1+\dfrac{1}{3}\right)}} = \frac{5}{8}$$

Hence gain margin = $1/d = 8/5 = 1.6 = K_g$

Let ω_c be the frequency at unity gain. Then from equation (3.11.5)

$$\frac{5}{\omega_c\sqrt{\left(1+\dfrac{\omega_c{}^2}{4}\right)\left(1+\dfrac{\omega_c^2}{30}\right)}} = 1$$

Form which ω_c = 2.79 rad/sec

The phase margin (γ) $= -90° - \tan^{-1}\dfrac{\omega_c}{2} - \tan^{-1}\dfrac{\omega_c}{6} + 180°$

$$= -90° - \tan^{-1}\frac{2.79}{2} - \tan^{-1}\frac{2.79}{6} + 180° = 10.36°.$$

Result: Phase margin (γ) = 10.36

Gain margin (K_g) = 1.6

EXAMPLE 5.13: Sketch the bode plot for the following transfer function determine phase margin and gain margin $G(s) = \dfrac{7.5(1+0.2s)}{s(s^2+16s+100)}$.

Solution: The sinusoidal transfer function $G(j\omega)$ is obtained replacing s by $j\omega$ in the given s-domain.

Given that $$G(s) = \frac{7.5(1+0.2s)}{s(s^2+16s+100)}$$

On comparing the quadratic factor of $G(s)$ with standard form of quadratic factor we can find ζ and w_n.

$$\therefore \quad s^2 + 16s + 100 = s^2 + 2\zeta\,\omega_n\, s + \omega_n^2$$

On comparing we get,

$$\omega_n^2 = 100$$

$$\therefore \quad \omega_n = 10$$

$$2\zeta\omega_n = 16$$

$$\therefore \quad \zeta = 16/2\,\omega_n$$

$$\zeta = .8$$

$$G(s) = \frac{75(1+0.2s)}{s\times 100\left(\dfrac{s^2}{100}+\dfrac{16s}{100}+1\right)} = \frac{.75(1+0.2s)}{s\left(0.01s^2+0.16s\right)}$$

$$\therefore \quad G(j\omega) = \frac{.75(1+0.2j\omega)}{j\omega\left(1+0.01(j\omega)^2+0.16j\omega\right)}$$

$$G(j\omega) = \frac{.75(1+j0.2\omega)}{j\omega\left(1-0.01\omega^2+j0.16\omega\right)}$$

Magnitude plot

The corner frequencies are $\omega_{c1} = \dfrac{1}{.2} = 5$ rad/sec and $\omega_{c2} = 10$ rad/sec.

The various terms of $G(j\omega)$ are listed in Table (5.13.1) in the increasing order to their corner frequencies. Also the table shows the slope contributed by each term and the change in slope at the corner frequency.

Choose a low frequency ω_l such that $\omega_l < \omega_{c1}$ and choose a high frequency ω_h such that $\omega_h > \omega_{c2}$

Let $\omega_l = .5$ rad/sec and $\omega_h = 20$ rad/sec

Let $A = |G(j\omega)|$ in db, let us calculate A at ω_l, ω_{c1}, ω_{c2} and ω_h

At, $\omega = \omega_l$,

$$A_1 = 20 \log |\text{first term}|$$

CHAPTER 5

Table 5.13.1

Term	Corner frequency rad/sec	Slope of each term *db*/dec	Change in slope from one term to another term *db*/dec
$\frac{.75}{j\omega}$	–	– 20	
$1 + j\,0.2\,\omega$	$\omega_{c1} = \frac{1}{.2} = 5$	+ 20	– 20 + 20 = 0
$\frac{1}{1 - .01\omega^2 + j0.16\omega}$	$\omega_{c2} = \omega_n = 10$	– 40	0 – 40 = – 40

i.e.,
$$A_1 = 20\log\left|\frac{.75}{j\omega}\right| = 20\log\frac{.75}{.5} = 3.5 = db$$

At $\omega = \omega_{c1}$,
$$A_2 = 20 \log |\text{first term}|$$

$$A = 20\log\left|\frac{.75}{j\omega}\right| = 20\log\frac{.75}{.5} = -16.5\,db$$

At $\omega = \omega_{c2}$, A_3 = [slope from one term to another term) × log $\left|\frac{\text{Present frequency}}{\text{Previous frequency}}\right|$ + Previous gain (A_2)

i.e.,
$$A_2 = \left[\text{slope from } \omega_{c1} \text{ to } \omega_{c2} \times \log\frac{\omega_{c2}}{\omega_{c1}}\right] + A_2\,(\text{at } \omega = \omega_{c1})$$

$$= 0 \times \log\frac{10}{5} + (-16.5) = -16.5\,db$$

At $\omega = \omega_h$, A_4 = [slope from one term to another term) × log $\left|\frac{\text{Present frequency}}{\text{Previous frequency}}\right|$ + Previous gain (A_3)

i.e.,
$$A_4 = \left[\text{slope from } \omega_{c2} \text{ to } \omega_h \times \log\frac{\omega_h}{\omega_{c2}}\right] + A_3\,(\text{at } \omega = \omega_{c2})$$

$$= -40 \times \log\frac{20}{10} + (-16.5) = -28.5\,db.$$

Let the points *a*, *b*, *c* and d be the points corresponding to frequencies ω_l, ω_{c1}, ω_{c2} and ω_h respectively on the magnitude plot.

Phase plot

The phase angle of $G(j\omega)$ as a function of ω is given by

$$\phi = \angle G(j\omega) = \tan^{-1} 0.2\omega - 90° - \tan^{-1} \frac{.16\omega}{1-.01\omega^2} \text{ for } \omega \le \omega_n$$

$$\phi = \angle G(j\omega) = \tan^{-1} 0.2\omega - 90° \left[\tan^{-1} \frac{.16\omega}{1-.01\omega^2} + 180°\right] \text{ for } \omega \le \omega_n$$

The phase angle of $G(j\omega)$ are calculated for various values of ω and listed in Table (5.13.2).

Table 5.13.2

ω rad/sec	**$\phi = \angle G(j\omega)$ deg**
0.5	– 88.9 ≈ – 88
1	– 87.9 ≈ – 88
5	– 91.8 ≈ – 92
10	– 116.6 ≈ – 116
20	– 147.3 ≈ 148
50	– 167.3 ≈ – 168
100	– 173.7 ≈ – 174

Note: In quadratic factors the angles varies from 0° to 180° But the calculator calculates $\tan^{-1}$ only between 0° to 90°. Hence a correction factor of 180° should be added to the phase angle after corner frequency.

On the same semilog sheet choose a scale of 1 unit = 20° on the y-axis on the right side of semi log sheet. Mark the calculated phase angle on the graph sheet. Join the points by a smooth curve.

The magnitude plot and the phase plot are shown in Figure (5.13.1). From the Figure (5.13.1) we find the phase angle at gain cross over frequency $(\omega_{gc}) = \phi_{gc} = 88°$.

∴ Phase margin = $180° + \phi_{gc} = 180° + \phi_{gc} = 180° - 88° = 92°$.

Here, K_g, Gain margin = + ∞

Note: The phase plot crosses – 180° only at infinite. The $|G(j\omega)|$ at infinity is –∞db. Hence the gain margin is +∞.

Result: Phaser margin, γ = 92°

Gain margin, $K_g = +\infty$

EXAMPLE 5.14: Given $G(s) = \dfrac{Ke^{-0.2s}}{s(s+2)(s+8)}$. Find K so that the system is stable under two conditions (*a*) gain margin equal to 6 db and (*b*) phase margin equal to 45°.

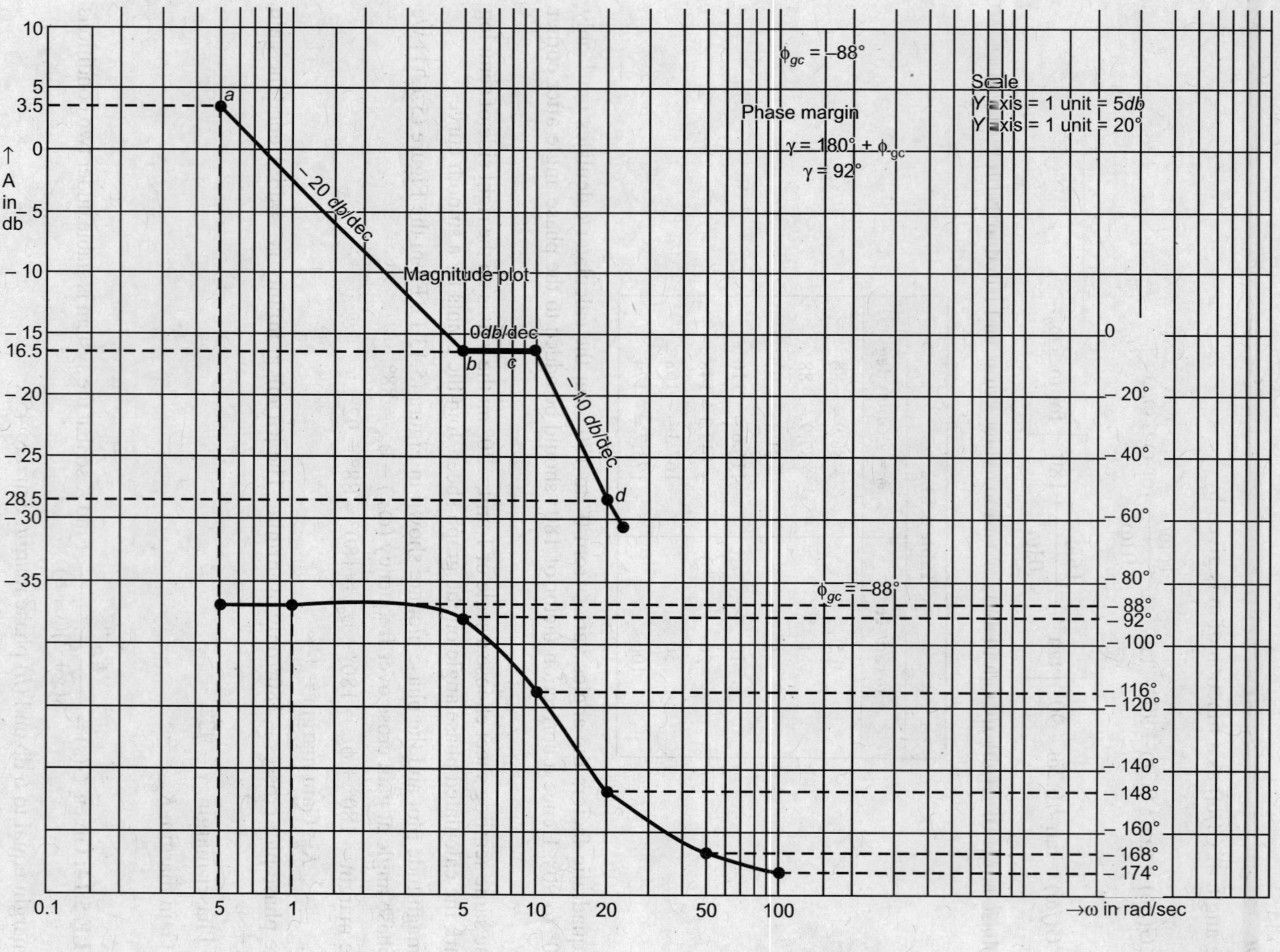

Fig. 5.13.1: Bode plot of transfer function $G(s) = \dfrac{75(1+0.2s)}{s(s^2+16s+100)}$.

Solution: Given that

$$G(s) = \frac{Ke^{-0.2s}}{s(s+2)(s+8)} = \frac{Ke^{-0.2s}}{s \times 2\left(1+\frac{s}{2}\right)\times\left(1+\frac{s}{8}\right)}$$

$$= \frac{0.0625ke^{-0.2s}}{s(1+0.5s)(1+.125s)}$$

Let $K = 1$

The sinusoidal transfer function $G(j\omega)$ is obtained by replacing s by $j\omega$ in the gives domain transfer function.

$$G(j\omega) = \frac{0.0625\ Ke^{-0.2j\omega}}{j\omega(1+j0.5\omega)\ (1+j0.125\omega)}$$

Note: $|.0625\ e^{-j0.2\omega}| = .0625$ rad

and $\angle e^{-j0.2\omega} = -0.2\ \omega$, radians

Magnitude plot

The corner frequencies are

$$\omega_{c1} = \frac{1}{0.5} = 2\,\text{rad/sec} \ \text{ and } \ \omega_{c2} = \frac{1}{.125} = 8\,\text{rad/sec}$$

The various terms of $G(j\omega)$ are listed in table 1 in the increasing order of their corner frequencies. Also the table shows the slope contributed by each term and the change in slope at the corner frequency.

Table 5.14.1

Term	Corner frequency rad/sec	Slope of each term *db*/dec	Change in slope from one term to another term *db*/sec
$\frac{0.0625}{j\omega}$	–	– 20	
$\frac{1}{1+j0.5\omega}$	$\omega_{c1} = \frac{1}{.5} = 2$	– 20	– 20 – 20 = – 40
$\frac{1}{1+j0.125\omega}$	$\omega_{c2} = \frac{1}{.125} = 8$	– 20	– 40 – 20 = – 60

Choose a low frequency ω_l such that $\omega_l = \omega_{c1}$ and choose a high frequency ω_h such that $\omega_h = \omega_{c2}$.

Let $\omega_l = 0.5$ rad/sec and $\omega_h = 50$ rad/sec

Let $A = |G(j\omega)|$ in db.

Let us calculated A at ω_l, ω_{c1}, ω_{c2} and ω_h

At $\omega = \omega_l$, $A_1 = 20\log\left|\frac{0.625}{j\omega}\right| = 20\log\frac{0.0625}{0.5} = -18\,db$ [A_1 = 20 log (first term)]

At $\omega = \omega_{c1}$, $A_2 = 20\log\left|\frac{0.625}{j\omega}\right| = 20\log\frac{0.0625}{0.5} = -30\,db$ [A_2 = 20 log |first term|]

At $\omega = \omega_{c2}$, A_3 = [slope from one term to another term] × log $\left|\frac{\text{Present frequency}}{\text{Previous frequency}}\right|$ + Previous gain

$$A_3 = [\text{slope from } \omega_{c1} \text{ to } \omega_{c2} \times \log\frac{\omega_{c2}}{\omega_{c1}}] + A_2 \text{ (at } \omega = \omega_{c1})$$

$$= -40 \times \log\frac{8}{2} + (-30) = -54 \text{ db.}$$

At $\omega = \omega_h$, A_4 = [slope from one term to another term] × log $\left|\frac{\text{Present frequency}}{\text{Previous frequency}}\right|$ + Previous gain

i.e.,
$$A_4 = \left[\text{slope from } \omega_{c2} \text{ to } \omega_h \times \log\frac{\omega_h}{\omega_{c2}}\right] + A_3 \text{ (at } \omega = \omega_{c2})$$

$$A_4 = -60 \times \log\frac{50}{8}\ (54) = -102 \text{ db.}$$

Let the points *a*, *b*, *c*, *d* be two points corresponding to frequencies ω_l, ω_{c1}, ω_{c2} and ω_h respectively on the magnitude plot.

Phase plot

The phase angle of $G(j\omega)$ as a function of ω is given by

$$\phi = -0.2\omega \times \frac{180}{\pi} - 90 - \tan^{-1} 0.5\omega - \tan^{-1} 0.125\omega$$

The phase angle of $G(j\omega)$ are calculated for various values of ω and listed in Table (5.14.2)

Table 5.14.2

ω, rad/sec	φ = deg
0.01	– 90.4 ≈ – 90
0.1	– 94.7 ≈ – 94
0.5	– 113.3 ≈ – 114
1	– 134.4 ≈ – 134
214	– 171.9 ≈ – 172
3	– 201.2 ≈ 202
4	– 225.8 ≈ 226

On the same semilog sheet choose a scale on the y-axis on the right side of the semilog sheet. Mark the calculated phase angle on the graph sheet. Join the points by smooth curve.

The magnitude and phase plot are shown in Figure. (5.14.1)

Calculate of K

Phase margin, $\gamma = 180 + \phi_{gc}$, where ϕ_{gc} is the phase of $G(j\omega)$ at $\omega = \omega_{gc}$.

where $\gamma = 45°$, $\phi_{gc} = \gamma - 180° = 45° - 180° = -135°$

With $K = 1$, the db gain at $\phi = -135°$ is -24 db.

This gain should be made zero to have to PM of 45°, hence to every point of magnitude plot a db gain of 24 db should be added. The corrected magnitude plot is obtained by shifting the plot with $K = 1$ by 24 db upwards. The magnitude correction is independent of frequency. hence the magnitude of 24 db is contributed by the term K. The value of K is calculated by equating 20 log be to 24 db.

$$\therefore \qquad 20 \log K = 24, K = 10^{24/10}, K = 15.84$$

with $K = 1$, the gain margin $= -(-.34) = 34$ db. But the required margin is 6 db. Hence to every point of magnitude plot a db gain of 28 db should be added. This addition of 28 db shifts the plot upwards. The magnitude correction is independent of frequency. Hence the magnitude of 28 db is contributed by the term k. The value of k is calculated by equating 20 log k to 28 db.

$$\therefore \qquad 20 \log K = 28, K = 10^{28/20}, K = 25.12$$

The magnitude plot with $K = 15.84$ and 25.12 are shown in Figure (5.14.1).

Result:

(*a*) K = 15.84

(b) K = 25.12

EXAMPLE 5.15: Plot the bode diagram for the following transfer function and obtain the gain and phase cross over frequencies $G(s) = 10/s\,(1 + 0.4s)\,(1 + 0.1s)$.

Solution: Given that $$G(s) = \frac{10}{s(1+0.4s)(1+0.1s)}$$

The sinusoidal transfer function of $G(j\omega)$ is obtained by replacing s by $j\omega$ in the given transfer function.

$$\therefore \qquad G(j\omega) = 10/j\omega\,(1 + 0.4\,j\omega)\,(1 + 0.1\,j\omega)$$

Magnitude Plot

The corner frequencies are,

$$\omega_{c1} = \frac{1}{0.4} = 2.5\,\text{rad/sec}$$

$$\omega_{c2} = \frac{1}{0.1} = 10\,\text{rad/sec}$$

The various terms of $G(j\omega)$ are listed in Table (5.15.1) in the increasing order of their corner frequencies. Also the table shows the slope contributed by each term and the change in slope at the corner frequency.

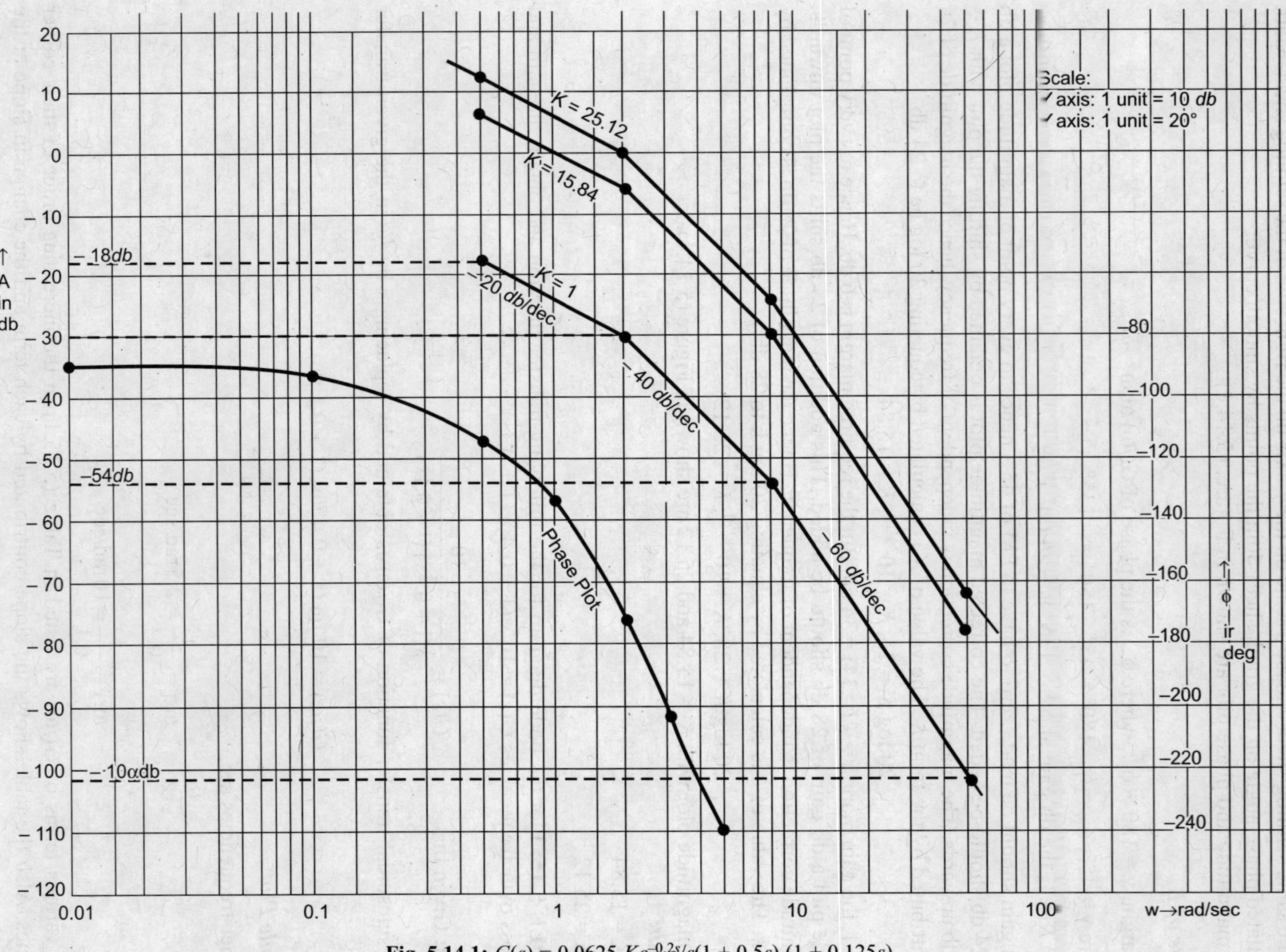

Fig. 5.14.1: $G(s) = 0.0625\ Ke^{-0.2s}/s(1 + 0.5s)(1 + 0.125s)$.

Table 5.15.1			
Term	**Corner frequency rad/sec**	**Slope of each term *db*/dec**	**Change in slope from one term to another term *db*/sec**
$\frac{10}{j\omega}$	–	– 20	
$\frac{1}{1+j0.4\omega}$	$\omega_{c_1} = \frac{1}{0.4} = 2.5$	– 20	– 20 – 20 = – 40
$\frac{1}{1+j0.1\omega}$	$\omega_{c2} = \frac{1}{0.1} = 10$	– 20	– 40 – 20 = – 60

Choose a low frequency ω_l such that $\omega_l < \omega_{c1}$ and choose a high frequency ω_h such that $\omega_h < \omega_{c1}$.

Let $\omega_l = 0.1$ rad/sec, and $\omega_h = 50$ rad/sec

Let $A = |G(j\omega)|$ in *db*

Let us calculate A at ω_l, ω_{c1}, ω_{c2} and ω_h

At $\omega = \omega_l$, $\quad A_1 = 20\log\left|\frac{10}{j\omega}\right| = 20\log\left|\frac{10}{0.1}\right| = 40\ db \quad [\because A_1 = 20 \log |\text{first term})]$

At $\omega = \omega_{c1}$ $\quad A = 20\log\left|\frac{10}{j\omega}\right| = 20\log\left|\frac{10}{2.5}\right| = 12\ db \quad [\because A_2 = 20 \log |\text{first term})]$

At $\omega = \omega_{c2}$, A_3 = [slope from one term to another term] × log $\left|\frac{\text{Present frequency}}{\text{Previous frequency}}\right|$ + Previous gain (A_2)

$$A_3 = \left[\text{slope from } \omega_{c1} \text{ to } \omega_{c2} \times \log \frac{\omega_{c2}}{\omega_{c1}}\right] + A_2 \text{ (at } \omega = \omega_{c1})$$

$$A_3 = -40 \times \log\frac{10}{2.5} + 12 = -12 \text{ db}$$

At $\omega = \omega_h$, A_4 = [slope from one term to another term] × log $\left|\frac{\text{Present frequency}}{\text{Previous frequency}}\right|$ + Previous gain (A_3)

$$A_4 = \left[\text{slope from } \omega_{c2} \text{ to } \omega_h \times \log \frac{\omega_h}{\omega_{c2}}\right] + A_3 (\text{at } \omega = \omega_{c2})$$

$$= -60 \times \log\frac{50}{10} + (-12) = -54 \text{ db}$$

Let the points *a*,*b*,*c*, and *d* be the points corresponding to frequencies ω_l, ω_{c1}, ω_{c2} and ω_h respectively on the magnitude plot.

Phase Plot

The phase angle of $G(j\omega)$ as a function of ω is given by

$$\phi = -90° - \tan^{-1} 0.4\,\omega - \tan^{-1} 0.1\,\omega$$

The phase angles of $G(j\omega)$ are calculated for various values of ω and listed in Table (5.15.2)

Table 5.15.2

ω rad/sec	φ, deg
0.1	$-92.86 \approx -92$
1	$-117.5 \approx -118$
2.5	$-149 \approx -150$
4	$-169.79 \approx -170$
10	$-210.96 \approx -210$
20	$-236.3 \approx -236$

on the same semilog sheet choose a scale on the *y*-axis on the right side of semilog sheet. Mark the calculated phase angle on the graph sheet. Join the points by a smooth curve.

The magnitude and phase plots are shown in Figure (5.15.1) from the graph the gain and phase cross over frequencies are found to be 5 rad/sec.

Result: Gain cross - over frequency = 5 rad/sec

Phase cross - over frequency = 5 rad/sec

EXAMPLE 5.16: For the following transfer function of raw bode plot and obtain gain cross-over frequency $G(s) = \dfrac{20}{s(1+3s)(1+4s)}$.

Solution: The sinusoidal transfer function of $G(j\omega)$ is obtained by replacing *s* by $j\omega$ in the given transfer function.

$$G(j\omega) = \frac{20}{j\omega(1+j3\omega)(1+j4\omega)}$$

Magnitude plot

The corner frequencies are $\omega_{c1} = \frac{1}{4} = .25$ rad/sec, $\omega_{c2} = \frac{1}{3} = .33$ rad/sec

The various terms of $G(j\omega)$ are listed in Table (5.16.1) in the increasing order of their corner frequencies. Also the table shows the slope contributed by each term and change in slope at the corner frequency.

Choose a frequency ω_l such that $\omega_l < \omega_{c1}$ and choose a high frequency ω_h such that $\omega_h > \omega_{c2}$.

Let $\omega_l = .15$ rad/sec, and $\omega_h = 1$ rad/sec

Let $A = |G(j\omega)|$ in db

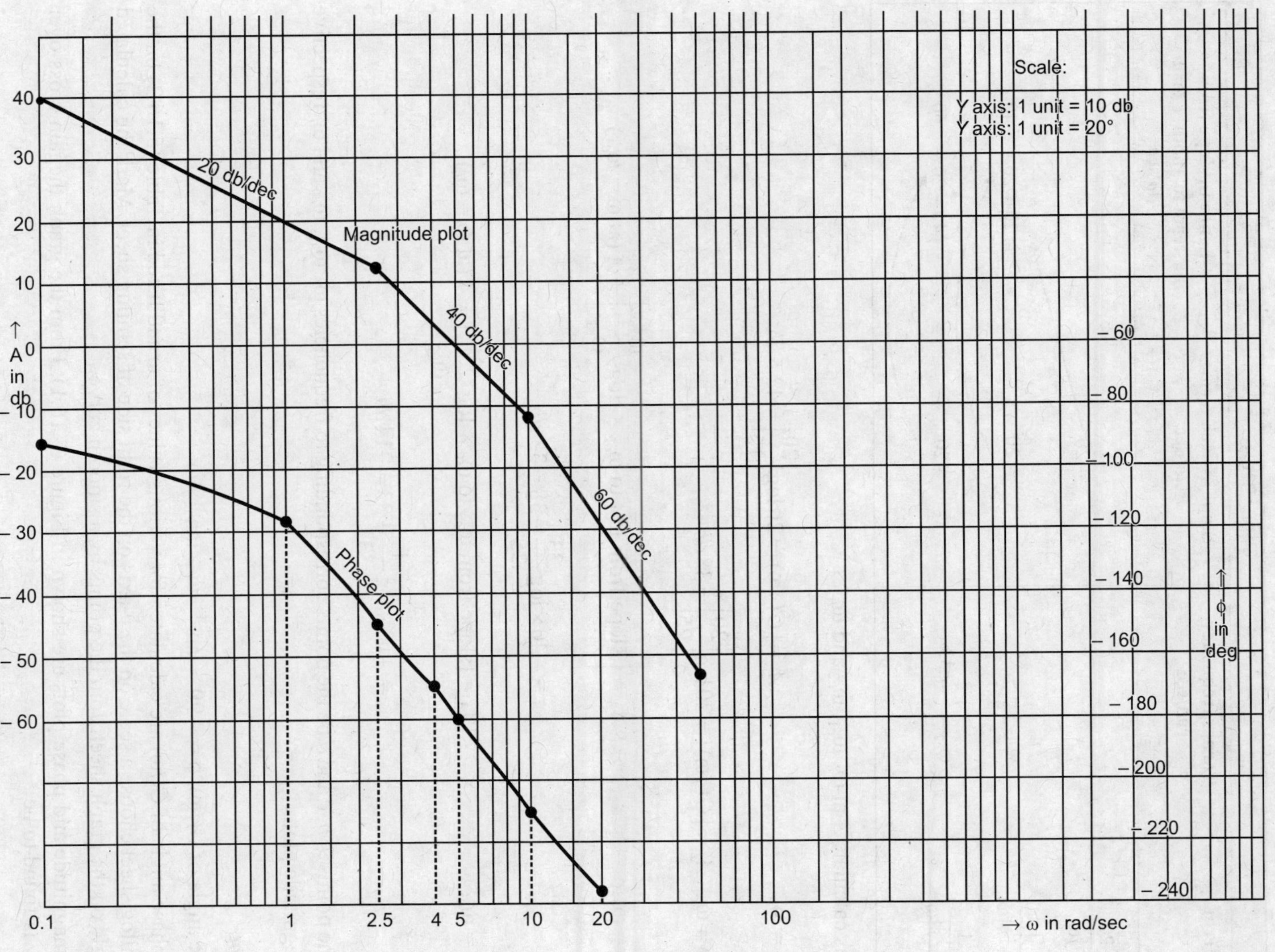

Fig. 5.15.1: Bode plot for $G(s) = 10/s(1 + 0.4s)\,(1 + 0.1s)$.

CHAPTER 5

Table 5.16.1			
Term	**Corner frequency rad/sec**	**Slope of each term *db*/dec**	**Change in slope from one term to another term *db*/sec**
$\frac{20}{j\omega}$	–	– 20	
$\frac{1}{1+j4\omega}$	$\omega_{c1} = \frac{1}{4} = .25$	– 20	– 20 – 20 = – 40
$\frac{1}{1+j3\omega}$	$\omega_{c2} = \frac{1}{3} = .33$	– 20	– 40 – 20 = – 60

Let us calculate A at ω_l, ω_{c1}, ω_{c2} and ω_h

At $\omega = \omega_l$, $$A_1 = |G(j\omega)| = 20\log\left|\frac{20}{.15}\right| = 42.5\,db$$

At $\omega = \omega_{c1}$, $A_2 = |G(j\omega)| = 20\log\left|\frac{20}{.25}\right| = 38\,db$

At $\omega = \omega_{c2}$, $$A_3 = \left[\text{slope from } \omega_{c1} \text{ to } \omega_{c2} \times \log\frac{\omega_{c2}}{\omega_{c1}}\right] + A\,(\text{at } \omega = \omega_{c1})$$

$$= -40 \times \log\frac{.33}{.25} + 38 = 33\,db$$

At $\omega = \omega_h$, $$A_4 = [\text{slope from } \omega_{c2} \text{ to } \omega_h \times \log\frac{\omega_h}{\omega_{c2}}] + A\,(\text{at } \omega = \omega_h)$$

$$= -60 \times \log\frac{1}{.33} + 33 = -14\,db$$

Let the points a, b, c and d be the points corresponding to frequencies ω_l, ω_{c1}, ω_{c2} and ω_h respective on the magnitude plot.

Phase plot

The phase angle of $G(j\omega)$, $\phi = -90 - \tan^{-1} 3\omega - \tan^{-1} 4\omega$

The phase angle of $G(j\omega)$ are calculated for various values of ω and listed in Table (5.16.2) on the same semilog sheet choose a scale on the y-axis on the right side of semilog sheet. Mark the calculated phase angle on the graph sheet. Join the points by a smooth curve.

The magnitude and phase plots are shown in Figure (5.16.1). From the graph thegain cross over frequency is found to be

$$\omega_{gc} = 1.1 \text{ rad/sec}$$

Result: Gain cross over frequency = 1.1 rad/sec

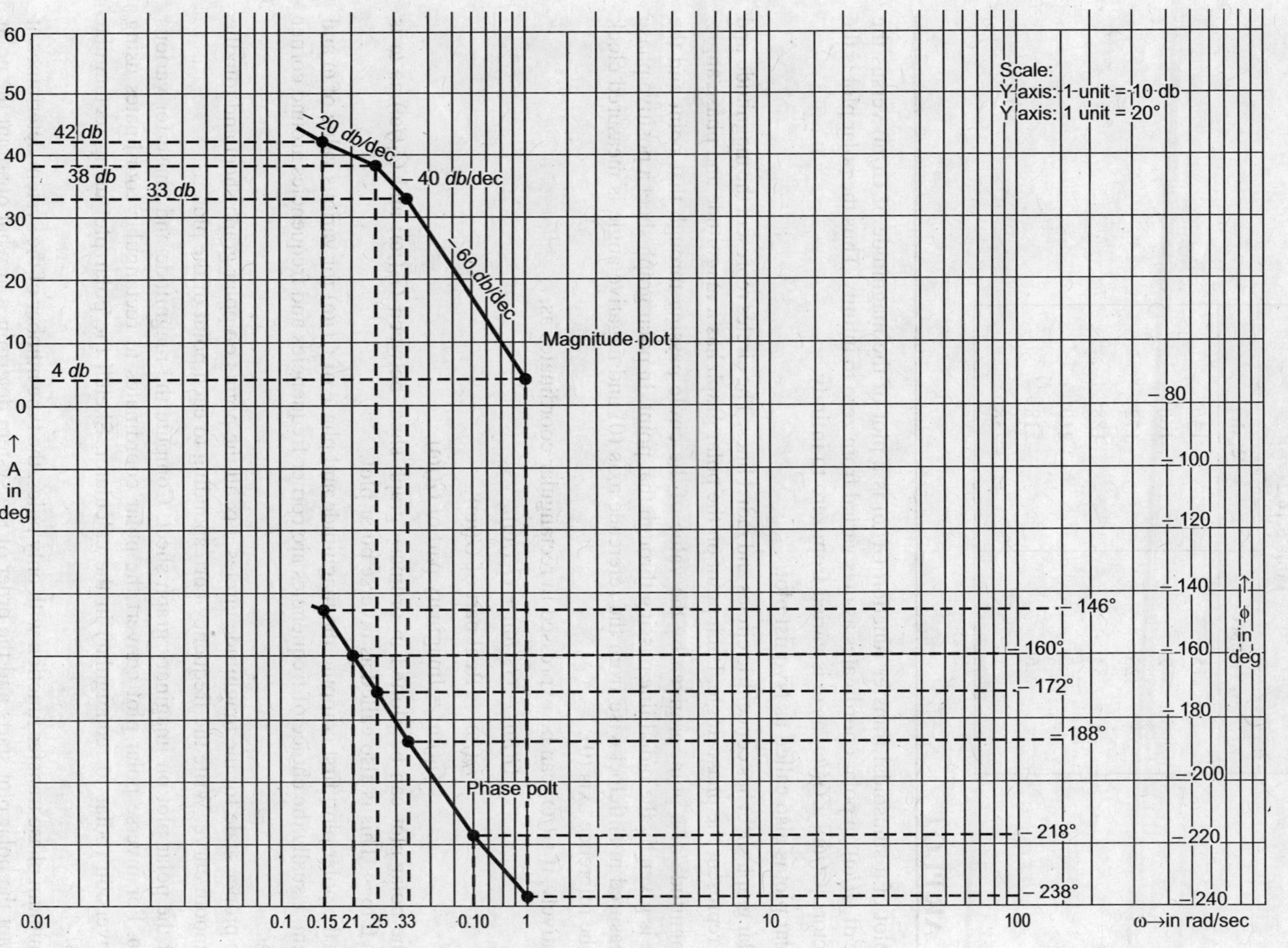

Fig. 5.16.1: Bode plot for 20s(1 + 3s) (1 + 4s).

Table 5.16.2

w, rad/sec	φ, deg
.15	−146°
.2	−160°
.25	−172°
.33	−188°
.5	210°
.6	−218°
1	−238°

5.6 POLAR PLOT

The polar plot of a sinusoidal transfer function $G(j\omega)$ is a plot of the magnitude of $G(j\omega)$ versus the phase angle of $G(j\omega)$ on polar coordinates as ω is varied from zero to intinity. Thus the polar plot is the locus of vectors $|G(j\omega)|\ \angle G(j\omega)$ as ω is varied from zero to infinity.

The polar plot is also called as Nyquist plot.

The polar graph sheet has concentric circles and radial lines. The circles represent the magnitude and radial lines represent the phase angle. Each point on the polar graph has a magnitude and phase angle.

The magnitude of a point is given by the value of the circle passing through that point and the phase angle is given by the radial line passing through that point. In polar graph sheet a positive phase angle is measured in anticlockwise from the reference axis (0) and negative angle is measured clock wise from the reference axis (0).

Alternatively, if $G(j\omega)$ can be expressed in rectangular coordinates as,

$$G(j\omega) = G_R(j\omega) + G_I\,(j\omega)$$

where, $G_R(j\omega)$ = Real part of $G(j\omega)$

and $G_I(j\omega)$ = Imaginary part of $G(j\omega)$,

Then the polar plot can be plotted in ordinary graph sheet between $G_R(j\omega)$ and $G_1(j\omega)$ as ω is varied from 0 to ∞. This is also called as inverse polar plot.

To plot the polar plot, first compute the magnitude and phase of $G(j\omega)$ for various value of ω and tabulate them. Usually the choice of frequencies are corner frequencies and frequencies around corner frequencies.

Choose proper scale for the magnitude circles. Fix all the points on polar graph sheet and join the points by smooth curve. Write the frequency corresponding to each point of the plot.

To plot the polar plot on imaginary graph sheet, Compute the magnitude and phase for various values of ω. For inverse polar plot convert the polar coordinates to rectangular coordinates using $P \to R$ conversion (polar to rectangular) in the calculator. Sketch the polar plot using rectangular coordinates.

For minimum phase transfer function with only poles, the type number of the system determines at what quadrant the polar plot starts and the order of the system determines at what quadrant the polar plot ends.

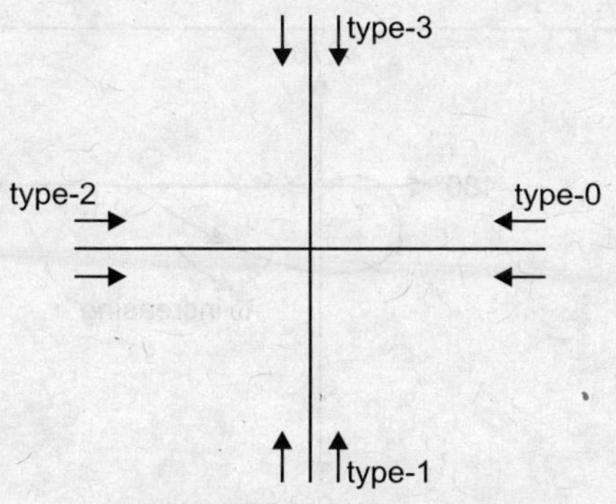

Fig. 5.30: Start of polar plot.

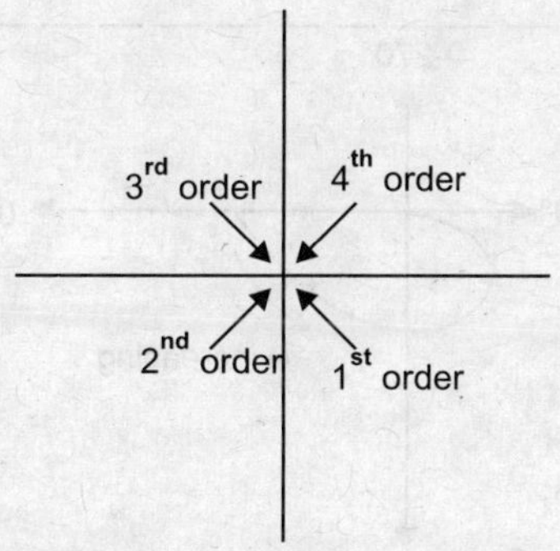

Fig. 5.31: End of polar plot.

Sketches of polar plot for different type and order of the system

Type: 0, Order: 1	***Type: 1, Order: 2***
$G(s) = \dfrac{1}{1+sT}$ $G(s) = \dfrac{1}{1+j\omega T}$ 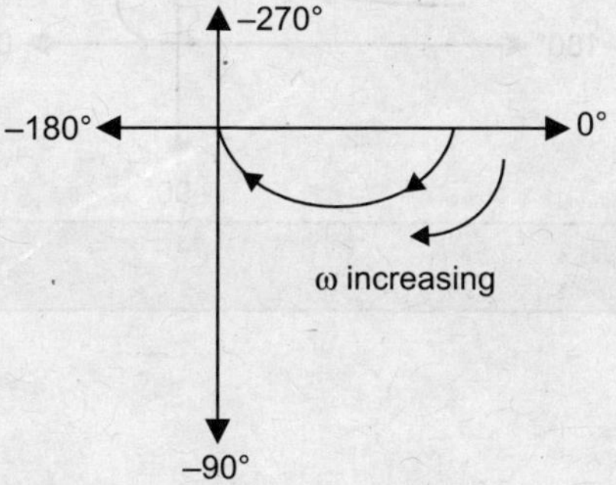	$G(s) = \dfrac{1}{s(1+sT)}$ $G(j\omega) = \dfrac{1}{j\omega(1+j\omega T)}$ 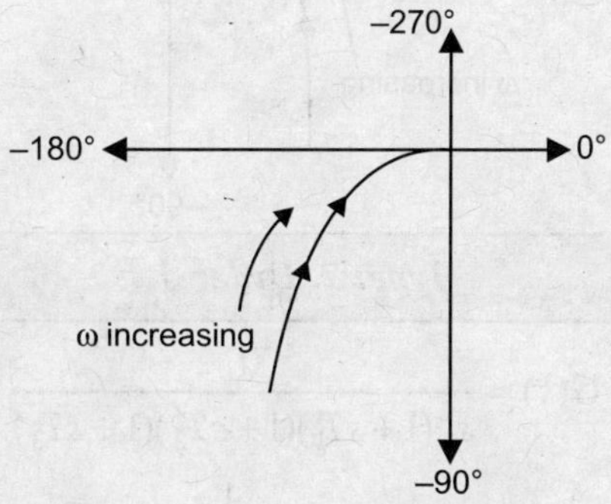
–270°, –180°, 0°, –90°, ω increasing	–270°, –180°, 0°, –90°, ω increasing
Type: 0, Order: 2	***Type: 0, Order: 3***
$G(s) = \dfrac{1}{(1+sT_1)(1+sT_2)}$ $G(j\omega) = \dfrac{1}{(1+j\omega T_1)\,(1+j\omega T_2)}$	$G(s) = \dfrac{1}{(1+sT_1)(1+sT_2)(1+sT_3)}$ $G(j\omega) = \dfrac{1}{(1+j\omega T_1)\,(1+j\omega T_2)\,(1+j\omega T_3)}$

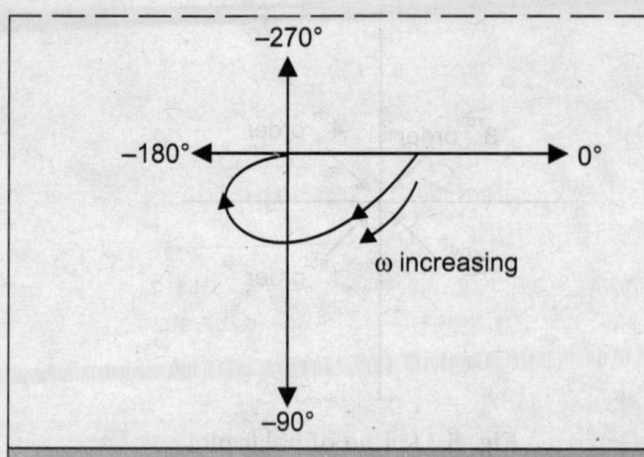	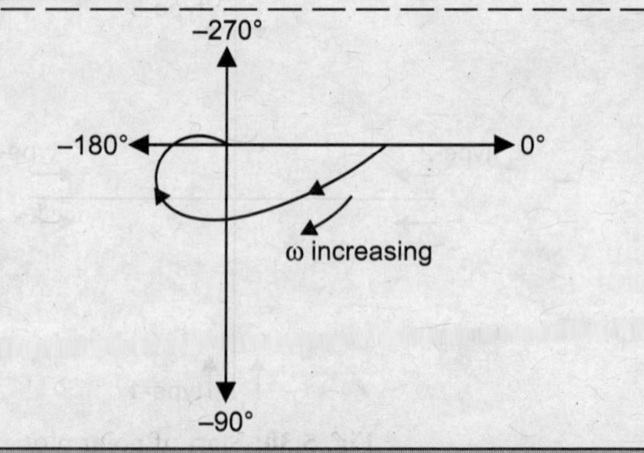
Type: 1, Order: 3	***Type : 2, Order : 4***
$G(s)=\dfrac{1}{s(1+sT_1)(1+sT_2)}$ $G(j\omega)=\dfrac{1}{j\omega(1+j\omega T_1)\,(1+j\omega T_2)}$ 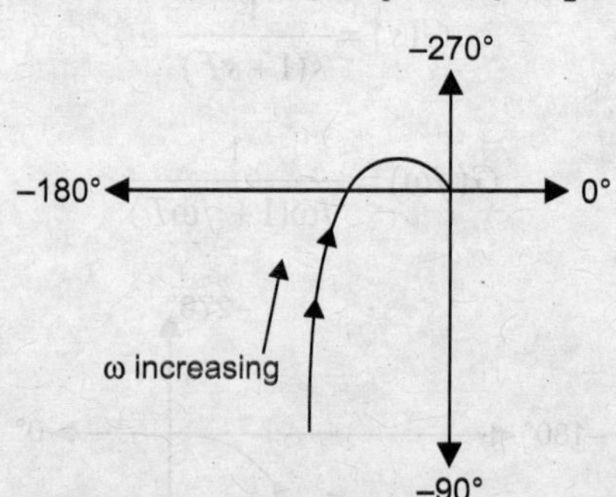	$G(s)=\dfrac{1}{s^2(1+sT_1)(1+sT_2)}$ $G(j\omega)=\dfrac{1}{(j\omega)^2\;(1+jT_1\omega)+(1+jT_2\omega)}$ 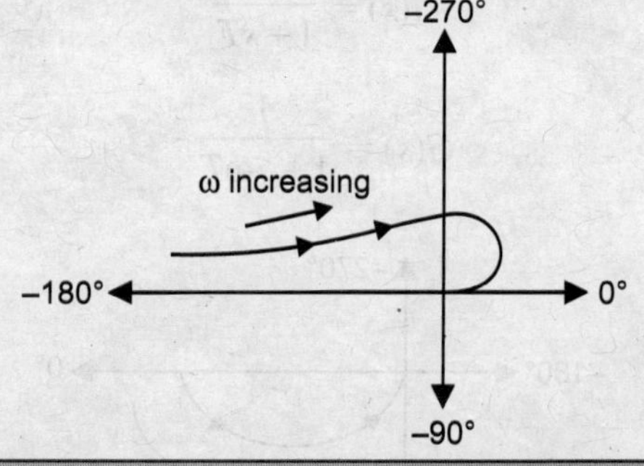
Type: 2, Order 5	
$G(s)=\dfrac{1}{s^2(1+sT_1)(1+sT_2)(1+sT_3)}$ $G(j\omega)=\dfrac{1}{j\omega^2(1+j\omega T_1)\,(1+j\omega T_2)\,(1+j\omega T_3)}$ 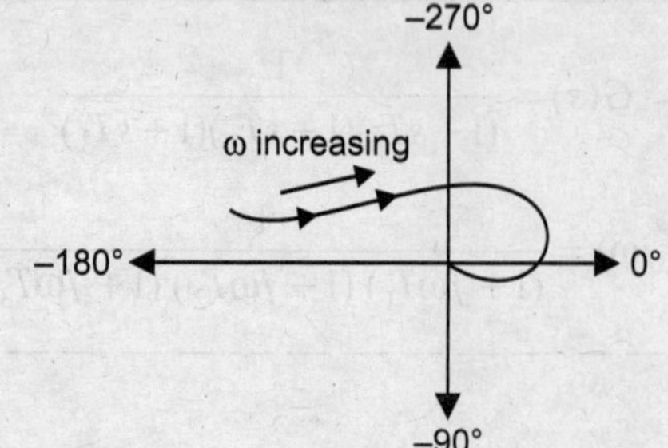	

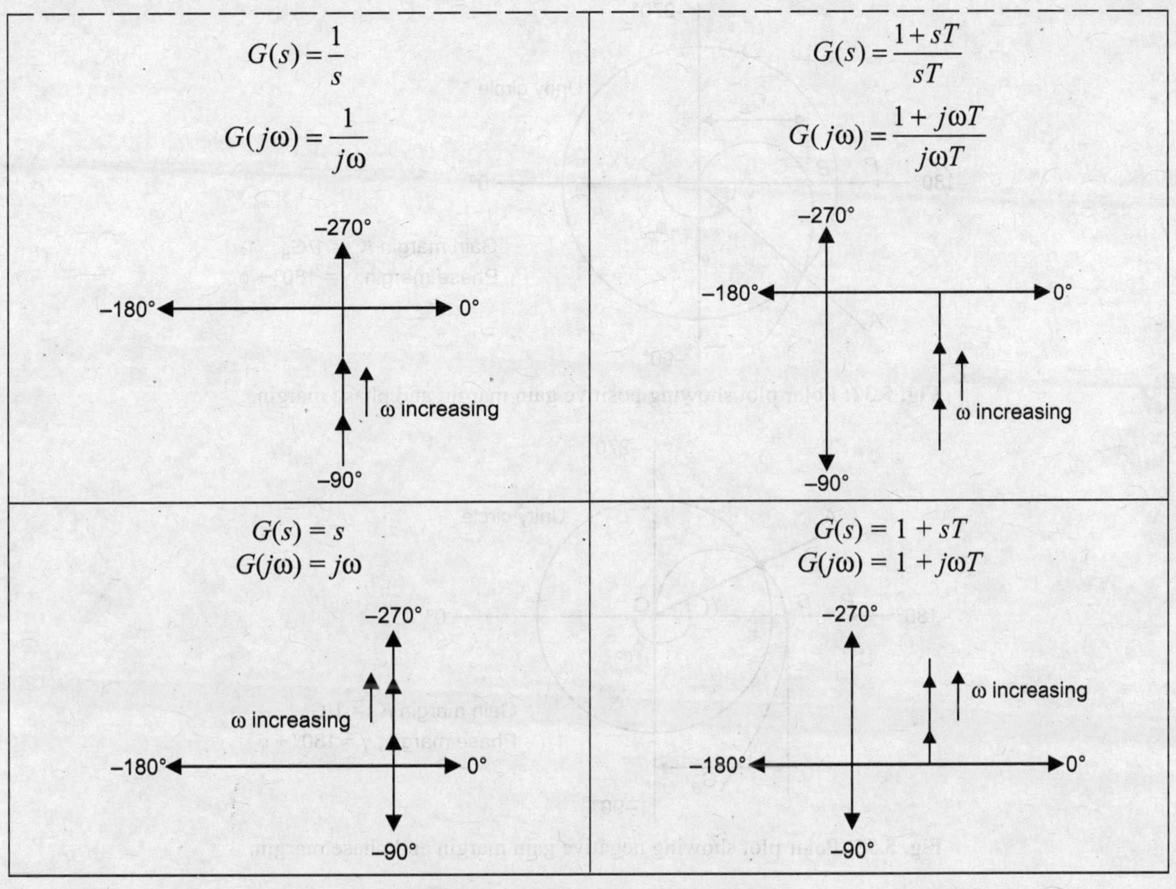

Determination of gain margin and phase margin from polar plot

The gain margin is defined as the inverse of the magnitude of $G(j\omega)$ at phase cross over frequency. The phase cross over frequency is the frequency at which the phase of $G(j\omega)$ is 180°.

Let the polar plot cut the 180° axis at point B and the magnitude circle passing through the point B be G_B. Now the gain margin $K_g = \frac{1}{G_B}$. If the point B lies within unity circle then the gain margin is positive otherwise negative. (If the polar plot is drawn in ordinary graph sheet using rectangular coordinates then the point B is the cutting point of $G(j\omega)$ locus with negative real axis and $K_g = \frac{1}{|G_B|}$ where G_B is the magnitude corresponding to point B).

The phase margin is defined as, phase margin $\gamma = 180 + \phi_{gc}$ where ϕ_{gc} is the phase angle of $G(j\omega)$ at gain cross over frequency. The gain cross over frequency is the frequency at which the magnitude of $G(j\omega)$ is unity.

Let the polar plot cut the unity circle at point A as shown in Figure (3.12.3) and (3.12.4). Now the phase margin, γ is given by $\angle AOP$. i.e. if $\angle AOP$ is below –180° axis then the phase margin is positive and if it is above – 180° axis then the phase margin is negative.

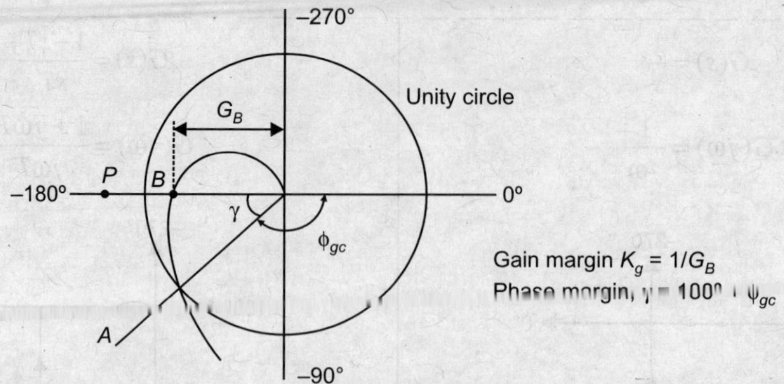

Fig. 5.32: Polar plot showing positive gain margin and phase margin.

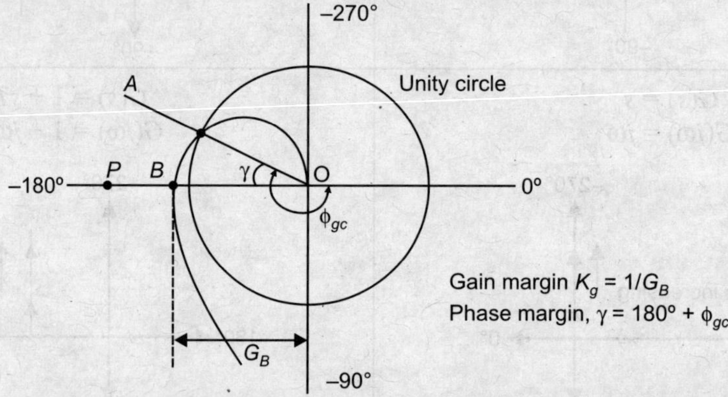

Fig. 5.33: Polar plot showing negative gain margin and phase margin.

Gain Adjustment using Polar Plot

To determine K for specified gain margin

Draw $G(j\omega)$ locus with $K = 1$. Let it cut the –180° axis at point B corresponding to again to G_B. Let the specified gain margin be x db. For this gain margin, the $G(j\omega)$ locus will cut –180° at point A whose magnitude is G_A.

Now,
$$20 \log \frac{1}{G_A} = x$$

$$\log \frac{1}{G_A} = \frac{x}{20}$$

$$\frac{1}{G_A} = 10^{\frac{x}{20}}$$

$$\therefore \quad G_A = \frac{1}{10^{\frac{x}{20}}}$$

Now the value of K is given by,

$$K = \frac{G_A}{G_B}$$

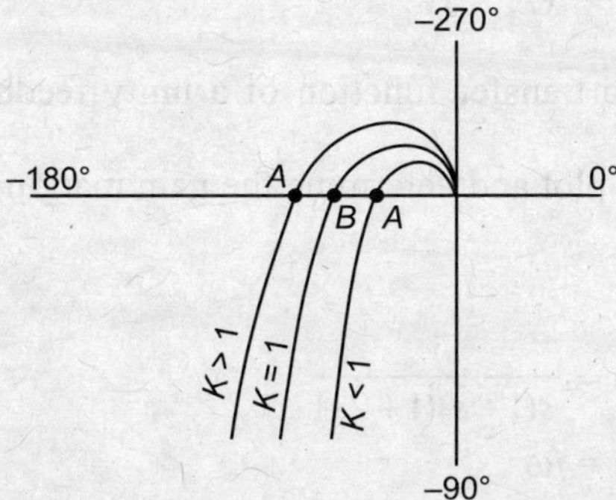

Fig. 5.24: Polar plot for different value of K.

If $K > 1$, then the system gain should be increased.

If $K < 1$, then the system gain should be decreased.

To determine K for specified phase margin

Draw $G(j\omega)$ locus with $K = 1$. Let it cut the unity circle at point B. (The gain at point B is G_B and equal to unity).

Let the specified phase margin be $x°$. For a phase margin of $x°$. Let ϕ_{gcx} be the phase angle of $G(j\omega)$ at gain cross over frequency.

$\therefore \quad x° = 180° + \phi_{gcx}$ and $\phi_{gcx} = x° - 180°$

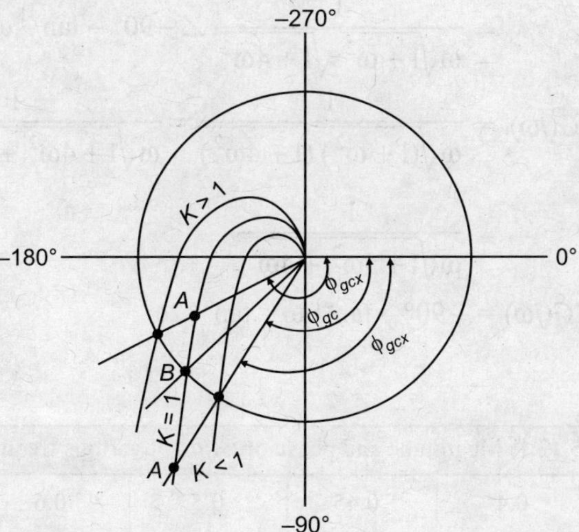

Fig. 5.35: Gain adjustment for required phase margin.

In the polar plot the radial line corresponding to ϕ_{gcx} will cut the locus of $G(j\omega)$ with $K=1$ at point A and the magnitude corresponding to the point be G_A.

Now, $$K = \frac{G_B}{G_A} = \frac{1}{G_A} \qquad [\because G_B = 1]$$

EXAMPLE 5.17: The open loop transfer function of a unity feedback system is given by $G(s) = \frac{1}{s(1+s)(1+2s)}$. Sketch the polar plot and determine the gain margin and phase margin.

Solution: Given that,

$$G(s) = \frac{1}{s(1+s)(1+2s)}$$

Put $$s = j\omega$$

The corner frequencies are $\omega_{c1} = \frac{1}{2} = .5$ rad/sec and $\omega_{c2} = 1$ rad/sec. The magnitude and phase angle of $G(j\omega)$ are calculated for the corner frequencies and for frequencies around corner frequencies and tabulated in Table (5.17.1). Using polar to rectangular conversion, the polar coordinates listed in Table (5.17.1) are converted to rectangular coordinates and tabulated in Table (5.17.2). The polar plot using polar coordinates is sketched on a polar graphs sheet as shown in Figure (5.17.1). The polar plot using rectangular coordinates is sketched on an ordinary graph sheet shown in Figure (5.17.2).

$$G(j\omega) = \frac{1}{j\omega(1+j\omega)(1+j2\omega)}$$

$$= \frac{1}{\omega}\angle 90° \sqrt{1+\omega^2}\ \tan^{-1}\omega \sqrt{1+4\omega^2}\ \angle \tan^{-1} 2\omega$$

$$= \frac{1}{\omega\sqrt{1+\omega^2}\ \sqrt{1+4\omega^2}} \angle -90° - \tan^{-1}\omega - \tan^{-1} 2\omega$$

$$\therefore \quad |G(j\omega)| = \frac{1}{\omega\sqrt{(1+\omega^2)(1+4\omega^2)}} = \frac{1}{\omega\sqrt{1+4\omega^2+\omega^2+4\omega^4}}$$

$$= \frac{1}{\omega\sqrt{1+5\omega^2+4\omega^4}}$$

$$\angle G(j\omega) = -90° - \tan^{-1}\omega - \tan^{-1} 2\omega$$

Polar Plot

Table 5.17.1: Magnitude and phase of $G(j\omega)$ at various frequencies

ω rad/sec	0.35	0.4	0.45	0.5	0.6	0.7	1.0
$\lvert G(j\omega)\rvert$	2.2	1.8	1.5	1.2	0.9	0.7	0.3
$\angle G(j\omega)$, deg	– 146	– 150	– 156	– 162	– 171	– 180	– 198

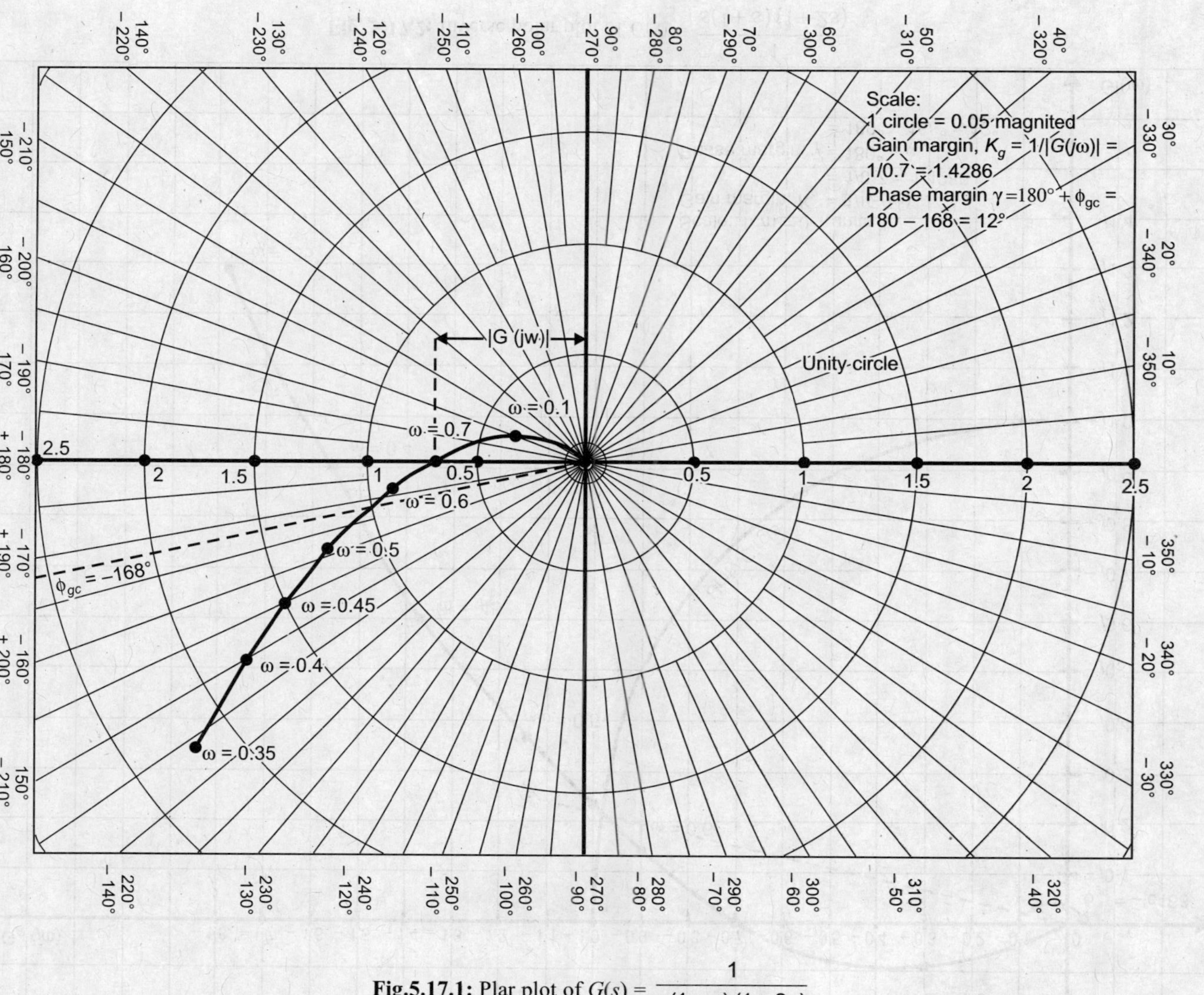

Fig.5.17.1: Plar plot of $G(s) = \dfrac{1}{s(1+s)(1+2s)}$.

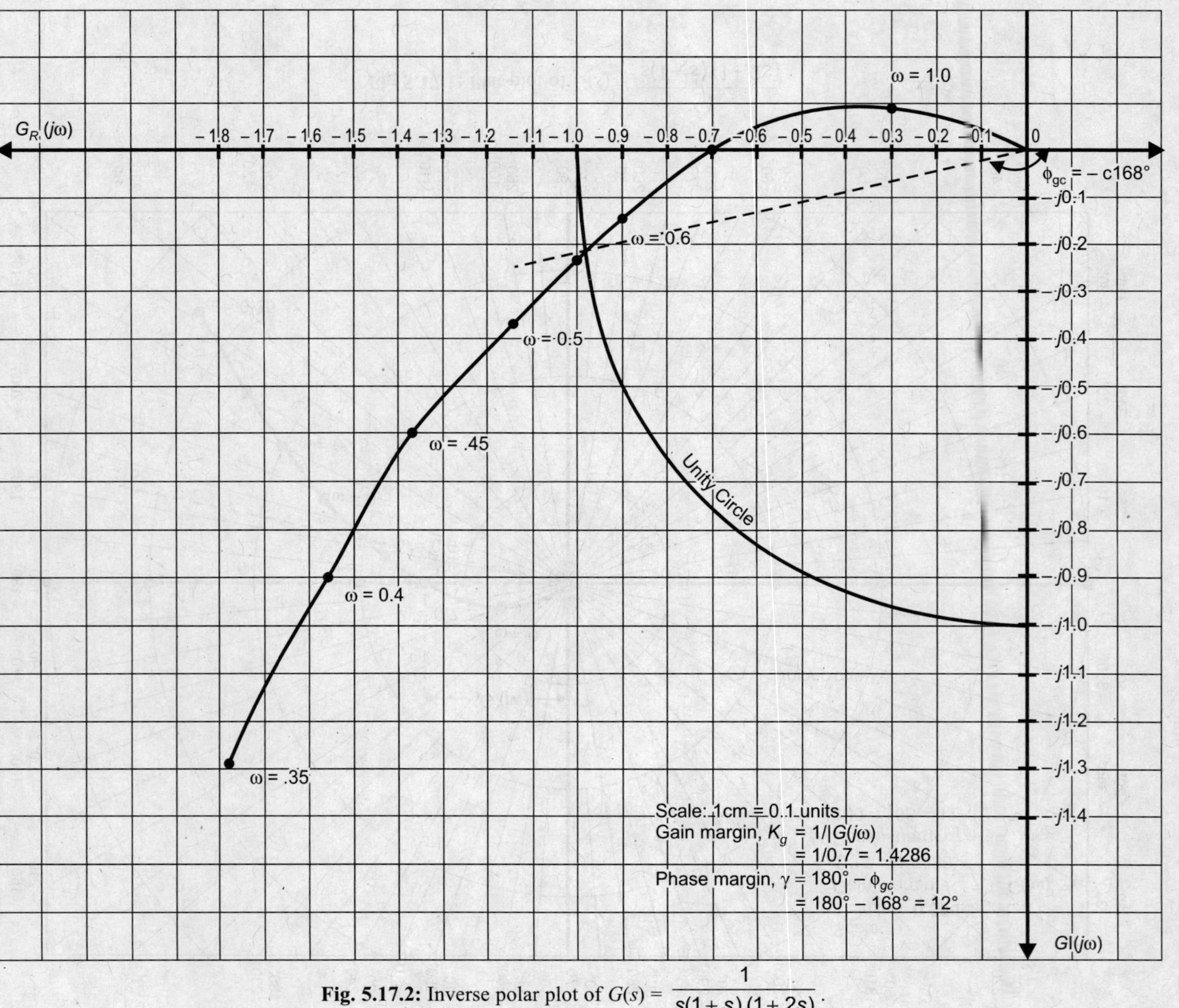

Fig. 5.17.2: Inverse polar plot of $G(s) = \dfrac{1}{s(1+s)(1+2s)}$.

Inverse Polar Plot

Table 5.17.2: Real and imaginary part of $G(j\omega)$ at various frequencies

ω rad/sec	0.35	0.4	0.45	0.5	0.6	0.7	1.0
$\lvert G(j\omega)\rvert$	– 1.56	– 1.56	– 1.37	– 1.14	– 0.89	– 0.7	– 0.29
$<G(j\omega)$ deg	– 1.29	– 0.9	– 0.61	– 0.37	– 0.14	– 0	0.09

Result: Gain margin, $K_g = 1.4286$

Phase margin, $\gamma = +12°$

EXAMPLE 5.18: The open loop transfer function of a unity feedback system is given by $G(s) = \dfrac{1}{s^2(1+s)(1+2s)}$. Sketch the polar plot and determine the gain margin and phase margin.

Solution: Given that,

$$G(s) = \frac{1}{s^2(1+s)(1+2s)}$$

Put

$$s = j\omega$$

$$G(j\omega) = \frac{1}{(j\omega)^2\,(1+j\omega)\,(1+2j\omega)}$$

The corner frequencies are $\omega_{c1} = 0.5$ rad/sec and $\omega_{c2} = 1$ rad/sec. The magnitude and phase angle of $G(j\omega)$ are calculated for the corner frequencies and frequencies around corner frequencies are tabulated in Table (5.18.1). The polar coordinates in Table (5.18.1) are converted to rectangular coordinates and tabulated in Table (5.18.2). The polar plot using polar coordinates is sketched on a polar graph sheet shown in Figure (5.18.1) and rectangular coordinates in sketched on an ordinary graph sheet as shown in Figure (5.18.2).

$$G(j\omega) = \frac{1}{(j\omega)^2\,(1+j\omega)\,(1+j2\omega)}$$

$$= \frac{1}{\omega^2\angle 180°\sqrt{1+\omega^2}\angle\tan^{-1}\omega\sqrt{1+4\omega^2}\angle\tan^{-1}2\omega}$$

$$G(j\omega) = \frac{1}{\omega^2\sqrt{1+\omega^2}\,\sqrt{1+4\omega^2}}\angle(-180° - \tan^{-1}\omega - \tan^{-1}2\omega)$$

$$|G(j\omega)| = \frac{1}{\omega^2\sqrt{1+\omega^2}\,\sqrt{1+4\omega^2}} = \frac{1}{\omega^2\sqrt{(1+\omega^2)\,(1+4\omega^2)}}$$

$$= \frac{1}{\omega^2\sqrt{1+5\omega^2+4\omega^4}}$$

$$<G(j\omega) = -180° - \tan^{-1}\omega - \tan^{-1}2\omega$$

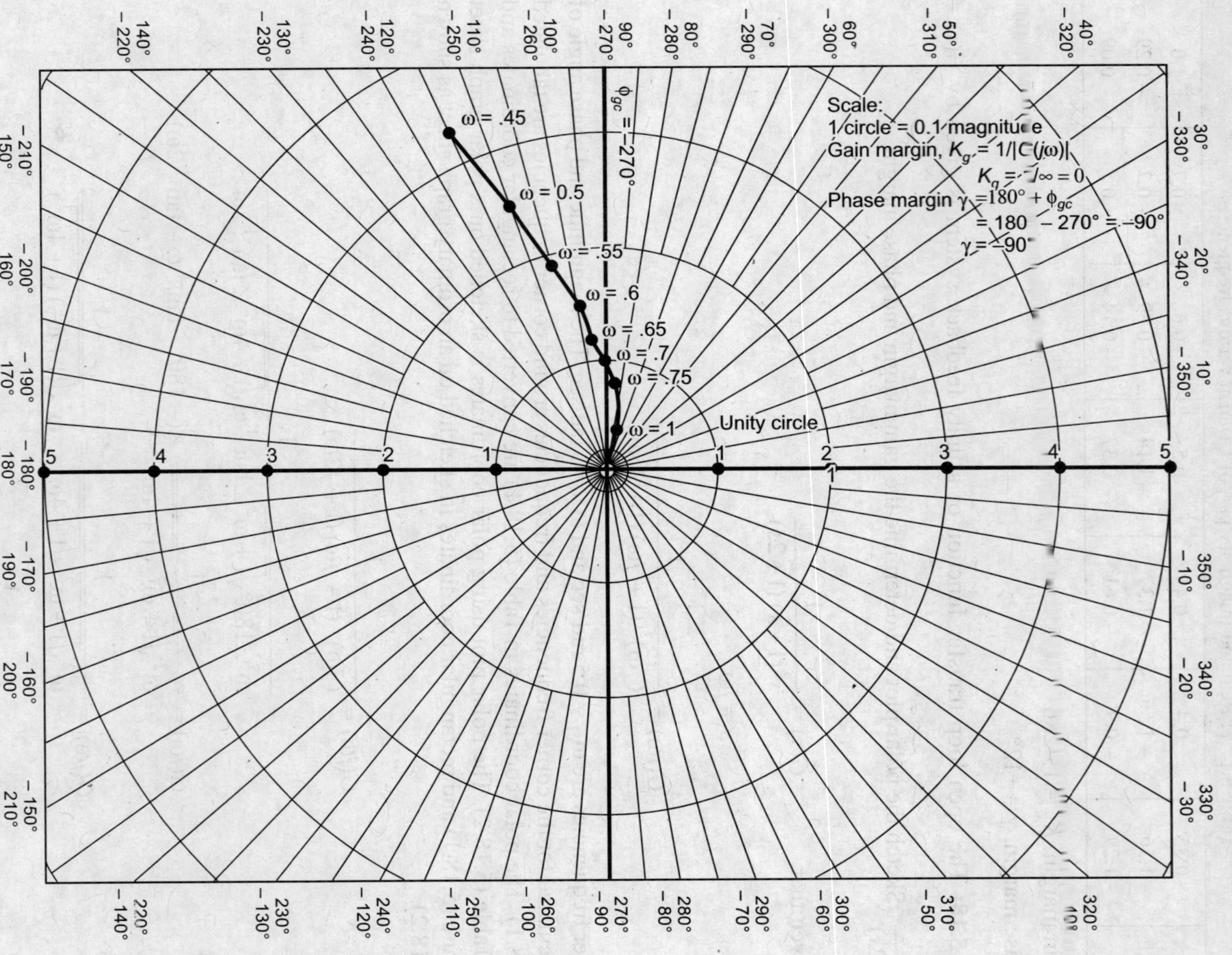

Fig. 5.18.1: Polar plot of $G(s) = 1/s^2\,(1 + s)\,(1 + 2s)$ (using polar coordinates).

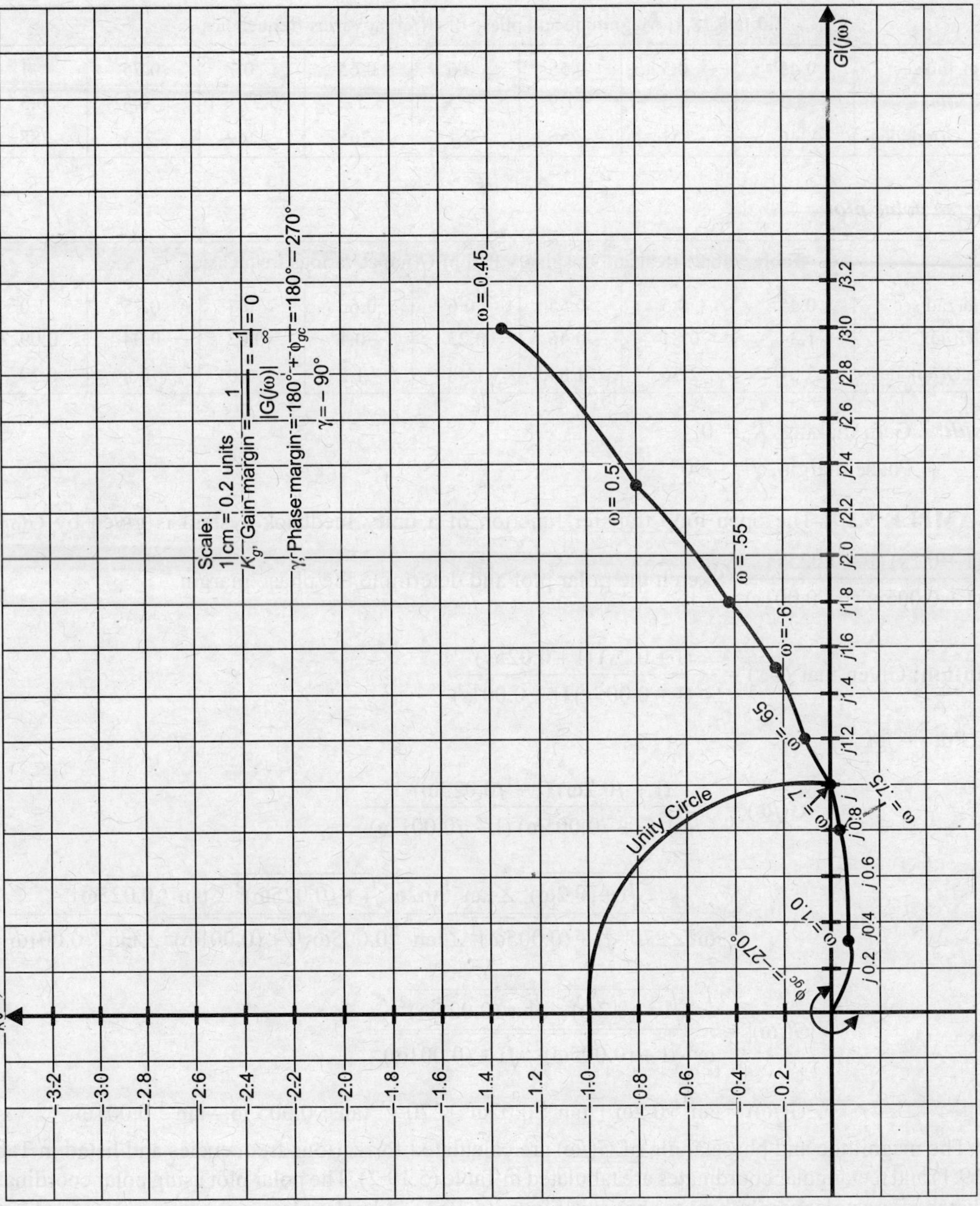

Scale:
1 cm = 0.2 units

K_g, Gain margin $= \frac{1}{|G(j\omega)|} = \frac{1}{\infty} = 0$

γ, Phase margin $= 180° + \phi_{gc} = 180° - 270°$

$\gamma = -90°$

Fig.5.18.2: Inverse polar plot of $G(s) = 1/s^2\,(1 + s)\,(1 + 2s)$ (using rectangular coordinates).

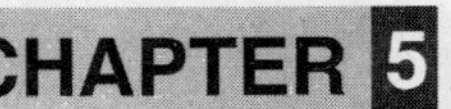

Polar plot

Table 5.18.1: Magnitude and phase of $G(j\omega)$ at varies frequencies								
ω, rad/sec	0.45	0.5	0.55	0.6	0.65	0.7	0.75	1
$\lvert G(j\omega)\rvert$	3.3	2.5	1.9	1.5	1.2	0.97 ≈ 1	0.8	0.3
$\angle G(j\omega)$, deg	–246	–251	–256	–261	–265	–269	–273	–288

Inverse polar plot

Table 5.18.2: Real and Imaginary Part of $G(j\omega)$ at various frequencies								
ω, rad/sec	0.45	0.5	0.55	0.6	0.65	0.7	0.75	1.0
$\lvert G(j\omega)\rvert$	–1.34	–0.81	–0.46	–0.23	–0.1	–0.02	–0.04	–0.09
$\angle G(j\omega)$, deg	3.01	2.36	1.84	1.48	1.2	1.0	0.8	0.29

Result: Gain making, $K_g = 0$

Phase margin, $\gamma = -90°$

EXAMPLE 5.19: The open loop transfer function of a unity feedback system is given by $G(s) = \dfrac{(1+0.2s)(1+0.025s)}{s^3(1+0.005s)(1+0.001s)}$. Sketch the polar plot and determine the phase margin.

Solution: Given that $G(s) = \dfrac{(1+0.2s)(1+0.025s)}{s^3(1+0.005s)(1+0.001s)}$

Put $s = j\omega$,

$$\therefore \qquad G(j\omega) = \frac{(1+j0.2\omega)(1+j0.025\omega)}{(j\omega)^3(1+j0.005\omega)(1+j0.001\,\omega)}$$

$$= \frac{\sqrt{1+(0.2\omega)^2}\angle\tan^{-1}0.2\omega\sqrt{1+(0.025\omega)^2}\angle\tan^{-1}0.025\omega}{\omega^3\angle 270°\sqrt{1+(0.005\omega)^2}\angle\tan^{-1}0.005\omega\sqrt{1+(0.001\omega)^2}\angle\tan^{-1}0.001\omega}$$

$$|G(j\omega)| = \frac{\sqrt{1+(0.2\omega)^2}\ \sqrt{1+(0.025\omega)^2}}{\omega^3\sqrt{1+(0.005\omega)^2}\ \sqrt{1+(0.001\omega)^2}}$$

$$\angle G(j\omega) = \tan^{-1}0.2\omega + \tan^{-1}0.02\omega - 270° - \tan^{-1}0.005\,\omega - \tan^{-1}0.001\omega$$

The magnitude and phase angle of $G(j\omega)$ are calculated for various frequencies and listed in Table (5.19.1) and rectangular coordinates are tabulated in Table (5.19.2). The polar plot using polar coordinates is sketched as a polar graph sheet as shown in Figure (5.19.1) and polar plot using rectangular coordinates is sketched on a ordinary graph sheet as shown in Figure (5.19.2).

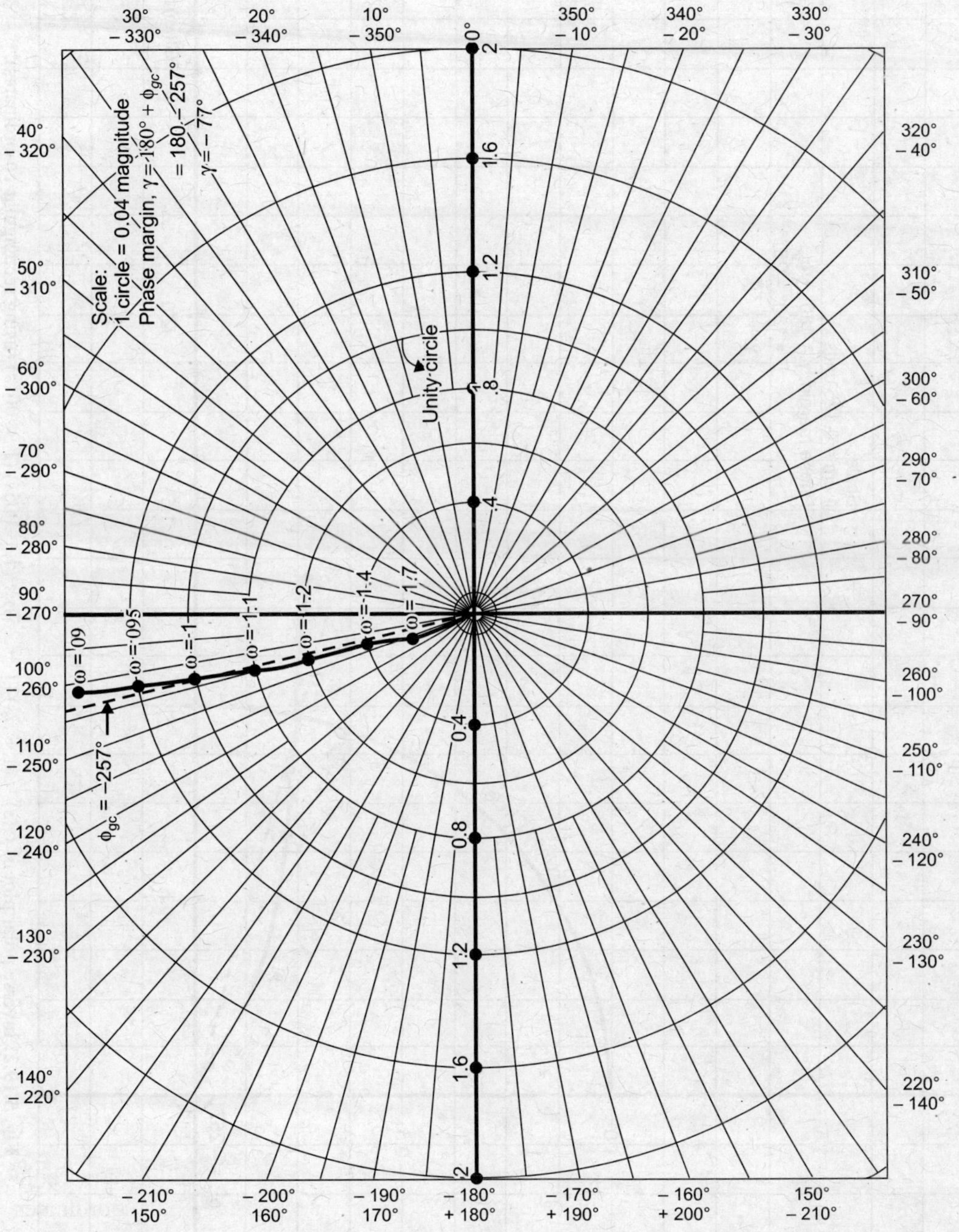

Fig. 5.19.1: Polar plot of $G(s) = (1 + .2s)\ (1 + .025s)/s3(1 + .005s)\ (1 + .001s)$ (using polar coordinates).

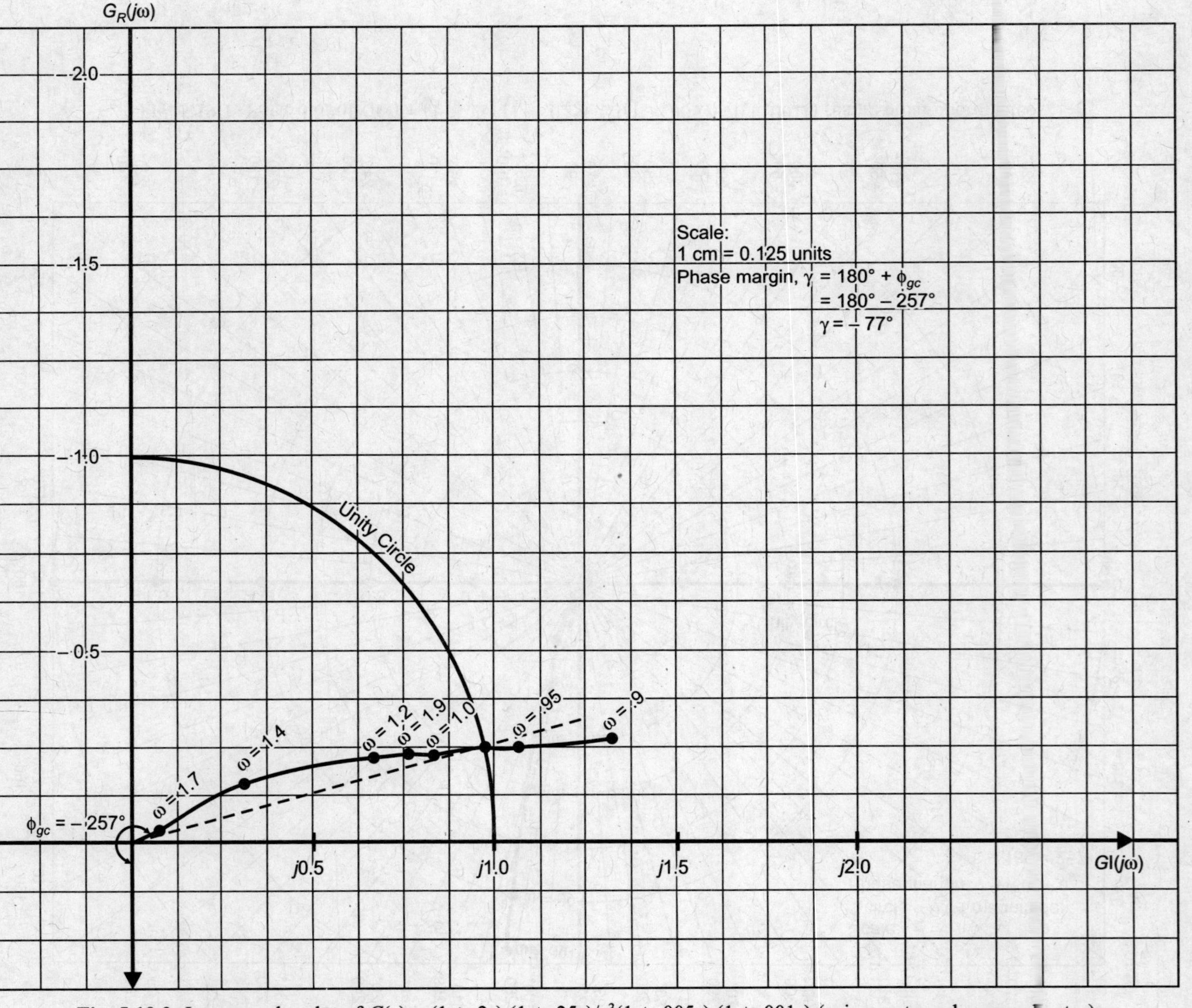

Fig. 5.19.2: Inverse polar plot of $G(s) = (1 + .2s)\,(1 + .25s)/s^3(1 + .005s)\,(1 + .001s)$ (using rectangular co-ordinates).

Polar plot

Table 5.19.1: Magnitude and phase of $G(j\omega)$							
ω, rad/sec	0.9	0.95	1	1.1	1.2	1.4	1.7
|G(jω)|	1.4	1.2	1	0.8	0.6	0.4	0.2
∠G(jω), deg	–259	–258	–257	–256	–255	–253	–249

Inverse polar plot

Table 5.19.2: Real and imaginary part of $G(j\omega)$							
ω, rad/sec	0.9	0.95	1	1.1	1.2	1.4	1.7
|G(jω)|	–0.27	–0.25	–0.22	–0.19	–0.16	–0.12	–0.07
∠G(jω), deg	1.37	1.17	0.97	0.78	0.58	0.38	0.19

Result: Phase margin $\gamma = -77°$

EXAMPLE 5.20: Consider a unity feedback system having an open loop transfer function $G(s) = \dfrac{K}{s(1+0.2s)(1+0.05s)}$. Sketch the polar plot and determine the value of K, so that

(i) Gain margin is 18 db

(ii) Phase margin is 60°.

Solution: Given that,

$$G(s) = \frac{K}{s(1+0.2s)(1+0.05s)}$$

The polar plot is sketched by taking $K = 1$.

∴ put $K = 1$ and $s = j\omega$ is $G(s)$

$$\therefore \quad G(j\omega) = \frac{K}{j\omega(1+j0.2\omega)(1+j0.05\omega)}$$

The corner frequencies are $\omega_{c1} = \dfrac{1}{0.2} = 5$ rad/sec and $\omega_{c2} = \dfrac{1}{0.05} = 20$ rad/sec. The magnitude and phase angle of $G(j\omega)$ are calculated for various frequencies and tabulated in Table (5.20.1). The rectangular coordinates and tabulated is Table (5.20.2). The polar plot using polar coordinates is sketched on a polar graph sheet as shown in Figure (5.20.1). The polar plot using rectangular coordinates is sketched on an ordinary graph sheet as shown in Figure (5.20.2).

$$G(j\omega) = \frac{1}{j\omega(1+j0.2\omega)(1+j0.05\omega)}$$

$$= \frac{1}{\omega\angle 90°\sqrt{1+(0.2\omega)^2}\ \angle\tan^{-1}0.2\omega\sqrt{1+(0.05\omega)^2}\angle\tan^{-1}0.05\omega}$$

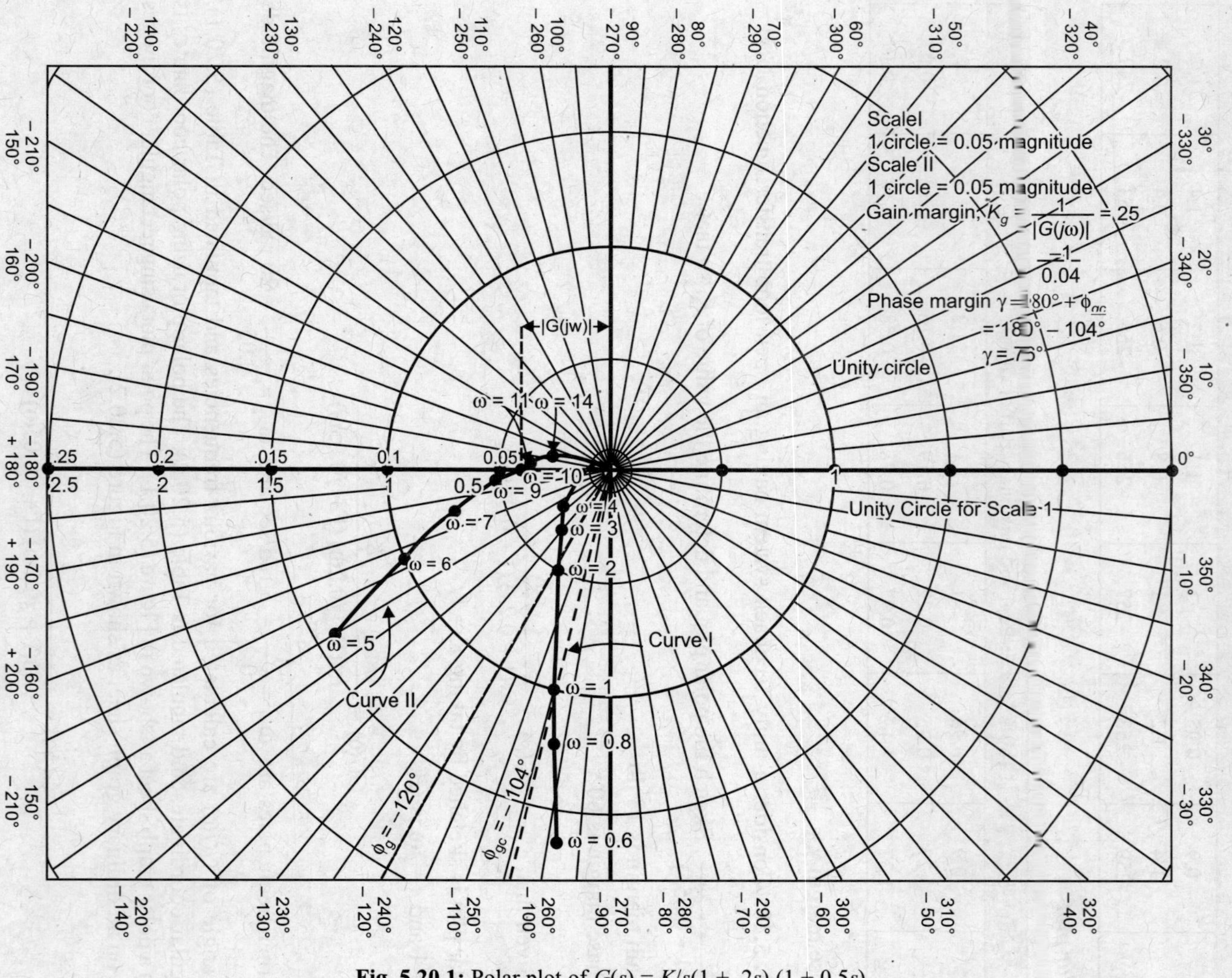

Fig. 5.20.1: Polar plot of $G(s) = K/s(1 + .2s)\,(1 + 0.5s)$.

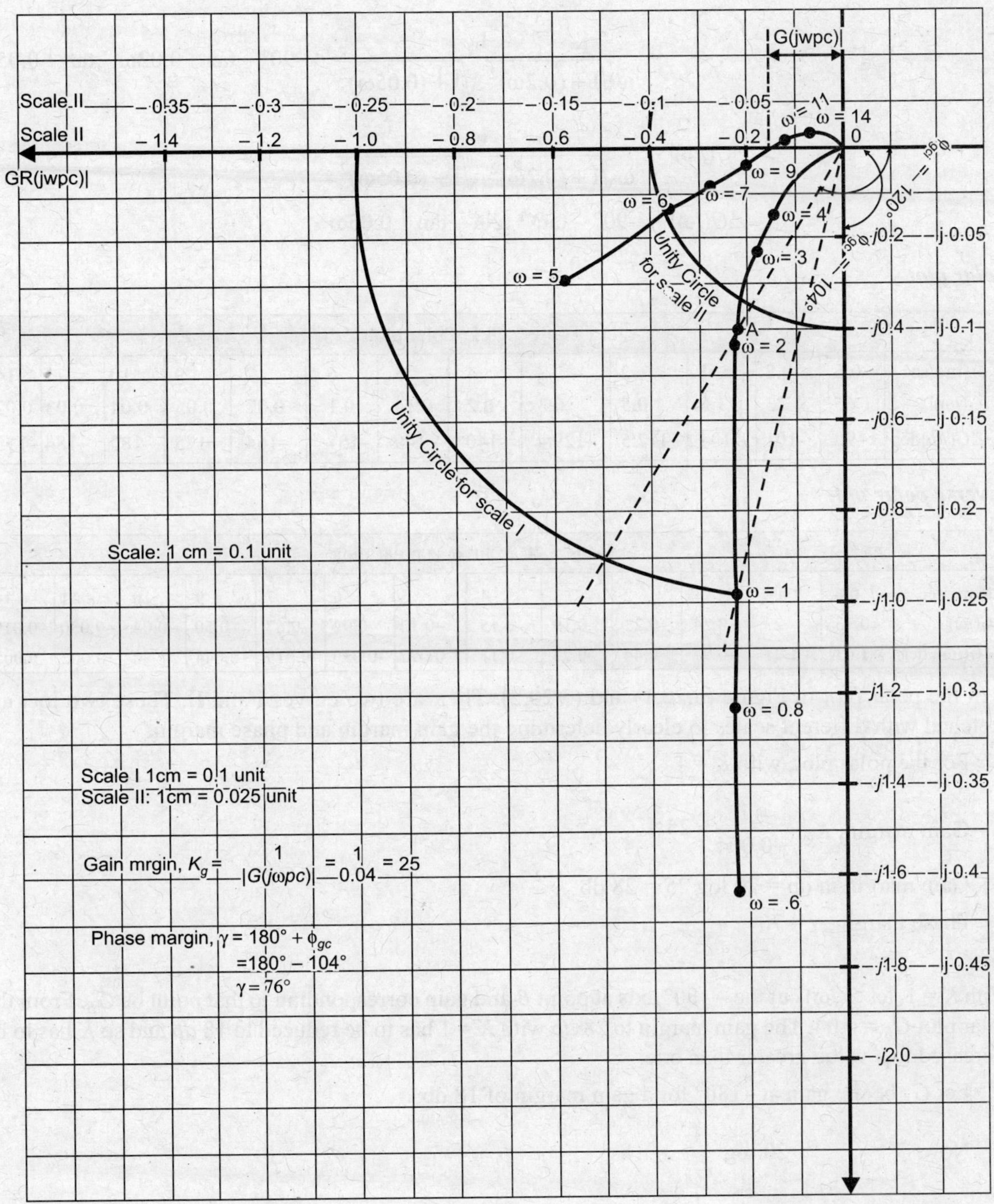

Fig. 5.20.2: Inverse polar plot of $G(s) = K/s(1 + .2s)(1 + 0.5s)$.

$$= \frac{1}{\omega\sqrt{1+(0.2\omega)^2}\ \sqrt{1+(0.05\omega)^2}} < -90° - \tan^{-1} 0.02\omega - \tan^{-1} 0.05\omega$$

$$\therefore \qquad |G(j\omega)| = \frac{1}{\omega\sqrt{1+(0.2\omega)^2}\ \sqrt{1+(0.05\omega)^2}}$$

$$\angle G(j\omega) = -90° - \tan^{-1} .2\omega - \tan^{-1} 0.05\omega$$

Polar plot

Table 5.20.1: Polar plot

ω, rad/sec	0.6	0.8	1	2	3	4	5	6	7	9	10	11	14
$\|G(j\omega)\|$	1.65	1.23	1.0	0.5	0.3	0.2	0.14	0.1	0.07	0.05	0.04	0.03	0.02
$\angle G(j\omega)$deg	–98	–101	–104	–117.5	–129.4	–140	–149	–157	–164	–175	–180	–184	–195

Inverse polar plot

Table 5.20.2: Inverse polar plot

ω, rad/sec	0.6	0.8	1	2	3	4	5	6	7	9	10	11	14
$\|G(j\omega)\|$	–0.23	–.23	–0.24	–0.23	–0.19	–0.15	–0.12	–0.92	–0.67	–0.50	–0.04	–0.030	–0.019
$\angle G(j\omega)$, deg	–1.63	–1.21	–0.97	–0.44	–0.23	–0.13	–0.072	–0.039	–0.019	–0.044	–0	0.002	0.005

In the polar plot in Figure (5.20.1) and (5.20.2). Their are two curves I and II. These two loci are sketched with different scales to clearly determine the gain margin and phase margin.

For the polar plot, with $K = 1$

$$\text{Gain margin, } K_g = \frac{1}{0.004} = 25$$

Gain margin in db = 20 log 25 = 28 db

Phase margin, $\gamma = 76°$

Case I

With $K = 1$, let $G(j\omega)$ cut the – 180° axis at point B and gain corresponding to that point be G_B. From the polar plot $G_B = 0.04$. The gain margin to 28 *db* with $K = 1$ has to be reduced to 18 *db* and so K has to be increased to a value greater than one.

Let G_A be the gain at –180° for a gain margin of 18 db.

$$\text{Now,} \qquad 20 \log \frac{1}{G_A} = 18$$

$$\log \frac{1}{G_A} = \frac{18}{20}$$

$$\frac{1}{G_A} = 10$$

$$\therefore \qquad G_A = \frac{1}{10^{\frac{18}{20}}} = 0.125$$

The value of k is given by $K = \frac{G_A}{G_B} = \frac{.125}{0.04} = 3.125$.

Case II

With $K = 1$, the phase margin is 76°. This has to be reduced be 60°. Hence gain has to be increased.

Let ϕ_{gc_2} be the phase of $G(j\omega)$ for a phase margin of 60°.

$$\therefore \qquad 60° = 180° + \phi_{gc_2}$$

$$\phi_{gc_2} = 60° - 180° = -120°$$

In the polar plot the 120° line cut the locus of $G(j\omega)$ at point A and cut the unity circle at point B.

Let G_A = Magnitude of $G(j\omega)$ at point A.

G_B = Magnitude of $G(j\omega)$ at point B.

From the polar plot, $G_A = 0.425$ and $G_B = 1$.

Now,
$$K = \frac{G_B}{G_A} = \frac{1}{0.425} = 2.353$$

Result:

1. When $K = 1$, gain margin, $K_g = 25$

   Gain margin is db = 28 *db*
2. When $K = 1$, phase margin, $\delta = 76°$
3. For a gain margin of 18 db, $K = 3.125$
4. For a phase margin of 60°, $K = 2.353$

EXAMPLE 5.21: Consider a unity feedback system having an open loop transfer function $G(s) = \frac{K}{s(1 + 0.5s)(1 + 4s)}$. Sketch the polar plot and determine the value of K so that

(i) Gain margin is 20 db and

(ii) Phase margin is 30°.

Solution: Given that,
$$G(s) = \frac{K}{s(1+0.5s)(1+4s)}$$

The polar plot is sketched by taking $K = 1$

Put $K = 1$ and $s = j\omega$ in $G(s)$

$$\therefore \qquad G(j\omega) = \frac{K}{j\omega(1 + j0.5\omega)(1 + j4\omega)}$$

CHAPTER 5

The corner frequencies are $\omega_{c1} = \frac{1}{4} = 0.25$ rad/sec and $\omega_{c2} = \frac{1}{0.5} = 2$ rad/sec. The magnitude and phase angle of $G(j\omega)$ are calculated for various frequencies and tabulated in table (5.21.1) and rectangular coordinates are tabulated in table (5.21.2). The polar plot using polar coordinates is sketched on a polar graph sheet as shown in Figure (5.21.1) and the polar plot using rectangular coordinates is sketched on an ordinary graph sheet as shown in Figure (5.21.2).

$$G(j\omega) = \frac{1}{j\omega(1 + j0.5\omega)\,(1 + j4\omega)}$$

$$= \frac{1}{\omega\angle 90°\sqrt{1+(0.5\omega)^2}\angle\tan^{-1}0.5\omega\sqrt{1+(4\omega)^2}\angle\tan^{-1}4\omega}$$

$$= \frac{1}{\omega\sqrt{1+0.25\omega^2}\,\sqrt{1+16\omega^2}}\angle(90° - \tan^{-1}0.5\,\omega - \tan^{-1}4\omega)$$

$$\therefore \quad |G(j\omega)| = \frac{1}{\omega\sqrt{1+0.25\omega^2}\,\sqrt{1+16\omega^2}}$$

$$\angle G(j\omega) = -\,90° - \tan^{-1}0.5\,\omega - \tan^{-1}4\omega$$

Polar plot

Table 5.21.1: Magnitude and Phase of $G(j\omega)$

ω rad/sec	0.3	0.4	0.5	0.6	0.8	1.0	1.2
$\|G(j\omega)\|$	2.11	1.3	0.87	0.61	0.35	0.22	0.15
$\angle G(j\omega)$, deg	–149	–159	–167	–174	–184	–193	–199

Inverse polar plot

Table 5.21.2: Real and Imaginary part of $G(j\omega)$

ω, rad/sec	0.3	0.4	0.5	0.6	0.8	1.0	1.2
$G_R(j\omega)$	–1.8	–1.21	–0.85	–0.61	–0.35	–0.21	–0.14
$G_I(j\omega)$	–1.09	–0.47	–0.2	–0.06	0.02	0.05	0.05

From the polar plot, with $K = 1$

Gain margin, $K_g = \frac{1}{0.44} = 2.27$

Gain margin in *db* = 20 log 2.27 = 7.12 db

Phase margin, $\gamma = 15°$

Case I

With $K = 1$, let $G(j\omega)$ cut the – 180° axis at point B and gain corresponding to that point be G_B. From the polar plot $G_B = 0.04$. The gain margin to 7.12 db with $K = 1$ has to be increased to 20 *db* and so K has to be decreased to a value less than one.

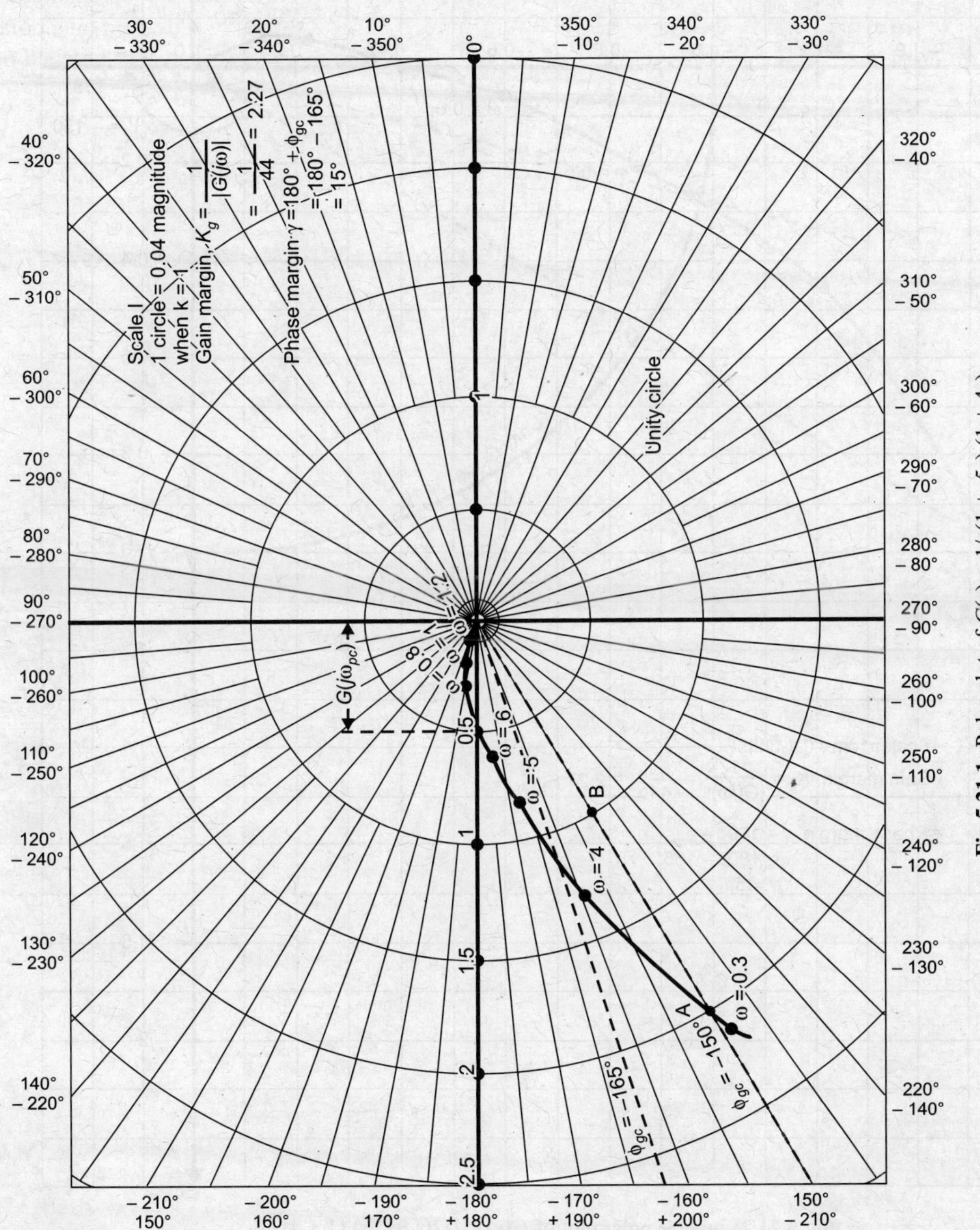

Fig.5.21.1: Polar plot of $G(s) = 1/s(1 + .5s)(1 + 4s)$.

CHAPTER 5

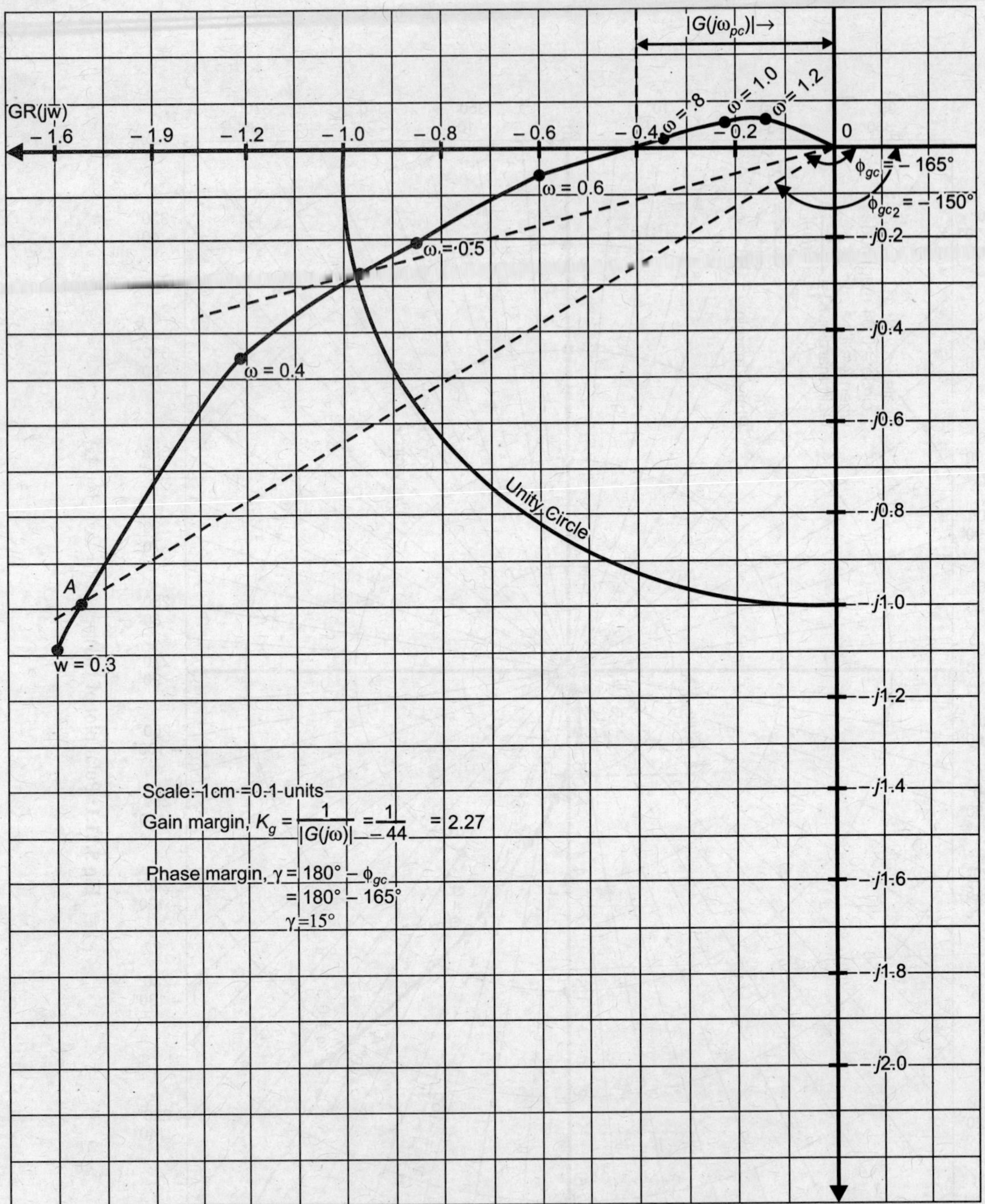

Fig. 5.21.2: Inverse polar plot of $G(s) = 1/2(1 + .5s)(1 + 4s)$.

Let G_A be the gain at –180° for a gain margin of 20 *db*.

Now, $$20 \log \frac{1}{G_A} = 20$$

$$\log \frac{1}{G_A} = \frac{20}{20} = 1$$

$$\frac{1}{G_A} = 10^{-1} = 10$$

∴ $$G_A = \frac{1}{10} = 0.1$$

The value of K is given by $K = \frac{G_A}{G_B} = \frac{0.1}{0.44} = 0.227$.

Case II

With $K = 1$, the phase margin is 15°. This has to be increased be 30°. Hence gain has to be decreased.

Let ϕ_{gc_2} be the phase of $G(j\omega)$ for a phase margin of 30°.

∴ $$30° = 180° + \phi_{gc2}$$

$$\phi_{gc2} = 30° - 180° = -150°$$

In the polar plot the 150° line cut the locus of $G(j\omega)$ at point A and cut the unity circle at point B.

Let G_A = Magnitude of $G(j\omega)$ at point A.

G_B = Magnitude of $G(j\omega)$ at point B.

From the polar plot, $G_A = 2.04$ and $G_B = 1$.

Now, $$K = \frac{G_B}{G_A} = \frac{1}{2.04} = 0.49.$$

Result:

1. When $K = 1$, gain margin, $K_g = 2.27$
2. Gain margin in *db* = 7.12 *db*
3. When $K = 1$, Phase margin, $\gamma = 15°$
4. For a gain margin of 20 *db*, $K = .227$
5. For a phase margin of 30°, $K = 0.49$

SOME WORKED EXAMPLES

Note: To make the students to understand some examples are given in very detail.

EXAMPLE 1: Sketch the bode plot of the following open loop transfer function and from the bode plot determine phase margin and gain margin $G(s) = \frac{100(1 + .1s)}{s(1 + .2s)(1 + .5s)}$

Solution:

Step 1: Put, $s = j\omega$ in $G(s)$

$$\therefore \quad G(j\omega) = \frac{100(1+.1j\omega)}{j\omega(1+.2j\omega)(1+.5j\omega)}$$

Step 2: Draw magnitude plot.

Table 1

Term	Corner frequency rad/sec	Slope of each term *db*/dec	Charge in slope from one term to another term *db*/dec
$\frac{100}{j\omega}$	–	–20	
$\frac{1}{1+.5j\omega}$	$\omega_{c1} = \frac{1}{.5} = 2$	–20	–20 –20 = 40
$\frac{1}{1+.2j\omega}$	$\omega_{c2} = \frac{1}{.2} = 5$	–20	–40 –20 = –60
$(1+.1j\omega)$	$\omega_{c3} = \frac{1}{-1} = 10$	+20	–60 + 20 = –40

Let, choose low and high frequencies

The condition for choosing low frequency is

$$\omega_l < \omega_{c1}$$

The condition for choosing high frequency is

$$\omega_h > \omega_{c3}$$

$\therefore$ Let, $\omega_l = 0.1$ rad/sec

$\omega_h = 100$ rad/sec

Let, $A = |G(j\omega)|$

at, $\omega = \omega l$

$$A_1 = 20 \log |\text{First term}|$$

$$A_1 = 20 \log \left|\frac{100}{j\omega}\right| = 20 \log \left|\frac{100}{1}\right|$$

$$\boxed{A_1 = 60\ db}$$

at, $\omega = \omega_{c1}$

$$A_2 = 20 \log |\text{First term}|$$

$$A_2 = 20 \log \left|\frac{100}{2}\right|$$

$$\boxed{A_2 = 34\ db}$$

at, $\omega = \omega_{c2}$

$$A_3 = \text{change in slope from one term to another term} \times \log\left|\frac{\text{Present frequency}}{\text{Previous frequency}}\right| + \text{Previous gain}$$

i.e., $$A_3 = \text{Change in slope from } \omega_{c1} \text{ to } \omega_{c2} \times \log\left|\frac{\omega c_2}{\omega c_1}\right| + A_2$$

$$A_3 = -40\log\left|\frac{5}{2}\right| + 34$$

$$\boxed{A_3 = +18\ db}$$

at, $\omega = \omega_{c3}$

$$A_4 = \text{change in slope from one term to another term} \times \log\left|\frac{\text{Present frequency}}{\text{Previous frequency}}\right| + \text{Previous gain}$$

i.e., $$A_4 = \text{Change in slope from } \omega_{c2} \text{ to } \omega_{c3} \times \log\left|\frac{\omega_{c3}}{\omega_{c2}}\right| + A_3$$

$$A_4 = -60\log\left|\frac{10}{5}\right| + 18$$

$$\boxed{A_4 = 0\ db}$$

at $\omega = \omega_h$

$$A_5 = \text{Change in slope from } \omega_{c3} \text{ to } \omega_h \times \log\left|\frac{\omega_h}{\omega_{c3}}\right| + A_4$$

$$A_5 = -40\log\left|\frac{100}{10}\right| + 0$$

$$\boxed{A_5 = -40\ db}$$

Table 2

ω, rad/sec	0.1	2	5	10	100
A in *db*	60	34	18	0	–40

Step 3: Draw phase plot

$$\phi = \angle G(j\omega)$$

$$\phi = -90^\circ + \tan^{-1} .1\omega - \tan^{-1} .2\omega - \tan^{-1} .5\omega$$

Table 3

ω, rad/sec	0.1	2	5	10	100
φ, in deg	–93.5°	–145.5°	–176.5°	–187°	–192°

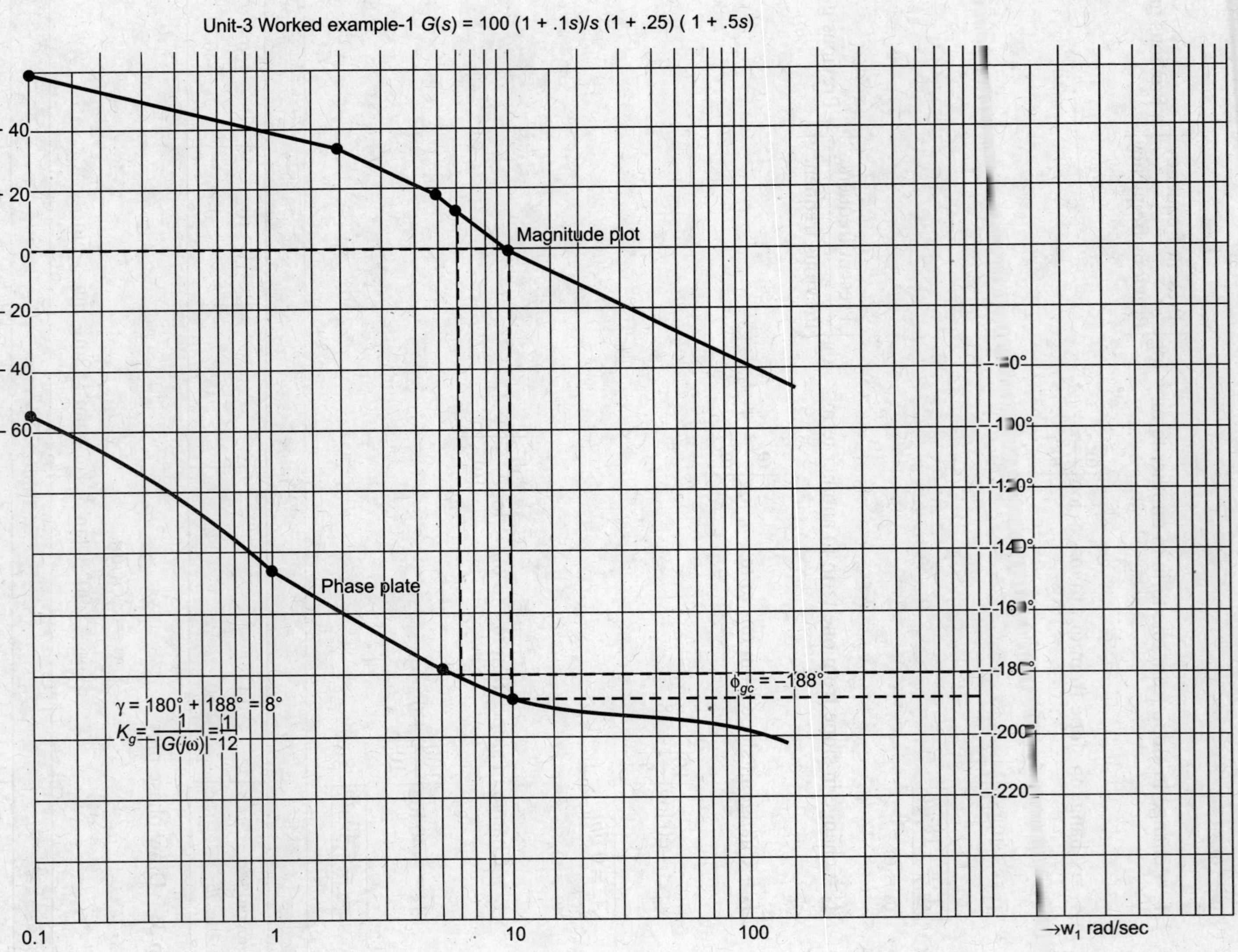
Unit-3 Worked example-1 G(s) = 100 (1 + .1s)/s (1 + .25) (1 + .5s)
Magnitude plot
Phase plate
γ = 180° + 188° = 8°
K_g = 1/|G(jω)| = 1/12
φ_gc = −188°
+ 40
+ 20
0
− 20
− 40
− 60
0.1
1
10
100
→w1 rad/sec
−180
−200
−220

From bode plot,

$$\text{Phase margin, } \gamma = 180^\circ + \phi_{gc} = 180^\circ - 188^\circ = 8^\circ$$

$$\text{Gain margin, } K_g = \frac{1}{|G(j\omega)|} = \frac{1}{12}$$

Result:

$$\gamma = 180^\circ$$

$$K_g = \frac{1}{12}$$

EXAMPLE 2: The open loop transfer function of a unity feedback system is given by $G(s) = \frac{(1+.2s)(1+.25s)}{s^2(1+.05s)(1+.01s)}$. Sketch the polar plot and determine the phase margin.

Solution: Given that

$$G(s) = \frac{(1+.2s)(1+.25s)}{s^2(1+.05s)(1+.01s)}$$

Put, $s = j\omega$ in $G(s)$

$$\therefore \quad G(j\omega) = \frac{(1+.2j\omega)(1+.25j\omega)}{(j\omega)^2(1+.05j\omega)(1+.01j\omega)} \quad (1)$$

From equation (1)

$$|G(j\omega)| = \frac{\sqrt{1+(.2\omega)^2}\sqrt{1+(.25\omega)^2}}{\omega^2\sqrt{1+(.05\omega)^2}\sqrt{(1+(.01\omega)^2}}$$

$$\angle G(j\omega) = \tan^{-1} .2\omega + \tan^{-1} .25\omega - 180^\circ - \tan^{-1} .05\omega - \tan^{-1} .01\omega$$

The magnitude and phase angle of $G(j\omega)$ are calculated for various frequencies and listed in table 1. The polar coordinates listed in table1 are converted to rectangular coordinates and tabulated in table 2. Polar plot using polar coordinates is sketched on a polar sheet and polar plot using rectangular coordinates is sketched on a ordinary graph sheet.

Table 1: Magnitude and phase of $G(j\omega)$

ω, rad/sec	0.9	0.95	1	1.2	1.4	1.7	2
$\|G(j\omega\|$	1.27	1.14	1	.74	.55	.39	.29
$\angle G(j\omega)$	–160°	–159°	–158°	–154°	–150°	–144°	–138.4°

Convert the above values given in table 1 to real and imaginary values i.e. convert the magnitude and phase to real and imaginary values and draw it on the ordinary graph sheet (refer example problems).

Result: Phase margin, $\gamma = 22^\circ$.

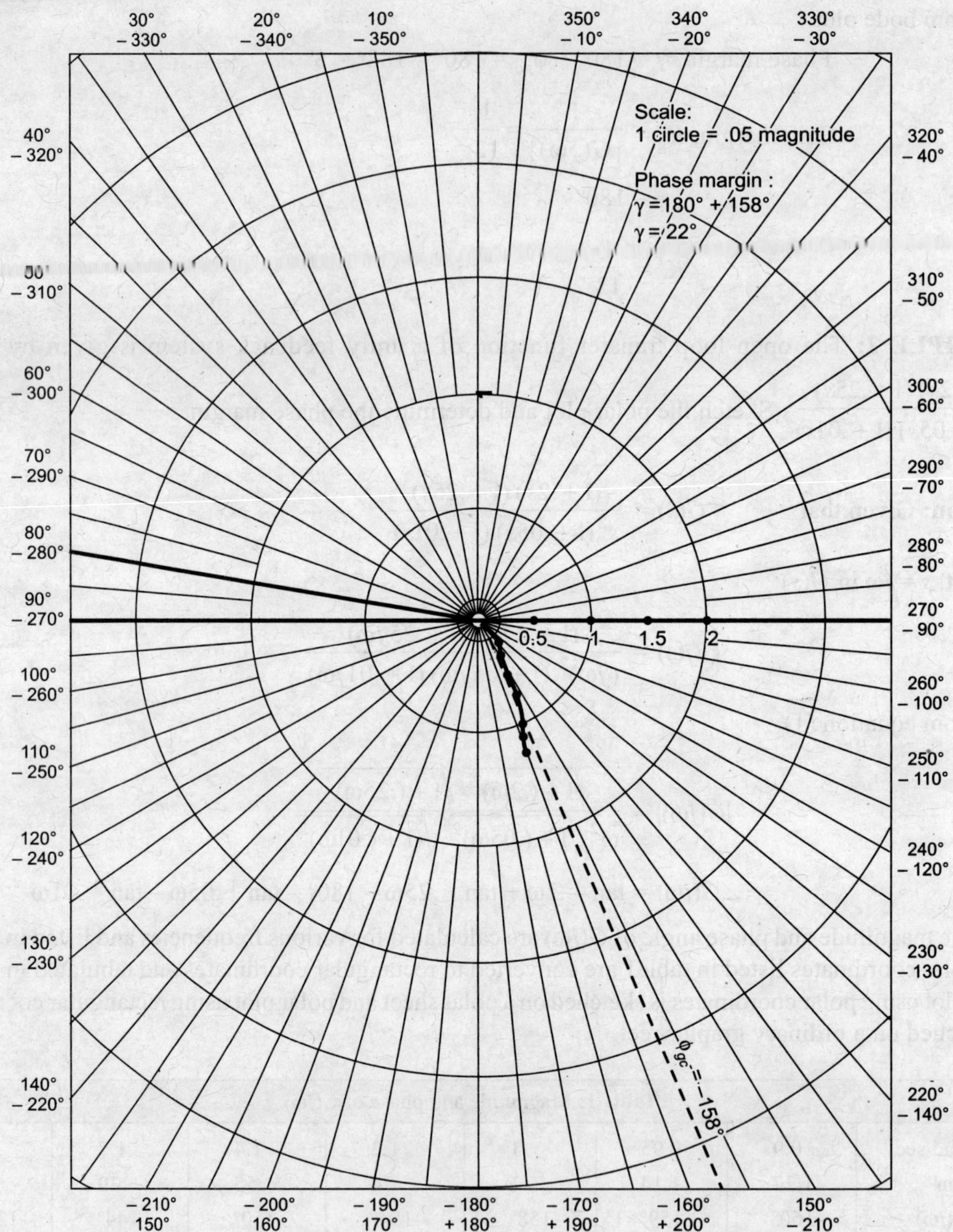

Worked example: Polar plot of $G(j\omega) = 1/j\omega\ (1 + j2\omega)$ (using polar coordinates).

EXAMPLE 3: Determine the gain margin of the system given below.

$$G(s) = 20/s(s + 1)(s^2 + 2s + 2)$$

Solution: Given that, $G(s) = \dfrac{20}{s(s+1)(s^2+2s+2)}$

Let, $s = j\omega$

$$\therefore \quad G(j\omega) = \frac{20}{j\omega(j\omega+1)[(j\omega)^2+2(j\omega)+2]}$$

$$= \frac{20}{j^4\omega^4 + 2j^3\omega^3 + 2j^2\omega^2 + j^3\omega^3 + 2j^2\omega^2 + 2j\omega}$$

$$= \frac{20}{\omega^4 - j2\omega^3 - 2\omega^2 - j\omega^3 - 2\omega^2 + 2j\omega}$$

$$= \frac{20}{\omega^4 - 4\omega^2 - 3j\omega^3 + 2j\omega}$$

$$= \frac{20}{\omega^2(\omega^2-4) + j\omega(2-3\omega^2)}$$

Let equate imaginary part to zero.

$$\omega_{pc}(2 - 3\omega_{pc}^2) = 0$$

Since $\omega_{pc} \neq 0$

$$2 - 3\omega_{pc}^2 = 0.$$

$$-3\omega_{pc}^2 = -2$$

$$3\omega_{pc}^2 = 2$$

$$\omega_{pc}^2 = 2/3$$

$$\omega_{pc} = \sqrt{2/3}$$

$\boxed{\omega_{pc} = .9}$ rad/sec

The gain margin, K_g is defined as the reciprocal of the magnitude of $G(j\omega)$ at phase cross over frequency.

Gain margin, $K_g = \dfrac{1}{|G(j\omega)|_{\omega=\omega_{pc}}}$

$$K_g = \frac{1}{1/|\omega^2(\omega^2-4)|_{\omega=\omega_{pc}}}$$

$$K_g = \omega^2(\omega^2 - 4) = 2.58$$

Gain margin in db $= 20 \log Kg = 20 \log 2.5$

$$K_g = 8\ db$$

Result: $K_g = 8\ db$.

CHAPTER 5

EXAMPLE 4: The open loop transfer function of a system in $G(s) = K/s(1 + .5s)$ (i) Determine the value of K os that gain margin is 6 db.

Solution: Given, that $G(s) = K/s(1 + .1s)\ (1 + s)$

Let, $s = j\omega$

$$\therefore \quad G(j\omega) = K/j\omega(1 + .5j\omega)\ (1 + j\omega)$$

$$= \frac{K}{j\omega\,(1 + j\omega + .5j\omega + .5j^2\omega^2)}$$

$$= \frac{K}{j\omega(1 + 1.5j\omega - .5\omega^2)}$$

$$= \frac{K}{j\omega(1 - .5\omega^2 + j\omega\ 1.5)} = \frac{K}{j\omega - .5j\omega^3 - 1.5\omega^2}$$

$$= \frac{K}{-1.5\omega^2 + j\omega - .5j\omega^3}$$

$$= \frac{K}{-1.5\omega^2 + j\omega\,(1 - .5\omega^2)}$$

At phase crossover frequency ω_{pc}, the $G(j\omega)$ is real and so equate the imaginary part to zero to solve for ω_{pc}.

At $\omega = \omega_{pc}$, $\quad \omega_{pc}(1 - 5\,\omega_{pc}^2) = 0$

$$\therefore \quad 1 - .5\,\omega_{pc}^2 = 0$$

$$-.5\ \omega_{pc}^2 = -1$$

$$\omega_{pc}^2 = 1/.5$$

$$\omega_{pc}^2 = 2 \Rightarrow \omega_{pc} = \sqrt{2}$$

$$\boxed{\omega_{pc} = 1.414} \text{ rad/sec}$$

$$|\,G(j\omega)\,|_{\omega = \omega_{pc}} = \left|\frac{K}{-1.5\omega^2}\right|_{\omega = \omega_{pc}} = \frac{K}{+1.5 \times (1.414)^2}$$

$$|G(j\omega)| = \frac{K}{3}$$

$$|G(j\omega)| = K\,(.33)$$

Given that $\quad K_g = 6$ db

$$\therefore \quad 20 \log K_g = 6$$

$$\log K_g = 6/20$$

$$K_g = 10^{6/20} = 1.9953$$

$$K_g = \frac{1}{|G(j\omega)|}$$

$$\therefore \quad 1.9953 = \frac{1}{.33K}$$

$$\therefore \quad K = \frac{1}{(.33)(1.9953)} = \frac{1}{.658}$$

$$\boxed{K = 1.51}$$

Result: The value of K when gain margin is 6 *db* is 1.51.

5.7 EFFECT OF P, PI AND PID CONTROLLER

The automatic controller determines the value of controlled variable and compare the actual value to the desired value (reference input), then determines the deviation and produces a control signal that will reduce the deviation to zero (or) to a smallest possible value. The method by which the automatic controller produces the control signal is called control action.

The control action may operate through either mechanical, hydraulic, pneumatic (or) electromechanical means i.e. controllers can be electrical, hydraulic, pneumatic electromechanical or electronics types.

5.7.1 Elements of Industrial Automatic Controller

Figure (5.35) is a block diagram of an industrial controller. The controller consists of an error detector and amplifier. The measuring element is a device which converts the output variable to another suitable variable such as displacement, pressure or electrical signals, which can be used for comparing the output to the reference input signal. The set point of a controller must be converted to reference input of the same units as feedback signal from the measuring element. From Fig. (5.35).

$$e = r - b$$

i.e., deviation is the difference between controlled variable and set point (reference input).

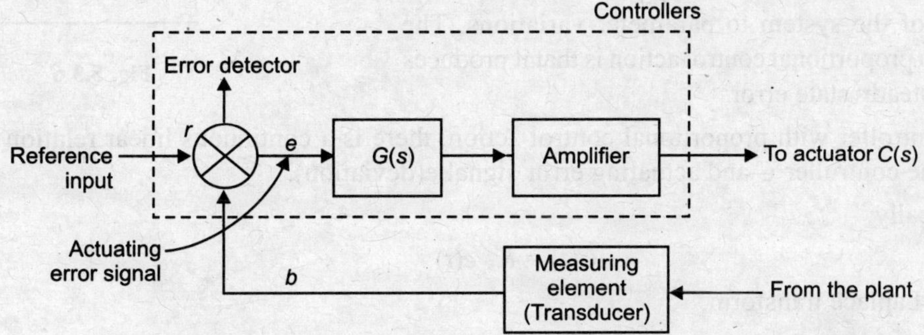

Fig. 5.35: Block diagram of industrial automatic controller.

CHAPTER 5

5.7.2 Classification of Controllers

Controllers are classified depending upon the type of controlling action used. Therefore, they can be classified as

1. Two position or ON–OFF controllers
2. Proportional controllers
3. Integral controllers
4. Derivative controller
5. Proportional-plus integral controllers
6. Proportional-plus derivative controllers
7. Proportional plus integral-plus derivative controllers.

They can also be classified according to the power source used for actuating mechanisms, such as electrical, electronics pneumatic and hydraulic controllers. Hydraulic controllers are used for controlling heavy loads, pneumatic controllers are suitable for shop floor applications.

The selection of a particular types of controller depends on the nature of plant, operating conditions such as safety, cost, availability, accuracy, weight and size.

Note: Derivative controller is not used induvidually it always combined with other controllers because the output will be zero for constant values for example denvative controller will give zero output when ste signal is given.

5.7.3 Effect of Proportional Controller

The proportional controller produces an output signal which is proportional to error signal. The transferfunction of proportional controller is K_p. The term K_p is called the gain of the controller. Hence the proportional controller amplifies the error signal and increases the loop gain of the system. The P controller has,

1. Steady state tracking accuracy.
2. Disturbance signal rejection.
3. Relative stability.

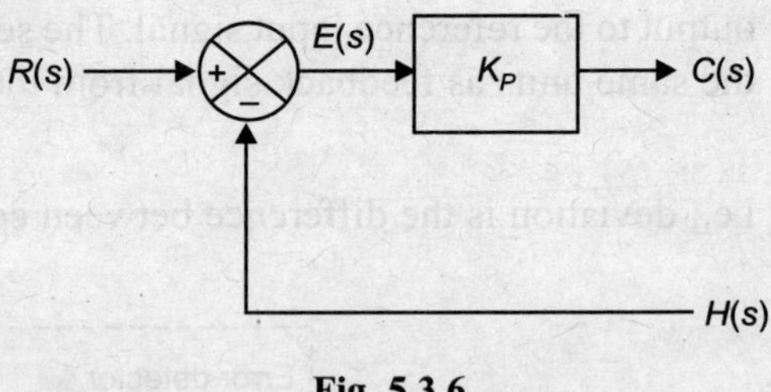

Fig. 5.3.6

In addition to increase in loop gain it decreases the sensitivity of the system to parameter variations. The drawback in proportional control action is that it produces a constant steady state error.

In a controller with proportional control action, there is a continuous linear relation between the output of the controller C and actuating error signal e(deviation).

Mathematically

$$C(t) = K_P\, e(t)$$

in terms of Laplace transform,

$$C(s) = K_P\, E(s)$$

$$K_P = \frac{C(s)}{E(s)} \tag{5.7}$$

5.7.4 Effect of Integral Controller [Re set Controller]

In a controller with integral control action, the output of the controller is changed at a rate which is proportional to the actuating error signal $e(t)$.

Mathematically,

$$\frac{d}{dt}C(t) = K_i e(t)$$

where $K_i \rightarrow$ Integral constant

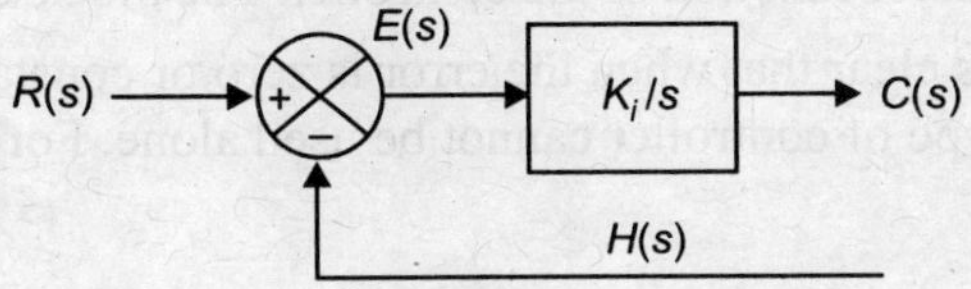

Fig. 5.37

$$C(t) = K_i \int e(t) \tag{5.8}$$

Laplace transform of equation (2.12.5.1)

$$C(s) = \frac{K_i}{s} E(s)$$

$$\frac{M(s)}{E(s)} = \frac{K_i}{s} \tag{5.9}$$

Note: $T_i = 1/K_i$

Equation (5.9) is the transfer function of integral controller. The block diagram of integral controller is shown in Fig. (5.37). The inverse of Ki is called integral time T_i and defined as the time of change of output caused by a unit change of actuating error signal.

5.7.5 Effect of Derivative Controller [Refer Controller]

In a controller with derivative control action, the output of the controller depends on the rate of change of actuating error signal $e(t)$.

Mathematically,

$$C(t) = K_d \frac{d}{dt} e(t) \tag{5.10}$$

where,

$K_d \rightarrow$ Derivative gain constant.

Laplace transform of Eq. (5.10)

$$C(s) = K_d \cdot s \cdot E(s)$$

$$\frac{C(s)}{E(s)} = K_d \cdot s \qquad (5.11)$$

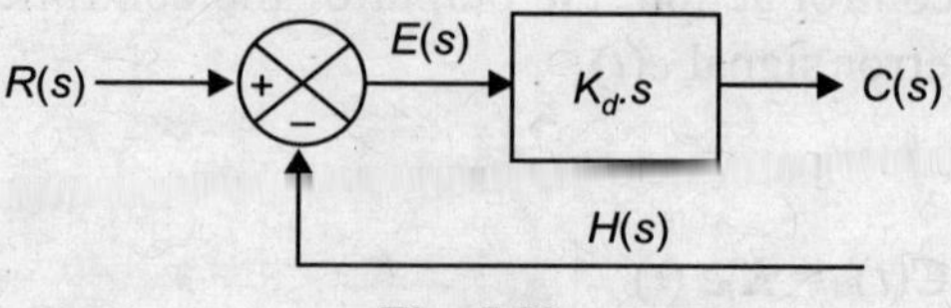

Fig. 5.38

Equation (5.11) is the transfer function of the controller. The block diagram is shown in Fig. (5.38).

From equation (5.19) it is clear that when the error is zero or constant, the output of the controller will be zero. Therefore this type of controller cannot be used alone. For this type of controller the gain should be small.

5.7.6 Effect of Proportional-Plus-Integral Controller [PI Controller]

The proportional plus integral controller (PI) produces an output signal consisting of two terms-one proportional to error signal and the other proportional to the integral of error signal.

$$C(t) = K_p\, e(t) + K_p K_i \int_0^t e(t)\, dt \qquad \text{Note: } T_i = \frac{1}{K_i} \text{ for } I \text{ controller}$$

$$C(t) = K_p\, e(t) + K_p\, \frac{1}{T_i} \int_0^t e(t)\, dt \qquad (5.12)$$

Laplace transform of equation (5.12)

$$C(s) = K_p\, E(s) + \frac{K_p}{sT_i} E(s)$$

$$C(s) = E(s) \left[1 + \frac{1}{sT_i}\right] K_p$$

$$\frac{C(s)}{E(s)} = \left[1 + \frac{1}{sT_i}\right] K_p \qquad (5.13)$$

Fig. 5.39

(or)

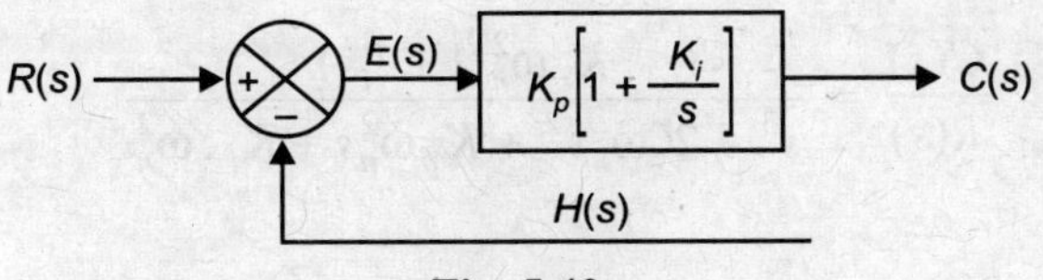

Fig. 5.40

Fig. 5.41

Let this PI controller is connected with one second order system. Under this condition let we see the effect of PI controller on the system.

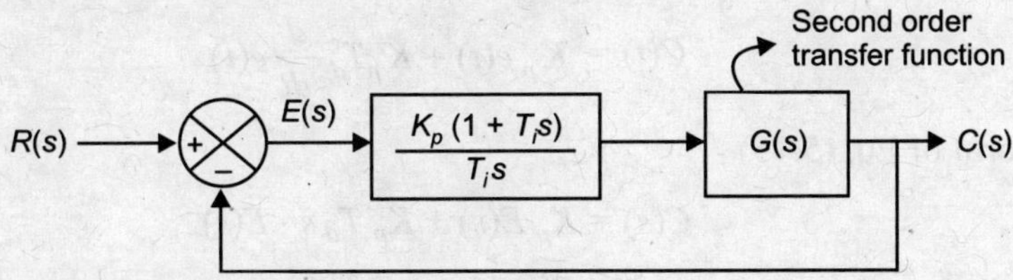

Fig. 5.42

Let $$G(s) = \frac{\omega_n^2}{s}[s + 2\zeta\omega_n]$$

Now the closed loop transfer function is given by

$$\frac{C(s)}{R(s)} = \frac{K_p\left[\frac{1+T_is}{T_is}\right]\left[\frac{\omega_n^2}{s[s+2\zeta\omega_n]}\right]}{1+K_p\left[\frac{1+T_is}{T_is}\right]\left[\frac{\omega_n^2}{s[s+2\zeta\omega_n]}\right]}$$

$$= \frac{K_p\omega_n^2(1+T_is)}{s^2T_i\,(s+2\zeta\omega_n)+K_p\omega_n^2\,(1+T_is)}$$

$$= \frac{K_p\omega_n^2\,(1+T_is)}{T_is^3+2\zeta\omega_nT_is^2+K_p\,\omega_n^2\,T_is+K_p\omega_n^2}$$

$$= \frac{(K_p/T_i)\omega_n^2\,(1+T_is)}{s^3+2\zeta\omega_ns^2+K_p\omega_n^2+\left(\frac{K_p}{T_i}\right)\omega_n^2}$$

$$\frac{C(s)}{R(s)} = \frac{K_i\,\omega_n^2(1+T_i s)}{s^3 + 2\zeta\omega_n s^2 + K_p\omega_n^2 s + K_i\,.\,\omega_n^2} \tag{5.14}$$

where, $K_i = K_p/T_i$

Note: $K_i = K_i/T_i$ for PI controller

From the closed loop transferfunction (equation 5.14) it is observed that the PI controller introduces a zero in the system and increases the order by one also the type number of open loop system is increased by one. The increase in type number results in reducing the steady state error.

The increase in order of the system results in a less stable system than the original one.

5.7.7 Effect of Proportional-Plus-Derivative Control Action (PD)

The proportional plus derivative controller produces an output signal consisting of two terms one proportional to error signal and the other proportional to the derivative of error signal.

$$C(t) = K_p\,e(t) + K_p T_d \frac{d}{dt} e(t) \tag{5.15}$$

Laplace transform of Eq. (5.15)

$$C(s) = K_p\,E(s) + K_p\,T_d\,s \cdot E(s)$$

$$= E(s)\,[1 + s\,T_d]\,K_p$$

$$\frac{C(s)}{E(s)} = K_p(1 + sT_d)$$

Fig. 5.43

T_d is known as derivative time constant. Derivative time is defined as the time interval by which the rate action advances the effect of the proportional control action.

Let we see the effect of PD controller on the second order system.

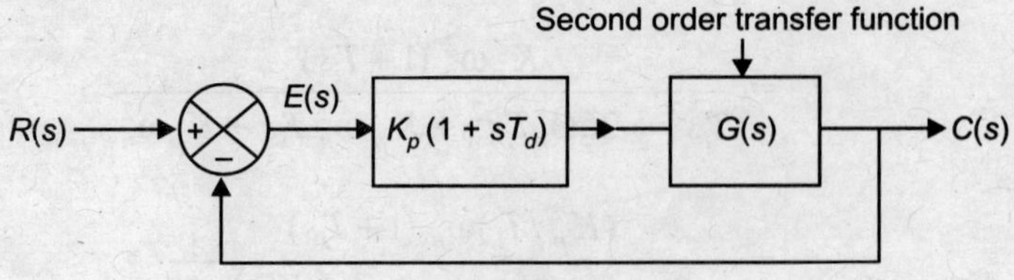

Fig. 5.44

$$\frac{C(s)}{R(s)} = \frac{K_p(1+T_d \cdot s)\cdot \dfrac{\omega_n^2}{s(s+2\zeta\omega_n)}}{1+K_p(1+T_d \cdot s)\cdot \dfrac{\omega_n^2}{s(s+2\zeta\omega_n)}}$$

$$= \frac{K_p\,\omega_n^2\,(1+T_d s)}{s(s+2\zeta\omega_n)+K_p\,\omega_n^2\,(1+T_d s)}$$

$$= \frac{K_p\,\omega_n^2\,(1+T_d s)}{s^2+2\zeta\omega_n\cdot s+K_p\cdot\omega_n^2+K_p\omega_n^2\,T_d s}$$

$$= \frac{K_p\,\omega_n^2\,(1+T_d\cdot s)}{s^2+(2\zeta\omega_n+K_p\,\omega_n^2\,T_d)s+K_p\omega_n^2}$$

$$\frac{C(s)}{R(s)} = \frac{\omega_n^2(K_p+K_d\cdot s)}{s^2+(2\zeta\omega_n+K_d\omega_n^2)\,s+K_p\omega_n^2}$$

where,

$$K_d = (K_p \cdot T_d) \tag{5.16}$$

From the closed loop transfer function it is observed that the PD controller introduces a zero in the system and increase the damping ratio. The addition of the zero may increase the peak overshoot and reduce the rise time. But the effect of increased damping ultimately reduces the peak overshoot. It has faster response and improves the bandwidth.

5.7.8 Effect of Proportional-Plus Integral-Plus Derivative Controller (PID Controller)

The combination of proportional, integral and derivative control action is called PID control action and the controller is called three action controller (or) PID controller.

Mathematically,

$$C(t) = K_p\,e(t) + K_p\frac{1}{T_i}\int e(t)\,dt + K_p\,T_d\frac{d}{dt}e(t) \tag{5.17}$$

Laplace transform of Eq. (2.12.9.1)

$$C(s) = K_p\,E(s) + \frac{K_p}{sT_i}E(s) + K_p\cdot T_d\cdot sE(s)$$

$$\frac{C(s)}{E(s)} = K_P\left(1+\frac{1}{sT_i}+sT_d\right) \tag{5.18}$$

K_p is the proportional gain, T_i is the integral time and T_d is the derivative time.

CHAPTER 5

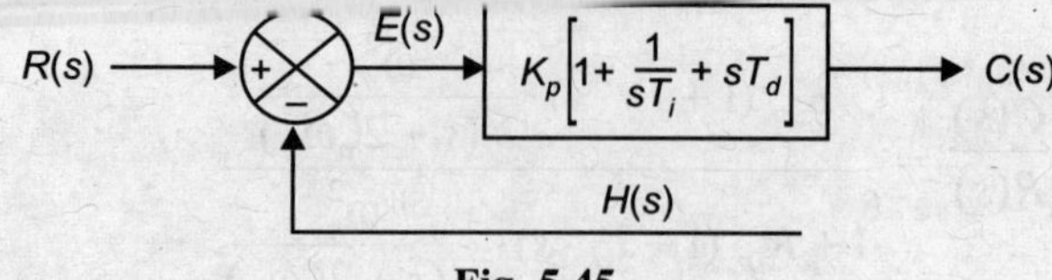

Fig. 5.45

Let this PID controller is connected with second order system. Under this condition let we see the effect of PID controller on the system.

Let $$G(s) = \frac{\omega_n^2}{s\,(s+2\zeta\omega_n)}$$

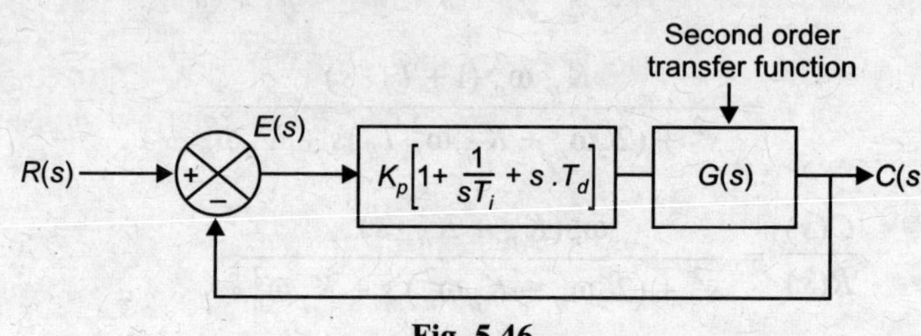

Fig. 5.46

Note: The transfer function of the Fig. (5.22.1) can be derived as same as previous cases.

After determining the closed loop transfer function. The following points are noted.

- The proportional controller stabilizes the gain but produces a steady state error.
- The integral controller reduces or eliminates the steady state error.
- The derivative controller reduces the rate of change of error. The combined effect of the three cannot be judged from the parameters K_p, K_i and K_d.

EXAMPLE 5.22: A closed loop control system with unity feed back is shown in Fig. (5.22.1) and Fig. (5.22.2). By using derivative control the damping ratio is to be made 0.7. Determine the value of T_d, also determine the rise time without derivative control and with derivative control. The input to the system is unit step.

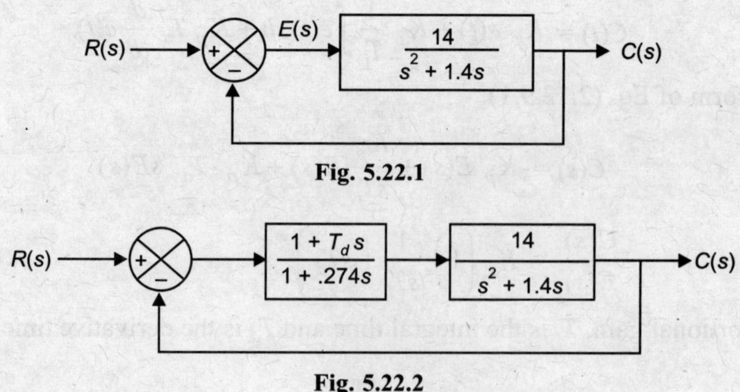

Fig. 5.22.2

Solution: The closed loop transfer function of the system,

From Fig. (5.22.1)
$$\frac{C(s)}{R(s)} = \frac{\frac{14}{s^2+1.4s}}{1+\frac{14}{s^2+1.4s}} = \frac{14}{s^2+1.4s+14}$$

The characteristic equation $s^2 + 1.4s + 14 = 0$

Compare with $s^2 + 2\zeta\omega_n s + \omega_n^2 = 0$

$$2\zeta\omega_n = 1.4$$

$$\omega_n^2 = 14 \Rightarrow \omega_n = 3.74 \text{ rad/sec}$$

$$2\zeta \times 3.74 = 1.4$$

$$\therefore \quad \zeta = .187$$

$$\zeta' = \zeta + \frac{\omega_n T_d}{2}$$

$$\zeta' = .187 + \frac{3.74\,(T_d)}{2}$$

$$.7 = .187 + \frac{3.74\,(T_d)}{2}$$

$$\therefore \quad T_d = .274 \text{ sec}$$

Rise time
$$t_r = \frac{\pi - \tan^{-1}\frac{\sqrt{1-\zeta^2}}{\zeta}}{\omega_n\sqrt{1-\zeta^2}}$$

$$= \frac{\pi - \tan^{-1}\frac{\sqrt{1-(.187)^2}}{.187}}{3.74\sqrt{1-(.187)^2}} = \frac{1.757}{3.67}$$

$$t_r = .478 \text{ sec}$$

Peak time,
$$t_p = \frac{\pi}{\omega_n\sqrt{1-\zeta^2}} = \frac{\pi}{3.74\sqrt{1-(.187)^2}} = .854 \text{ sec}$$

$$t_p = .854 \text{ sec}$$

Percentage maximum ovrshoot

$$M_p = e^{\frac{-\zeta\pi}{e\sqrt{1-\zeta^2}}} \times 100$$

CHAPTER 5

$$= e^{-\frac{\pi(.187)}{\sqrt{1-(.187)^2}}} \times 100$$

$$M_p = 55\%$$

The overall transferfunction with derivative control

For Fig. (5.22.2), $$\frac{C(s)}{R(s)} = \frac{14\,(1+.274s)}{s^2+1.4s+14+3.836s}$$

Note: Find transfer function of Fig. (2.27.2) using the formula $G/1 + GH$.

$$\frac{C(s)}{R(s)} = \frac{14\,(1+.274s)}{s^2+5.236s+14}$$

Input $R(s) = 1/s$ (for step input)

$$C(s) = \frac{1}{s}\cdot\frac{14\,(1+.274s)}{s^2+5.236s+14} = \frac{1}{s}\left[\frac{14+3.836s}{s^2+5.236s+14}\right]$$

$$= \frac{1}{s}\cdot\frac{14}{s^2+5.236s+14} + \frac{1}{s}\cdot\frac{3.836s}{s^2+5.236s+14}$$

$$C(s) = \frac{A}{s} + \frac{Bs+C}{s^2+5.236s+14} + \frac{3.836s}{s^2+5.23s+14}$$

Now calculate the coefficients

$$14\,(1 + .274s) = A\,(s^2 + 5.236s + 14) + (B + C)s + 3.836s$$

Equating the coefficient of s^2, s and constant terms

$$A + B = 0$$

$$5.236A + C + 3.836 = 3.836$$

$$14A = 14 \Rightarrow A = 1 \Rightarrow B = -1$$

$$C = -\,5.236$$

Substitute the A, B and C values in $C(s)$

$$C(s) = \frac{1}{s} - \frac{s+5.236}{s^2+5.236s+14} + \frac{3.836}{s^2+5.236s+14}$$

$$= \frac{1}{s} - \frac{s+5.236}{(s+2.618)^2+(14-2.618)^2} + \frac{3.836}{(s+2.618)^2+(14-2.678)^2}$$

$$= \frac{1}{s} - \frac{s+5.236}{(s+2.618)^2+(2.673)^2} + \frac{3.836}{(s+2.618)^2+(2.673)^2}$$

$$C(s) = \frac{1}{s} - \frac{+(s+2.618+2.618)}{(s+2.618)^2+(2.673)^2} + \frac{3.386}{(s+2.618)^2+(2.673)^2}$$

$$= \frac{1}{s} - \frac{s+2.618}{(s+2.618)^2+(2.673)^2} + \frac{1.218}{(s+2.618)^2+(2.673)^2}$$

$$C(s) = \frac{1}{s} - \frac{s+2.618}{(s+2.618)^2+(2.673)^2} + \frac{2.673}{(5+2.618)^2+(2.673)^2}\left(\frac{1.218}{2.673}\right)$$

on taking inverse Laplace transform

$$C(t) = 1 - e^{-2.618t}\cos 2.673t + .455\, e^{-2.618t}\sin 2.673t$$

For rise time $C(t) = 1$

$$1 = 1 - e^{-2.618t}\cos 2.673t + 0.455\, e^{-2.618t}\sin 2.673t \quad (2.27.1)$$

$$0.455 \sin 2.673t = \cos 2.673t$$

$$\tan 2.673t = 2.19$$

$$t = .427$$

$$t_r = .427 \text{ sec}$$

Result:

(i) Without derivative control

$t_r = .478$ sec

(ii) With Derivative control

$t_r = .427$ sec

(iii) $T_d = .274$ sec

EXAMPLE 5.23: Fig. (5.23.1) shows a system employing proportional plus error control. Determine the value of error rate factor K_e so that the damping ratio is 0.5. Determine the value of settling time, maximum overshoot and steady state error (for unit ramp input) with and without error rate control.

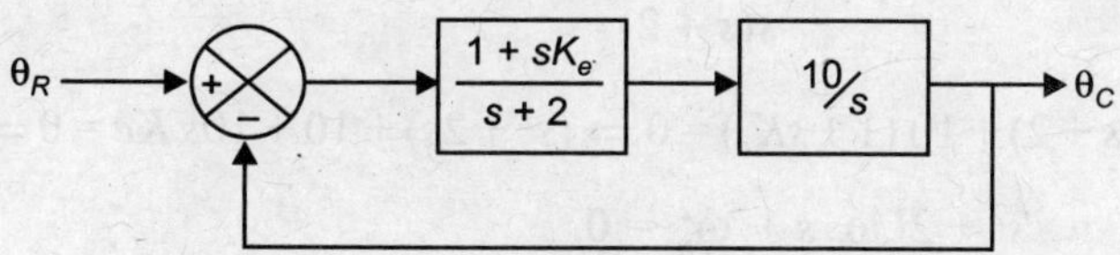

Fig. 5.23.1

Solution: Without error rate control i.e. $K_e = 0$ if $K_e = 0 \Rightarrow \left(\frac{1}{s+2}\right)\left(\frac{10}{s}\right) = \frac{10}{s(s+2)} = G(s)$

The overall transfer function $\frac{C(s)}{R(s)}$ will be $G(s)/1 + G(s)\,H(s)$

i.e.,
$$\frac{C(s)}{R(s)} = \frac{\frac{10}{s(s+2)}}{1+\frac{10}{s(s+2)}} = \frac{10}{s^2+2s+10}.$$

Characteristic equation $s^2 + 2s + 10 = 0$

CHAPTER 5

Compare without $s^2 + 2\zeta\omega_n s + \omega_n^2 = 0$

$$\omega_n^2 = 10$$

$$\therefore \quad \omega_n = 3.16 \text{ rad/sec}$$

$$2\zeta\omega_n = 2$$

$$2 \times \zeta \times 3.16 = 2$$

$$\zeta = .316$$

Maximum overshoot $M_p = e^{-\pi\zeta/\sqrt{1-\zeta^2}} \times 100$

$$= e^{\frac{(-\pi \times .316)}{\sqrt{1-(.316)^2}}} \times 100$$

$$\% M_p = 35.14\%$$

Settling time $t_s = \dfrac{4}{\zeta\omega_n} = \dfrac{4}{0.316 \times 3.16} = 4 \text{ sec}$

steady state error

$$e_{ss} = \frac{2\zeta}{\omega_n} = \frac{2 \times .316}{3.16}$$

$$e_{ss} = 0.2 \text{ rad}$$

With error control

The overall transferfunction of Fig. (5.23.1)

$$\frac{C(s)}{R(s)} = \frac{\dfrac{10\,(1+sK_e)}{s(s+2)}}{1+\dfrac{10\,(1+sK_e)}{s(s+2)}} = \frac{10\,(1+sK_e)}{s(s+2)+10\,(1+sK_e)}$$

characteristic equation $s(s+2) + 10\,(1+sK_e) = 0 \Rightarrow (s^2+2s) + 10 + 10s\,Ke = 0 \Rightarrow s^2 + s(10\,Ke + 2) + 10$

compare with $s^2 + 2\zeta\omega_n\, s + \omega_n^2 = 0$

$$\omega_n^2 = 10$$

$$\therefore \quad \omega_n = 3.16 \text{ rad/sec} \Rightarrow 2\zeta\omega_n = 2 + 10\, K_e$$

$$= 2 \times .5 \times 3.16 = 2 + 10\, \text{K}_\text{e}$$

$$\therefore \quad K_e = .116$$

The characteristic equation with $Ke = .116$ is

$$s^2 + 3.16s + 10 = 0$$

Compare with $s^2 + 2\zeta\omega_n s + \omega_n^2 = 0$

$$2\zeta\omega_n = 3.16$$

$$2\zeta(3.16) = 3.16$$

$$2\zeta = 1$$

$$\zeta = 1/2 = .5$$

Maximum overshoot, $M_p = e^{-\pi\zeta/\sqrt{1-\zeta^2}} \times 100$

$$M_p = e^{\frac{-\pi(0.5)}{\sqrt{1-(0.5)^2}}} \times 100 = 16.3\%$$

$$\%M_p = 16.3\%$$

settling time

$$t_s = \frac{4}{\zeta\omega_n} = \frac{4}{0.5 \times 3.16} = 2.53 \text{ sec}$$

$$t_s = 2.53 \text{ sec}$$

Steady state error $e_{ss} = \underset{s \to 0}{\text{Lt}}\ sE(s)$

$$E(s) = \frac{1}{1+G(s)} \cdot R(s) \qquad [\because H(s) = 1]$$

$$e_{ss} = \underset{s \to 0}{\text{Lt}}\ s \cdot \frac{1}{s^2} \cdot \frac{s(s+2)}{10 + 1.16s + s(s+2)} \quad [\because R(s) = 1/s^2 \text{ for unit ramp i/p}]$$

$$e_{ss} = 0.2 \text{ rad}$$

Results:

(i) Without error control
  (i) $\%M_p = 35.14\%$
  (ii) $t_s = 4$ sec
  (iii) $e_{ss} = 0.2$ rad

(ii) With error control
  (i) $M_p = 16.3\%$
  (ii) $t_s = 2.53$ sec
  (iii) $e_{ss} = 0.2$ rad

EXAMPLE 5.24: A feedback system employing output-rate damping shown in Fig. (5.24.1) In the absence of derivative feedback ($K_0 = 0$). (a) Determine the damping factor and natural frequency of the system. What is the steady state error resulting from unit ramp state error resulting from unit ramp input. (b) Determine the derivative feedback constant K_0, which will increase the damping factor of the system to 0.6. What is the steady state error resulting from unit ramp input with this settling of the derivative feedback constant?

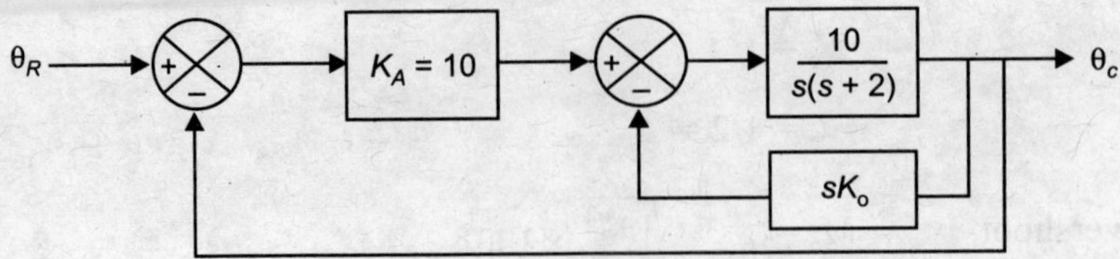

Fig. 5.24.1

Solution: (a) *Without derivative feedbacK*

(a) When $K_0 = 0$

$$G(s) = \frac{10}{s(s+2)} = \frac{\frac{10}{s(s+2)}}{1+\frac{10}{s(s+2)}} \qquad [\because H(s)=1]$$

$$= \frac{\frac{10}{s(s+2)}}{\frac{s(s+2)+10}{s(s+2)}} = \frac{10}{s(s+2)+10}$$

$$= \frac{10}{s^2+2s+10}$$

Compare characteristic equation $s^2 + 2s + 10 = 0$ with $s^2 + 2\zeta\omega_n s + \omega_n^2 = 0$

$$\omega_n^2 = 0 \quad \omega_n = 3.16 \text{ rad/sec}$$

$$2\zeta\omega_n = 2$$

$$2 \times \zeta \times 3.16 = 2$$

$$\zeta = .316$$

Steady state error e_{ss} $\qquad e_{ss} = \frac{2\zeta}{\omega_n} = \frac{2 \times .316}{3.16}$

$$e_{ss} = 0.2 \text{ rad}$$

(b) *With derivative feedback*

The overall transferfunction of Fig. (2.29),

$$\frac{C(s)}{R(s)} = \frac{G(s)}{1+G(s)\,H(s)} = \frac{\frac{10}{s(s+2)}}{1+\frac{10}{s(s+2)}(sK_0)}$$

$$\frac{C(s)}{R(s)} = \frac{10}{s^2+(2+K_0)\,s+10}$$

Characteristic equation $s^2 + (2 + K_0)\, s + 10$ compare with $s^2 + 2\zeta\omega_n s + \omega_n^2$

$$\omega_n^2 = 10$$

$$\therefore \quad \omega_n = 3.16 \text{ rad/sec}$$

$$2\zeta\omega_n = 2 + K_0$$

$$2 \times 0.6 \times 3.16 = 2 + K_0$$

$$K_0 = 1.792$$

steady state error (for unit ramp input)

$$e_{ss} = \underset{s\to 0}{\text{Lt}}\; s \cdot E(s)$$

$$E(s) = \frac{R(s)}{1 + G(s)\,H(s)} \qquad [\because R(s) = 1/s^2 \text{ for unit ramp input}]$$

$$e_{ss} = \underset{s\to 0}{\text{Lt}}\; s\,\frac{1}{s^2}\,\frac{s(s+2)+1.8s}{s(s+2)+1.8s+10}$$

$$e_{ss} = \underset{s\to 0}{\text{Lt}}\; \frac{s+2+1.8}{10} = 0.38$$

$$e_{ss} = 0.38 \text{ rad}$$

Results:

(i) Without derivative feedback
   (i) $\zeta = 0.316$
   (ii) $\omega_n = 3.16$ rad/sec
   (iii) $e_{ss} = 0.2$ rad

(ii) With derivative feedback
   (i) $K_0 = 1.792$
   (ii) $\omega_n = 3.16$ rad/sec
   (iii) $e_{ss} = 0.38$ rad

SOME QUESTION AND ANSWERS

Q.1 What is frequency response?

Ans. It is a steady state output of the system, when the input is a sinusoidal signal.

Q.2 What are the advantages of frequency response analysis?

Ans. 1. Estimation of absolute and relative stability fo the closed loop system can be done.
2. The transfer function of complicated functions can be determined.
3. Corrective measure for parameter variation can be easily carried.

Q.3 What are frequency domain specifications?

Ans. 1. Resonant peak

2. Resonant frequency
3. Bandwidth
4. Cut-off rate
5. Gain margin
6. Phase margin

Q.4 What is bode plot?

Ans. The bode plot is a frequency response plot of the transfer function of a system. It consist of magnitude and phase plots. The magnitude plot is a graph between magnitude and frequency and phase plot is a graph between phase angle and frequency.

Q.5 Define corner frequency?

Ans. The magnitude plot can be approximated by asymptotic straight lines. The frequencies corresponding to the meeting point of asymptotes are called corner frequency.

Q.6 What are the advantages of bode plot?

Ans. 1. Frequency domain specifications can be easily determined.

2. It is used to analyse both open loop and closed loop system.

Q.7 What is polar plot?

Ans. The polar plot is a sinusoidal transformation $G(j\omega)$ is a plot of the magnitude of $G(j\omega)$ on polar or rectangular coordinates as w is varied from zero to infinity.

Q.8 Define BIBO stability?

Ans. A linear relaxed system is said to have BIBO stability if every bounded input results in a bounded output.

Q.9 What is the requirement of BIBO stability?

Ans. $\int_0^\infty m(\tau)d\tau < \infty$, where $m(\tau)$ is the impulse response of the system.

Q.10 What is impulse response?

Ans. It is the inverse laplace transform of the system transfer function.

Q.11 What is characteristics equation?

Ans. The denominator polynomial of $C(s)/R(s)$ is the characteristic equation of the system.

Q.12 How the roots of characteristic equation are related to stability?

Ans. If the roots of characteristic equation has positive real part then the impulse response of the system is not bounded. Hence the system will be unstable. If the roots has negative real parts then the impulse response is bounded. Hence the system will be stable.

Q.13 What is the necessary and sufficient conditions for stability?

Ans. 1. The necessary condition for stability is that all the coefficients of the characteristic polynomial be positive.

2. All the elements in the 1st coloumn of the routh array should be positive.

Q.14 What is the raltion between stability and coefficinet of characteristic polynomial?

Ans. If the coefficients of characteristic polynomial are negative or zero, then some of roots lies on right half of *s*-plane. Hence the system is unstable. If the coefficnets of characteristic polynomial are positive and if no coefficient is zero then there is a possibility of the system to be stable provided all the roots are lying on left half of *s*-plane.

Q.15 What is routh stability criterion?

Ans. Routh criterion states that the necessary and sufficient conditions for stability is that all of the elements in the first column of the routh array be positive. If this condition is met, the system is unstable and the number of sign changes is the elements of the first column of routh array corresponds to the number of roots of characteristic equation is the right half of the *s*-plane.

Q.16 What is auxiliary polynomial?

Ans. In the construction of routh array a row of all zero indicates the existence of an even polynomial as a factor of the given characteristic equation. In an even polynomial the exponents of *s* are integers or zero only. This even polynomial factor is called auxiliary polynomial. The coefficients of auxiliary polynomial are given by the elements of the row just above the row of all zeros.

Q.17 What is limitedly stable system?

Ans. For a bounded input signal, if the output has constant amplitude oscillations then the system may be stable or unstable undersome constraints. Such a system is called limitedly stable system.

Q.18 What is root locus?

Ans. The path taken be a root of characteristic equation where open loop gain K is varied from 0 to ∞ is called locus.

Q.19 What are asymptotes and what is centroid?

Ans. 1. Asymptotes are straight lines which are parallel to root locus going to infinity and meet the root locus at infinity.

2. The meeting point of asymptotes with real axis called centroid.

Q.20 What is breakaway and breaking points?

Ans.
- The point where the root locus breaks from the real axis to enter into the complex plane is called breakaway point.
- The point where the root locus enters the real axis from the complex plane is called break in point.

Q.21 What is dominant pole?

Ans. The dominant pole is a pair of complex conjugate pole which decides transient response of the system.

In higher order systems the dominant poles are very close to origin and all other poles of the system are widely separated and so they have less effect on transient response of the system.

Q.22 What is the function of controller?

Ans. It determines the value of controlled variable and compare the actual value to the desired value, then determines the deviation and produces a control signal that will reduce the deviation to zero.

Q.23 What are the elements of controller?

Ans. 1. Error detector

2. Measuring element
3. Amplifier

Q.24 What are the types of controllers?

Ans. 1. Two position (or) ON-OFF controllers

2. Proportional controllers of (*P* controller)
3. Integral controllers (*I* controller)
4. Derivative controller (*D* controller)
5. *PI* controller
6. *PID* controller

Q.25 What is the advantages and disadvantages of *P* controller?

Ans. *Advantages*

- Loop gain increases
- Less sensitive to parameter variation

Disadvantages

It produces constant steady state error.

Q.26 What is the effect of *PI* controller?

Ans.
- It introduces a zero to the system so, order of the system increased by one.
- It also increases the type of the system by one reduced steady state error.
- Less stable.

Q.27 What is the effect of *PD* controller?

Ans.
- It also introduces zero in the system as *PI*.
- It increases damping ration.
- Peak overshoot increases.
- Rise time get reduced.

- Improved bandwidth.
- Fast response

Q.28 What is the effect of *PID* controller?

Ans. It has all effects of *P*, *I* and controllers.

Note:

All the points of *P*, *I* and controllers.

(or)

Combine the points of *PI* and *PD* controllers.

EXERCISE

1. Sketch the bode plot of the following open loop transfer functions and from the plot determine the phase margin and gain margin.

(a) $G(s) = s^2(s + 10)/(s + 5)^2\ (s + 0.1)$

(b) $G(s) = 10(1 + s)e^{-0.1s}/s\ (1 + 0.2s)$

(c) $G(s) = 30/(1 + 0.1s)/s\ (1 + .01s)\ (1 + s)$

2. Sketch the polar plot of the following transfer functions and from the plot, determine the phase margin and gain margin.

(a) $G(s) = 10(s + 1)/(s + 10)^2$

(b) $G(s) = 200(s + 2)/s(s^2 + 10s + 100)$

(c) $G(s) = 1/s\ (s + 4)\ (s + 8)$

3. The open loop transfer function of a system is given by $G(s) = K/s(s^2 + s + 4)$ using polar plot, determine the value of *K*, so that phase margin is 50°. What is the corresponding value of gain margin?

4. The open loop transfer functions fo certain unity feedback systems are given below. Sketch the root locus of each system.

(a) $G(s) = K(s + 2)/(s + 3)^2\ (s^2 + 2s + 17)$

(b) $G(s)\ K(s + 4)/s(s + .5)\ (s + 2)$

(c) $G(s) = K/s(s + 3)\ (s^2 + s + 2)$

(d) $G(s) = K/s(s + 2)\ (s^2 + 2s + s)$

(e) $G(s) = K(s + 1)/s^2(s + 12)$

(f) $G(s) = K(s^2 + 1)/s(s + 2)$

5. A unity feedback system has an open loop transfer function, $G(s) = K/s(s^2 + 8s + 32)$. Sketch the root locus and determine the dominant closed loop poles with $\zeta = .5$. Determine the value of *K* at this point.

- Improved bandwidth.
- Fast response.

Q.28. What is the effect of PD controller?

Ans. [illegible] effects of PD controllers.

Note: [illegible]

All the points of PI and controllers [illegible]

[illegible]

combine the points of PI and PD controllers.

[illegible]

1. Sketch the bode plot of the following open loop transfer functions and from the plot determine the phase margin and gain margin.

   (a) [illegible]

   (b) [illegible]

   (c) [illegible]

2. Sketch the polar plot of the following transfer functions and from the plot, determine the phase margin and gain margin.

   (a) [illegible]

   (b) [illegible]

   (c) [illegible]

3. The open loop transfer function of a system is given by [illegible]. Using polar plot, determine the value of K so that phase margin is 20°. What is the corresponding value of gain margin?

4. The open loop transfer functions of unity feedback systems are given below. Sketch the root loci of each system.

   (a) [illegible]

   (b) [illegible]

   (c) [illegible]

   (d) [illegible]

   (e) [illegible]

   (f) [illegible]

5. A unity feedback system has an open loop transfer function [illegible]. Sketch the root locus and determine the dominant closed loop poles with [illegible]. Determine the value of K at this point.